Green Energy and Technology

More information about this series at http://www.springer.com/series/8059

Roberto Álvarez Fernández
Sergio Zubelzu · Rodrigo Martínez
Editors

Carbon Footprint and the Industrial Life Cycle

From Urban Planning to Recycling

Springer

Editors
Roberto Álvarez Fernández
Nebrija University
Madrid
Spain

Rodrigo Martínez
University of Deusto
Bilbao
Spain

Sergio Zubelzu
Technical University of Madrid
Madrid
Spain

ISSN 1865-3529 ISSN 1865-3537 (electronic)
Green Energy and Technology
ISBN 978-3-319-54983-5 ISBN 978-3-319-54984-2 (eBook)
DOI 10.1007/978-3-319-54984-2

Library of Congress Control Number: 2017938533

Printed on acid-free paper

This Springer imprint is published by Springer Nature
The registered company is Springer International Publishing AG
The registered company address is: Gewerbestrasse 11, 6330 Cham, Switzerland

Contents

About the Editors

Roberto Álvarez Fernández is an Electrical Engineer and Ph.D. in manufacturing process engineering. Currently, he heads the Mechanical Engineering Degree in Nebrija University (Madrid-Spain). He has over fourteen years of experience as a Lecturer in electric machines, manufacturing processes, quality systems and lean management, environmental science, and electric vehicles. He acts as an expert on these topics for various companies. He is the author of numerous articles and contributions to scientific conferences related to lean manufacturing, energy and greenhouse gas emissions. High-profile journal outlets include Energy Policy, Energy, Land Use Policy, Journal of Automobile Engineering and Journal of Cleaner Production.

Sergio Zubelzu is an Agricultural Engineer and Ph.D. in Environment and Sustainable development. Currently, he works as an Assistant Professor in the Technical University of Madrid. He has over ten years of experience as a Lecturer in statistics, operation research, and hydraulic as well as a consultant engineer. His research activities mainly concern the methodologies to simulate physical and environmental phenomena. He has authored and coauthored several articles, and has contributed to scientific congress and books.

Rodrigo Martínez is an Industrial Design Engineer and Ph.D. in industrial design. Currently, he works at Design Department in Deusto University (Bilbao-Spain). His experience is centered on the development of design projects as a Lecturer in Projects of industrial design, design and environment or eco-design. He has developed and coordinated projects with several companies and agencies, such as Ecoembes, Lékué, Ikea, Retecork, or Finsa. He is the author of numerous papers and contributions related to the methodological process of design, including the importance of a life cycle in the conception of products and solutions. He has also served as a Reviewer for relevant journals in this area. He collaborates as an expert with the main agencies of design in Spain, as RED-AEDE or READ.

Introduction

As Elvis Presley sung: “a little less conversation, a little more action please”, in this book we have tried to show the reader some conversation and some action. Because this book has coincided in time with a series of circumstances that are promoting actions that will be determinant in the near future. For the first time, the road traffic in Madrid has been restricted due to the pollution levels. For many (inhabitants like us, or not, of the city of Madrid), this act can be seen as an anecdote more, an informative flash, and a warning sign. For others, this year 2017 represents the beginning of the actions.

In this book, we have selected authors from different countries (Spain, Japan, South Africa, France, Colombia, Sweden, Holland, Italy, etc.), aiming to reflect the globalism of the problem and trying to reflect this global magnitude through different perspectives.

Part I, sustainability and urban planning, includes contributions from UNESCO, Nebrija University (Madrid), Spanish nonprofit association Asociación Sostenibilidad y Arquitectura (ASA), Kagawa University and Osaka University in Japan, University Autónoma of Madrid (Spain), and Subdirección General de Planificación Regional (Madrid, Spain).

Part II explains the relation between urban planning and policy. It contains contributions from Valencia Institute of Building (Spain), Oslo and Akershus University College of Applied Sciences (Norway), Roskilde University (Denmark), University of Castilla-La Mancha (Spain) with University of Los Andes (Colombia), University of Maryland (USA) and Montpellier Business School (France), Rey Juan Carlos University (Spain) and Public Policy Institute (Belgrade, Serbia).

Part III displays the metrics for sustainability focusing on infrastructure. Works from University of Castilla-La Mancha (Spain), University of Hong Kong and University of Johannesburg (South Africa) explain methodologies and case estudies in order to clarify this subject.

In Part IV an interesting work from Valladolid University (Spain) displays innovative biotechnologies applied to CH_4 and N_2O removal, and assesses their potential as well as their current limitations.

Part V and Part VI compile works about eco-cities and eco-planning from University IUAV of Venice (Italy), Universidad Europea (Spain), University of Nottingham (UK), and Universidad de Alcalá (Spain) and two powerful chapters from the Faculty of Engineering and the Built Environment Johannesburg (South Africa) about the characterization of emissions and urban planning and low carbon cities.

Note that this book includes contributions from four continents, aiming to analyze from as many as possible points of view and different approaches the sustainability problem that affects this globalized world in we are living.

We, the editors, are unquestionably indebted to all the authors who have contributed with their work and efforts to make this book possible.

Roberto Álvarez Fernández
Sergio Zubelzu
Rodrigo Martínez

Part I
Sustainability and Urban Planning

Chapter 1
Urban Planning in Developing World: Which Alternative for Poor Cities?

Jean-Claude Bolay

Abstract The efforts made to plan cities in emerging and developing countries are confronted to multiple issues, especially in small and middle-sized cities, which can be considered as poor through several criteria: socio-economic level of majority of population; low levels of public investments, weak quality of local administration, and large dependence of external donors. Following several authors, one of the main reason is that philosophy and methods of urban planning applied to these specific contexts are directly reproduced from a Western tradition, which does not correspond to the local and national context in terms of needs, priorities and organization of the financial resources. The cases of Koudougou, a medium-sized city in one of the poorest countries in the world, Burkina Faso, with a population of 115,000 inhabitants, and of Montes Claros, an industrial blooming city of 360,000 inhabitants in Brazil, one of the most dynamic emerging countries in the world, will give the opportunity to make comparisons in order to understand concretely which and how these deficiencies are translated in an urban context. And foresee, more globally, alternative models of urban planning better adapted to medium-sized cities, focusing on the intermediation with their environment, in the perspective to offer new instruments of urban planning able to tackle in an efficient way the main constraints of their urbanization: growing population; territorial extension and fragmentation; environmental contamination and health; poverty and social exclusion, urban governance.

1 Between Theories and Praxis, Is Urban Planning More Than an Instrument of Management?

Urban planning is both a technique and a method of the observation and analysis of spatial, material and human reality. It is also a vision of what the city will be in the near and distant future. It is known above all by its projective and operational

J.-C. Bolay (✉)
Swiss Federal Institute of Technology, Lausanne, Switzerland
e-mail: jean-claude.bolay@epfl.ch

R. Álvarez Fernández et al. (eds.), *Carbon Footprint and the Industrial Life Cycle*,
Green Energy and Technology, DOI 10.1007/978-3-319-54984-2_1

actions. Beyond that reality, indirectly, urban planning is also under the influence of theories and reflections which have, along the decades, consider and analyzed the city transformation and the urban mutation of the world. Discourses evolved, in the same manner that reality changes, going from the city to the urban, according to Choay (1999). Paquot et al. (2000) underlined the apparent contradiction between the normativity of a globalized urbanization, following similar models, and yet in the complex ambiguity of a gigantic heterogeneity, born of distinct local and national histories, of social practices, and of natural, climatic, geographic environments. This complexity must be taken into account in our discourse on the city and urbanization. In the words of Harvey (2012), cities are the place where people of all kinds and all classes gather, willingly or unwillingly, to produce a constantly changing common life. These perspectives, from this point of view, allow us to keep some generic elements, which also apply to the South: The city is first and foremost a social system (Kilcullen 2012), characteristic of modernity made of societal interactions and techniques (Bolay and Kern 2011). And the city is changing under the influence of social, economic and technological transformations, forming an environment that is both natural and constructed, a form of urban ecology, confronted by endogenous and exogenous conflicts of interest. Its socio-spatial dimensions are polymorphic, variable and dynamic (Brenner and Schmid 2014), making it impossible to give it an unambiguous definition, accepted by all. As stated in Scott and Storper (2014), the debate is endless because, based on a multi-dimensional subject and on continuous transformation, different currents of thought confront each other, some to deny the prevalence of an urban singularity, others to decipher the different characters: global city, neo-liberal city, creative city, ordinary city, and post-modern city. Certain general trends must nevertheless be remembered; as this is a global phenomenon that now massively impacts emerging and developing countries. It is first the synergy that exists between economic development and urban development, cities being, at the global level, high-tech hubs, driving forces of contemporary economics (OECD 2006). Although these trends may be more diffused in highly industrialized countries with dense and diverse networks, they remain significant in many developing countries, with a city, usually the capital, sometimes accompanied by two or three large agglomerations, which polarize all the elements of modern technology and economic attractiveness. Infrastructure density and the sophistication of services are all assets, regardless of the country or regional context, which reinforce the centrality of the city, its dynamism and its social interactions.

This theoretical debate tends to delineate the scope of knowledge about the city, its past and present (Wheeler and Beatly 2014), and will inevitably affect the level of planning, based on the precept that urban planning marks the translation of concepts in approaches and methods, with impact when the transfer of these theories, essentially of western origin, are applied to “other societies” (Endensor and Jayne 2012).

Urban planning should be understood as the prospective achievement of theories aiming for real transformation in its material as well as social, economic, environmental, and political dimensions. Planning itself is not considered a science but rather a method applied through technologies adapted to needs in the field, based on precepts often not clearly defined, but guided by instruments capable of spatially and materially organizing the distribution of individuals, their activities, goods, services, facilities and equipment in a territory that is clearly identified and limited for geographical and administrative reasons. Urban planning takes into account the potential and limitations of the natural (spatial and environmental) and human entity in question, including in its analysis the causes and impacts of the dynamics that affect the transformation of the city and its people.

The difficulty encountered with urban planning is that it is based more or less explicitly on different disciplines (architecture, urbanism, engineering, economics, sociology, public management, etc.), without rigorous obligations to refer to, and with many professional practices generally used as a basis in periodic and repeated exercises. A contrasting critical reading, forwards and backwards, is needed to understand: first, explicit or implicit links with various theories relative to city and urban societies; secondly, the anchoring points between (a) the intentions highlighted by urban planning, the operations planned and the resulting implementation, and (b) the procedures followed and the instruments used for their application. Using real concrete cases, in specific local, regional, and national contexts, the results of such planning processes will reveal their direct and indirect impacts on urban society, and the influence that these methods, instruments and activities have in the configuration of territorial and societal transformations. Examining the implementation and outcomes of real cases of urban planning should lead us to review urban planning, in its precepts, its methods and its applications, based on social, spatial and environmental reality, so as to make available to urban stakeholders useful tools that will help to provide sustainable solutions to urban problems that confront urban populations, in their entirety (as these problems are repeated in almost all towns in developing countries) and in their characteristics (as these problems are expressed in a particular way in each context, depending on the history of places and people, of constraints, and of current and future potential).

2 Global South, Urbanization and Planning

First, it appears that the results of urban planning, as applied in emerging and developing countries, whether in Africa, Latin America, or Asia, only partially address the real problems facing urban populations. It is partial on the territorial level, covering only certain parts of the city, generally abandoning precarious housing neighborhoods, areas poorly regulated by law, and peripheries. It is biased on the socio-economic level, focusing primarily on the areas invested in by the privileged social actors of urban society, on the basis of their financial status, their relations to power, or even their community or ethnicity. Second, it appears that there is a caesura between urban policy makers, planners and residents. And this is largely because the fundamentals of planning were built in the North and then transferred to the South, without having been redesigned on a clear basis, but simply adapted to other physical, political and financial constraints.

Devas (2001) concluded from a comparative study conducted in nine cities of the South that in these cases overall, the standard chosen in infrastructure, equipment and buildings were totally unsuitable to the conditions of the poor, but that they formed a perfectly workable system of regulation by individuals with power, whether economic and/or political. And this is explained, according to Edensor and Jayne, in their introduction to the book "Urban theory beyond the West" (2012), by the fact that the methodological and technical discussions are dominated by Europe and North America. In this perspective, cities of the South are almost always considered less "modern" than Western cities. Yet cities in the South are gradually, and at an extremely rapid pace, inserted in the world of economic globalization. This economic and political integration of countries less marginalized than 20–30 years ago had the effect of commodifying even more the relationship between urban society and its territory, by investment priorities (public budgets facilitating the emergence and the strengthening of private operators), by the privatization of many collective services (water, energy, transportation, culture, public spaces, to name only the most obvious sectors). For Watson (2009), demographic and territorial growth of cities in the South inevitably causes a concentration of poverty and social, economic, but also spatial inequalities in cities. And urban planning, as applied in South Africa and in many other emerging countries, is not able to apprehend with anticipation and to solve a multitude of intertwined problems, between on the one hand the local needs, of each family, of each community, of each neighborhood, and on the other hand the production by the specialists of planning based on the entire urban territory, but with inexplicit priorities.

Precarious living conditions, at different levels, are reflected by a continuous expansion of slums on the fringes of the models adapted by policy makers and planners, but central to urban issues, by the increasing number of people living there, by the key problems to be addressed to vision and implement a more inclusive and coherent city. As expressed by Roy, the slums reflect both a territorial exclusion, through the lack of equipped spaces accessible to the poor, and the tensions arising from the occupation by the poor of land that sometimes becomes very attractive for the market development of the city (Roy 2004). Which does not prevent, with very few planning and realizations, the inhabitants of the slums to greatly contribute to the economic and social dynamics of the city: job creation, income generation, community organization, and social and political participation. Also according to Watson, the question is twofold, first the models that inspire these professional practices come from the North, based on totally different socio-spatial contexts; on the other hand, planning is primarily viewed as an implementation of future territorial interventions, in a purely technical posture, with little concern for local urban history, the players involved, the interests at stake and, more generally, a societal vision taking into account all of the urban community, in its various components, especially the poor who, as is often the case in cities in South, represent the majority of citizens. The result is often that investments that are often poorly targeted and do not address the crucial questions that the majority of urban dwellers are facing. The question arises in terms of equipment choices to focus on, but also, and most often, in terms of conditions of accessibility, as costs are not adapted to the financial conditions of the most underprivileged segments of the population. These changes have induced increasingly marked urban land fragmentation, the gentrification of neighborhoods according to their level of equipment, and a socio-economic segmentation of the functions and uses of the city. The alternative, according to Yftachel (2006) would be to question five key dimensions of urban development: land-use and its allocation criteria; the policies put in place to fight against segregation; decision-making procedures, in a way that adapted forms of social participation fight against exclusion of the urban poor; the consideration of the socio-economic conditions of urban dwellers as a whole and particularly groups in the disadvantaged population; as well as the financial impact of urban transformations, especially in terms of increasing the value of land and property of rehabilitated neighborhoods. Classic planning procedures, as believed in the 60s, are no longer appropriate. It can no longer be designed in a linear and progressive perspective, insofar as it applies to rapidly changing territories, whose evolution is often misunderstood. It must, on the contrary, accept a certain level of uncertainties, which, to be understood and accepted, need to be integrated into the planning process itself, by a communication and collaboration effort between stakeholders (Woltjer 2000).

Bearing in mind the specificities of the majority of cities in the South, the first variable to consider in appropriate planning in these contexts is the fight against urban poverty (UN-Habitat 2010) in its different facets and consequences at various levels. Tannerfeldt and Ljung (2006) emphasize that urban poverty is relative, considering the local context and the social and economic inequalities. Financial and economic criteria are not the only ways to characterize urban poverty. In addition, indicators based on health, education, environmental quality, violence and insecurity, represent a multitude of risks that these poor families must confront. Tacoli et al. (2015) designate urban sectors where the poor are systematically marginalized: insecure tenure, poor quality housing and a lack of public provision for infrastructure. Which explains that, faced with these shortcomings, residents of precarious districts must cope alone to solve these basic needs. A significant number of variables must be included in the planning process, taking into account interactions between social, ecological, economic, political and administrative dimensions. This should lead us to think of urban planning as a creative approach, innovative, ready to face the unexpected, and not as a routine approach doomed to failure (Grunau and Schönwandt 2010).

3 Poverty and Urban Management

This reflection is posed urgently and with high priority when it comes to cities located in emerging and developing countries; in particular with respect to small and medium-sized cities (Bolay and Kern 2016). Overall, these are the areas that suffer the highest rates of population growth; it is their authorities who suffer most from the lack of financial and human resources to be able to anticipate and address these issues. And these are the inhabitants, at local and regional levels, who suffer the consequences in terms of human and material precariousness, contamination of natural resources, informality of economic activities, a malfunction in the process of decision-making and governance. And it is this urbanization that will in the coming decades be put under increasing pressure, knowing that 95% of urban growth will primarily impact emerging and developing countries, and first of all in intermediate cities. Depending on the regions of the world, the process of a rapid urbanization is directly linked with a growth of poverty and more socio-economic disparities. For Davis (2006), slums are the prominent feature of contemporary urbanization. He focuses on its negative aspects such as violence, insecurity, informality and poverty, which, in his opinion, are the result of the economic power relations of a globalised world. Around one billion of people in the world live in this kind of poor settlements (Bolay et al. 2016). The figures collected by the United Nations on this issue in 2011 (UN-HABITAT 2010) show that their expansion varies widely according to regions of the worlds. If, on a global level, about 32.7% of the world urban population live in slums in 2010, this concerns 61.7% of the sub-Saharan population in Africa, 35% in Southern Asia, 31% in South Eastern Asia, when 23% in Latin America and the Caribbean, and 13.3% in Northern Africa. With, as demonstrated by Mboup (2004), the relative significance of the individual identification parameters of slums on a global level: lack of secure tenure 70%; lack of durable housing 65%; lack of sufficient living space 60%; lack of improved sanitation 50%; lack of improved water 20%.

All these characteristics allow us to speak about "poor cities", meaning cities in Southern countries confronted with different forms of precarity, having a high percentage of dwellers living in sub-standards conditions (in terms of incomes as in terms of habitat and access to services). The situation is particularly critical in intermediate cities, between 20,000 and 500,000 individuals. On one side because they include around 50% of the whole urban population in the world (United Nations 2014), on the other side because they represent the fringe of cities with the higher level of urban growth. In front of that, theses "ordinary cities" (Robinson 2006; Parnell and Robinson 2012) are generally not under the political attention of the central government, and their public budget are relatively low to tackle all the urgent questions to resolve and investing for the future. And priorities of investments, as we shall see in the 2 case studies, are not decided based on a rigorous diagnostic involving a process of urban planning at long term, but more in relation with opportunities or under pressure of powerful stakeholders, inside the city or as powerful outsiders.

4 Urban Planning in Practice: The Cases of Koudougou, Burkina Faso, and Montes Claros, Brasil

At first view, these two cities have few points in common. Koudougou, a little bit more than 100,000 dwellers, is located in Burkina Faso, one of the poorest countries in Africa (Bolay 2015); when Montes Claros, 400,000 inhabitants, is considered as an intermediate city in the State of Minas Gerais in Brazil, one of the most powerful emerging countries (despite the crisis) in the world (Bolay 2016).

But some convergences exist between the 2 cities, too. Both of them play an important and growing role of intermediation linking the city, their hinterland, the region, and other cities in the country, even at international level (more than all through industry in Montes Claros, and with a strong support of external agencies of cooperation in Koudougou).

Koudougou is the capital of the province of Boulkiemdé and represents a pole of activities for this regions, with its big city's market, shops, banks, province and local administrations, and number of delegations of national ministries, university, among other assets. The city is a real platform of exchange, with the agriculture producers of the region, and with Ouagadougou, capital of the Burkina Faso, only 100 km from there with one of the rare excellent roads in the country. These elements explain in large part that urban growth of Koudougou stays largely dominated by a continuous flow of rural immigrants. In turns, the city of Montes Claros, even not capital of the State of Minas Gerais, is the main urban pole in the Northern part of this State. Having during decades be the center of distribution of agriculture products from this region, these activities continue, but gradually overtaken, since the 1970', by the establishment of many industrial enterprises, national and foreign branches, and more and more by new private and public services for the whole population of Northern Minas Gerais (universities, hospitals, shopping centers, banks and so on …) (Figs. 1, 2).

Fig. 1 Suburb of Koudougou (*Photo* Bolay 2014)

Fig. 2 New social housing settlement in Montes Claros (*Photo* Bolay 2015)

4.1 Lessons Learned from Koudougou

There have been many plans drawn up for the city of Koudougou over the past fifteen years—some with an overview as are the local development plans or master plans—or more sectoral as the strategic sanitation plan or strategic household waste plan. Their advantage is that they give a picture of the investments to be made to improve the situation in the municipality. Their great weakness is that they are not executed because they are out of step with the financial means and the competencies of the municipal administration. At best, they serve to reassure donors during financial negotiations. We face this fundamental contradiction as we face the objectives of urban planning. What do the stakeholders say? What are the intentions of this urban planning that has been largely developed in Burkina Faso since the 2000s, in line with the strategies of international donors that support the government in its development efforts?

The plans do not meet the traditional goals to develop the future of Koudougou, but mark the ambitions of the authorities facing the organization of the territory. And it is clear that challenged by the multitude of problems to solve, it is difficult, both politically and technically, to set priorities in terms of areas and sectors. Everything immediately becomes a priority, without criteria that justifies choices made. And consultation frameworks between policy makers, operators and population—if they are desired and recognized useful—are gradually set aside, for lack of resources and available time. In fact, much focus is put on the tool, implementing a proven technicality (that of consulting firms mandated to do this), and on the indicative outcomes—and very little on the approach and the objectives of the application of this instrumentation.

The second very large problem encountered is related to the conditions of the production of urban development plans, in Koudougou as in other cities of Burkina Faso. Three stumbling blocks: the first is that these local plans are decided by the national government and "imposed" on the municipalities; the second comes from the fact that the municipal administration of Koudougou, like many Burkinabe cities, is under endowed with competent personnel, and is thus unable to participate in the design, the supervision and the monitoring of this production of local plans. Neither does the glaring lack of financial resources allow the Municipality to implement this planning that takes more than wishful thinking and is primarily used for advocacy with funders, rather than as a guide in an assured control of local urban development.

And the conclusion of all the experts agree that planning tools exist but are not used as such, and prove to date to be unnecessary. Indeed, it is revealed by the study, they are diverted from their original purpose and become an object for urban marketing and communication with donors, since all donors think it essential that each city where they intervene includes planning. In the words of one amused speaker: the plan is a catalog of all that must be done in the municipality, donors choose what they want to finance! Two rationales conflict with the overall interests of coherent and sustainable urban planning: first, the aspirations of foreign donors, to which national and local authorities will submit, are a priority and guide investments; second, the will of national and local political leaders is to make their mark on the territory by occasional symbolic "gestures" of their presence in power, rather than stewardship over the long term.

4.2 Lessons Learned from Montes Claros

Montes Claros is a perfect example of an intermediate Brazilian city. While it would be incorrect to describe it as a poor city, the urban population is highly segregated; one third of the city's population is poor, and the wealthy—who represent 20% of the population—control 66% of the local wealth. This fragmentation of the social fabric is also reflected in the more than 140 unregulated neighborhoods and 50,000 people living in makeshift housing conditions.

In addition to these socio-economic and territorial disparities, two other major problems exist. The first is the depletion of natural resources due to advanced deforestation and periodic flooding of central neighborhoods, a problem that largely results from the construction of new housing developments (social and luxury), wherein the environmental impact of these forms of urbanization, which have become increasingly popular in the past 20 years, are not fully considered. The second is the lack of regional integration of the municipality's rural hinterlands. These areas, where services and telecommunications are still rudimentary, are gradually being abandoned as purely agricultural region and progressively transformed in new peri-urban settlements.

To address this situation, the local authorities are in the process of developing a new master plan. As such, the plan aims to create a sustainable city, including a right to urban land, housing, infrastructure, services, transport and employment for everyone.

The critiques lean in two directions:

First, the implementing of decisions made by the council falls on technical teams and administration members who have neither a guaranteed job in the long term, nor attractive salaries. Secondly, political changes in the city's administration have direct political repercussions on the continuity of urban alternatives, with each new mayoral team seeking to set itself apart from its predecessors and to "leave its mark".

The current process raises many issues. To begin, it is impossible to discern the authorities' medium- and long-term vision, which would be helpful in streamlining the urban planning. Furthermore, the authorities' perception of Montes Claros is biased as it only takes into account the dense areas in the city center. The rapidly growing outskirts and suburban areas are still disconnected from the rest of the city and are poorly integrated in this prospective exercise. In the future, Montes Claros must be seen as an urban hub for northern Minas Gerais, in a kind of urban/rural/interurban interplay that involves environmental, social, educational and housing issues. All neighborhoods and people cannot be dealt with in the same way; specific needs and priorities must be taken into account. While there is undeniably cooperation with national and regional authorities in the implementation of the current plan, it is not a participatory master plan in the eyes of many respondents. Considering the current process is that, once completed, the master plan should be supported by enabling legislation. Without it, the plan cannot be fully functional.

This ongoing process may be can be seen as an associative process, but not a veritably participatory one. City Hall has indeed teamed up with private agencies and representatives from different sectors (universities, representatives from the economic community, and the ad hoc committees of the City Council), but without including the public in the initial phase of the process. Instead of collecting opinions and request of the population, it is only now in the final phase that the provisional results, which will determine the guidelines for future planning, have been presented and discussed in the more urbanized areas of the city. This exclusive approach has received considerable criticism from "outsiders", particularly volunteer organizations and social groups, who were not invited to participate in the process from the beginning. For them, this denotes a certain authoritarianism on the part of the public authorities, who have not seized the opportunity the new plan affords to engage in an open dialogue with the public to better understands its needs, desires and vision of their urban future.

Though Montes Claros cannot be described as a "poor city", it can be described as an intermediate city that, like many other cities in Brazil, is experiencing the kind of unbalanced growth that is typical of what one can see in cities around the world. Brazil has become one of the most segregated countries in the world, an emblem of economic globalization in a growing struggle between global regions and cities (nationally and internationally) and strategies to improve social cohesion and cultural/economic integration, reflected politically in terms of urban planning.

5 From and Towards the South: Rethinking Urban Planning

Urbanization in emerging and developing countries is today an irreversible trend of the transformations that shape the world, and this regardless of the country where one looks, and whatever the city concerned. Driven by mass migration of rural populations to urban centers and by natural growth that is still very strong, sustainable urban development is a question that confronts all cities, in priority in Africa and Asia, and in all small and medium-sized cities of Southern countries, with identical problems: organizing the planning of territory in continuous demographic and spatial expansion, and, in parallel, doing its best with limited financial and human resources. If the very large cities, national capitals or economic and political hubs enjoy comparatively more marked attention from governments and funders, the smaller sized agglomerations remain neglected and almost invariably face a multitude of glaring necessities, with no real means to respond to social demands for investing in infrastructure and community facilities that match the identified needs. These questions remain unanswered, addressed mainly in an emergency or according to the allocations provided sporadically by national governments or by foreign financiers.

This urban context of great insecurity and uncertainty about the future of these small and medium-sized cities allows us to speak of "poor cities", not only because a large part of their citizens actually live on the edge of destitution, but also because urban authorities are poorly equipped in means to assume the investment that would be needed to improve the daily life of all residents.

In this context, urban planning must be completely rethought and taken out of the patterns of development models that were designed for completely different environments, being for the most part the simple reproduction of defined standards and rules implemented in Western countries by specialists whose credentials are totally foreign to global South. The major risk in this situation is that of projecting these cities on a basis that will serve only the interests of a minority of citizens, favoring the most advantaged in economic and territorial development, and leaving the modest population on the margins, those living in informality and in the most underserved neighborhoods. But it also represents an extraordinary opportunity to think about the future from what exists, taking into account the real resources, not only financial but also social, to devise and implement urban planning with the goal to fight against poverty and invest in equipment that has a sustainable impact on the living conditions of the poor.

As such, the analysis carried out in Koudougou, Burkina Faso, is very instructive. Provincial capital at 100 km from Ouagadougou, Koudougou is a commercial and political center of a large rural area, but it has also long been considered as a rebellious city, unwilling to obey the requirements of the Burkinabe central government. In Koudougou, as in many intermediate African cities, the urban planning process is exogenous, not really consistent with the requests of the people, nor with the human, material and financial resources of the city, and therefore rarely applied.

This is easily explained when we know that urban planning in its design, is initiated as part of a collaborative framework between the central government and foreign donors. The initial diagnosis is made by quality professionals but who are disconnected from local administrative and social realities. In fact, it is a census of all needs to be met, but without a guidance manual! How then the facilities to be created whose costs are more than ten times that of the annual municipal budget reserves? In fact, plans produced in this context do not serve to guide local authorities in the current and future development of the urban territory. Neither are they an instrument of dialogue between the said authorities and the population. On the contrary, any consultation with the community that does not result in expected and desired deliverables will strengthen the distrust, or even defiance towards public, political, and administrative powers. At best, the plans, losing their principal essence, become promotional tools, pure marketing products, a catalogue of intentions of penniless communities at the mercy of the donors' desideratum, whether they be State or foreign cooperation agencies.

With a population of nearly 400,000 inhabitants Montes Claros, in Brazil, with much more resources than in Koudougou, cannot be described as a poor city, but the urban population is highly segregated; one third of the city's population is poor. This fragmentation of the social fabric is also reflected in the more than 140 unregulated neighborhoods. In addition to these socio-economic and territorial disparities, a major problem exists: the lack of regional integration of the municipality's rural hinterlands. These areas, where services and telecommunications are still rudimentary, are gradually being excluded of urban and regional planning.

Today, the municipality is facing the issue of social exclusion, with an increasing proportion of the urban population living in economic insecurity and relative marginalization, both geographically and socially. These are key issues for a comprehensive urban plan whose aim is to not only act spatially. This ongoing process, which will determine the guidelines for future planning, is primarily focused on the more urbanized areas of the city.

This distortion of urban planning is dangerous, since it destroys any coherence in a more regional perspective, both in establishing priorities in infrastructure and equipment to realize, in the economic and social sectors to be favored, as in the implementation timeframes, without any possible guarantee that things will be done on time, potentially creating more long term disorganization than anything else.

Urban planning in developing and emerging countries must be entirely reconsidered. The essential point—too often overlooked—is to begin from a participatory diagnosis in which the actual situation of the city is examined in its various dimensions, both demographic and spatial, infrastructural, but also economic, social and environmental, permitting all the stakeholders to position themselves. This information, cartographic, as well as documentary and anthropological, will serve as the foundation for the establishment of a database that can then be fed in real time, facilitating the monitoring of "urban development" and a collaborative, up-to-date decision-making process. In parallel is the question of establishing priorities, in terms of structures to be built, but also in terms of standards, rules and plans tailored to the context, with regard to the needs identified by specialists, to requests from the different social actors, as well as the available resources, both local and from external sources. Two principles should guide this work: first, that urban investments be directly or indirectly involved in the fight against poverty; second, that an overall coherence guide the specific actions in the short, medium and long term. These precepts can only be applied if the framework conditions are respected: local and regional governments must be given the human competencies and financial resources enabling them to act. And it is not impossible if the political will is there, and is based on a legitimacy in the eyes of the population. And this inevitably involves consultation frameworks that will animate the dialogue between representatives of the population, public administration, political powers, industry professionals and other special interest groups (private sector; social, religious, and political groups; NGOs; etc.). Training plays a key role, as does communication and dialogue. And it is these same guidelines that should guide implementation. Here too there is room for innovation, starting with the social practices and deployed human dynamics, beyond any formalism, on the local and regional levels. While it is clear that it takes the technical know-how of experts of the city and businesses, we must also remember that the own inhabitants did not wait to take the place of these players—too often absent—to build their houses, build community facilities, to better manage their neighborhoods.

These vital forces should neither be overlooked, nor marginalized; they are the center of a participatory process that is not limited to the consultation but goes from conception to action. They must be integrated into the planning process and thus contribute concretely to the implementation of decisions taken collectively. Communication is also a key issue. How to learn from other cities via the Internet and increasingly frequent global exchanges on urban matters. Whether we look at the frequent international summits on these issues by the United Nations or at visits of municipal delegation between continents. If, as indicated by Campbell (2012), we learn from near and far, and learning is no longer unilateral from North to South, but also from South to South and South to North. There nevertheless remain three reservations in this regard: first, urban technological innovations, even from emerging countries, are first found in the largest and richest agglomerations; in return, small and medium sized cities remain on the margins of these innovative processes and rarely have the opportunity to apply them, for lack of means; the question of precarious housing and urban poverty is generally treated as a problem, without utilizing and applying in similar contexts the lessons learned from this situation. These views are not so arrogant as to believe that all problems will now magically be solved, but rather that we are breaking out of a vicious circle in which urban planning is not playing its role, totally disconnected from a complex and changing reality. And put forward, as an innovative alternative, a more realistic vision, more pragmatic, based on what exists and focusing the efforts of all the citizens in favor of a gradual improvement in well-being for all, giving priority to the most deprived urban citizens.

References

Bolay, J.-C. (2016). Prosperity and social inequalities: Montes Claros, how to plan an intermediary City in Brazil. *Current Urban Studies, 4,* 175–194. doi:10.4236/cus.2016.42013

Bolay, J.-C., Kern, A. L. (to be published in 2016). Intermediary cities. In *Wiley-Blackwell encyclopedia of urban and regional studies*. Hoboten: Wiley.

Bolay, J.-C., Chenal, J., & Pedrazzini, Y. (2016). *Learning from the slums: the habitat of the urban poor in the making of emerging cities*. London: Springer. doi:10.1007/978-3-319-31794-6

Bolay, J.-C. (2015). Urban Planning in Africa: Which alternative for poor cities? The case of Koudougou in Burkina Faso. *Current Urban Studies, 3,* 413–431. doi:10.4236/cus.2015.34033

Bolay, J.-C., & Kern, A. (2011). Technology and cities: What type of development is appropriate for cities of the South? *Journal of Urban Technology, 18*(3), 25–43. doi:10.1080/10630732.2011.615563

Brenner, N., & Schmid, C. (2014). The "urban age" in question. In N. Brenner (Ed.), *Implosions/Explosions. Towards a study of planetary urbanization*. Berlin: Verlag.

Campbell, T. (2012). *Beyond smart cities. How cities network, learn, and innovate*. London: Earthscan.

Choay, F. (1999). De la ville à l'urbain. *Urbanisme, 309,* 6–8.

Davis, M. (2006). *Planet of slums: Urban involution and the informal working class*. London: Verso.

Devas, N. (2001). Does city governance matter for the urban poor? *International Planning Studies, 6*(4), 393–408.

Endensor, T., & Jayne, M. (2012). *Urban theory beyond the West. A world of cities*. New York: Routledge.

Grunau, J.-P., & Schönwandt, W. L. (2010). Dealing with society's "Big Messes". In G. De Roo & E. A. Silva (Eds.), *A planner's encounter with complexity* (pp. 41–62). Surrey: Ashgate.

Harvey, D. (2012). *Rebel cities. From the right to the city to the urban revolution*. New York: Verso.

Kilcullen, D. J. (2012). The city as a system: Future conflict and urban resilience. *Fletcher Forum of World Affairs, 36*(2), 19–39.

Mboup G. (2004). Slum dwellers estimates country level. In *Training workshop on data and indicators to monitoring progress towards the MDGs and the habitat agenda*, 1–5 November, Bangkok.

OECD. (2006). *Territorial reviews: Competitive cities in the global Economy*. Paris: OECD.

Paquot, T., Lussault, M., & Body-Gendrot, S. (2000). *La ville et l'urbain, l'état des savoirs*. Paris: La Découverte.

Parnell, S., & Robinson, J. (2012). (Re)theorising cities from the global south: Looking beyond neoliberalism. *Urban Geography, 33*(4), 593–617.

Robinson, J. (2006). *Ordinary cities: Between modernity and development*. London: Routledge.

Roy, A. (2004). Transnational trespassings. The geopolitics of urban informality. In A. Roy, N. Alsayad (Eds.), *Urban informality. Transnational perspectives from the Middle East, Latin America and South Asia* (pp. 289–317). Oxford: Lexington Books.

Scott, A. J., & Storper, M. (2014). The nature of cities: The scope and limits of urban theory. *International Journal of Urban and Regional Research, 39*(1), 1–15.

Tacoli, C., McGranahan, G., & Satterthwaite, D. (2015). *Urbanisation, rural-urban migration and urban poverty*. London: IIED.

Tannerfeldt, G., & Ljung, P. (2006). *More urban less poor. An introduction to urban development and management*. London: Sida & Earthscan.
United Nations, Department of Economic and Social Affairs, Population Division (2014). *World urbanization prospects: The 2014 revision, highlights*. New York: United Nations. https://esa.un.org/unpd/wup/publications/files/wup2014-highlights.Pdf. Accessed December 22, 2016.
UN-Habitat (2010). *State of the world's cities 2010/2011. Bridging the urban divide*. London: UN-Habitat & Earthscan.
Watson, V. (2009). Seeing from the South: Refocusing urban planning on the Globe's central issues. *Urban Studies, 2009*(46), 2259–2275.
Wheeler, S. M., & Beatley, T. (2014). *Sustainable urban development reader (Routledge Urban Reader Series)* (3rd ed.). London: Routledge.
Woltjer, J. (2000). *Consensus planning: the relevance of communicative planning theory in Dutch infrastructure development*. UK: Ashgate.
Yftachel, O. (2006). *Ethnocracy, land and identity politics in Israel/Palestine*. Philadelphia: University of Pennsylvania Press.

Chapter 2
Social and Economic Management of Sustainable Neighborhoods Regeneration Projects

Alejandro Bosqued, Francisco Javier Gonzalez and Susana Moreno

Abstract The aim of this paper will be to analyze how social and economic management affects on sustainable neighborhood regeneration projects. Specifically, will go in depth of processes of social and economic management in Spanish neighbourhood regeneration processes, where environmental and social benefits are the main goal of the intervention, but finance is, nowadays, an obstacle. Therefore, this study of real cases will offer a view on which are the environmental objectives, how these projects are funded in Spain and what is the role of neighbors in the process. On the other hand, will help to identify different possibilities of funding to get this type of projects through successfully. A mapping of significant existing projects will be necessary to reach the main objective of the research, which will produce unpublished documentation, useful for future research on this matter, identifying cases of success and failures.

1 Introduction

1.1 Evolution of Intervention Practice. Last Experiences of Urban Regeneration in Spain

During the years of economic prosperity, the public administration promoted a wide range of experiences of urban regeneration. If we had to do a quick typing, we could point out some cases and kinds of intervention as an example exposure.

A. Bosqued (✉)
Department of Architecture, Universidad Nebrija, Madrid, Spain
e-mail: abosqued@nebrija.es

F.J. Gonzalez · S. Moreno
School of Architecture, Engineering & Design,
Universidad Europea de Madrid, Madrid, Spain
e-mail: franciscojavier.gonzalez@universidadeuropea.es

S. Moreno
e-mail: susana.moreno@universidadeuropea.es

R. Álvarez Fernández et al. (eds.), *Carbon Footprint and the Industrial Life Cycle*,
Green Energy and Technology, DOI 10.1007/978-3-319-54984-2_2

1.2 The Last Generation of Interventions in Historic Centers

Following the conservationist success of the plans for the protection of historic centers during the 1980s and 1990s, the proposals have pointed in recent years to combine the tourist and commercial capacity of their heritage by restoring these urban spaces, which lost vitality, both functional and demographic.

In Santiago de Compostela case of study (MAP 1) there are programs to link public housing to public rent with the refurbishment of buildings or the implementation of "green strategies" to improve urban metabolism of these tissues, with new points of view on the use of water, naturalized spaces and energy.

In Lavapiés, Madrid (MAP 2), public investment has focused on the re-qualification of public spaces and facilities as strategy for integrating the cultural diversity of a neighborhood that has 42% of foreign immigrant population belonging to 113 different places of Origin. Without abandoning the rehabilitation of housing, the investment in equipment of metropolitan rank has not meant the deployment of a mass gentrification process and nevertheless has allowed to developing a way of coexistence.

Similar experiences, adjusted to the local peculiarities, have been developed in many Spanish cities, among which we could highlight the rehabilitation of the Casc Antic and the discussed renovation operations in the Raval of Barcelona (MAP 3), the performances in the squares of the Historic center of Seville (MAP 4) and the recovery of plots as squares, in a remarkable performance for the recovery of the center of Zaragoza (MAP 5), to highlight some examples (Fig. 1).

1.3 Retrofitting Developmentalism Peripheral Neighborhoods of the Twentieth Century

Entering into use of urban regeneration peripheral neighborhoods, has marked the way to follow on which will be the activity in the years to come.

Zaragoza has been one of the pioneer cities to consider the set of neighborhoods built since the 1940s in a massive way, with typologies mostly of exempt dwelling blocks from collective housing.

Improvements in energy efficiency of buildings, have begun to incorporate as a fundamental part of the menu of actions, in these rehabilitations as update of housing.

Along the same lines, San Cristóbal de los Angeles neighborhood, in Madrid, has begun to experiment with bioclimatic restorations in some of its buildings. Another of the outstanding experiences of this example, is the organization of a Citizen Action Plan as a participation mechanism linked to the regeneration of the neighborhood, although the expectations of this process have not been satisfactorily fulfilled until its last objectives.

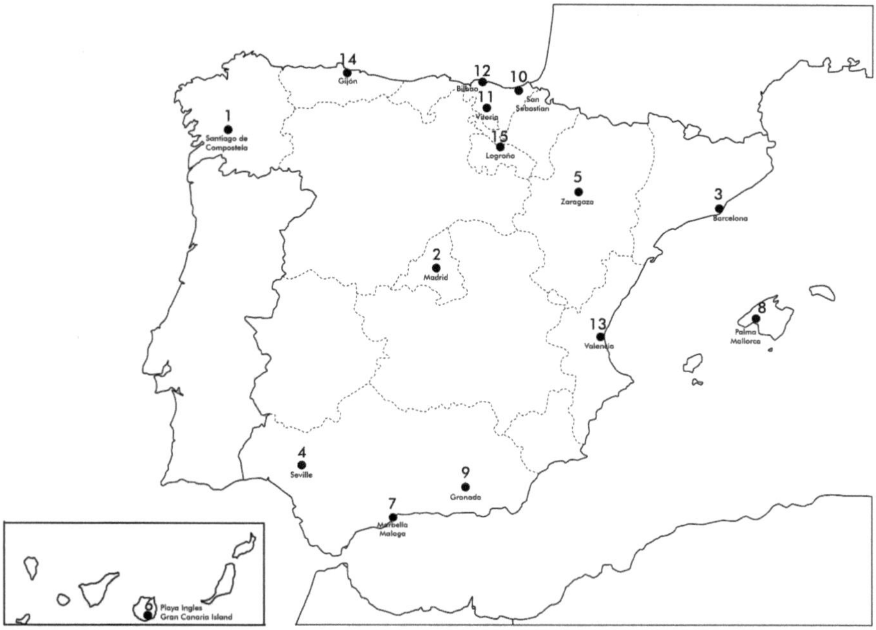

Fig. 1 Map of main urban regeneration actions in Spain

1.4 Addressing Social Exclusion from Urban Regeneration

These attempts to satisfy social demands in the processes of urban regeneration have been more successful in the case of the La Mina neighborhood in Barcelona. Problems of social exclusion of gypsy origin population in the neighborhood have been considered central in the process.

The renovation and physical reorganization of the neighborhood is only a part of the actions, with the most important investment item being the social policy package (15%) together with the re-qualification of public facilities and facilities (25% among two).

The axis of the urbanistic proposal transformation, resides in the construction of an urban rack in the middle of the neighborhood, as a new central space, a backbone that connects new buildings and existing ones. A place of relationship for new civic, social, economic and cultural activities.

With these elements as a starting point, actions such as the incorporation of new housing—both protection and free market- to the neighborhood, as well as the implementation of a University of Barcelona site, seeking to increase social diversity.

1.5 Taking Action in Urban Tourist Tissues

In recent years there has been a need to retrofit the tourist space. With lots of secondary residences, many of them with pathologies appeared because of the time elapsed since its construction and its lack of adaptation to current tourism and residential quality standards. These "seasonal neighborhoods" are numerous in the Mediterranean area and in the Canary Islands, and it is here that public-private consortia have been set up to regenerate some representative beaches Playa del Inglés on Gran Canaria Island (MAP 6), the beaches of Marbella and Málaga (MAP 7) or Playa de Palma in Mallorca (MAP 8).

This last one is the one that has managed to finalize a regeneration project in which the novelties go through the search for a metabolism of "Carbon zero" and the fixation of the resident population throughout the year in the neighborhood.

1.6 Strategic and Sectorial Regenerations

These kind of actions are the ones that have proliferated the most over the last 20 years in Spanish cities. There are different actions, and to mention some of them, we can point out several referent examples:

- **Sustainable mobility and regeneration of public space actions**.
  The first sustainable mobility plans had the cities of Granada (MAP 9) and San Sebastián (MAP 10) as pioneering scenarios. With the generalization of public aid, this type of plans has proliferated in most cities with over 50,000 inhabitants. In recent years the use of tram and bicycle have been used as spur to deeper regeneration processes, like Vitoria (MAP 11), Parla in Madrid or Seville.
- **Recovery actions of maritime fronts and river bank spaces in cities**.
  After the Olympic experience in Barcelona and the Ría de Bilbao (MAP 12) with the Guggenheim Museum as a star, many cities have reorganized their port area looking out for an exit to the sea for the city, which previous infrastructures denied.
  The result is the recovery of an urban beach space and regeneration of the traditional neighborhoods linked to the ports in cities such as Valencia (MAP 13), Málaga, Gijón (MAP 14) and many others.
  In an analogous sense, the recovery of urban contact with rivers has led to operations like Zaragoza Expo or Madrid-Rio linear park, which is related to a previous burial of an urban motorway and the growth prospect of 20,000 homes to be regenerated in the neighborhoods in contact with the linear park in Madrid.
- **Actions to regenerate urban natural spaces**.
  In these years, a deepening treatment of urban green areas has been pursued in order to increase the diversity of their biomass, their operational role in the urban metabolism and to convert them into areas of real contact with nature.

Periurban, riverbeds and banks has been widespread in recent years. Perhaps the most successful example is the green ring of Vitoria, which main strategy was recovering a natural space of high ecological value right next to the city, such as the lagoons of Salburúa.

- **Reinforcement actions or generation of urban centrality**.
  These types of operations have been very numerous. Many of them have been carried out over old industrial or railway installations or on military spaces that have left the city. Examples like Prolongación de la Castellana operations in Madrid or the establishment of Ave stations in cities such as Logroño (MAP 15), Valencia, etc.
  An interesting case, the 22@ neighborhood in Barcelona, because it incorporates in an obsolete industrial tissue an advanced tertiary space, to accommodate economic activities linked to both business and university innovation, maintaining more than 4000 pre-existing homes; All incorporating criteria of sustainability that have allowed to reformulate the patterns of mobility of the zone and the use of the energy.

2 Urban Regeneration Model: Environmental, Social and Economic Aspects. The Specific Case of Madrid

For the case of urban regeneration in Spain, speaking of an intervention instrument is somehow elusive. There are no one, but several and clearly little articulated among them: urban planning, sectoral policies of equipment or public space, laws and technical standards that affect the built and its property, etc. This will be the scenario without entering into social care from minority integration policies and gender policies, resisting against unemployment, youth, or care for dependency, among others.

Instruments that intervene in urban regeneration can be resumed pointing out three columns: the first one, formed by financing rehabilitation policies and social housing, understood as a redistributive mechanism.

Among them are Rehabilitation Integrated Areas (ARIs), urban spaces delimited with operational criteria and where public aid to the private rehabilitation of housing have more intensity than in other areas of the city. The last column is made up of various legislation, the instruments of which are not connected to one another, but which affect rehabilitation processes, such as horizontal property law, historical heritage law, Leasing laws, etc. These three columns should be completed with all kind of sectoral urban policies that affect, in a subsidiary way, the planning and intervention in regenerated neighborhoods, such as mobility policies, urban sustainability, specific social policies or regulating labour market, etc. Almost always external factors to neighborhood processes itself. However, we can say that there is

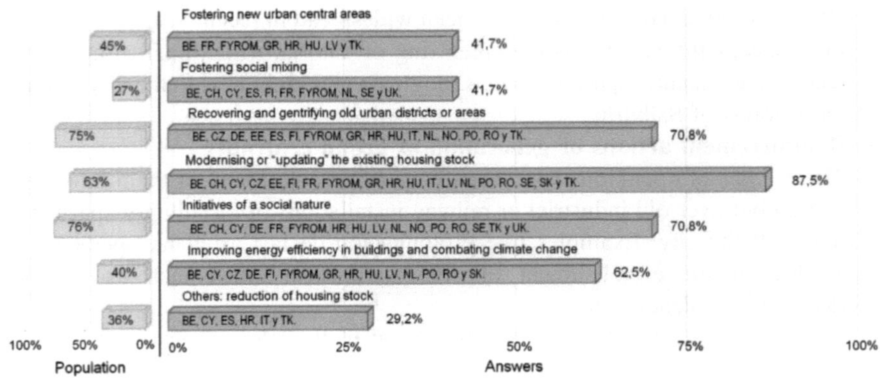

Fig. 2 In your country, does urban regeneration correspond to a specific policy or is it associated with or an integral part of other policies? (Alvarez Mora et al. 2010)

a conceptual model, theoretical, but emerged and fed from institutional practice, which guides how things should be done and understood as a set of knowledge and attitudes shared by agents involved in the processes. It is a model that we have commissioned the regeneration of our neighborhoods culturally and that has a vocation of integral action.

As a conclusion: in this model of action, the integral part is what it considers to act on the whole of the urban structure and the citizen as the actions beneficiary, almost all of a physical nature, because they improve their incorporation into society as a whole or, at least, allows to keep its share capital. To this end, the actions focus on four pillars: habitability in housing and in building, improving or eliminating por or substandard housing, but also combating energy poverty or incorporating vertical accessibility of elevators as something that can not be waived.

The second pillar is access to equipment and other urban services.

The third is improvement of collective meeting spaces to strengthen social capital and, finally, promotion of local economic activities.

It is evident that the practice does not accompany the model in Alfonso Álvarez Mora and Fernando Roch work (Álvarez et al. 2010), on European urban regeneration, where the results of a survey of technicians belonging to public administrations in EU countries are given. In Spain, regenerative activity is not associated with either social or environmental policies, but with patrimonial protection (Fig. 2).

For this purpose, the discussion is to see if the processes of urban regeneration in Madrid have considered these premises by which social cohesion is shed.

2.1 *Coincidences and Instruments in a Rather Diffuse Model in the Situation of Madrid. ARIs—Integration Rehabilitation Areas*

In line with the Spanish and European situation, in Madrid there has not been an explicit and unified instrument, either legal or technical, covering all aspects of interventions, i.e. their integrality of a deep meaning.

It is true that, on the positive side of the balance, the first actions in rehabilitation areas are able to agree on the three levels of administration, and even considering financing social programs together with actions on buildings and public space, including matching practices subsidy with those of urban planning.

RD 2329/83 of July 28, not only establishes a general framework for collaboration between administrations, matching the state funding to the municipal and the financing of the private rehabilitation of buildings, but for the first time are targeted solutions within of a global performance (EMV 1999). With this decree the areas of Integrated Rehabilitation are defined for the first time. The dependence of the determinations of urban planning for its application is striking, a condition that does not include the rest of the ARI formulations that have been developed by the following decrees, both state and regional. Thus, for the ARI declaration, it was established that: (a) that the area is affected by urban planning that contains and develops criteria for protection, conservation and integrated rehabilitation of the same. (b) In the absence of such urban planning, the Directorate General of Territorial Action and Urbanism may collaborate for the drafting of the same, without prejudice to the legally established powers for its processing and approval. (c) Rehabilitation studies that, in accordance with urban planning criteria, contain information on the area, analyze their conditions and formulate proposals for action.

Other legislation that affect the rehabilitation activity are Law 2/2011, of March 4, on Sustainable Economy; The measures to promote rehabilitation activity contained in Royal Decree-Law 8/2011, of 1 July, on measures to support mortgage debtors, control of public expenditure and cancellation of debts with companies and self-employed persons contracted by the entities Promotion of business activity and promotion of rehabilitation and administrative simplification; RD 1113/84 of 22 of February on actions of remodeling and relocation in certain neighborhoods.

On other occasions, there have been spatial coincidences of urban, economic and social policies in which apply European structural funds. But the truth is that, on many occasions, sectoral approaches have been imposed and have developed powerful but biased instruments, which have ended the integrated approach and the mere rehabilitation of housing and epithelial improvement of public space has become the center of the interventions.

In fact, the evolution of events has made the housing subsidy policies to be rehabilitated in the most generalized instrument, since the Integrated Rehabilitation Areas (and the derived denominations) have been established from them, with the logic of channeling the investments there where they were most needed.

This shifting of the mode of intervention towards the financing has affected very specific aspects, such as the basic habitability of housing. The very logic of an instrument such as the housing plan, without underestimating its importance, has limited the integral vision, since it has been, above all, a means of financing actions in building, since what is outside the plan, financing is not done.

The management of urban regeneration policies and interventions more closely linked to an integrative model are those developed from the city council, while the management of the Community of Madrid has been atomized with emphasis on financing parts of the building envelope, for example.

As for both institutions this has been an evolutionary process it is advisable to go deeper into this by identifying the chronological stages with which the Rehabilitation Areas have been developed and which districts may be more representative of each one of them.

2.2 *Chronology and Proposal of Stages in Regeneration of Neighborhoods in Madrid*

From Preferential Rehabilitation Areas (ARP) to Integrated Rehabilitation Zones (ZRI). Proposal of delimitation of stages. We can identify several stages in this evolution of regeneration of neighborhoods in Madrid according to:

- The confluence of rehabilitating ways between City Council and Community of Madrid.
- Legislative production linked to housing plans and ingredients taken into account in rehabilitation.
- And finally, moments in which declarations of rehabilitation areas are encouraged.

There have been identified three periods: from 1994 to 2001, from 2002 to 2008, with two sub-stages from 2001 to 2005 and from 2006 to 2008 and, finally, from 2009 to 2012, which closes most areas of rehabilitation.

From 1994 until 2001, we find a stage marked by the convergence of rehabilitation policies and the clear institutional and sectoral collaboration between administrations. It is possibly the moment in which integral approach of the actions is carried out with more fidelity.

Actually the novelty is twofold. On the one hand, it is recognized that social problems are among the main reasons for granting rehabilitation and, on the other, it is not only considered in rehabilitation policies that these problems have a spatial dimension with which counting for.

The results of investment efforts are more effective in concentrating efforts on specific urban spaces, but this is done without the spatial support of urban planning. That is to say, the policy of subsidies assumes the rehabilitation effort. The investment effort is over 50,000 million pesetas (300 million €), allowing the results to be displayed in the public space and not only in the building.

From 2002 to 2008, we can say that the model of previous period is given continuity from two interpretations, despite the fact that a very similar legal framework is available in state and regional housing plans. On the one hand we have the rehabilitation of the City Council, which, based on the state housing plans and the Integrated Rehabilitation Areas (ARI), maintains an articulation with urban programs, neighborhood action plans, etc. As well as financing for improvements in infrastructure and public space. On the other hand, from the Community of Madrid emphasizes the rehabilitation of the building in a philosophy that leaves to a large extent the integrated approaches, producing a displacement, according to the years, that facilitates the access to the financing of a fragmentary way, even with regard to what is eligible within the same building. Thus, the plans to "renew" boilers, windows and elevators become the axes of the rehabilitation policy from the Community, which does not prevent a large number of Integrated Rehabilitation Zones (ZRIs) from being produced.

The state housing plan 2009–2012 opens a new period in the midst of economic crisis, with the reinforced idea that we must consider neighborhoods as a unit of action, which is an advance of what is reflected in the Charter of Toledo 2010 (Fig. 3).[1]

From the quantitative point of view, dwellings involved in securable buildings have been varying for each period described. As shown in Fig. 4, which also includes the stages, the houses that are part of the quota that can be protected by the initiative of the City Council and the Community of Madrid, as well as its peripheral or central location in the city, the rehabilitation effort carried out and planned since 1994 concerns approximately 111,000 homes.

What is clear and, is a very interesting aspect of how rehabilitation of neighborhoods in Madrid has been, is that there seems to be a certain role distribution between the City and the Community of Madrid, especially when the rehabilitation policies acquired certain independence from year 2001.

This is a division of space roles. While the EMVS has focused its efforts in the Historic Center and the Community of Madrid has specialized in the peripheral neighborhoods. As can be seen in the Table, the City Council has been involved in the rehabilitation of 28,572 homes in the center for only 7190 in the periphery, this is 80% compared to 20% of all its interventions. In the case of the Community, 2503 homes have been considered in the center, compared to 72,709 in peripheral neighborhoods of the ZRI subsidies, a negligible 3.5% compared to an overwhelming 96.5%. The proportion is incontestable. On the other hand, in the total

[1]Other legislation that affect the rehabilitation activity are Law 2/2011, of March 4, on Sustainable Economy; The measures to promote rehabilitation activity contained in Royal Decree-Law 8/2011, of 1 July, on measures to support mortgage debtors, control of public expenditure and cancellation of debts with companies and self-employed persons contracted by the entities Promotion of business activity and promotion of rehabilitation and administrative simplification; RD 1113/84 of 22 of February on actions of remodeling and relocation in certain neighborhoods.

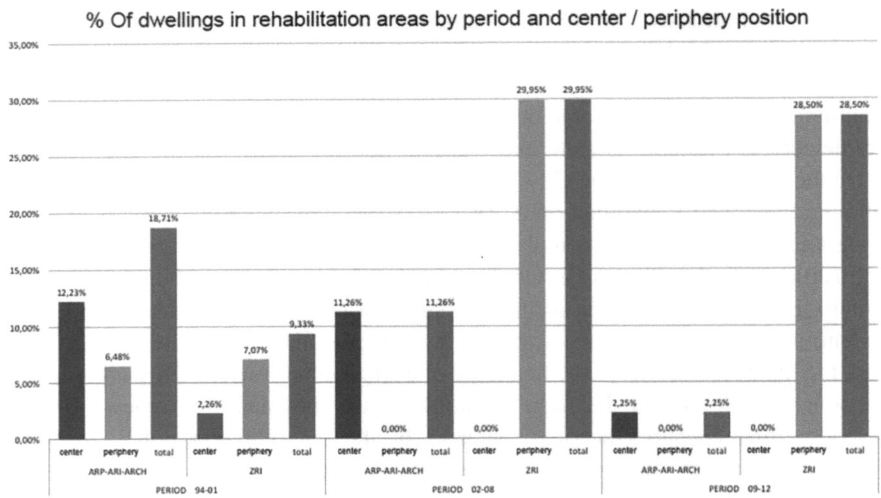

Fig. 3 Center and periphery. Percentage of households declared by rehabilitation area

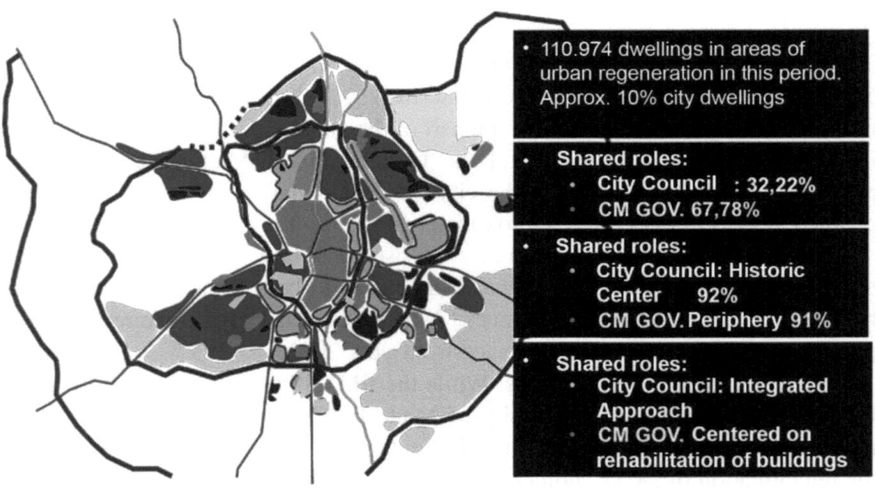

Fig. 4 Roles in urban regenerated areas

computation, the City Council has taken over 92% of the Housing of the Center and the Community of 91% houses of the periphery that have been incorporated in rehabilitation areas (Fig. 5).

Therefore, there are not only variations in the intervention model depending on what has been considered as protected intervention, but in turn there has been a certain spatial specialization. This is an interesting fact when crossing the rehabilitation experience with the processes of change of the social contents in neighborhoods.

	ARP-ARI-ARCH			ZRI			
PERIOD 94-01							
# dwellings	center	periphery	total	center	periphery	total	total
	13.575	7.190	20.765	2.503	7.851	10.354	31.119
PERIOD 02-08							
# dwellings	center	periphery	total	center	periphery	total	
	12.497	-	12.497	-	33.235	33.235	45.732
PERIOD 09-12							
# dwellings	center	periphery	total	center	periphery	total	
	2.500	-	2.500	-	31.623	31.623	34.123
total	28.572	7.190	35.762	2.503	72.709	75.212	110.974

Fig. 5 Household totals by rehabilitation areas and periods

From the actions promoter point of view, in Fig. 5, we see that the City Council has considered approximately 30% of total subsidized housing in the ARP, ARI and ARCH rehabilitation, while the ZRI' S of the Community have considered subsidies for the remaining 70%.

2.3 *End of Cycle and New Legal Framework. The Law 3/8 of 2013*

Everything indicates that a change of model looms from 2011 and its consolidation is visualized in 2013. There are several reasons: on the one hand, the Rehabilitation Areas have been closed, leaving only the City of Angels ZRI, and some of the latest statements from the community have not come to fruition.

The new housing plan 2013–2015 and the approval of the Urban Rehabilitation, Regeneration and Renovation Law[2] clearly moves within the framework of the Toledo Charter to recognize in Integrated Urban Regeneration the conceptual reference in which to develop new operations.

At the same time, given the financing difficulties foreseen for the coming years, the law sets out a model for the implementation of urban regeneration areas, which seems to be aimed at the presence of professionalized private investment in interventions. Therefore, it seems that we are facing a cycle that has closed and the opening of a new context that we will see what praxis entails (Fig. 6).

[2]Law 8/2013, of June 26, on urban rehabilitation, regeneration and renovation.

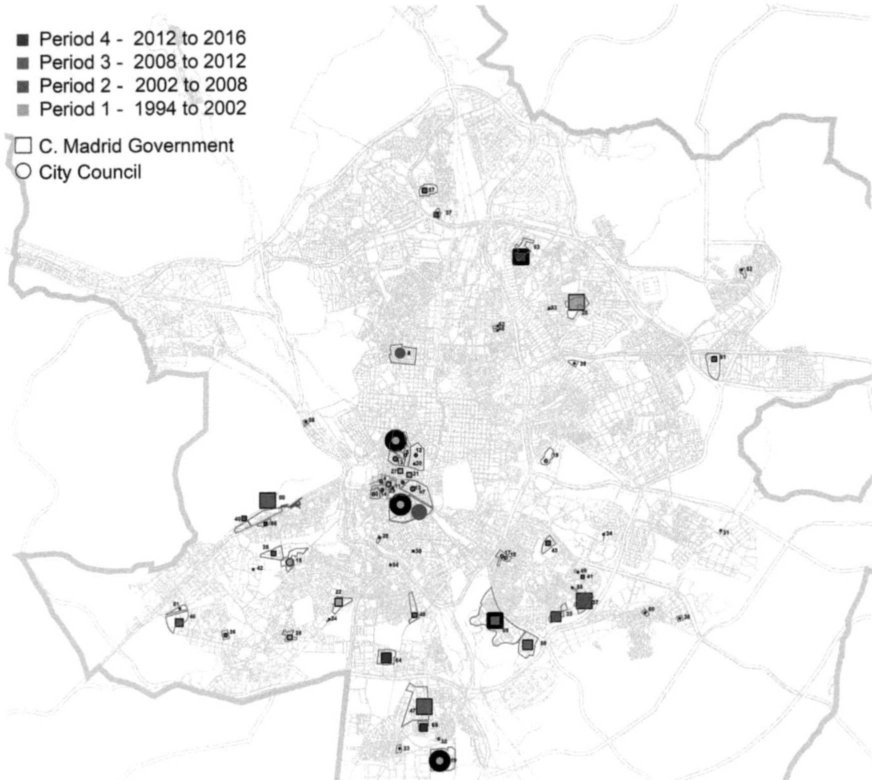

Fig. 6 ARIs and ZRIs evolution

3 Conclusions

3.1 Present Situation and Future Prospects Reveal Urban Regeneration as a Necessary Practice to Achieve the Challenges of the Present

The growth of Spanish cities in the last decade has been accompanied by the improvement of the existing city. However, in the next few years, production pace of the city[3] does not appear to be repeated, and of course from the point of view of territorial and urban sustainability it is far away from desirable. It opens a stage in which urban assets that we have, are the base, the wicker with which cities will be more livable. In this sense, Sustainable Urban Regeneration is considered as a desirable and possible urban practice. This new stage that is being opened

[3]Approximately 800,000 homes annually at their peak according to CSCAE.

nowadays presents, from the point of view of its strategic contents, the following challenges:

- **The situation of the current residential park**.
  Spanish built park standing is in synthesis: 70% of the buildings were built after 1960, a 16% in the first European post-war, being the remaining 14% before the civil war.
  If we take into account the growth prospects of the report "A Vision-Country for the building sector in Spain. Road map for a new housing sector" (Cuchí et al. 2012).
  The residential park eligible for rehabilitation between 2012 and 2050 in Spain is approximately 14 million homes. The lack of adaptation to service standards in housing is also a clear deficiency,[4] as well as the lack of thermal comfort and energy inefficiency, being a good part of the park prior to the use of standards such as the Spanish Technical Building Code.
- **Urban vulnerability is another factor to take into account in regenerative practices of socio-economic or socio-demographic origin**.
  It is concentrated in four large cities: Barcelona, Madrid, Seville and Valencia. The affected population is proportionally greater in historical centers, where it has come to concentrate immigrants and/or old population.
- **The need to activate the construction sector**.[5]
  Which before the crisis amounted about a 14% of GDP in the country and which in recent years has collapsed to only 5% of its activity a few years ago, may represent an opportunity of re-founding an economic sector around refurbishment.
- **Urban sustainability has become a prerequisite**.
  The improvement of the neighborhood's metabolism nowadays, has its strong point over energy treatment (rational use of private vehicles and energy efficiency in building) and the sustainable use of water, both in buildings and at neighborhood level.
  Most of the current efforts in the rehabilitation of housing are being focused on this line.
  In European context, efforts have been made to coordinate the implementation of transpositions from the energy efficiency directives which have resulted in the Concerted Action EPBD in 2005, promoting dialogue and exchange of good practices between them.
  The current review provides insights on how the 2010/31/EC Directive on recasting is being applied, the evolution towards new and rehabilitated near zero-energy buildings by 2020 and the application of an optimal cost

[4]To give an example, if we look at the situation of the residential park in France, Germany or Italy more than 92% of it has a bathroom or shower, hot water and some type of heating. In Spain, only 63.8% of the existing dwellings have these basic habitability elements.

[5]GTR estimates that housing refurbishment, including energy rehabilitation, can generate between 110,000 and 130,000 jobs annually.

methodology to determine the minimum requirements for both the build envelope and the conditioning systems mechanical facilities.

3.2 *Urban Sustainability in the Way of Been Implemented*

This last conclusions block, is reduced to a small set of cities representative of all actions cited in this paper. The selection is made, because they are actions of an important size, and because they are located in different areas of the Iberian Peninsula with different climate types, such as mediterranean, continental, and oceanic.

Finally, a more detailed study of the situation in the city of Madrid is carried out, since it is the residence of the authors and the one where data have been obtained more easily or where have been directly participated in the proceedings. A summary table of analysis is presented, which indicates the sustainability criteria that contemplate the actions in the different analyzed cases (Fig. 7).

Given this sample of experiences some guidelines of intervention can indicated:

- All regenerated neighborhoods install lifts as one of the indispensable measures. The sensitivity of the population towards these types of improvements in the accessibility of housing is very strong. Sometimes it is determined by the urgency of solving difficult human situations of people who can not go down the street, and the social consensus that this is a situation below the acceptable standards of habitability. Other times it is understood that the installation of an elevator allows to revalue the property, which without this utility is even out of the market.
- Energy efficiency is another generalized characteristic in those neighborhoods that have been intervened in the last 10 years. Only Lavapies or La Mina have been given out of this type of performance. The impulse always comes from the public energy policies associated with building and the population demand for this type of improvement is retained by very strong inertia. In some cases, such

Neighborhoods	City	Housing habitability improvement. Vertical Accessibility	Energy Efficiency	Sustainable Mobility sostenible	Sustainable management of water	Renaturalization (urban gardens, etc).	Requalification. New community facilities	Public space improvent as resting space
Lavapiés	Madrid	Elevator not in all occasions. Unresolved poor housing problems	No	Residential priority areas	No	Yes . New park	New self-managed equipment. Theaters, libraries	Remodeling of all squares
Ciudad de los Ángeles	Madrid	Elevators Dwellings enlargement	Yes	No	No	No	No	Handicapped Accessibility
Balsas Viejas de Rio Ebro	Zaragoza	Elevators	Yes	Residential priority areas	Yes. Storm pond	Yes. Tio Jorge park renovation	Improvements in two public schools and a Day & Home senior Center	Adaptation for public use of children's games of water tank covers
La Mina	Barcelona	Elevators Dwellings enlargement	No	central tram	No	Besos river recovery	New university headquarters	Central Boulevard
Simancas	Gandía	Elevators. 140 dwellings renovation	Yes	Withdrawal of parking spaces from the street	No	No	Social center	Yes. New square
Coronación	Vitoria	Elevators. Building pathologies	Yes	Pedestrian space improvement	No	No	District heating	Currently to be defined by the participation process

Fig. 7 Sample neighborhoods analisys data

as Ciudad de los Angeles, after the modernization of the former blocks, the advantages of energy saving and improved comfort are becoming known thanks to the diffusion of neighborhood associations in the neighborhood. In the neighborhood of Coronación in Vitoria, the effective diffusion of the advantages is combined with the offer of subsidies only if within the rehabilitation of the buildings is considered the energetic improvement and also they are connected to the new network of centralized neighborhood heating included in the proposal. In Andalucía (Granada, Marbella, Malaga and Seville), energy efficiency of facilities have been considered as well. In the very case of Simancas, Gandía (Valencia), housing with B energy efficiency certification, are not energy rehabilitation but complete replacement by new housing.

- The actions of sustainable mobility appear in an unequal way. In all cases, except in Ciudad de los Angeles, the regenerated areas are considered as a pedestrian priority, but often without deep physical changes that accompany it, leaving everything to private vehicle traffic regulation. In the case of Simancas in Gandía, these measures manage to reduce parking spaces in the street, although new underground garages of residents are proposed. Only in the neighborhood of La Mina in Barcelona, the intervention does involve complex infrastructures, such as the new tram that crosses the boulevard. The bicycle is not explicitly contemplated although pedestrian priority areas can function as spaces of coexistence with the pedestrian in those neighborhoods in which the single platform has been contemplated.
- No neighborhood have considered measures for saving, collecting and treating local water, except for Balsas-Viejas in Zaragoza that views an storm pond for collection rain water.
- The renaturalization of parks or public spaces is also not widely considered. The improvement of the trees in the streets is the most common measure, but does not go beyond a treatment close to the ornamental. Only the improvement of Besós, in La Mina and the incorporation of urban gardens in a new park in the historical fabric (Tabacalera) of Lavapiés are actions that can be classified in this category. The actions on streets and squares of Coronation neighborhood, in Vitoria, are the current object of debate on the possibilities of naturalization in a city that has extensive experience in this type of practices, developed since the implementation of its “Green Ring”.
- The incorporation of new facilities in neighborhoods depends very much on each case. In a city as well equipped as Vitoria, the novelty is the proposal of a new centralized district heating. The use of low buildings, replacing houses and dedicating space to community facilities, public or private is a strategy that is developed in the neighborhood of Simancas in Gandía and is not very common because of difficulties of managing a stock of homes with individual owners up to 85%, which is normal in Spanish neighborhoods. The most ambitious strategies are those of Lavapiés and La Mina, aimed at discouraging these two neighborhoods. The most symbolic equipment that joins the neighborhood is the University, going beyond the efforts to re-qualify these neighborhoods supplying the deficits of everyday urban services. It is a way of connecting the

neighborhood with the population of the rest of the city. We have already commented on the novelty of incorporating the programs of central equipment of community heating, as in the case of Coronation in Vitoria.

- Remodeling of existing squares or the creation of new ones, as well as boulevards and parks, is relatively frequent in regenerated neighborhoods, especially those that were intervened during the first decade of this century. Understanding that the improvement of social cohesion goes through the improvement of collective public space is a simplification with which the social agenda of these neighborhoods has been approached in some cases. In any case it is true that socialization spaces improvement, if not replaced, helps to generate social capital.

As the reader can see, conclusions lead us to understand that within the concept of integrality, it is necessary to influence the agenda of sustainable actions from the point of improving the metabolism of these neighborhoods.

Issues such as energy production with renewable energies in neighborhoods are paralyzed by state regulations of production and distribution in a country with a large amount of hours of sunshine and wind.

Water and its sustainable management in a context of increasing aridity of much of the Spanish territory, incomprehensibly is out of the focus of all actions.

Recycling of garbage is centralized and its management does not rest in any way in districts, nor even for the composting of the residues to produce fertilizers of urban orchards, that on the other hand are beginning to turn into a citizen demand of districts more aware about ecological crisis that cross our cities.

Energy efficiency, sustainable mobility and reconquered basic habitability are the central axes of the most immediate past or currently under development urban regeneration actions.

References

Álvarez Mora, A., Roch Peña, F., et al. (2010). *Regeneración urbana integrada en Europa. Documento de Síntesis*. Valladolid, Castilla-León: Instituto Universitario de Urbanística de la Universidad de Valladolid.

Campesino Fernández, A. J. (1989). *La rehabilitación integrada de los centros históricos: el reto del urbanismo de finales de los ochenta*. Alicante, C. Valenciana: Reunión científica La utilización de los estudios geográficos en la planificación del territorio. Departamento de Geografía Humana. Universidad de Alicante.

Cuchi, A., & Sweatman, P. (2012). *Informe GTR 2012 - Una visión-país para el sector de la edificación en España: hoja de ruta para un nuevo sector de la vivienda*. Madrid: European Climate Fundation - Fundación Complutense CCEIM.

de Terán Troyano, F. (1984). *Teoría e intervención en la ciudad, balance de un período. Estado de la cuestión. Perspectivas. Ciudad y Territorio 59/60*. Madrid: IEAL.

EMVS. (1999). *Áreas de Rehabilitación Preferente. Intervenciones en el centro histórico y barrios periféricos de Madrid 1994–1999*. Madrid: EMVS.

Moreno Soriano, S., González, F. J., et al. (2010). *Mesa Redonda Bloque temático: La formación Universitaria en Rehabilitación Sostenible*. Barcelona: Congreso Internacional Rehabilitación y Sostenibilidad. El futuro es posible.
Moreno Soriano, S., González, F. J., & Gómez, A. (2013). La regeneración urbana en España Revisión crítica para un estudio de caso de regeneración sostenible. Revista De Arquitectura n.27 Facultad de Arquitectura y Urbanismo de la Universidad de Chile. pp. 6–14.
Rubio del Val, J. (2011). Rehabilitación urbana en España 1989–2010. Barreras actuales y sugerencias para su eliminación. Informes de la Construcción, Vol 63, Extra 5.

Chapter 3
Urban Planning Research in the Climate Change Era: Transdisciplinary Approach Toward Sustainable Cities

Masanobu Kii, Kenji Doi and Kazuki Nakamura

Abstract This article discusses the direction of urban planning policies toward sustainable cities. Sustainable development goals (SDGs) adopted by United Nations General Assembly range from poverty eradication, improvements in education to protection of global assets including oceans and climates. Achieving these wide ranging goals requires holistic and transdisciplinary approaches. Urban planning is one of the effective measures for SDGs because those goals are closely related with the urban activities. In this article, we first summarize a conceptual framework of the interaction between urbanization and climate change, and forecast the prospective trajectory of rapid global urbanization in this century and the consequent sustainability problems. Then, we discuss the direction of urban planning analysis to tackle the sustainability problems as sustainable science, and propose the cross-assessment approach for vision-led urban planning. The cross-assessment approach aims to explore synergistic solutions combining different value systems by assessing the impacts on a range of outcome indices of measures pursuing each value factor. The case studies for the cross assessment are taken from Japanese public transport policies. Finally, we discuss the impacts on urban expansion from socioeconomic changes and technological progress, especially in the fields of transport, construction, and communication.

M. Kii (✉)
Faculty of Engineering, Kagawa University, Takamatsu, Kagawa, Japan
e-mail: kii@eng.kagawa-u.ac.jp

K. Doi
Department of Global Architecture, Graduate School of Engineering, Osaka University, Osaka, Japan
e-mail: doi@civil.eng.osaka-u.ac.jp

K. Nakamura
Faculty of Science and Engineering, Meijo University, Nagoya, Aichi, Japan
e-mail: knaka@meijo-u.ac.jp

R. Álvarez Fernández et al. (eds.), *Carbon Footprint and the Industrial Life Cycle*,
Green Energy and Technology, DOI 10.1007/978-3-319-54984-2_3

1 Introduction

Both urbanization and climate change are expected to greatly proceed in the 21st century and to have significant mutual influence as they interact. Technological revolutions, especially in the field of construction and transport, have raised the productivity and quality of life in megacities, which are not only massive but are the cities continuously increasing in population. While the improvement of productivity in cities fulfils people's needs, the increase in population and concentration and economic growth have also increased energy consumption and greenhouse gas (GHG) emissions in cities. Climate change is anticipated to have a wide range of impacts on human activities, with cities in developing countries considered to be especially vulnerable to extreme weather and meteorological disasters.

The fifth assessment report (AR5) by the Intergovernmental Panel on Climate Change (IPCC) (IPCC 2014) indicates potential impacts of climate change on urban areas. Those climate risks influence key resources, including water, food and energy supply, public health, and infrastructure. Further, it is estimated that more than 70% of global CO_2 emissions are caused by urban activities (Marcotullio et al. 2013; Grübler et al. 2012). Therefore, we need to understand the interactive effects between urban activities and climate change. Adaptation to climate change is considered an opportunity for growth and revolution of urban areas, and at the same time, the structural change of urban systems to meet adaptation needs will have enormous impacts on the efficiency of urban activities and quality of life of residents. In other words, appropriate approaches to sustainable urban planning can contribute to realising the co-benefits of reduction of climate disaster risks and upgrading urban systems. Moreover, if we do not pay enough attention to the mechanisms of urban activities or citizen values, mitigation or adaptation measures for climate change may curb urban activities to the point of decline in the quality of life.

In this article, we discuss the framework of urban planning research for co-evolution of climate policy with urban planning. While each city has specific urban planning issues, the overall GHG emissions of all cities affect the climate on a global scale. To understand the interaction between urban activities and climate change, we need to conduct an urban analysis of all cities across the globe with consideration of several individual cities. In addition, climate change has extremely long-term impacts, far beyond the scope of past urban planning research. The necessary research methods for incorporating both the long-term and global scales have not been established, and therefore, we think a new framework is needed.

The structure of this article is as follows. In the next section, we discuss the conceptual framework of interaction between urbanisation and climate change. Section 3 examines a prospective problem of long-term and global-scale urbanisation under IPCC scenarios. Section 4 describes how sustainability science has become the new direction of urban planning research and discusses the needs for transdisciplinary approaches. Section 5 discusses cross-assessment analysis using

Japan's public transport strategy to show how vision-led planning incorporates stakeholders' sense of values. Section 6 discusses the impact of technological innovation on urban activities and urban planning.

2 Urbanisation and Climate Change Interaction

Urban areas are the primary place of economic activities. In turn, these economic activities cause GHG emissions, requiring mitigation and adaptation to climate change. Some adaptation measures may increase GHG emissions; for example, massive infrastructure to counter climate disasters may yield significant emissions. Thus, climate change and urban activities interact with each other. Climate change may shift the spatial and temporal patterns of temperature and precipitation that affect the demand for heating and cooling, cost for utilising water resources and food supply, and public health. Sea level rise is also expected to increase the risk of high tide, which may deteriorate both public and private fixed capitals in coastal cities and degrade land resources. All of those impacts can increase the cost of urban activities; however, impacts are unevenly distributed across the globe, and differ among cities. There are various adaptation measures to climate change, but their efficiency varies, and some measures may not only increase the costs of urban activities but also GHG emissions.

Research for cities in developing countries is no less valuable than that for developed countries. Many of the cities in developing countries are expected to experience urbanisation on a massive scale in the future. Urban structure affects GHG emissions and it usually takes a long time to modify the structure once it is constructed. Therefore, cities where future growth is expected are the effective target for climate policy. Urban planning will induce different trajectories of urban growth: some plans may lead to environmentally friendly urban structures and the others may not. Planning also affects the efficiency of urban activities and quality of life; therefore, we need to consider the various factors in consideration with the long-term impact of climate change.

There have been many studies for climate change impact and adaptation measures in urban areas which were reviewed in the IPCC-AR5, Working group II, Chap. 8 (Revi et al. 2014). The report stresses the importance of considering climate change impacts in urban areas: including increased risk of sea level rise, high tide, heat wave, rainstorm, flood, landslide, drought, water shortages, and air pollution. All of these risks are considered to have the potential to severely affect water and energy supplies, public health, social infrastructure and construction, and ecosystems. Further, they also pointed out that adaptation measures against these risks could provide the opportunity for urban growth and the promotion of innovation.

However, the pathways climate change impacts on urban activities are complicated. For instance, the shift of precipitation patterns may cause drought in some area, which may affect water resources in urban areas, as well as agricultural productivity. The latter may have an impact on food price and security, and that may also possibly affect migration from rural to urban areas (Vairavamoorthy et al. 2008; Herrfahrdt-Pähle 2010; Farley et al. 2011). The sea level rise will increase the scale and frequency of high tide, which will affect residents, fixed assets, ecosystems, as well as damage port facilities and industrial plants along coastlines.

Generalising the effect of adaptation measures is considered to be difficult because it varies due to the conditions of climate impacts and economic levels of cities. Less developed countries are usually not able to afford investments in water resource facilities or climate disaster prevention structures, so they are considered more vulnerable (IPCC 2014). If it is affordable, tide embankment is considered a cost effective measure to prevent climate disaster along coastal areas (Hinkel et al. 2013); however, construction of these embankments may emit large volumes of GHGs. Some country water supplies consist of desalinated seawater, which requires energy intensive systems and may contribute to global GHG emissions (McEvoy and Wilder 2012).

On the other hand, the impact of urban policies on climate change was reported in Working Group III, Chap. 12 (Seto et al. 2014). Currently half of the global population lives in urban areas, while CO_2 equivalent emission caused by urban activities is more than 70% of global emissions. This means that urban areas are more intensive in terms of CO_2 emissions per capita than rural areas. Some recent literatures for local studies also suggest the results (Heinonen and Junnila 2014; Zhang et al. 2016). However the report also indicates that CO_2 emissions per capita vary not only among countries but also among cities within the same country. It has also been mentioned that urbanisation and its accompanied massive infrastructure development will be a major factor in intersectional GHG emissions, and that the factors of urban structure affecting GHG emissions are density, land use mix, connectivity, and accessibility. Gudipudi et al. (2016) indicates the lower urban density causes higher per-capita emission based on land use/land cover map and spatial CO_2 emission datasets in United States circa 2000. The speed of urban area expansion is estimated to be twice that of the speed of urban population increase, and it is anticipated that the new development of urban areas will be greater in the next three decades than what has been urbanised up until now. This means that if this urbanisation trend continues, the population density will decrease, urban areas will become dispersed, and infrastructure—roads, sewer, water, electricity, gas—needs to be dispersed as well, which is costly and energy intensive. Cities will consume even more energy, and emit more environmental loads, especially in developing countries.

Combining these findings, it is clear how climate change and urban activities interact with each other. It is necessary to conduct an analysis of this interaction in order to create climate and urban policies consistent with each other that can also obtain co-beneficial outcomes for human activities and the environment. Some theoretical or empirical studies demonstrate that urban compactness will contribute to the reduction of GHG emissions; however, simple land use regulations may significantly increase the cost of living and deteriorate quality of life. On the other hand, coordination of infrastructure development and land use management in growing cities may restrain both increases to living costs and environmental loads, leading to better quality of life in total. The trajectory of urban growth and its sustainability will differ depending on urban and environmental policies.

Naturally, integrated urban planning, considering both urban activities and climate change, is desirable; however, no method is established to frame and evaluate such a plan. Climate change is a global scale phenomenon and the impact of mitigation measures in one single city is negligible. Therefore in the urban planning of a single city, the climate must be an exogenous condition because urban planning for a single city cannot control the climate. Further, the total GHG emissions of all the cities across the globe obviously affect the climate and of course the sum total is not negligible. On the other hand, climate change impacts differ according to city and the effect of urban policies on GHG emission reductions also varies between cities. Therefore, to comprehend the interaction between climate change and urban activities, the analysis must take cities on an individual and sum total basis into account.

One possible analytical framework of the interaction is shown in Fig. 1. In the long term, urban activities are naturally affected by socioeconomic factors, including urban population, economic growth, as well as technological innovations. For example, the innovation of transport technologies typified by automobiles is recognised as significantly contributing to urban expansion over the past 100 years. Building technologies have allowed us to build skyscrapers, which has increased urban density at the centre of metropolises where there is a need for higher affordability in rental units. In addition, national legal systems (which may affect land use regulations), and local geographical features, historical events, and trajectories, may affect the individuality of a city. Furthermore, GHGs emitted by urban activities as a whole causing climate change, do not have a uniform impact but rather impacts vary among regions and cities. Climate change will bring various climate disasters, as well as a shift to precipitation patterns and sea level rise. These may affect natural resources including water, land, and energy. As a result, these disasters and resource degradation affect urban activities. Both private and public investment is needed in order to adapt to these changes: private investment includes air conditioning facilities and public investment includes water control and

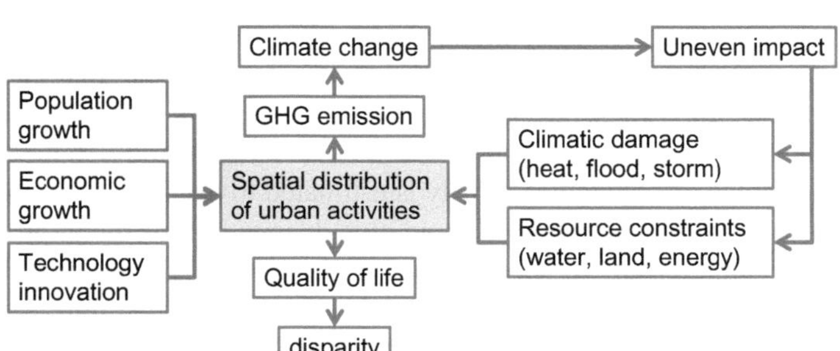

Fig. 1 Interaction between urban activities and climate change *Source* Prepared by authors

utilisation, tide embankment, land elevation, and possibly relocation of a part or whole city. The public investment required for adaptation could be massive and could also significantly contribute to GHG emissions.

Climate change is expected to have an impact on people's lives, while, at the same time, mitigation and adaptation policies will also have an effect on people. If mitigation policies severely regulate emissions resulting from urban activities, the opportunity costs may be high enough to stifle the economy. If all cities do not reduce GHG emissions, they will pay the economic price when having to address severe climate disasters, such as enormous investments into climate change adaptation. Therefore, there is trade-off between the economy and the environment in terms of GHG emissions, and the problem is not isolated within a single city. This requires analysis of policy impacts on individual city emissions and economies as well as analysis of the aggregated effects on climate change and the feedback to the individual cities. This framework is a new challenge for urban research, and further study is needed in the field of environmental policy. The problem cannot be solved according to individual study fields, thus a transdisciplinary approach will be needed to address this challenge.

3 Long-term Global Urbanisation and Environmental Challenge

It is essential to understand prospective global urbanisation in order to consider the interaction between urban activities and climate change. Long-term socio economic situations throughout this century are unpredictable, so climate policy studies assume several socio economic scenarios. The Special Report on Emissions Scenarios (SRES) (Nakicenovic and Swart 2000; IPCC 2000) have been commonly used for the studies cited in the IPCC report, and currently Shared Socio Economic pathways scenarios (SSPs) (Ebi et al. 2013; IIASA 2015) have been developed for the studies to be used in the next IPCC report.

A few studies estimate that by 2030, the global urban area (Seto et al. 2012; Angel et al. 2011) under the SRES scenario will be two-to three-fold the size of 2000. We also estimated urban expansion and population density under the SRES and SSP scenarios using a simple urban model (Kii et al. 2016a). Our estimates for 2030 show an increase of urban areas by 1.8–2.4 times under the SRES and 1.9–2.2 times under the SSPs: these are lower estimates compared with past studies. We estimate that the urban area will grow 2.3–3.6 times by 2050 and 3–6 times in 2100. The global average of population density is estimated to decline during this century under most scenarios. The decline of population density reflects the improvement of mobility and suburbanisation, accompanied by economic growth even though the population increases in most scenarios. Some research on urban transport clarified the negative correlation between urban density and transport energy consumption per capita (Kenworthy and Laube 1999, 2001), indicating that the global decline of urban population density may accelerate GHG emissions in the urban transport sector.

Looking at the urban area according to regions, Asian cities are prominent in urban expansion for all scenarios. In 2100, Asia's urban areas are expected to make up 40% of the world's urbanisation. In the SSPs except SSP3, it is assumed that Asia's urban population will peak during this century. The resulting economic growth may make higher transport costs more affordable, which can induce suburbanisation of residential areas, resulting in a decline in population density. However in Africa, some scenarios assume higher population growth and lower economic growth. For example, SSP3 represents a fragmented world inducing extremely high population density, which estimates African cities having an average of more than 20,000 people/km^2. These cities may possibly emit smaller amounts of GHGs due to shorter travel distances within the city; however, this ultra-high density will cause other severe urban problems, such as public health and congestion.

Urban expansion will have various impacts on the environment, including GHG emissions and deterioration of ecosystems by land use change. At the same time, urban expansion will affect quality of life and the value of assets and property in the city. Infrastructure and buildings offer decades or possibly centuries of use, so spatial planning and management should consider future socioeconomic situations within a climate change context. Sensitivity analysis in our study indicates that the urban area can be varied, even under the same population and GDP scenario depending on the transport and construction costs: as a result the total travel length is also different. When transport costs decrease, the urban area increases. Contrarily, when construction costs of high-rise buildings decrease, the urban area decreases. This implies that the policy mix integrating transport and land use may induce eco-friendly urban structures.

4 Urban Planning Toward the Sustainability Science

To comprehend the perspective of interaction between urban activities and climate change, we need further transdisciplinary study. A holistic approach toward the construction of a sustainable society is called sustainability science.

Sustainability science is a research field defined by its goal of addressing problems, but not by employing its disciplines (Clark 2007). Problems include challenges never faced before by our society, such as sustainable water supply, clean energy, human health, food security, rapid urbanisation, concerns for environmental and natural resources, poverty reduction, global climate, conserving ecosystems, and biodiversity. Many of us have become more concerned about the complexities and dilemma of cities regarding these problems (Batty 2013). Aiming to solve pressing global challenges with a tremendously wide scope, sustainability science tries to fuse natural science with social science and humanities (Komiyama and Takeuchi 2006). However, efforts regarding this fusion are, as yet, considered superficial and a transdisciplinary approach is required. The review panel of the International Council for Science discussed the idea that the transdisciplinary approach seeks not only to transcend boundaries among disciplines but also to create fusion, and to establish connections with stakeholders outside academia to solve global problems (Takeuchi 2014).

Modern urban planning has been, by nature, required to be problem-oriented and to take a transdisciplinary approach. It emerged just after the industrial revolution when concerns began to emerge regarding public health and the industrialisation of cities in England. Since then, planning has progressed by integrating the knowledge of various fields, including architecture, civil engineering, public health, economics, and politics. Currently, urban planning is its own discipline established from various fields, such as architecture, civil engineering and geography, independently. Of course, study on the particular problems within the discipline is indispensable. However, urban planning has a wide range of impacts on society and the environment. A transdisciplinary research approach will be required to contribute to the urban and climate policy for a sustainable future, instead of restricting study to a closed discipline. Kajiwara et al. (2014) indicated that the field of "urban and transportation systems" was the seventh largest research cluster in sustainability research in 2013. This field is closely related to nature and society, and therefore comprises crosscutting study topics with wide variation.

Even though the transdisciplinary approach is effective, dealing with the problems on various levels, all at once, will make the problem over-complex, and that effort may not provide any useful findings. Lee (1973) has already pointed out the flaw of large-scale models for urban analysis as a tool for policy-making. Since then, urban models had been upgraded through employing economic theory that makes the process of political impact on society intelligible, and those models are utilised in the practice of urban policy-making in some countries. These models represent the urban activities based on the locational and travel behaviours of actors in the city, and they analyse the impact of target policies through the mechanism of urban activities. The urban model is also applicable for some urban climate policies; however, the model usually requires spatial data of urban activities. Therefore, it is still difficult to apply the detailed urban model to the all of the global cities for long-term evaluation of urban climate policy due to the data constraint and cost for model building. For the long-term and global analysis of urbanisation and climate change interaction, we need a simple and reasonable model that is operable with less datasets. Further, urban models are required for the application to other urban problems, including social exclusion or community formation related to housing and transport policies. Those requirements dictate a more detailed and complex model be used to represent the interaction among various actors in the city. This social aspect may also be referable to climate policies. Therefore, the research of urban models will have two directions: simplification in global analysis and complication in local analysis. Figure 2 indicates the conceptual frame of co-evolution of these two research directions and its contribution to the sustainability science (Kii et al. 2016b).

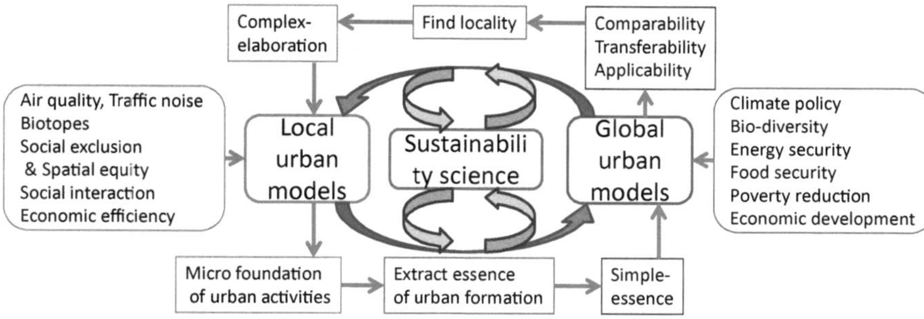

Fig. 2 Cross-reference between local and global urban modelling studies *Source* Kii et al. (2016b)

The proposed combination of two different research directions, complication and simplification, or local modelling and global modelling can be interpreted as a concept to avoid Lee's problems by discussing the balance between details of reality and abstraction of theory to lead the collective knowledge. The objective of sustainability science is to comprehend the wide spectrum of problems that may have an interrelationship and may thus provide solutions to the problems. However, it is difficult to solve the problems all at once because of the range and complexity of the system. Therefore, piecemeal approaches have to be taken to solve the problems. We consider that our concept, shown in Fig. 2, provides insight into the revolution of deeper piecemeal research on urban planning and its integration into holistic knowledge for sustainability science.

5 The Cross-assessment Approach: A Tool for Vision-led Planning

As discussed above, there is a trade-off between promotion of urban activity and GHG emission reductions. The balancing point between them may depend on the evaluation standard. The standard may differ even by person, depending on his/her position or ideology. People may make a judgement based solely on their own standard without referencing other standards. To better integrate strategies, a wide range of stakeholders with different values should be encouraged to fully participate in strategy formulation. This enables the development of a common understanding of policy objectives and a shared vision for the sustainable city. To make it possible, we need to reconcile conflicting objectives among stakeholders by clarifying the pros and cons of respective strategies and meet the requirements of a sustainable society.

As a tool for supporting the shared vision formation, we have developed a strategic cross-assessment model (Doi and Kii 2012). The cross-assessment combines the essential elements of multi-criteria analysis and conflict analysis (Keeney and Raiffa 1993; Toman 1994; Knoflacher and Himanen 1991; Dodgson et al. 2000; Minken 2002). It aims to explore synergistic solutions combining different value systems by assessing the impact on all outcome factors of measures pursuing each value factor, as shown in Fig. 3. An iterative feedback in the figure indicates that an appropriate weighting and combination of three value factors should be examined based on the results of the cross-assessment. Decisions are usually made based on consensus among stakeholders whose value systems are different from each other. Each of the three strategies in this study is based on a particular value factor, and is evaluated for the impacts on all of the outcomes.

We set the triple bottom lines of sustainability as economy, society, and environment, assuming the following three strategic targets in transport policy: (1) Profit maximisation of the public transport operator; (2) Net benefit maximisation; and (3) CO_2 emissions minimisation in the transport sector. The first target is equal to minimisation of subsidy by the government. In the second, net benefits are defined as the sum of user benefits and operator profits. Based on these targets, we set three outcome indices; operator profits, user benefits, and CO_2 emissions. The outcomes of these three strategies are shown in Fig. 4.

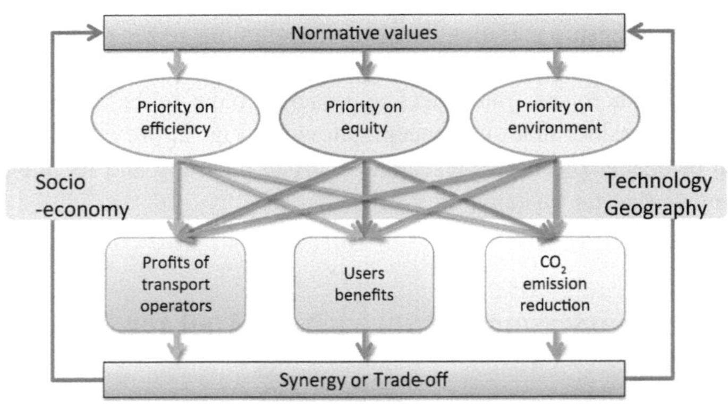

Fig. 3 Concept of cross-assessment in this study *Source* prepared by authors based on Doi and Kii (2012)

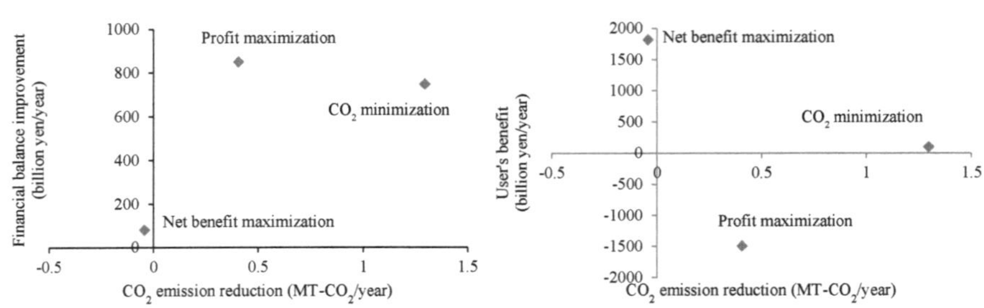

Fig. 4 Comparison of estimated outcomes of strategies *Source* prepared by authors based on Doi and Kii (2012)

Comparing with the "business as usual" (BAU) case in 2030, the profit maximisation strategy is estimated to largely decrease the public transport operation deficit and CO_2 emissions as well; however, the user's benefit will be largely negative because of the decline of level of service of public transport This strategy will decrease the services offered by unprofitable operators. That contributes to a reduction in CO_2 emissions because unprofitable operators cease bus and train services where there is little ridership.

The CO_2 emission minimisation strategy largely reduces the emissions, and also results in an improved financial balance of the public transport operation. In addition, the user's benefit is slightly positive. Different from the profit maximisation strategy, some public transport operators may increase their service level to shift the demand from car to public transport, which reduces CO_2 emissions. Of course, some operators cease service at lower demand routes, but in an aggregated figure, the user's benefit is slightly positive.

The net benefit maximisation strategy largely increases the user's benefit, but the financial balance of the public transport cannot be improved, and furthermore, CO_2 emissions slightly increase. This indicates that prioritising the user's convenience justifies inefficient operation, allowing its service to continue. When evaluating these three strategies, the CO_2 minimisation strategy has the most balanced effect on the three outcome indices.

The cross-assessment framework evaluates optimal strategies based on various value standards using the outcome indices corresponding to the value standards. This method makes it possible to comprehend the synergy and trade-off among the strategies, and it can contribute to building a shared vision and consensus among stakeholders with different value standards, resulting in balanced policy formation.

6 The Impact of Technological Innovation on Urban Sustainability

Technological innovation changes people's lifestyles as well as the urban structure. The information revolution and various automated technologies inspire people's activities. Artificial intelligence and virtual reality technologies expand the cognitive world, which would contribute to changing the normative value. These technologies in total make it possible to integrally manage the social infrastructure, to promote consensus among stakeholders with different interests, and the integration of policy and strategies across various sectors, including environment, health, and disaster prevention.

For example, fully automated vehicle technologies may enable all people to enjoy freedom of movement. People can use their time in the vehicle for pursuits other than driving, which means the opportunity cost for driving is reduced. Self-driving cars improve road safety, which is beneficial to society. Furthermore, due to improved technology and optimised driving, the resulting energy efficiency will alleviate environmental constraints. All of these benefits may reduce the need for urban compactness and possibly lead to more dispersed and low-density urban structure that may require further investment in road infrastructure. Further, if the automated technologies are integrated with sharing technologies (Fagnant and Kockelman 2014; Liang et al. 2016), urban compactness will improve the operational efficiency of vehicle stocks and the cost per user could be reduced. That implies the impact of technological innovation can vary depending on the combination of technologies and socioeconomic situations. Therefore, we should consider how to utilise those innovative technologies in urban planning to achieve a desirable vision for society.

Developing countries are expected to utilise those technologies for leapfrogging to bypass the industry-intensive stage that developed countries experienced (Perkins 2003). For this integration to work it is essential to obtain a shared vision and consensus among stakeholders (May 2005). Simulation technologies in climate research and urban studies can conceptualise a future vision. Virtual reality or augmented reality may be effective for intelligible and friendly presentations of the future vision of urban strategies and policy options. It can also be effective for consensus building. However, friendly presentations and interest sharing techniques must be exercised with caution because they are closely linked with manipulation of public opinion (Shteynberg et al. 2016).

Technological innovation has the potential to change the urban structure, and the urban space is an incubator for this innovation. Universities are recognised as knowledge network hub and are the centre of science and technology innovations (Youtie and Shapira 2008). The knowledge accumulation and high-density communication in a city are considered to generate the seeds of innovation and the ensuing spillover effect on productivity (Carlino et al. 2007). The conditions by which cities promote innovation, however, are not sufficiently clear, but cities with various types of industries and human resources may provide more opportunities for connecting innovative ideas and enterprises. If so, megacities may have more chance of realizing innovation, and some studies explored this concept (Florida et al. 2008). Moreover, the progress of information communication and transportation technologies could enable such opportunities for technological innovations in smaller cities, because the agglomeration effect of megacities may reach to the small cities. Local cities have advantages over megacities, with regard to the richness of liveability, closeness to the natural environment, and original culture, which could possibly attract some innovators. Hitherto, the study of urban planning does not pay enough attention to the impact of technological innovations on urban space and its liveability, as well as the effect of urban features on innovation. Of course innovation is also considered in connection with climate change for sustainable future. These issues can be new challenge for urban studies as well.

7 Conclusions

In this article, we discussed the framework of urban planning research for the co-evolution of climate policy. We need both a microscopic view to analyse individual cities and a macroscopic view to capture the impact of world cities on the climate as a whole. Climate change is an immensely long-term phenomenon, compared with the time scale of conventional urban planning. We proposed a cross-assessment framework to coordinate the normative values in society in order to inform political decisions in urban and climate issues. It is expected that this approach will harmonise the different value standards and lead to a shared vision and consensus among stakeholders. Climate change and urbanisation have a wide range of impacts, which are affected by various factors, thus requiring a transdisciplinary research approach to capture the interaction. Sustainability science can be a research field to study this interaction. Finally, the relationship between technological innovation and urban sustainability was discussed. While technological innovation provides solutions to specific problems, depending on how these innovations are combined with the other technologies or urban policies, it may result in a totally different unintended impact on urban space and living. This means technological innovation can contribute to solve urban and climate problems, but only if we have the shared vision of the future urban system and the consensus among stakeholders.

Acknowledgements This research was supported by MEXT KAKENHI (Grant Number 15H02869) and Japan Science and Technology Agency (JST), e-ASIA Joint Research Program, e-ASIA JRP.

References

Angel, S., Parent, J., Civco, D. L., Blei, A., & Potere, D. (2011). The dimensions of global urban expansion: Estimates and projections for all countries, 2000–2050. *Progress in Planning, 75,* 53–107.

Batty, M. (2013). *The new science of cities*. Cambridge: MIT Press.

Carlino, G. A., Chatterjee, S., & Hunt, R. M. (2007). Urban density and the rate of invention. *Journal of Urban Economics, 61*(3), 389–419. doi:10.1016/j.jue.2006.08.003.

Clark, W. C. (2007). Sustainability science: A room of its own. *Proceedings of the National Academy of Sciences of the United States of America, 104*(6), 1737–1738.

Dodgson, J., Spackman, M., Pearman, A., & Phillips, L. (2000). *Multi-criteria analysis: A manual*. London: Department of the Environment Transport and the Regions.

Doi, K., & Kii, M. (2012). Looking at sustainable urban mobility through a cross-assessment model within the framework of land-use and transport integration. *IATSS Research, 35,* 62–70.

Ebi, K., Hallegatte, S., Kram, T., Arnell, N., Carter, T., Edmonds, J., et al. (2013). A new scenario framework for climate change research: Background, process, and future directions. *Climatic Change, 122,* 363–372.

Fagnant, D. J., & Kockelman, K. M. (2014). The travel and environmental implications of shared autonomous vehicles, using agent-based model scenarios. *Transportation Research Part C: Emerging Technologies, 40,* 1–13.

Farley, K. A., Tague, C., & Grant, G. E. (2011). Vulnerability of water supply from the Oregon Cascades to changing climate: Linking science to users and policy. *Global Environmental Change, 21*(1), 110–122.

Florida, R., Gulden, T., & Mellander, C. (2008). The rise of the mega-region. *Cambridge Journal of Regions, Economy and Society, 1*(3), 459–476. doi:10.1093/cjres/rsn018.

Grübler, A., Bai, X., Buettner, T., Dhakal, S., Fisk, D., Ichinose, T., et al. (2012). Urban energy systems. *Global energy assessment: Toward a sustainable future* (pp. 1307–1400). Cambridge: Cambridge University Press.

Gudipudi, R., Fluschnik, T., Ros, A. G. C., Walther, C., & Kropp, J. P. (2016). City density and CO_2 efficiency. *Energy Policy, 91,* 352–361. doi:10.1016/j.enpol.2016.01.015.

Heinonen, J., & Junnila, S. (2014). Residential energy consumption patterns and the overall housing energy requirements of urban and rural households in Finland. *Energy and Buildings, 76,* 295–303. doi:10.1016/j.enbuild.2014.02.079.

Herrfahrdt-Pähle, E. (2010). South African water governance between administrative and hydrological boundaries. *Climate and Development, 2*(2), 111–127.

Hinkel, J., van Vuuren, D. P., Nicholls, R. J., & Klein, R. J. T. (2013). The effects of mitigation and adaptation on coastal impacts in the 21st century. An application of the DIVA and IMAGE models. *Climatic Change, 117*(4), 783–794.

IIASA. (2015). *SSP database* (version 0.93) available online at https://secure.iiasa.ac.at/web-apps/ene/SspDb. Accessed March 29, 2016.

IPCC. (2000). *Special report on emissions scenarios: A special report of working group III of the intergovernmental panel on climate change*. Cambridge: Cambridge University Press.

IPCC. (2014). Summary for policymakers. In C. B. Field, V. R. Barros, D. J. Dokken, K. J. Mach, M. D. Mastrandrea, T. E. Bilir, et al. (Eds.), *Climate change 2014: Impacts, adaptation, and vulnerability. Part A: Global and sectoral aspects. Contribution of working group II to the fifth assessment report of the intergovernmental panel on climate change* (pp. 1–32). Cambridge: Cambridge University Press.

Kajiwara, Y., Tacoa, F., & Yamaguchi, K. (2014). Sustainability science: The changing landscape of sustainability research. *Sustainability Science, 9,* 431–438.

Keeney, R. L., & Raiffa, H. (1993). *Decisions with multiple objectives: Preferences and value trade-offs reprinted.* Cambridge: Cambridge University Press.

Kenworthy, J., & Laube, F. (1999). *An international sourcebook of automobile dependence in cities, 1960–1990.* Boulder: Boulder University Press of Colorado.

Kenworthy, J., & Laube, F. (2001). *Millennium cities database for sustainable transport*. Belgium: UITP.

Kii, M., Nakamura, K., & Morita, H. (2016a). Long term estimation of global urbanization under SRES and SSPS scenarios. *Journal of Japan Society of Civil Engineers, Ser. G (Environmental Research), 72*(6): I_187–194. (in Japanese).

Kii, M., Nakanishi, H., Nakamura, K., & Doi, K. (2016b). Transportation and spatial development: An overview and a future direction. *Transport Policy, 49,* 148–158.

Knoflacher, H., & Himanen, V. (1991). *Transport policy between economy and ecology, VTT Research Notes, 1221.* Espoo: VTT.

Komiyama, H., & Takeuchi, K. (2006). Sustainability science: Building a new discipline. *Sustainability Science, 1,* 1–6.

Lee, D. B. (1973). Requiem for large-scale urban models. *Journal of the American Institute of Planners, 39,* 163–178.

Liang, X., Correia, G. H. D. A., & van Arem, B. (2016). Optimizing the service area and trip selection of an electric automated taxi system used for the last mile of train trips. *Transportation Research Part E: Logistics and Transportation Review, 93,* 115–129. doi:10.1016/j.tre.2016.05.006.

Marcotullio P. J., Sarzynski A., Albrecht J., Schulz N., & Garcia, J. (2013). The geography of global urban greenhouse gas emissions: An exploratory analysis. *Climatic Change, 121,* 621–634. doi:10.1007/s10584-013-0977-z. ISSN: 0165-0009, 1573–1480.

May, A. D. (2005). *A decision makers' guidebook—developing sustainable urban land use and transport strategies*. Prospects European Commission Deliverable, N. 15.

McEvoy, J., & Wilder, M. (2012). Discourse and desalination: potential impacts of proposed climate change adaptation interventions in the Arizona-Sonora border region. *Global Environmental Change: Human and Policy Dimensions, 22*(2), 353–363.

Minken, H. (2002). *A framework for the evaluation of urban transport and land use strategies with respect to sustainability, the 6th Workshop of the Transport*. Haugesund: Land Use and Environment Network.

Nakicenovic, N., & Swart, R. (Eds.). (2000). *Special report on emissions scenarios*. Cambridge: Cambridge University Press.

Perkins, R. (2003). Environmental leapfrogging in developing countries: A critical assessment and reconstruction. *Natural Resources Forum, 27*(3), 177–188.

Revi, A., Satterthwaite, D. E., Aragón-Durand, F., Corfee-Morlot, J., Kiunsi, R., et al. (2014). Urban areas. In C. B. Field, V. R. Barros, D. J. Dokken, K. J. Mach, M. D. Mastrandrea, T. E. Bilir et al. (Eds.), *Climate change 2014: Impacts, adaptation, and vulnerability. Part A: Global and sectoral aspects. Contribution of working group II to the fifth assessment report of the intergovernmental panel on climate change* (pp. 535–612). Cambridge: Cambridge University Press.

Seto, K. C., Dhakal, S., Bigio, A., Blanco, H., Delgado, G. C., Dewar, D., et al. (2014). Human settlements, infrastructure and spatial planning. In O. Edenhofer, R. Pichs-Madruga, Y. Sokona, E. Farahani, S. Kadner, K. Seyboth, A. Adler, I. Baum, S. Brunner, P. Eickemeier, B. Kriemann, J. Savolainen, S. Schlömer, C. von Stechow, T. Zwickel, & J. C. Minx (Eds.), *Climate change 2014: Mitigation of climate change, contribution of working group III to the fifth assessment report of the intergovernmental panel on climate change* (pp. 923–1000). Cambridge: Cambridge University Press.

Seto, K. C., Güneralp, B., & Hutyra, L. R. (2012). Global forecasts of urban expansion to 2030 and direct impacts on biodiversity and carbon pools. *Proceedings of the National Academy of Sciences, 109,* 16083–16088.

Shteynberg, G., Bramlett, J. M., Fles, E. H., & Cameron, J. (2016). The broadcast of shared attention and its impact on political persuasion. *Journal of Personality and Social Psychology, 111*(5), 665–673. doi:10.1037/pspa0000065.

Takeuchi, K. (2014). The ideal form of transdisciplinary research as seen from the perspective of sustainability science, considering the future development of IATSS. *IATSS Research, 38,* 2–6.

Toman, M. A. (1994). Economics and sustainability: Balancing trade-offs and imperatives. *Land Economics, 70*(4), 399–413.

Vairavamoorthy, K., Gorantiwar, S. D., & Pathirana, A. (2008). Managing urban water supplies in developing countries—climate change and water scarcity scenarios. *Physics and Chemistry of the Earth, Parts A/B/C., 33*(5), 330–339.

Youtie, J., & Shapira, P. (2008). Building an innovation hub: A case study of the transformation of university roles in regional technological and economic development. *Research Policy, 37*(8), 1188–1204. doi:10.1016/j.respol.2008.04.012.

Zhang, C., Cao, X., & Ramaswami, A. (2016). A novel analysis of consumption-based carbon footprints in china: Unpacking the effects of urban settlement and rural-to-urban migration. *Global Environmental Change, 39,* 285–293. doi:10.1016/j.gloenvcha.2016.06.003.

Chapter 4
Edge Open Spaces in Madrid and Its Metropolitan Area (Spain), Sustainable Urban Planning and Environmental Values

Fernando Allende, Elia Canosa, Nieves López and Gillian Gómez

Abstract This contribution is focused on the valuation of the spaces that remain vacant at the city border, as part of a future urban Green Infrastructure. To that end, they have been identified and characterized. Furthermore, a specific GIS has been developed for Madrid, the largest urban metropolitan area in Spain. This chapter is organised in five parts and the final conclusions. First of all, in the introduction, the general objectives are exposed and the concept and importance of the open spaces in the city are studied. Secondly, the methodology describes the design of the research and the development of the GIS. Afterwards, the urban and territorial context, where these pieces have their origin, are characterized. This is a synthetic approach which also includes the recent evolution of the legislation as well as that of the city. Through this evolution it is possible to see how these pieces have been framed into a failed green belt since the middle of the last century. Fourthly, a presentation of these open spaces, their urban regime and their land uses is carried out. Subsequently, their valuation in a potential network of open spaces in the Madrilenian metropolitan area is performed. The basic criteria for this valuation is the land adjacent to each of these open spaces. Finally, in the conclusions, we set out the evidence arising from this research, requiring the need for having accurate tools, such as GIS, in order to reach solid proposals that deal with the challenge of Green Infrastructures. We also believe that the conservation and integration of open spaces should be a priority in general strategies of urban development for mitigation Climate Change and to reduce the effects of Global Warming.

F. Allende (✉) · E. Canosa · N. López · G. Gómez
Department of Geography, Faculty of Humanities,
University Autónoma of Madrid, Madrid, Spain
e-mail: fernando.allende@uam.es

E. Canosa
e-mail: elia.canosa@uam.es

N. López
e-mail: nieves.lopez@uam.es

G. Gómez
e-mail: gillian.gomez@uam.es

R. Álvarez Fernández et al. (eds.), *Carbon Footprint and the Industrial Life Cycle*,
Green Energy and Technology, DOI 10.1007/978-3-319-54984-2_4

1 Introduction

The term Open Spaces, was probably used for the first time around 1833 by the "Select Comittee on Public Walks" of London (Turner 1992). Over the last few decades Open Spaces have gone on to form part of the Green Infrastructures. These are defined by the European Council as "a network of natural and semi-natural areas and other environmental elements, that delivers a wide range of ecosystem services". The European Commission has developed a Green Infrastructure Strategy (Council of Europe 2013a) of high significance in the territorial planning and natural resources conservation. This strategy sets out the potential components of a Green Infrastructure: legally protected natural areas, healthy ecosystems and areas with high ecological value outside protected areas, natural landscape features, restored habitats, artificial features such as eco-ducts or eco-bridges designed to enhance wildlife movements, multi-functional zones where compatible land uses can help to maintain or restore healthy ecosystems against other incompatible activities, areas that are managed sustainably and help improve the general ecological quality and permeability of the landscape to biodiversity, urban features such as green parks that help ecosystems to work and deliver their services through the connection of urban, metropolitan and rural areas, and features for the adaptation and mitigation of climate change (Council of Europe 2013b). Land uses transformations change drastically the conditions of the near-surface atmosphere (increment of temperatures) over the cities and peri-urban areas and increase the impacts of climate change. Measures for their mitigation are necessary in urban areas and adaptation actions have been proposed from European Environmental Agency (2012) according to data of Intergovernmental Panel on Climate Change. Open Spaces and Green Infrastructures play an essential role in the urban adaptation to climate change and the sustainability of the urban environment.

The case that has been studied is the city of Madrid, where the current proposal of a Green Infrastructure would include some of the components mentioned above, such as: urban green spaces or protected natural areas designated at the regional level. However, in this chapter, a proposal is made so that some of the categories of peripheral lands, either non-urban or urban land but without urban development yet, can form part of a network of free spaces from the interior to the edges of the city.

The free spaces that have been recognized can be grouped into two major types: *urban green land* and *edge open space*, formed by peripheral lands that remain as unoccupied. In this second type, it is possible to distinguish the following sub-types: *Protected Natural Areas, Wasteland* (urbanization works without starting) and *Future Open Urban Land* (areas in transformation towards General Systems as roads, equipment and public spaces).

The roles of the free spaces are usually classified into two main categories: those that provide recreation and other services to society, such as health (Tzoulas et al. 2007), and those whose main objective is the conservation of natural values (Maruani and Amit-Cohen 2007). These two approaches start from the idea that open spaces with recreational function are located in the metropolitan areas while those that pose the conservation of natural values are in peripheral areas and with rural dominance. This work offers a third option in which some of the open spaces that remain within the urban and metropolitan space, under specific planning figures, can develop a function of conservation of natural, agricultural, landscape, visual and aesthetic values as well as of reduction of urban densities, reducing the recreational and landscaped role with an intensive use, which has been much more frequent in this type of spaces.

In different European areas (Koomen et al. 2008) a model is recognized, which is formed by a triangle whose vertex would be "agricultural landscapes—versus—Nature—versus—urban development". Land use changes in this triangle do not experience the same force and dynamic: it is much easier and more frequent to change from agricultural lands to urban lands than from natural areas to urban or agricultural lands. This lack of resistance to change of the agricultural areas against the urbanization is interesting and vital to understand the evolution that this spaces have followed and its scarce extension in the present time. This is what Maruani and Amit-Cohen consider a "market failure", the market value of these lands and their speculative value are much higher than the value of the land that is conserved as open space (Koomen et al. 2005, 2008). In some cases, the incorporation of a restrictive zoning has been proved as efficient tool in the preservation of open spaces, arriving at the purchase of land for the maintenance of agricultural activities (Koomen et al. 2008). It is at this point where planning takes on an essential role, allowing us to find iconic examples on this question in Europe (Randstad in the Netherlands and the Berlin Greenbelt) (Kühn 2003), as well as research that has made comparative analyses in relation to the management of urban green spaces (Baycan-Levent and Nijkamp 2009).

Furthermore, the ESDP recommendations (European Spatial Development Perspective) about the urban-rural gradient planning are focused on: the promotion of the agrarian restructuration acceleration, the diversification of rural economies, the mobilization and the increment of natural resources production through the valuation and of cultural services, and on the promotion of sustainable development in metropolitan areas and other urbanized regions (European Commission 1999).

The European urbanization process has been intensively studied (Champion 2001; Van der Berg et al. 1982; Antrop 2004). Different cycles and stages have been recognized. In them, the new features and structures overlap the traditional landscapes, which keep but fragmented and lose their identity, changing to functionally homogeneous areas. In these urbanization processes, transport routes have had an essential role, being accessibility the main factor of landscape transformation, not only because of the impact of the transport route itself, but also because of the indirect effects it provokes. The recognized phases are: suburbanization, counterurbanization, and reurbanization (Antrop 2004). For the southern European region, Antrop (2004) establishes the step from urbanization phase to suburabanization phase was occurred in the nineteen-eighties. The case of Madrid could be framed in this period. These drastic changes in the land use threatened the landscapes and their patrimonial values, leading to changes in the legal categories (zoning or planning control) and in urban planning, which leave the adjacent areas of the city without protection, under the "urban—shadow". Furthermore, the research focused on the evaluation of the natural potential that these lands have, are very numerous (Weber et al. 2006; Beer 2005), although almost always from the ecological perspective and with extensive areas, being scarcer the researches focused on the key role that these pieces can play in the landscape quality and its agrarian functionality.

In the case of Madrid, our study areas represent one of the versions of the "Opportunistic Model", (Maruani and Amit-Cohen 2007) in which the remaining space after the urban planning is represented as residual plots, usually small, irregulars or inaccessible for other uses.

As we have already pointed, the current planning in Madrid has left fragments of open spaces on the edges of the municipal area. These are not always protected, and could be converted, with those already catalogued as green areas and other areas included as General Systems, into the network of the green infrastructure that the city needs.

The objective of this analysis is to value the opportunity this kind of land offers in the achievement of that network. Throughout this chapter, the lose opportunities during the development of the previous planning processes will be reviewed. Furthermore, the evolution of the regulations of 1997th Plan will be examined critically. Since 1997, these regulations have allowed the occupation of some sectors. After that, we will characterize the lands that we have called *Edge Open Spaces* and are under these categories: Protected Non Developable Land, Common Non Developable Land (since 2001, this is equivalent to Developable Non-Sectorized Land), Non-Programmed Developable Land and Specific Planning Areas (Urban Land). Finally, a proposal of valuation of these spaces will be presented, with the aim of forming a network that, with the internal spaces of the city, connects with the open spaces outside of Madrid city, in the metropolitan area, multiplying its natural and landscape importance. The protection of these areas should be included in the guidelines for reduction the effects of climate change and open spaces should be considered in local policies (Ayuntamiento de Madrid 2014).

2 Methodology

The areas of interest and their buffer zones are located in Madrid municipal borough (Fig. 1). These zones were defined according to the development of the General Plan of Madrid (GPM 1997), and in particular, the areas included in the adjustment of GPM to the Land Law of Madrid (9/2001) had taken into account. Mostly are located in the border of municipal area (Edge Open Spaces, EOS), as well as their adjacent areas that were considered in a buffer of 700 m around external margin of the polygon. Four methodological approaches were used: search of digital and bibliographical info, field work, analysis and validation of thematic info to evaluate their socio-environmental interest and, finally, the implementation of all this information in a Geographical Information System.

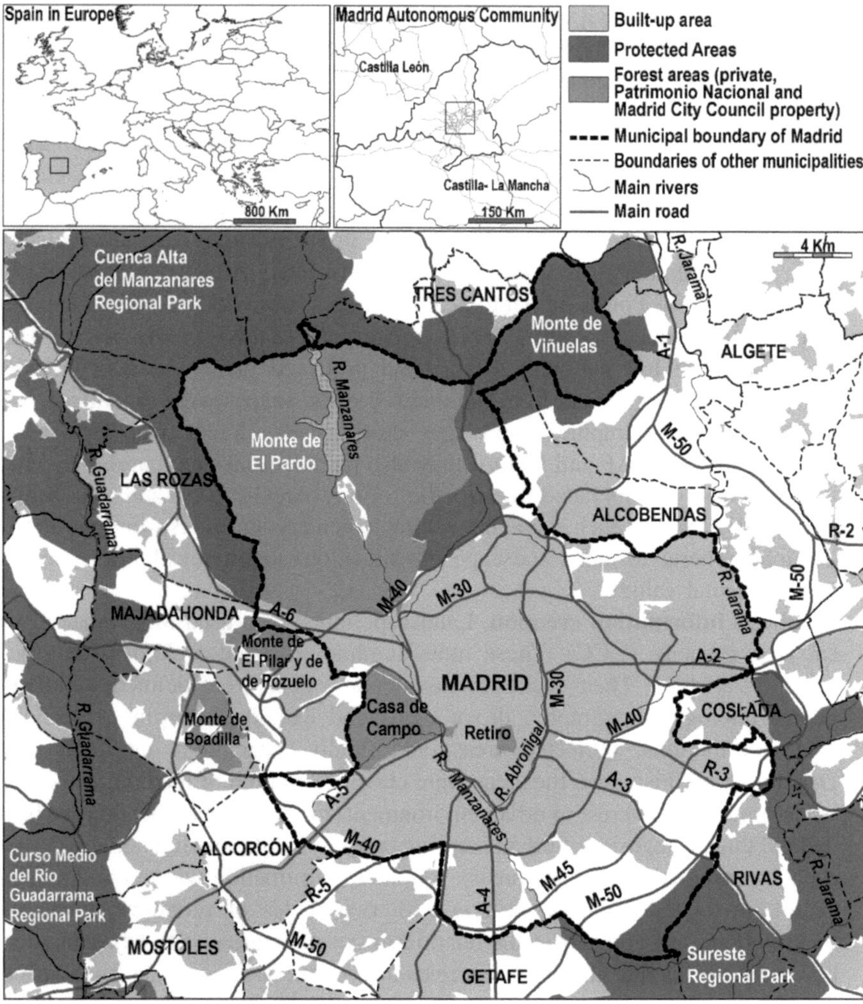

Fig. 1 Madrid and nearby municipalities (2016). *Source* Compiled by Community of Madrid. Territorial information and regional cartography

- **Digital and bibliographical information**. Data of development have been obtained from Planning, Housing and Infrastructure Area of Madrid City Council in *shapefile* and *dgn* formats. All the other geo-referenced information was developed for this research. Open spaces, green infrastructures and city dynamics were the main thematic areas of the bibliographic information.
- **Field work**. 22 basic tracks were designed in *shapefile* format using orthophotographies. For each track different thematic information was registered and they were geopositioned by DGPS Trimble Nomad 6G (elements of interest and terrestrial photography). A second assessment of the information was

developed, and in case it was necessary, new tracks for complete the information were designed. The shape and number of the tracks are linked to the attribute table. All tracks were coded with associated information in line and point format (*shapefile, kmz and gpx*).

- **Analysis and validation of thematic info using Geographic Information Systems**. Valuable social-environmental areas and areas protected by current laws were considered green spaces and, these last ones, preserved of development projects. These areas excluded from urban planning include protected areas of Community of Madrid Network Protected Areas (Regional Parks, Preserved Forest, Points of Geological Interest), Natura 2000 Network (Special Conservation Area, Special Protection Areas for Birds) and Important Bird Areas (Birdlife International). Rivers and protected plains are also included through a buffer of 100 m from the thalweg. New shapes were digitalized over PNOA orthophotographies: landscape, drove roads (drawn over 1956 aerial photography of USAF and PNOA, they also include protected margins by law). All variables were included in shapefile format (ArcGis 10.x) and kmz format (Google Earth). For each shape three complementary attributes were considered in alphanumeric format (landscape classification, toponymical reference and environmental value).
- **Synthesis information creation**. Landscape units were used to estimate urban carrying capacity (UCC). These units summarize social-ecological and geographic variables. Their social-environmental values were defined: lithology, geomorphology, topography, drove roads and excluded margins, rivers and excluded margins, protected areas, and finally, vegetation and land uses. Depending on this value, the areas were classified in high or medium ecological value or area for interest in urban environment. From this classification we have defined three categories of urban carrying capacity: high, medium and low urban capacity (proposed protected areas). Synthesis information was classified using singularity and distribution criteria with two scales of reference: local and metropolitan. For both scales the relationship of these areas with adjacent zones has been considered, therefore the relevance of these areas in an open spaces network has been analysed.

The highly detailed analysis allows a better assessment of the social-environmental values of Edge Open Spaces. Without this detailed view it would be an impossible task to identify the real value of the EOS and the results and conclusions of this chapter.

3 Open Spaces Evolution in Madrid During the Recent Municipal General Plans (1963–1997)

Planning of open spaces around Madrid municipal boundary is the chronical of a continued failure. In the nineteen fifties began the legally binding urban planning. The first three General Plans established successive green belts that were consistently broken. The fourth one, approved in 1997 and currently in force, propose the complete urban development for all parts of municipal area. The *Spanish housing bubble* (1996–2006) promoted this expansionist policy of urbanizing nearly without limitation, always supported by municipal government. Nowadays Madrid has maintained only a small representation of undeveloped edge open spaces. The footprint of these small areas in the edge of the city is higher thanks to their continuity with historically protected forests.

Following the characterization of the recent evolution of the city, which includes the origin and structure of the more relevant open spaces, we will deeply analyze the content of each General Plan and their vision of these spaces. Then, the destruction of the green compromise for a *habitable and greener city* will be discussed. The end of this compromise starts with the partial invasion of projected open spaces to the definitive disappearance with the neoliberal urban model.

3.1 Recent Evolution: Spanish Housing Bubble

The spatial configuration of the city of Madrid is determined by its role as the State capital and because of that, as the residence of the monarchy, since the second half of the XVI century. At the West of the city, over natural interest areas, Spanish monarchs created protected spaces for leisure activities. These preserved properties around Madrid were the forests of El Pardo (15,821 ha, 1000 of them are open to general public), Casa de Campo (1722 ha), Viñuelas (3000 ha) and the forests of Boadilla and Pozuelo (800 ha each one). These forests cover the northwest area of the municipality over an environment of great quality, base of the residential urban developments for high social classes. In opposition, the southwest, with the lack of these qualities, is the place where working classes are located. The Casa de Campo is open to general public since the 2nd Republic when the state gave the enjoyment to the city of Madrid. In the General Plan of 1997 this area was incorporated as a consolidated Specific Planning Area and not as a peripheral open space.

After the Spanish Civil War (1936–39) Franco's regime revitalized Madrid as the state center. This has been essential in the whole configuration of the city. In the early twentieth century a period of slow but continuous population migration from rural areas to the city began and, after the civil war, it drastically increased. Government policies for economic modernization of the country allowed the secondary and tertiary sectors to take off, increasing the hope of having a job and improving the future of rural immigration. From 1900 to 1930 the population doubled to about one million. Thereafter, when the war was finished, the population increased by more than 4% a year until 1950 and then by 5.7% until 1965, when the city reached 2,800,000 inhabitants. Among the urban development strategies of Franco was *The Great Madrid*. Its aim was to manage the city directly, especially the metropolitan urban growth. Remember that from 1947 to 1954, thirteen municipalities with 325,000 people were absorbed by Madrid and its area increased from 6676 to 60,437 ha.

A slower but regular urban growth led to a population of 3,228,057 in 1975. During the 1970's there began a process of metropolization with a movement to its bordering municipalities with cheap land and lower urban control for industry and housing. The municipality of Madrid had a negative demographic growth with a loss of almost 400,000 citizens, while the nearby municipalities had a fast increase. The 22 municipalities included in the *Madrid Metropolitan Area* between 1964 and 1980 grew from 101,180 inhabitants in 1960 to 323,140 in 1970, reaching 820,765 in 1981 and 1,242,051 in 1991. The current ratio in growth rates still shows a great increase in the municipalities within the Madrid Community, but outside the city of Madrid, although in practice the Madrid Community is considered a single urban area.

Only during the *Spanish housing bubble* was there a demographic growth in the city of Madrid driven by economic growth, the continued expansion of credit and the influx of 600,000 foreign immigrants (55% of the total for the Madrid

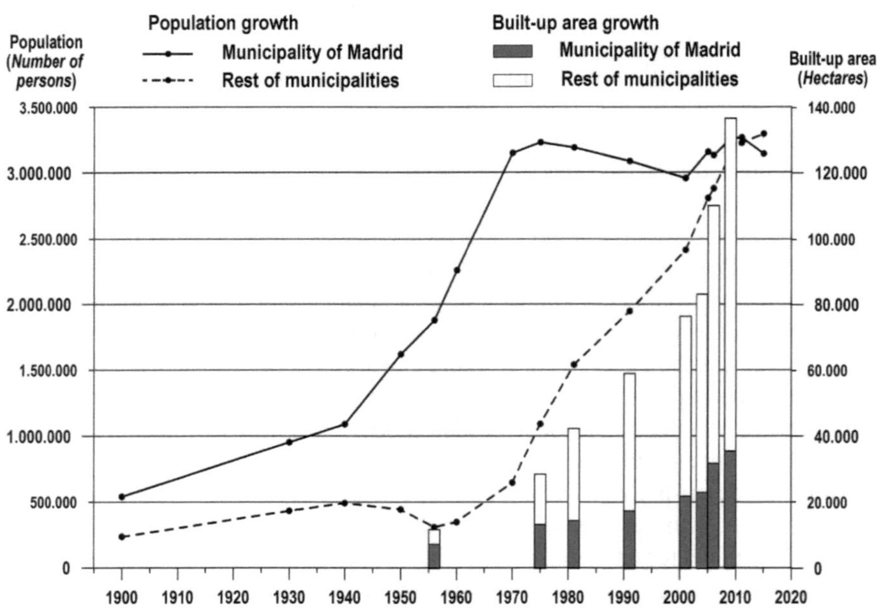

Fig. 2 Population of the municipality and Madrid Region. *Source* Compiled by INE, 1900–1997 collection of statistical year-books; 1997–2015. Municipal Register of Inhabitants and population censuses; land use: 1956–2005, General Direction of Urbanism (2006), Corine Land Cover (2006) and SIOSE (2009), Ministry of Development, System of Urban Information (SIU)

Community). The growth rates became positive again between 2000 and 2011 at 1%, although lower than the figure of 2.9% in the rest of the municipalities in Madrid (Fig. 2).

This demographic growth was accompanied by disproportionate urban sprawl, in the classic model form of *oil stain*, following the city's road network. From 1956 to 1975 Madrid city doubled its urban area from 7400 to 13,304 ha. At that time urban land of Madrid city was equivalent to the urban land of the rest of 178 municipalities of Madrid Community. During the subsequent decade urban grew moderately with 269 ha a year from 1995 to 2001 and more than 300 ha a year in the following five years. This increment of urban land has not any correspondence with the demographic growth. From 1995 to 2005, population increased in 125,600 new inhabitants while the urban sprawl increased in 3753 ha. In other words, an increment of 4% of the population does not endorse an increment of 20% of the urbanized area. The growth model implanted in Madrid in the last few years, based on the property development and speculation, has promoted the urban sprawl over new neighborhoods that in many cases are not finished yet. The economic and financial crisis has endangered the future of these areas, being very difficult to re-plan them.

Suburban railways, underground, motorways, national roads have hyper-connected the city and their regional area, in addition, the new section of high-speed rail (AVE) and the expansion of Madrid airport have increased national and international connections, essential support for urban sprawl. Taking benefit of this accessibility, public and private urban development initiatives were implemented over large areas.

In the last 20 years 360 new kilometers of motorways have been built for the access to Madrid city: ring roads M-40, M-45 y M-50 and tolled motorways (R-2, R-3, R-4 and R-5). These big rings were proposed by the national and municipal planning of infrastructures since the middle of the past century. First ring road (M-30) has been underground and its traffic volume has increased. Road network of high capacity in Autonomous Community of Madrid has grown from 319 to 1000 km. (1985–2004) and total road system reached from 2728 to 3492 km (De Santiago Rodríguez 2012).

3.2 General Plan of 1946. The First Green Belt of Madrid

In this context of rapid urban growth but no so rapid demographic one, the urban planning has accompanied, but not led, the different periods of the configuration of Madrid city. Each of the four General Plans that have been approved since the middle of the last century summarizes in some extent, the aspirations of the public policies in the city. Nevertheless, interests of dominant economic groups and the lack of political willing has prevented the real application of these urbanistic regulations.

We have analyzed their treatment of edge open spaces, their identification and their proposed actions within the general urban planning. From this analysis, it should be noted that two types of open spaces can be distinguished. The first one includes green areas inside the city, in urban land. In other words, parks and gardens. *Casa de Campo* and the *Parque del Retiro*, are two examples of these green areas. We will only consider them at the end of this work, forming part of the green corridors network. The second one includes all the green belts or forestry belts that since 1946 were designed around the urban border with a double purpose: as a limit of urban growth and as an interface between urban and rural areas. These green belts were mentioned into the General Plans until 1985 and they were thought as naturalized areas that should been reforested and also where other uses could appear complementarily: agrarian uses, urban farms, etc. With this design the urban planning in Madrid was linked to the European tendency of green belt solutions: London, Frankfurt, Berlin or Vienna after the Second World War (Amati 2016; Kühn 2003; Werquin et al. 2005).

The 1941 General Plan, approved in 1946, proposed a city model, as far as green areas are concerned, based on two main pillars. Firstly, the new scale, a regional scale. This Plan went further the municipal limits of Madrid city. The city area increased with the join of adjacent municipalities and the Plan could set up the proposals of the city decongestion as well as the green belts development. Furthermore, the creation of an autonomous agency depending by the State with the objective of controlling the development and enforcement of the Plan (*Comisaría General para la Ordenación Urbana de Madrid y sus Alrededores*) reflects the will expressed by Franco regime to create The Grand Madrid as a city representative of the new Regime and to control the municipalities in the urbanization development process (Fig. 3).

The second pillar of the 1946 General Plan, the green belt, is the result of the combination of the congestion necessity of the city and the worries about the incorporation of environmental values. Therefore, the Plan proposes a concentrated and limited urban structure by a system of three green belts connected with transversal green wedges (Sambricio 2015; Sassi 2016).

The first green belt surrounds the central city and separates Madrid from the suburbs already developed and the future satellite districts (Fig. 4) that will be created for the relief of the city. These future neighborhoods were isolated using green wedges and limited by a second belt connecting with the previous one. The third green belt, delimited the urban influence area and included the main forests already mentioned: *Montes del Pardo y Viñuelas* at the north area and new areas that will be reforested along the river plains that surround the city: *Guadarrama* river (West), *Jarama* river (East) and *Culebro* stream (South).

These first proposed green belts can be characterized for their limited width with the exception of the forest areas already existent. They seemed to be designed for the embellishment the accesses of the city more than a true worry about the creation of green areas. Other characteristic of these green belts is that they were designed such as natural areas, and not as urban parks. As it was said in the official magazine of *Gran Madrid* (Planeamiento Urbanístico 1953, 15), “they should be pieces of Nature adapted to their urban vicinity”.

During that period, the land purchase policy and the reforestation programs were not effective because twofold reasons: Spain was suffering an economical dearth

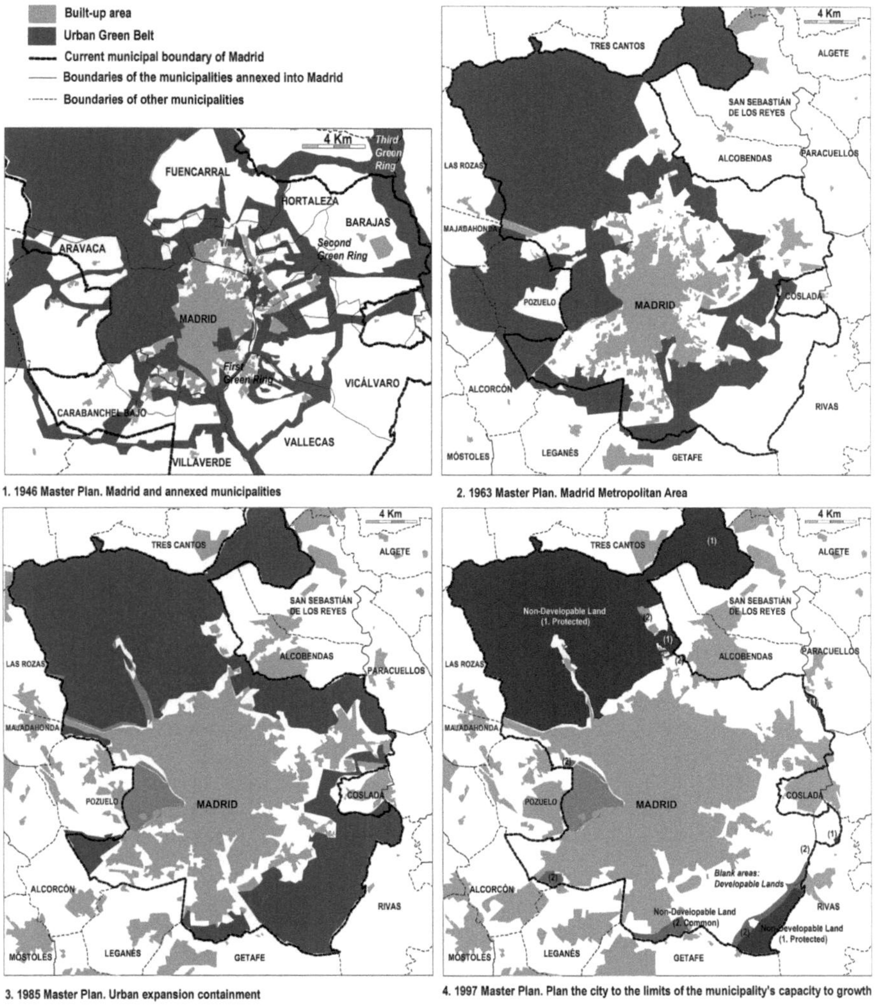

Fig. 3 Evolution of Green Belt models. *Source* Compiled by General Direction of Urbanism, 1990 and 2014. Territorial information and regional cartography

and the real-state sector was being created with landowners that do not accept any restrictions in their construction rights. From 1946 to 1963 only eight green areas were created, four of them over the green belt and *wedge forestry areas* (only 146 ha in 17 years). In 1952 four forestry parks were included as a part of a program of reforestation over the green belts (106.2 ha), two of them were located in cleaned up streams where urbanization was forbidden (Canosa Zamora and García Carballo 2013).

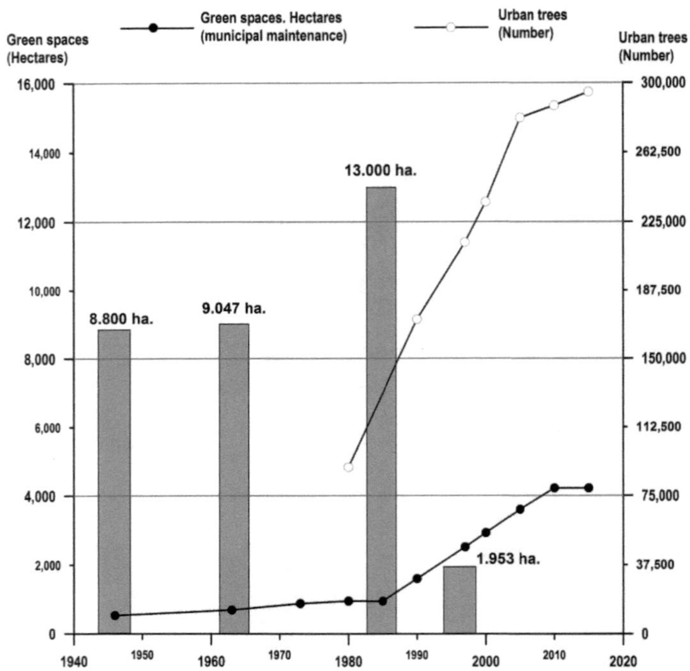

Fig. 4 Municipal green areas (gardens and trees) and surface planning in General Plan (green belts). *Source* Compiled by parks and trees. Municipality of Madrid. Statistical Information Area. Database forestry areas and gardens owned by National Heritage are not included: Campo del Moro (20 ha), El Pardo (15,821 ha) y la Casa de Campo (1722 ha) and Tres Cantos forestry park (225 ha owned by Madrid municipality from 1941). Green belt areas compiled by aerial photography and Jiménez (2015). El Pardo, Viñuelas and Casa de Campo were not considered (urban parks)

Other element that would definitely hindered the green belts enforcement was the scarcity of affordable housings for new arrivals with low wages. From 1940 to 1960, more than 750,000 immigrants had arrived to the city. In 1960 there were more than 58,530 shanties as well as numerous self-built neighborhoods without basic infrastructure and equipment (Otero Carvajal 2010). The pressure of laborers and entrepreneurs together with the fear of a social riot allowed a development of public housing policy from municipal and state authorities. Expropriated properties for the second green belt were used (more than 2000 ha until 1960) to build social housing promotions. Barely, 100 ha were reforested, most of them located in the edges of the access roads to the city with the intention of improving the visual quality.

3.3 General Plan of 1963. Metropolitan Green Belt Approach

The 1963 General Plan offers a strengthening of the proposals more linked to our objectives: it designs a Grand Green Belt surrounding the city and has a metropolitan approach, planning beyond municipal boundaries. This Plan includes for the first time the concept of *Metropolitan Area* with competences in 23 adjacent municipalities. Furthermore, it creates the Planning and Coordination Commission of Madrid Metropolitan Area (COPLACO), disappeared in 1981. Despite the wishes of intervention in this new metropolitan frame and with the aim of decongesting the city, the Plan was unable to manage the urban development of Madrid over the following decades.

The 1963 General Plan, re-shapes green belts designed in the previous period. Green areas of the first green belt were eliminated and interstitial-empty spaces were occupied by communication infrastructure and residential areas. Nevertheless, a single green belt is kept in peripheral area amplifying the second Green Belt designed in 1946 General Plan. This area was considered a forestry belt with *green wedges* that complete the borders of the city. Once these lands were reforested they should serve as leisure places.

For this large green belt, 14,860 ha of rural and forestry lands were closed, 9047 of them within municipality of Madrid. These areas were thought as connection nodes with natural areas around the city (*El Pardo* at the North, and the river plains of *Jarama-Manzanares* at the South). Westward the Plan changes the green belt design of the previous plan to adapt to the important private forest areas (*Quercus ilex* subsp. *rotundifolia* and *Pinus* ssp.) in the municipalities of Boadilla and Pozuelo.

Green Belt of 1963 was a multi-functional system. From the more classic functions, urban, landscape or social functions, to the newer ones, as environmental benefits (Jiménez Garcinuño 2015, 63). The Plan of 1963 highlighted the necessity of the improvement the urban environment through the regeneration of the vegetal cover.

The conservation of this forestry areas was reflected in the regulations of the Plan but compulsory expropriations were not considered by the Plan; mainly because of the economic crisis which made difficult to allocate public funds to these proposes. Once more time, the preservation of lands for open spaces was not possible. More than one third of the green belt would be built during the following 20 years, until the next General Plan endorsement. Most of the lands were used for residential purposes, although 25% were used for healthy, military and educational equipment and only 290 ha were dedicated for forest reserve (Jiménez Garcinuño 2015, 174).

3.4 General Plan of 1985. The Short-Lived Enlargement of Protected Areas

With the arrival of democracy, a new territorial and political order changed the urban planning and land use policies. The land planning and management competencies were transferred to the regional governments (in our case, Madrid Autonomous Community). In addition, local government acquired new competencies regarding urbanism such as drafting, approval and management of General Plans. These Plans must be ratified by Autonomous Communities and coordinated with regional and sub-regional planning strategies (Tomás 2009, 48). In conclusion, the old government structure disappears and the State loses its competencies as mediator between the different urban actors. Instead, from 1979 and for 10 years, a left-wing coalition would govern Madrid.

This new government designed a General Plan appropriated for an economic and demographic stagnation and for an urban model committed to social and territorial debt. The main objectives of the new municipal government will be focused on what is called "finishing the city" through the development of service facilities and equipments in disadvantaged areas, taking use of the empty spaces that the urban sprawl had left. Not only the peripheral area was requalified, also the city center and the numerous slums settlements were remodeled by the local government, characterized by a strong public interventionism. Regarding open spaces, parks, forest and agricultural lands, their policies were linked to the municipal strategies. In particular, a green areas network was considered especially appropriate to improve urban quality of life.

Almost 13,000 ha were classified by the 1985 General Plan as "non-developable land", distinguishing different categories, against the "rural land" which was the category recognized in the previous Plans. These non-developable lands are very heterogeneous areas, with natural, landscape, environmental, agricultural, or mining values (Ayuntamiento de Madrid 1988, 69). Hence, this Plan makes a distinction between common, agricultural and ecological non-developable land, and for the two last categories, a new sub-category named *special protection land* is created. These areas were going to constitute a new green belt that was going to surround the municipal boundary and limit the growth of the city.

Nevertheless, planning results were below expectations. A new highway bypass (M-40) was built on the Green Belt of 1963 and, in practice, completed residential areas on the perimeter of the city. This period must be considered a *green period*: more than 600 ha of gardens and urban parks were created in 10 years while in the past 40 years only 400 ha of green areas were developed. Nevertheless, the General Plan targets were not achieved, owing to the change of economic dynamics and political government, that altered the expected orientation completely.

3.5 General Plan of 1997. The Decrease and Fragmentation of the Green Belt. Edge Open Space Configuration

Municipal urban development strategies and policies changed from 1991 when, for the first time, right-wing governed Madrid municipality. With the *economic boom* Madrid becomes the Spanish core of the neoliberal urban model. The system collapsed with the economic crisis and the *bursting of the housing bubble*. The keys of this model were analyzed by, among others, Burriel (2016), Romero et al. (2015) and Naredo (2013). In summary, 1997 General Plan summaries the will of urbanization, an urban growth until the limit of urban capacity without temporal period of development (Rodríguez Avial 1997). This will be traduced in scarcely 36% of lands were catalogued as non-developable land (Ayuntamiento de Madrid 2012a, 23).

The municipal government proposal of urbanizing the whole municipality of Madrid was implemented in two stages. Firstly, the increment of the lands under the category "Developable Lands" through one-time modifications and the subsequent drafting of the Plan. This increment was achieved at the expense of the remaining "rustic land" and declassifying a large area of protected lands by the General Plan of 1985. Thus, from 1991 to 1997, 3344 ha were illegally got for new residential developments (Brandis 2014). The second stage started after approval of the Land Law of Madrid (9/2001). This law deals the regulations laid down in the State Land Law of 1998. For the first time, all the non-urban and non-protected lands were considered developable land. As a result of the former, *Common Non-Developable* lands in 1997 General Plan (almost 1250 ha), became *Non Sectorized Developable* lands with the exception of areas protected by different codes. In other words, these lands could be converted into urban land with a zoning plan even if the General Plan had not foreseen their development (Calvo et al. 2007, 229).

A large part of residential areas, a total of 21 on 9000 ha, were planned on protected land by the General Plan of 1985 because of their natural or agrarian values. Among these areas are six of the eight great urban developments designed in the General Plan of 1997. In total, more than 110,000 housings on 4400 ha. Their construction, as well as the construction of the 13 remaining developments, has been troubled by the ecologist groups and municipal opposition appeal to the Court. It has been 19 years since the annulment of the reclassification (2003) and four additional legal proceedings, sentences, until the final resolution. In 2012 municipal government reviewed in detail the planning, arguing in each case, the reclassification of protected lands into developable lands. In 2016, a definitive sentence has endorsed that review and unblocked the enforcement of the new neighborhoods, preventing from recovering large open spaces for the city.

Due to financial and economic crisis house building sector had stopped and no urban development work has started in large city areas of the Southeast of the city.

However, although the preservation of these areas is not possible, the reality shows that in large part of these areas, especially at the Southwest of the city, the development works have not started yet. Not only because of the slowness of legal process, but also because of the economic crisis.

Increasing urbanization of this period was possible because of the emergence of right-wing into the municipal and regional government. Madrid Autonomous Community has been a driving of this expansion, through the construction of road infrastructures and helping the immediate land occupation (Mata Olmo 2007). During economic boom the highway M-45 was built, and the M-50 has been finished, among other new radial highways. New urban developments are located between M-40, M-45 and M-50, and over the empty spaces between the different highways built since 1999 (Gómez Mendoza 2013, 23).

The final result cannot be bleaker. Not only for because of the created scenery, but also because of the lost opportunities. Madrid urban planning has not really promoted the open spaces creation, and their current total area is scarce compared with the initially proposed area. Furthermore, the current pieces of open spaces have neither the connection nor the reforested image that was expected at the beginning. It clearly seems that there has not been a political will for their enforcement, therefore, the needed strategies or tools for their development were never created (Fig. 5).

There has been an alliance between all the urban agents, public authorities, land owners, banks and property developers which has allowed an indiscriminate and excessive construction in the city of Madrid. Even the citizens have not shown any worry for the lack of parks and green areas, except in specific and late periods. Only the preserved areas (formerly monarchy properties), are currently the dreamt large extent of forests as a kind of Green Belts for the city.

Green Belt concept has lost its significance in the last project of "Forest Green Belt of Madrid" designed in 2009. This new and perimetral Green Belt is a discontinuous narrow ring band, which is made up of already existent parks and small protected areas located on both sides of the M-40 highway. These areas are completed with 100 new hectares adjacent to the highway or isolated by the roads (Ayuntamiento de Madrid 2012b, 68).

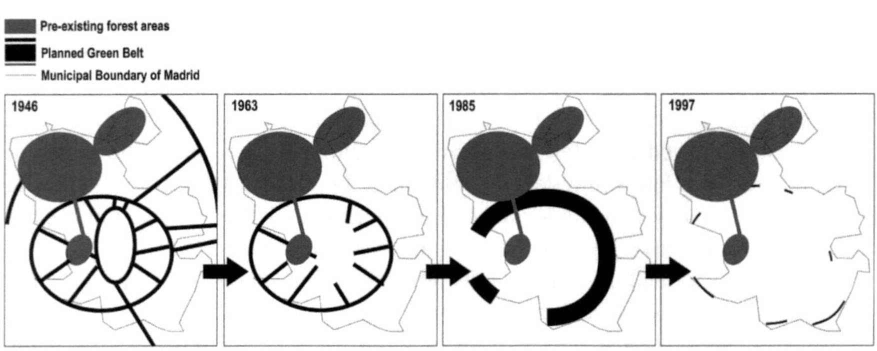

Fig. 5 Schematic representations of the structures of the successive Green Belts

Since the 1970's as a mitigation measure, municipal governments have promoted the tree planting on pavements, boulevards, central reservations and other pedestrian spaces. Through these actions Madrid, after Tokyo, is the second city of the world in number of trees on its streets (Calderón et al. 2009, 12).

4 Edge Open Spaces Evolution in Madrid and Intervention Proposals

4.1 Types of Edge Open Spaces in Madrid City

The city planning regulations of the General Plan of 1997, have been updated several times over the years because of the modification works in the Plan, the agreements of the General Plan Monitoring Committee, the regulations and the sentences which affect these city planning regulations.

As it has been pointed out before, there are fewer and fewer open spaces defined by the urban planning. They are included in the following categories: Protected Non Developable Land, Common Non Developable Land (as it has been said before, it has changed to Non Sectorized Developable Land, and therefore, it could be subject to development), Non-programmed Developable Land and Specific Planning Areas that in practise are General Systems, within Urban Land.

The General Plan includes as Non Developable Land *those lands that, according to the chosen territorial model, should be preserved of the development processes,* as well as, *lands with higher natural values that advise against urban development, being protected by a specific regime according to the autonomic legislation* (Ayuntamiento de Madrid 2015).

The General Plan distinguishes two categories. The first one, **Protected Non Developable Land**, includes different kind of protections: Special Protection, Ecological Protection, Forest Protection; Protection of River beds and banks, Drove Roads Protection and Protection of Infrastructure and Equipment.

The second one, **Common Non Developable Land**, disappears as such, after de approval of the Madrilenian Land Law of 2001. This Law establishes that all the non-urban and non-protected land can turn into Developable Land. The lands included in this category at that moment have had differing success: either they have been protected if they were affected by a specific protection or they have been changed into Non programmed developable land.

It has been analysed areas of all the categories mentioned above, except: Protected Non Developable Land (Infrastructure and Equipment).

Regarding **Non Programmed Developable Land**, the General Plan says *it is formed by those areas that, being suitable to be urbanised, they could be only incorporated to the urban development through the approval of the Specific Planning Areas* (Compendio de las Normas Urbanísticas del PGOUM-97, 2015, 91). Several study areas are under this category.

Finally, the **Specific Planning Areas** that are part of Urban Land. The difference between them and Common Urban Land is that those *constitute independent management unities, either because they have a specific protection regime of edification or because they have urban conditions that justify their individual treatment* (Ayuntamiento de Madrid 2015). The study areas included in this category are open spaces because they are green areas that were not developed in the General Plan of 1997.

As it has been already pointed out, some of the open spaces studied in this work, have seen how their land use classification was modified because of the enforcement of the sentences of High Court of Justice of Madrid, 27 February 2003, and of the Supreme Court, 3 July 2007, 28 September 2012, 1 August 2013 and 6 September 2016.

These sentences modify some of the study areas. For example: *El Monte de El Pilar (Aravaca—La Escorzonera-Monte de El Pilar)*, classified as Non Programmed Developable Land in 1997, has become Protected Non Developable Land (Forest Protection). With the *Área de Ordenación Especial de Valdegrulla* has happened the same. Nevertheless, there have been changes in the inverse sense. It is the case of the old Non Programmed Developable Land, *Solana de Valdebebas,* that nowadays is considered Sectorized Developable Land.

In Table 1 it is shown all the areas that have been studied and their classification according to the General Plan of 1997 and the current General Plan (Fig. 6).

At present, edge open spaces in Madrid are groups of disconnected patches, and in some cases of reduced extent. We can assure that Madrid and its immediate metropolitan border are held up due to the city sprawl promoted by the public policies. In many cases the open spaces of social and environmental interest have been disappeared in favour of the partial or total development of land. Open spaces free of development in the planning of 1997 maintain to a great extent their urban classification in subsequent revisions. That implies that in some cases the open space has disappeared. It is true that the aggressive neoliberal ideology of 1997 General Plan has been softened, but once the possibilities of the territory are almost exhausted.

Within this scarcity and dispersion already mentioned, it should be pointed out that the majority of the open spaces of Madrid are protected areas (more than 80%), and only more than 12% are wastelands, with forest, agrarian and leisure potentiality. The remaining 5% are included in the General Systems of Urban Land (Fig. 7).

4.2 Territorial Value of Edge Open Spaces in Regional Context

Beyond their dimensions and the natural features of the group, it is essential for the valuation of the pieces, the interconnection that it is stablished with the immediate

Table 1 Categories of land in the study areas

Categories of land	Number on the map	Study areas	Land regime (master plan 1997)		Land regime (land law 9/2001)	Edge open spaces (typology)
Non-developable land	1	Monte de El Pardo	Protected	Special protection	Without changes	Protected areas
	2	Valdemingómez-Cumbres de Vallecas		Ecological protection		
	3	Viñuelas-Fuencarral		Ecological protection		
	4	Los Ladrones		Ecological protection		
	5	Valdelatas		Forest protection		
	6	Cañada Galiana		Drove roads protection		
	7	Jarama		Protection of river beds and banks	Urban land. Built into Barajas Plan to enlargement of the Madrid-Barajas Airport	Restricted access. Incorporated in the airport grounds
	8	El Goloso				
	9	Valdegrulla 1	Common		Application of the regime of developable non-sectorized land	Wasteland
	10	Mina del Cazador				
	11	Entorno de Valdemingómez				
	12	Valle de las Mimbreras				
	13	Villaverde				

(continued)

Table 1 (continued)

Categories of land	Number on the map	Study areas	Land regime (master plan 1997)	Land regime (land law 9/2001)	Edge open spaces (typology)
Developable land	14	Los Cerros (Desarrollo del Este-Ensanche de San Fernando de Henares)	Non-programmed	Non-Sectorized	
	15	Campamento (Remate del Suroeste)			
	16	El Encinar de los Reyes		Sectorized. In process of development: Sectorization Plan with initial or definitive approval	Wasteland (Urbanization works without starting)
	17	Mesa de Rejas-El Moral-Vicalvaro (Nueva centralidad del Este)			
	19	Solana de Valdebebas			
	18	Monte de El Pilar (Aravaca - La Escorzonera-Monte de El Pilar)		Application of the regime of non-developable land. Forest protection	Protected areas
	20	Área de Ordenacón Especial de Valdegrulla			
Urban land	21	Manzanares Norte	Specific Planning Areas	Land divided. Part of the area is considered non-developable	Areas in transformation towards General systems (roads)

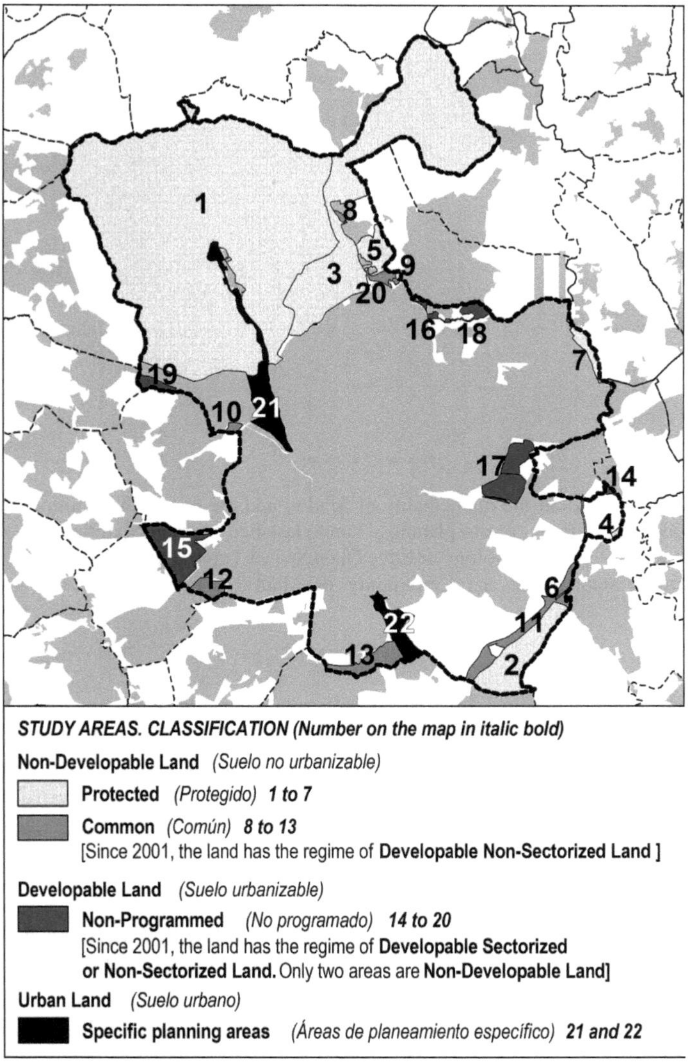

Fig. 6 Study areas classification

territories, and therefore, the connectivity in local and metropolitan scales. The final objective is to set their role in the definition of a Madrilenian green infrastructure.

Five basic types of contacts have been defined (Table 2 and Fig. 8). They include all the existing interconnections between EOS and immediate areas. First of all, the contact between protected EOS with other protected spaces. These protected EOS can have contacts also with urban areas (second typology) and with wastelands (third typology). The fourth type of contact is between wastelands. And finally, the last typology is the contact between wastelands and urban land.

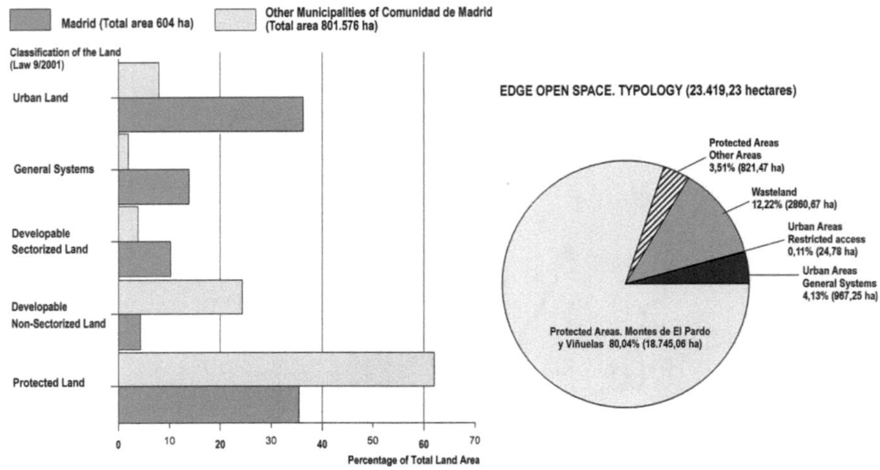

Fig. 7 Major land uses in the municipality of Madrid and the rest of the Community of Madrid municipalities according to current planning. Land classified by municipalities for each planning (according to Law 9/2001). Typology of Edge Open Spaces (*right*). *Source* Compiled by data from the Institute of Statistics of the Community of Madrid and our own Edge Open Spaces quantification

Table 2 Type of contacts between study areas and their nearest surroundings

Type of contacts		
Type 1	Protected EOS	Protected EOS
Type 2	Protected EOS	Urban land
Type 3	Protected EOS	Wastelands
Type 4	Wastelands	Wastelands
Type 5	Wastelands	Urban land

Type 1. **Contact between Protected EOS and other protected spaces**. In many cases, this protection is associated to the existence of protected land by national laws (ZEC, ZEPA), regional laws (Regional Parks and public forests) and even by local laws (protected non-developable land). Objectively, this is the contact with the highest repercussion, as it means the union between spaces of recognized quality in the region. It should be noted, again, the singular presence of the two main forests of the capital, the forests of El Pardo and Viñuelas (northwest sector). Thanks to them and their secular existence, Madrid reaches a high position in the ranking of European Green Cities. Both of them are close to each other and they have a connection with a protected space in large part of their perimeter, Cuenca Alta del Manzanares Regional Park, which multiplies their potentialities. Other five protected EOS enjoy the connection with other protected areas as well, although to a lesser extent and significance. Even so, they increase, apart from their scale, areas of special interest.

Type 2. **Contact between Protected EOS and consolidated urban land**. Type of contact we consider especially unfortunate. The Northwest—Southeast social

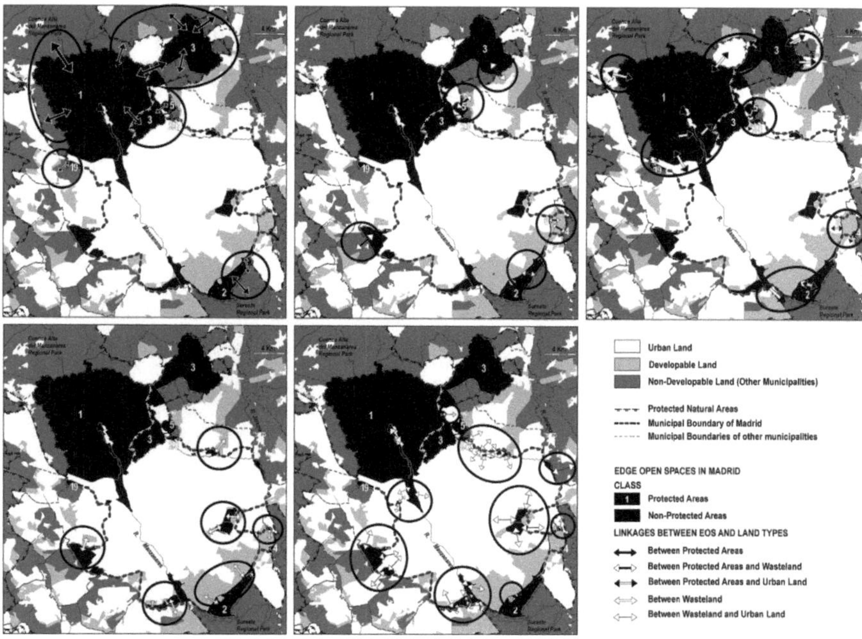

Fig. 8 Typology of contacts between Edge Open Spaces and bordering land uses

and spatial segregation in Madrid has its perfect reflection in the type of uses that shades this connection. Next to El Pardo, Viñuelas and El Pilar forests (numbers 1, 3 y 19) upper class neighbourhoods appear in the end of the seventies of last century. Whilst, around Valdemingómez (number 2 and 11), there is a garbage treatment plant inside the technological complex of urban garbage treatment located in its surroundings. This typology has a lesser territorial representation than the previous one (Fig. 9).

Type 3. Contact between Protected EOS and wastelands, territories without protection and without development. The higher environmental sensitivity at present should have driven these zones away from the property development. But the reality is the contrary and almost all the protected spaces have in their perimeter some sector of wasteland. This third alternative should be considered as a threat, but also as an opportunity. Although these are non-protected lands, therefore developable lands by the Madrilenian Land Law of 2001, the economic crisis has stopped the property developments: For that reason, these wastelands maintain residual but valuable agrarian uses and most of all, with the right aptitude for its reformulation towards a more conservative approach or at least towards the partial insertion of these zones into the corridors.

The case of *Cañada Real Galiana* (number 6) deserves special mention. It was a protected area in the Plan of 1997 and in the following legislation as drove away for the movement of cattle. The urban pressure has been devastating in this case and

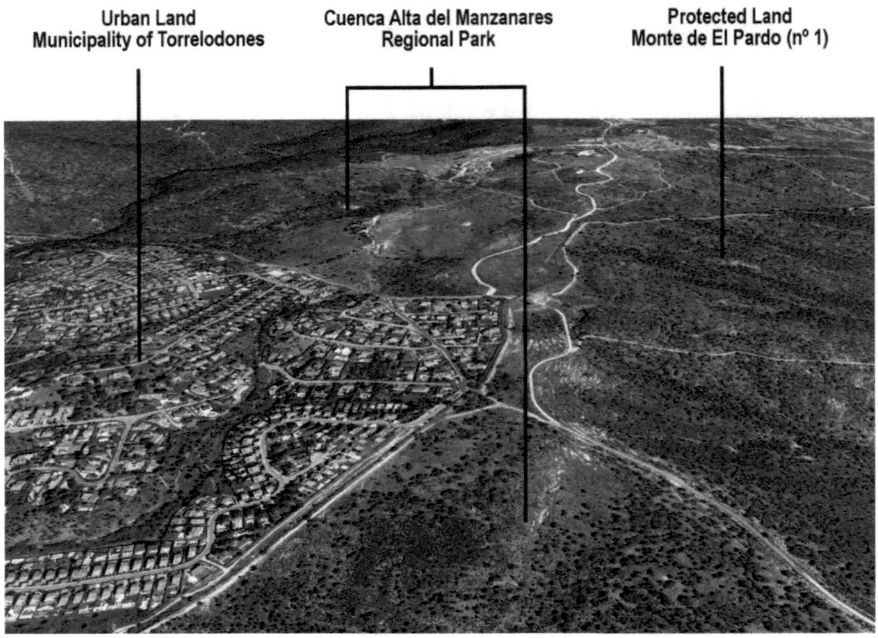

Fig. 9 EOS contacts at the Northeast of Madrid (EOS identified by their reference numbers in Table 1). *Source* Own elaboration based on Google Earth

finally it has ended with its disaffection in 2011, when loses its regime of public domain to be incorporated into the city. The problem of its illegal occupation since the seventies of last century for the construction of substandard housings has been tried to be solved with this procedure. This area has sections adjacent to urban areas, new developable areas and even in its meridional margin to the protected land of Valdemingomez. Precisely in this sector there is a serious problem of extreme poverty, ethnical segregation and delinquency. In total, with its almost 14 km of length, 72 m of width and more than 8000 dwellers, it is one of the biggest informal settlement of Europe. Recovering the Cañada doesn't seem compatible with the regularization plans of the sections more consolidated, located adjacent to developable areas or urban areas (Fig. 10).

Type 4. **Contact between wastelands**. This contact appears in areas where the open space is very fragmented and generally is formed by a mosaic of planning features with high socio-environmental interest and urban lands. Six of our study areas have this contact. Only two of them (numbers 17 and 18) have been sectorized, therefore they have their action zone well delimited. In the remaining areas, without urban development, it could be possible, at least temporally, to propose alternative uses until their future will be defined (Fig. 11).

Type 5. **Contact between wastelands and urban land**. In this typology the areas with social meaning, with or without environmental value, stand out. This contact is the most numerous quantitatively, although, as in the previous one, it

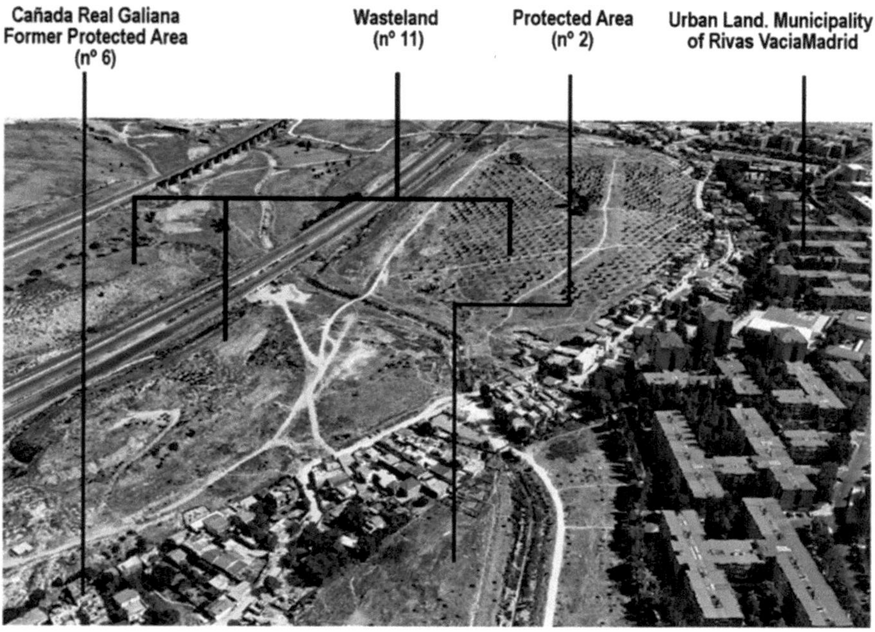

Fig. 10 EOS contacts at the southeast of Madrid (EOS identified by their reference numbers in Table 1). *Source* Own elaboration based on Google Earth

Fig. 11 EOS contacts at the East of Madrid (EOS identified by their reference numbers in Table 1)

Fig. 12 EOS contacts at North of Madrid (EOS identified by their reference numbers in Table 1). *Source* Own elaboration based on Google Earth

affects small pieces. Fourteen over the twenty-two study areas have this contact. The relevance of these spaces is very high because they can serve as the continuity of green spaces, urban farms, vacant lots, green roofs and all urban areas with this ecological and social characteristics and services for the citizens. It can be distinguished three situations in function of size and position of the area. First of all, there are small spaces at the Northeast just near to high class neighbourhood (numbers 9, 16 y 18), and at the South, surrounding by dense urban working-class neighbourhoods. In both cases, their future is clearly unsure (Fig. 12).

Secondly, around the *Manzanares* river, there are two areas located at the North and South of Madrid, that form two fringes of variable width. Their conversion to urban park and public equipment has been agreed, it is only a question of political will and appropriate finance. Their interest consists in the continuity with *Casa de Campo*, real lung of the city, with the already developed *Manzanares* park and, through footbridges, with Madrid Rio, result of the M-30 highway burying in the city centre, and the recovery of the *Manzanares* river itself. These two areas will contribute to the consolidation of a green diagonal that will connect the protected spaces from the Northwest to the Southeast.

Finally, it should be noted the large pieces at the Southwest and East, *Campamento* y *Mesa de Rejas-El Moral*. Both with projected developments with economic and political importance. In the case of the *Campamento* development, in 2016, the High Court of Justice cleared up the doubts about the legality of the

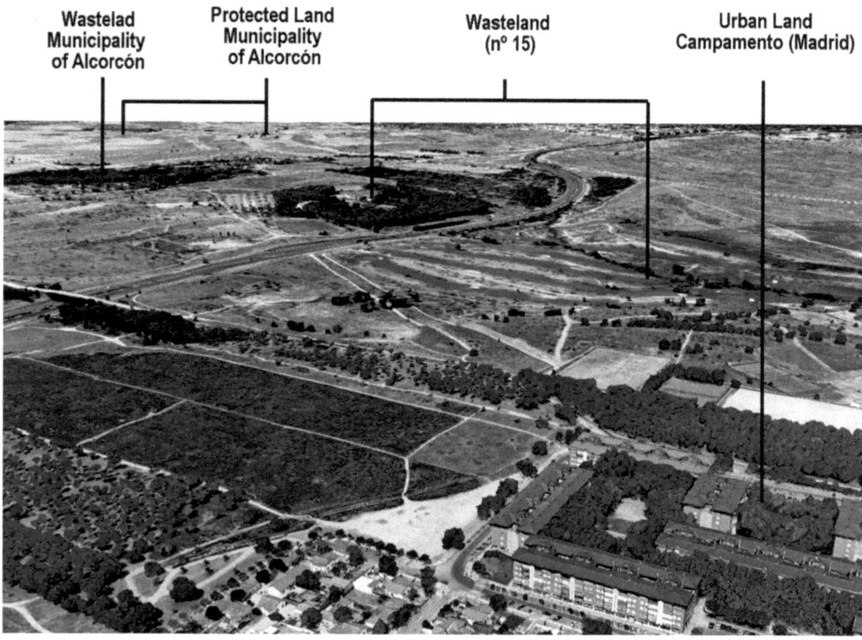

Fig. 13 EOS contacts at Southwest of Madrid (EOS identified by their reference numbers in Table 1). *Source* Own elaboration based on Google Earth

reclassification of the 98% of the land that was considered protected land in 1995 Plan. Predictably, large part of this land is going to be sold by the Defence Ministry, main owner of the land. This has been considered one of the last possible residential urbanistic action in Madrid. The area located at the East of the city will be also dedicated to the development of the *Nueva Centralidad del Este*, where there will be housings and offices. Initially, in this area was expected the development of the *Villa Olimpica*, within the failed Olympic Games project in Madrid. Nowadays there are mining activities still functional whose profitability has served to stop the expectation of changes in the land use. The current context allows the reconversion of these two proposals to other less compact and also the improvement of the continuity and complexity of open spaces (Fig. 13).

4.3 Proposals for the EOS Insertion in Regional Green Infrastructure Planning

As we have seen, residual spaces on which this research is focused on, remained as functional pieces of an agrarian system based on non-irrigated crops (wheat and barley), until almost half of the last century. Besides, this was combined with the use of stubbles by sheep and goats. This last use was disappearing little by little, in the same way that urban expansion was more patent and nearer. Currently, these spaces have lost their rural function completely, although some of them, the minority, are used eventually as temporal pastures by goat and sheep shepherds. However, nowadays, rural-agrarian landscape is perceived as cultural landscape that should be conserved because of their visual and inherited values (Maruani and Amit-Cohen 2007; Thompson 2002; Beer 2005). For that, an important issue is the proposal of applying regenerative and rehabilitant measures that improve the quality of these spaces (Maruani and Amit-Cohen 2007). In this case, the measures could be focused on the maintenance of the agrarian activities that have been their engine until few years ago.

In this context, it is complex to propose a perimeter connection system that creates a peripheral green belt and its continuity with adjacent municipal areas. So, there is no other option than to plan from within the law, recovering or maintaining areas through different strategies. For example, the different codes include specific treatments of the open spaces. In other words, the drove road code or the public water code, among others, include, in a generic manner, references to the treatment of the open spaces within the limits of the municipal area, including when possible, their incorporation in the current urban planning or zoning.

Once all the spaces available for protection have actually been protected, then it would be possible to implement other types of actions that improve this Green Infrastructure. In this sense, at least 5 interventions could be implemented:

- To facilitate and reinforce the green permeability between bordering spaces as a binding rule, especially in the contacts with natural protected areas (Cuenca Alta del Manzanares or Sureste Regional Park), lands that have been protected by local planning, or in perimeter sectors of contact between urbanized areas.
- To minimize the impact of the non-consolidated urban limits, reducing to the minimum the artificial areas.
- To preserve drove roads, wastelands or fallow lands, considered or not as Non Developable Land by the in-force local regulations, and to preserve these spaces as free spaces of social interest within the consolidated developments.
- To protect the areas with natural or naturalized vegetation paying attention to fluvial areas.
- To avoid modifications in the hydrography and physiography thus maintaining the hydromorphic diversity.

Even more, specific interventions would be advisable in order to recover open spaces. Such interventions could be: one-off re-plantations (reinforcements), generation of connection vectors between disconnected areas, widening of riparian forests, increment of the area of non-irrigated agriculture, both fallow and active, reaffirmation of their role as open spaces of social and environmental interest, promotion of the adaptation of the degraded open spaces on the municipal border that promote flows and minimize the potential existence of obstacles, and finally, maintenance and rehabilitation of thalwegs (northern, eastern and western arkosic sectors) and slopes and scarps (southern gypsum sector) of the municipal limits, promoting, in necessary cases, core reforestations with possibility of interconnection trough lineal reforestation.

5 Conclusions

As final conclusions, the following issues should be pointed out:

First of all, there has been a clear change in the ethical and aesthetical approaches that defined the city model projected by the public authorities through their different General Plans. This is deduced not only from the ultraliberal focus of the General Plan of 1997, but also from the *2020 Madrid City Future* strategy. The document that should have guided the conservative municipal policy banked on improving mobility and improving its own smart city management, but it forgot the issues related with green infrastructure. Unlike previous General Plans, it did not seek to create a Green City. Therefore, there is a radical contrast to other European cities determined to reinforce their green belts such as Oresund, the metropolitan region of Copenhagen, recognized as European Green Capital by the European Commission in 2014, as well as Vitoria-Gasteiz in Spain or Hamburg in Germany, winning cities in other years.

In second place, even when aware of the limitations that the Madrid territorial model of open spaces has, it is the value of what already exists that should be promoted. The opportunity of reordering the territory is very important in very dense and dynamic cities such as Madrid. In this sense, intervention over the wastelands is essential, the areas that tend to deterioration because of the rural abandonment but still with landscape, visual, natural and strategic values. They can be useful as the base of the articulation of regional open spaces.

In third place, we insist on a higher political commitment that will be able to reach an agreement on a new territorial model where the open spaces play an important role as environmental services. The important challenge that these spaces mean for urban planning has already been mentioned. Nevertheless, these areas are still undervalued in the planning. The success of this planning depends to a great extent on the proportion of green spaces, on the recent changes of their total area, on the number of departments responsible for these spaces and on the experience with citizen participation.

In fourth place, conservation and integration of open spaces in a network of green infrastructures have been considered as a priority action for mitigation climate change and we believe that their inclusion in general strategies of urban development it is necessary (Fernández et al. 2016).

Finally, as a vital idea, it has been shown that the design of GIS tools and the technical capacity of the local authorities are key points for the development of an efficient management of information. Without a detailed study of the EOS it is unlikely that an effective strategy of intervention can be presented. This is the essential contribution of the work we present here.

References

Amati, M. (Ed.). (2016). *Urban green belts in the twenty-first century*. London: Urban Planning and Environment, Routledge.

Antrop, M. (2004). Landscape change and the urbanization process in Europe. *Landscape and Urban Planning, 67*(1), 9–26. doi:10.1016/S0169-2046(03)00026-4

Ayuntamiento de Madrid. (1988). *Plan General de Ordenación Urbana de Madrid 1985. Normas Urbanísticas* (2ª edición). Madrid: Ayuntamiento de Madrid.

Ayuntamiento de Madrid. (2012a). *Revisión del Plan. Preavance. Evaluación del Plan General de 1997*. Madrid: Dirección General de Revisión del Plan General.

Ayuntamiento de Madrid. (2012). *Memoria de Actividades 2012*. Madrid: Dirección General de Patrimonio Verde.

Ayuntamiento de Madrid. (2014). *Plan de Energía y Cambio Climático de la ciudad de Madrid - Horizonte 2020*. Madrid: Ayuntamiento de Madrid.

Ayuntamiento de Madrid. (2015). Compendio de las Normas Urbanísticas del PGOUM-97, p. 93.

Baycan-Levent, T., & Nijkamp, P. (2009). Planning and management of urban green spaces in Europe: Comparative analysis. *Journal of Urban Planning and Development, 135*(1), 1–12. doi:10.1061/(ASCE)0733-9488(2009)135:1(1)

Beer, A. (2005). The green structure of Sheffield. In A. C. Werquin et al (Eds.) (2005). *Greenstructure and urban planning*. Final Report, European Commission, COST Action C11, Luxembourg, Office for Official Publications of the European Communities.

Brandis, D. (2014). La periferia inacabada madrileña. El penúltimo capítulo de una historia interminable. Proceedings of the XII Coloquio y Trabajos de Campo del Grupo de Geografía Urbana (AGE). Madrid y Castilla La Mancha, 69–80.

Burriel, E. L. (2016). Empty urbanism: the bursting of the Spanish housing bubble. *Urban Research & Practice, 9*(2), 158–180. doi:10.1080/17535069.2015.1110196

Calderón, C., Saiz de Omeñaca, J. A. & Günthardt-Goerg, M. S. (2009). Contribución del arbolado urbano y periurbano del municipio de Madrid en la mejora de la calidad del aire y sumidero de contaminantes atmosféricos como beneficio para la sociedad. *Proceedings of the 5º Congreso Forestal Español*, Ávila, Sociedad Española de Ciencias Forestales, 2009.

Calvo, R. et al. (2007). La explosión urbana de la conurbación madrileña. In Observatorio metropolitano de Madrid (Ed.) Madrid: ¿la suma de todos? Globalización, territorio, desigualdad, Madrid, Madrid: Traficantes de Sueños, 223–325.

Canosa Zamora, E. & García Carballo, Á. (2013). Lutte de voisinage et patrimonialisation d'espaces naturels résiduels à Madrid. *VertigO-la revue électronique en sciences de l'environnement*, 16. doi:10.4000/vertigo.13690

Champion, T. (2001). Urbanization, suburbanization, counterurbanization and reurbanization. In R. Padison (Ed.), *Handbook of urban studies*. London: SAGE Publication.

Commission, European. (1999). *European Spatial Development Perspective (ESDP)*. Luxembourg: Office for Official Publications of the European Communities.

Council of Europe. (2013a). Communication from the Commission to the European Parliament, the Council, the European Economic and Social Committee and the Committee of the Regions. Green Infrastructure (GI)—Enhancing Europe's Natural Capital.

Council of Europe. (2013b). Technical information on Green Infrastructure (GI). Accompanying the document: Communication from the Commission to the European Parliament, the Council, the European Economic and Social Committee and the Committee of the Regions. Green Infrastructure (GI)—Enhancing Europe's Natural Capital.

De Santiago Rodríguez, E. (2012). Una lectura de las políticas de suelo y los modelos urbanísticos madrileños desde mediados de los años 1990: de la liberalización a la resaca inmobiliaria. *Geopolítica (s) Revista de estudios sobre espacio y poder* 3(1), 83–116. doi:10.5209/rev_GEOP.2012.v3.n1.39303

European Environmental Agency. (2012). *Urban adaptation to climate change in Europe. Challenges and opportunities forcities together with supportive national and European policies*. EEA Report No 2/2012.
Fernández, F., Allende, F., Rasilla, D., Martilli, A. & Alcaide. J. (2016). Estudio de detalle del clima urbano de Madrid. Unpuplished report for Madrid City Council (Government Department of the Environment and Mobility).
Gómez Mendoza, J. (2013). El caso de Madrid: 1997–2012. Del urbanismo explosivo al inane. Sacar Lecciones de la crisis. *Cuadernos y Posiciones, Cuadernos del Círculo Cívico de Opinión, 9,* 21–28.
Jiménez Garcinuño, L. (2015). Dinámicas de ocupación urbana del Anillo Verde metropolitano, desde sus orígenes en el Plan General del Área Metropolitana de Madrid de 1963 hasta el Plan General de Ordenación Urbana de 1997. Doctoral dissertation, Universidad Politécnica de Madrid.
Koomen, E., Dekkers, J., Koetse, M., Rietveld, P. & Scholten, H. (2005). *Valuation of metropolitan open space; presenting the research framework.* Paper presented at the 45th Congress of the European Regional Science Association, 2005.
Koomen, E., Dekkers, J., & Van Dijk, T. (2008). Open-space preservation in the Netherlands: Planning, practice and prospects. *Land Use Policy, 25*(3), 361–377. doi:10.1016/j.landusepol.2007.09.004
Kühn, M. (2003). Greenbelt and Green Heart: separating and integrating landscapes in European city regions. *Landscape and Urban Planning, 64*(1), 19–27. doi:10.1016/S0169-2046(02)00198-6
Maruani, T., & Amit-Cohen, I. (2007). Open space planning models: A review of approaches and methods. *Landscape and Urban Planning, 81*(1), 1–13. doi:10.1016/j.landurbplan.2007.01.003
Mata Olmo, R. (2007). Madrid: environment. Case study. In Comunidad de Madrid (Ed.), *Capital regions: planning and sustainable development of metropolitan capital regions* (pp. 393–411). Madrid: Consejería de Medio Ambiente y Ordenación del Territorio.
Naredo, J. M. (2013). El modelo inmobiliario español y sus consecuencias. In J. M. Naredo & C. Taibo (Eds.), *De la burbuja inmobiliaria al decrecimiento: causas, efectos y perspectivas de la crisis* (pp. 9–68). Madrid: Fundación Coloquio Jurídico Europeo.
Otero Carvajal, L. E. (2010). Madrid, de capital a metrópoli. In C. Priego (Ed.), *Sociedad y espacio urbano de Madrid en el siglo XX* (pp. 10–33). Madrid: Ayuntamiento de Madrid.
Planeamiento urbanístico de Madrid. (1953). *Gran Madrid: Boletin informativo de la Comisaría General para la ordenación urbana de Madrid y sus alrededores, 23*, 3–26.
Rodríguez Avial, L. (1997). New master plan for the city of Madrid. *Urban, 1,* 110–123.
Romero, J., Brandis, D., & Melo, C. (2015). The neoliberal turn of city policies in Spain. The examples of Madrid and Valencia. *Boletín de la Asociación de Geógrafos Españoles, 69,* 573–577.
Sambricio, C. (2015). On urbanism in the early years of francoism. In H. Bodenschatz, P. Sassi, & W. G. Max (Eds.), *Urbanism and dictatorship: A European perspective* (pp. 117–153). Basel: Birkhäuser Verlag.
Sassi, P. (2016). A new master plan for the "Gran Madrid Capital de España" after the Civil War. *International Planning History Society Proceedings, 17*(4), 347–358. doi:10.7480/iphs.2016.4
Thompson, C. W. (2002). Urban open space in the 21st century. *Landscape and urban planning, 60*(2), 59–72. doi:10.1016/S0169-2046(02)00059-2
Tomás, M. (2009). Building metropolitan governance in Spain: Madrid and Barcelona. In H. Heinelt & D. Kübler (Eds.), *Metropolitan governance in the 21st century: Capacity, democracy and the dynamics of place* (pp. 47–62). Routledge/Ecpr Studies in European Political Science: London and New York.
Turner, T. (1992). Open space planning in London: From standards per 1000 to Green strategy. *Town Planning Review, 63*(4), 365–386.
Tzoulas, K., Korpela, K., Venn, S., Yli-Pelkonen, V., Kaźmierczak, A., Niemela, J., et al. (2007). Promoting ecosystem and human health in urban areas using Green Infrastructure: A literature

review. *Landscape and Urban Planning, 81*(3), 167–178. doi:10.1016/j.landurbplan.2007.02.001
Van der Berg, L., Drewett, R., Klaassens, L. H., Rossi, A., Vijverberg, C. H. T. (1982): *Urban Europe. A study of growth and decline* (Vol. 1). Oxford: Pergamon Press.
Weber, T., Sloan, A., & Wolf, J. (2006). Maryland's green infrastructure assessment: development of a comprehensive approach to land conservation. *Landscape and Urban Planning, 77*(1), 94–110. doi:10.1016/j.landurbplan.2005.02.002
Werquin, A. C. et al (Ed.) (2005). *Greenstructure and urban planning*. Final Report, European Commission, COST Action C11, Luxembourg, Office for Official Publications of the European Communities.

Chapter 5
Controlled Landscapes or Building Sustainability in Public Spaces. Case of Studies of Padova and Moscow

Elena Merino Gómez and Fernando Moral Andrés

Abstract In the present chapter, a comparative exercise of insertions of green areas in two European urban centers of great historical value is realized: Padova (Italy) and Moscow (Russia). Its urban fabrics, with a high degree of consolidation in their central areas, are the object of interventions that contribute to alleviate the congestion that they suffer in our days. More than two hundred years separate the intervention of Andrea Memmo, known as *Prato della Valle* from the ambitious project meant for the center of Moscow, next to the Red Square, for the plot of Zarydaye. Besides the temporal distance, there is a great difference in the intentions that animate both projects, however, the results obtained in the Italian case and expected in the Russian intervention participate in current sustainable values with full validity.

This work is based on the observation of the concomitance between two urban interventions in which vegetation and the treatment of water-related spaces play a fundamental role in shaping the consolidated urban centers of Padua (Italy) and Moscow (Russia). The present relations between two interventions as far back in time as the neoclassical *Prato della Valle* of Andrea Memmo, in Padua, and the still in process of implantation Zaryadye Park, in Moscow, are connected with the eternal return to the classic idea of *rus in urbe*, that, although preserving the ideological essence of landscaping the cities, has served different purposes throughout history: from the consolation of nostalgia for Roman peasant origin (Albardonedo 2015) to the consideration of m^2 Green areas per inhabitant as a healthy urban indicator by the World Health Organization (2012). The attenuation of the heat island effect of densely populated nuclei or the capacity of CO_2 sequestration mitigates, by processes which involve age-old precepts, fully contemporary questions.

E.M. Gómez (✉) · F.M. Andrés
Department of Architecture, Universidad Nebrija, Madrid, Spain
e-mail: emerino@nebrija.es

F.M. Andrés
e-mail: fmoral@nebrija.es

R. Álvarez Fernández et al. (eds.), *Carbon Footprint and the Industrial Life Cycle*, Green Energy and Technology, DOI 10.1007/978-3-319-54984-2_5

The existence of public spaces in urban territory finds far precedents in the Rome of the Empire (García Lorca 1989), when rulers and benefactors of diverse nature made public promotion of their prodigality through works enjoyable by all citizens. Throughout the Middle Ages, the confinement of urban layouts within the walls, the scarcity of open public spaces, beyond those in which markets were celebrated, coupled with the consequences of the distribution of property inherent to feudal structure, prevented, among other causes, the planning of public areas of a certain extent that would alleviate the city's congestion of that period. The green areas of the medieval city were restricted to the domestic gardens inside some privileged dwellings and to small-scale gardens cloistered inside convents and palaces (Rodríguez-Avial Llardent 1980).

It will be necessary to bridge over the medieval interlude to find new examples of spaces for the enjoyment of the common people on European soil. The dominant medieval tendency that tended to fill the cities survived even in the ideas of egregious Renaissance architects such as Alberti or Palladio, who continued to conceive the city as a "completely built place" (Lawrence 2008).

The existence of green spaces open to the public in urban environments, as we know them today, is a relatively modern phenomenon, since the first examples on European soil go back to the middle of the sixteenth century, related to the arboreal conditioning in areas tangential to the urban nuclei. In these first interventions, the utilitarian ends prevail over the sustainable ones and have nothing to do with the exercises of the later romantic urbanism that will purposely pursue, as a fundamental objective, the introduction of the nature in the city.

A pioneer in the second half of the XVI century was the intervention on the wall of Lucca (Italy) (Martinelli and Parmini): shortly after being erected, on the parapet walk, poplars were planted (Fig. 1). Their main mission was far from the contemporary ends with which are projected today the green areas. The grove was planted to achieve, through its roots, the consolidation of the land that integrated the walls. The proximity of the place and its attractive aspect made it possible for the people of Lucca to immediately take over the place, turning it into a leisure destination (Lawrence 2008).

In the image of the Luquean intervention, we find in other places of European geography linear plantations of groves, although they cannot yet be considered relief operations within compact urban layouts. These are actions in places adjacent to high density nuclei related to the existence of problems of unhealthiness or degradation. The landscaping of public spaces for merely recreational purposes, different from utilitarian ones, arises in Spain in the last third of the sixteenth century (Albardonedo Freire 2015).

Thus, in 1570, at the request of Philip II, king of Spain, was planned in Madrid the layout of what later would become the present Paseo del Prado (Jiménez Garnica 2002). Its main intention was to provide the city with an organized recreational meeting place according to the canons of the prevailing mannerism (Jiménez Garnica 2002) in a place remote and alternative to the busy areas of Plaza Mayor and Calle Mayor (Lopezosa Aparicio 2009), both located in central areas of the urban fabric. In a similar way was carried out, between 1573 and 1574,

Fig. 1 Merian (1668) "Luca". Bella veduta a volo d'uccello della città e delle mura fortificate

the *Alameda de Hércules* in Seville, planned in a degraded and relatively peripheral area which required hydraulic sanitation works for its adaptation (Albardonedo Freire 1998). The intervention bequeathed to the city an open environment with vegetal presence that has survived until our days.

The most primitive examples of the presence of green areas in the Early Modern Period consist of tangential interventions to the urban nuclei with a marked linearity, related to the presence of walls or the conditioning of river banks (Lawrence 1988). The irruption of nature in consolidated urban fabrics, linked to the purposes of decongestion or improvement of the urban landscape continues to be an idea alien to the Renaissance and Baroque approaches of the city. The opening of the Place des Vosges at the beginning of the 17th century in Paris is a commemorative proposal (Morriset and Noppen 2003) that will have to wait until 1670 to be covered with green and have exported to the rest of Europe the model of gardening insertion of public character in the inhabited urban centers (Lawrence 2008).

Functionalism linked to the conception of urban spaces in the context of the illustrated precepts of the eighteenth century reaches its maximum level in the intervention that Andrea Memmo performed in the city of Padua (Italy) in 1775. The Paduan space known since medieval times as *Valle del Mercato*—Valley of the Market-(Azzi Visentini 2005) was compromised by the presence of stagnant waters. The problem was aggravated by the recurrent floods that affected it from one of the deviations of the Brenta River which struck at the beginning of the fourteenth century (Gennari 1796). Located in the southeastern part of the city, in an area that once housed the *Patavium Roman* theater, the place required an urgent intervention to solve the problem of frequent avenues (Stratico 1795) that rendered the place a space muddy and unhealthy.

The vitiated constitution, in Patte's words (1769), of cities could be remedied by the application of rational principles to urban planning. Memmo's proposal for the conditioning of the *Valle* is fully in line with the functionalist ideology spread by Patte in France (Zaggia 2010) closely related to the eighteenth-century diffusion of hygienist ideas.

After a severe flood in 1722 (Neveu 2011) began the renovation of a space that had been dedicated to different purposes since Roman times: once the amphitheater disappeared, playful and commercial activities of diverse nature took place. The two main agricultural fairs in Padua were held in June and September (Zaggia 2010), a period in which water frequently threatened to undermine commercial activity (Azzi Visentini 2005).

The *idea*, as his pretended personal secretary (Neveu 2011), Vincenzo Radicchio (1786) refers to, meant one of the most innovative interventions in the creation of urban public spaces, not only for its functional and formal aspects, but for the ideology underlying the conception itself. In addition to solving the problems of periodic flooding, the project sought to create an urban garden island where commercial activities could be developed, housed in a series of ephemeral constructions that could be assembled during the fair and dismantled once the fair had finished (Williamson 1996).

The rest of the year, in the terms of the explanation at the foot of the engraving by Francesco Piranesi (Fig. 2), commissioned to illustrate the *Descrizione* (Radicchio 1786), the "paludoso terreno contenente lo spazio di 974,012 piedi quadranti" would be enjoyable as a "pubblico passeggio, pinacoteca e lapidaria che

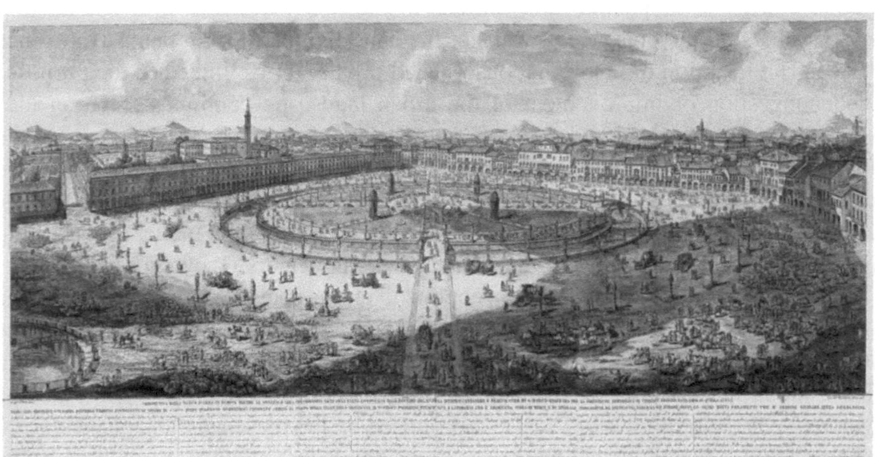

Fig. 2 Piranesi (1786) Prospettiva della Nuova Piazza dietro la generale idea già concepita ed in gran parte efettuata dall Eccemo Sig. Andrea Memmo

la adornano, […] indicazioni di spettacoli, bosco, lago, strade, ed altri molti ornamenti che si possono rilevare senza spiegazione[1]".

The island, surrounded by a canal whose purpose was to redirect the muddy waters to drain the space of the *Prato*, is conceived as a wooded place for collective enjoyment. The author of the project takes advantage of the opportunity to go far beyond what would be a mere hygiene and sanitation operation, approaching an unprecedented intervention of urban qualification: the execution of the "most beautiful square" (Goethe 1891) in which the *Prato della Valle* would eventually become. The conscious design of a scenery that would modify the urban landscape transcends in the words with which Memmo inspires the description of his idea:

> Ognuno sa per esperienza, che due filare sole di alberi per quanto folti siano, non bastano a ripararci del sole caminando tra essi; ve ne vogliono quattro almeno, perchè il viale di mezzo sia tutto ombroso […]. Quel che somministra l'ombra al di dentro esclude la vista al di fuori […].
>
> Non si potrebbe negare che una tale idea non fosse bellissima in un quadro di Claudio di Lorena, o Gaspare Pussino; ma nel Prato, come come mai sarebbe eseguibile quando appunto que' viali così ben dipinti toglierebbero la vista?"[2] (Radicchio 1786)

Memmo's interest in the visual impact of his work is linked to an early awareness of air quality, at a time when the premises of an environmental nature were not yet part of the overall design intentions. Memmo is worried about the irrespirable humidity of an excessively wooded and shady environment next to the water, which would affect the quality of the air: "Imprigionata l'aria già umida in quel basso sito, non si respirerebbe certamente colà la più leggiera, e sana. […][3]" (Radicchio 1786).

The validity of the proposal in all its propositional aspects survives today. Pragmatic aspects of the idea, some of them as incipient for the time as the tourist attraction, were announced in a visionary way by Memmo when he emphasized that the project program would be capable of "render deliziosa la città per maggior

---

[1]"The muddy land contained in the space of 974,012 square feet" would be enjoyable as a "public walk, art gallery and statuary that adorn it, […] indications of spectacles, forest, lake, streets and many other ornaments that can be observed without need of explanation" (Radicchio 1786). Translation from Italian provided by the authors of this paper.

[2]"Everyone knows from experience that only two rows of trees, however leafy they may be, are not enough to protect us from the sun walking between them; at least four are required for the central path to have shadow […].That which provides shade to the interior limits the views of the exterior […].

It could not be denied that an idea of this type was not beautiful for a painting by Claudio de Lorena or Gaspare Pussino; but in the *Prato*, how could something like this be realized, considering that those well-painted paths would take away the views?" (Radicchio 1786). Translation from Italian provided by the authors of this paper.

[3]"The damp air, trapped in that sunken place, would not be breathed precisely as if it were the healthiest and lightest […]. (Radicchio 1786). Translation from Italian provided by the authors of this paper.

attrattiva de' forestieri[4]" (Memmo 1775 cited by Zaggia 2010). The environmental value of the idea demonstrates how the citizen present welfare observed in its vicinity is directly related to it, as well as the direct beneficial effects for the revaluation of the soil, public and private assets, and the indirect ones related to the tourist attraction of the area, which even nowadays contributes to the overall income of the whole city of Padua (Amrusch 2007).

The 88,620 m^2 (Azzi Visentini 2005), occupied by the *Prato della Valle*, make it the second largest open public space in Europe, only surpassed by that of Red Square in Moscow (Noro et al. 2014b). The interest of the possible positive environmental repercussions for the city in which such space is inserted is reflected in the works developed by Noro et al. (2014a, b) at the University of Padua.

The green areas and the channels that make up the *Prato* predict a decrease of the temperature in its environs with mitigating consequences of the heat island effect suffered by the city of Padua. However, in the measurements carried out during the month of August 2010, higher than expected temperature values were recorded. The authors of the work attribute it to the fact that the instrument used collected data mainly in the perimeter of the green zone and hardly in the interior. It is noteworthy that the temperatures measured on the lawn and next to the central pond of the ellipse that defines the intervention turned out to be inferior in something more than 1 °C with respect to those collected in the rest of external points of the *Prato*. The data would be in line with the heat-reducing effects, favored by the presence of vegetation and water. The thermal inertia of the materials that make up the hard pavements of the environment will possibly be the origin of data different from those expected in the area of influence of the *Isola Memmiana*.[5] The authors propose future research in the area to obtain conclusive data (Noro et al. 2014b), which demonstrates the current validity of the proposal and its renewed interest in aspects that affect the particular sustainability of the intervention and its general repercussion over the city of Padua.

Between the illustrated intervention of Memmo in Padua and the contemporary urban operations to introduce nature in the city there are innumerable actions of public character adapted to the idiosyncrasy of the different urban nuclei that, from the Industrial Revolution, begin to undergo a new densification. For the first time, public health issues are being linked to the planning of green areas in cities that were beginning to suffer the consequences of incipient industrialization (Gordon and Shirley 2002).

Ebenezer Howard's (1902) visionary proposals about *Garden Cities* or the underlying ideology in the Park Movement (Bluestone 1987), whose purportedly naturalistic aims actually responded to an urbanized and domesticated perception

[4]"Render the city delicious and make it more attractive to foreigners" (Memmo, 1775 cited by Zaggia 2010). Translation from Italian provided by the authors of this paper.

[5]*Isola Memmiana*, meaning *Memmo's Island*, is one of the most famous names given to the intervention.

Fig. 3 Diller Scofidio + Renfro (2013) Zaryadye Park Competition, Winner Project by Diller Scofidio + Renfro, view from Bolshoy Moskvoretsky Bridge

more than to the sincere Horacian remembrance of *beatus ille*, connect directly with examples of contemporary green interventions of medium and large scale.

The project for Zaryadye Park (Fig. 3), in Moscow, according to the plan of the New York office, Diller Scofidio + Renfro, dating from 2013, bursts into the scene of contemporary green spaces as a banner of what has come to be called *Wild Urbanism*. The intention to spontaneously populate determined areas of urban environments is related to the identification of authentically natural spaces that survive in or around the city (Box and Harrison 1993). However, the novelty of *Wild Urbanism* lies in the high planning of the actions that arise, so that the appearance of spontaneous or survivor nature is in reality the result of a complex sophistication, both in design and in the processes of implementation and maintenance of scheduled sites.

The main park of Moscow up to the present is the "Park of Culture and Rest in memory of Maxim Gorky", 120 Ha intended to endorse the values that emanate from his name, a place where Muscovites can enjoy their leisure. Its layout was defined by Konstantin Melnikov in 1929 in the estate of Prince Trubetskói, where the gardens, characterized by their roundabouts, of the old Golitsin Hospital, founded in 1802 according to the project of the architect Matvey Kasakov, and those of the Neskuchny Palace, from the XVth century and under project of Dmitry Ukhtomsky. It occupies a privileged position in the city. Its characteristic boundaries today are the Moskva River and Lenin Avenue. It is very close to emblematic places of the metropolis such as the Tretiakov Gallery or the Kremlin. In the same project a varied structure of constructions is articulated by buildings of different historical and architectonic value, along with diverse equipments. A place that has hosted from medicine, nobility, exhibitions, fairs, attractions to the "Cosmic Experience" linked to the shuttle Buran. Also quieter activities such as ice skating on winter floodplains or playing sports such as tennis or football (Horn 2014). All in a context where different gardens, ponds or grasslands articulate this great green area of Moscow (Park-Gorkogo 2015).

In 1957 the last great park of Moscow, the one of the "Amistad (Druzhbi)", is designed, coinciding with the celebration of the VI World Festival of the Youth. It lies outside the central circle of the city but in the immediate vicinity of the Leningrad- St. Petersburg highway. The water has a special accent on the whole: it has several ponds and is very close to the Moscow Canal. In general terms it presents a more or less picturesque structure with vegetation based on groves, prairies and floral flower beds. In any case the distinctive symbol by which it is characterized is by the group of sculptures of different origin and style installed in different places of the set such as the two pieces of the iconic Vera Mukhina or the monument to Cervantes donated by Spain (Morley 2016).

These two antecedents are fundamental to understand the effective weight of this urban form in the structuring of the city of Moscow but also to contextualize the construction of the park that will fill the most privileged position of the city, next to the Red Square, behind the back of the St. Basil's Cathedral, tangent to the Moskva River and with a dimension of 13 Ha: Zaryadye Park will be one of the operations in progress, of greater importance and which aspires to become one of the fundamental areas for the redefinition of the city in this XXIst century. A project aimed at restoring an extended plot, which between 2007 and 2013 has been a large abandoned site (Fig. 4).

Only a complex sociopolitical history, such as the Russian one, can explain how this urban vacuum has reached our days. The germ of this future park was a walled neighborhood, also known as Kitay-Gorod, which was demolished between 1930 and 1960. This place housed the first commercial district outside the walls of the

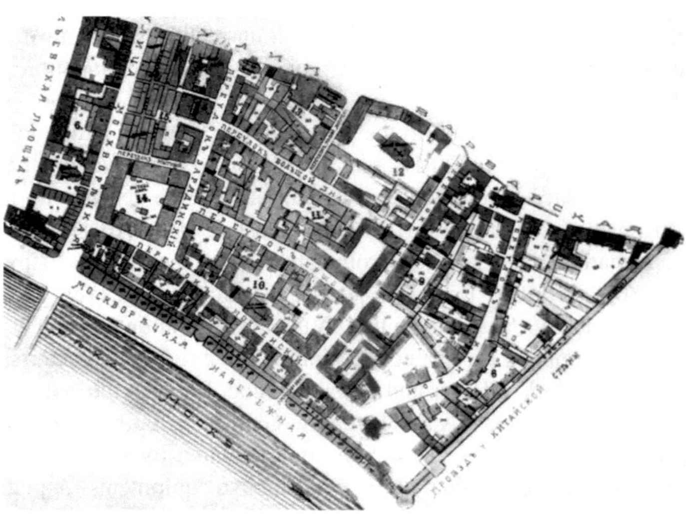

Fig. 4 Unknown (1881) Map of Zaryadye District

Kremlin. Its buildings were articulated around blocks with interior courtyards of different shape and size while making an irregular road network. Only the main historical buildings, almost all in the vicinity of Varvarka Street, were saved: Cathedral of the Sign (1679–84), Church of All Saints (1680s), St. George's Church on Pskov Hill (1657), St. Maksim's Church (1698), St. Anna's Church at the Corner (1510s), St. Barbara's Church (1796–1804), the Old English Embassy (1550s), and the 16th century Romanov boyar residence. They compose an odd set of buildings if we attend to the lack of context that explains and allows a full understanding. A kind of *cadavre exquis* that would not be outside the plans of reconstruction of Stalinist era. In the 30's of the XXth century, it was the proposed site to locate the new headquarters of the Commissariat of the Heavy Industry of the USSR, Narkomtiazhprom (NKTP). Various projects signed by some of the architects of the Constructivist movement, such as the Vesnin Brothers and Ivan Leonidov, were presented. Subsequently, in the 1950s, the Zaryadye Administrative Building was planned and built. It was the eighth tower of the *Seven Sisters* (Stalin skyscrapers), with heights between 133 and 240 m, under the project and direction of Dimitry Checulin, in line similar to the rest of skyscraper-landmark of the time. This proposal was paralyzed after the foundation and execution of part of the structure of the first floors. Stalin had died and with him, the conclusion of the project (Citizenarcane 2005; Strelka 2013).

Taking advantage of part of the already executed and with a new project of the same architect would be built the Hotel Rossiya in 1967, of smaller height, but of colossal dimensions that made of it the biggest hotel of the world at the time and that maintained its record in Europe until the year of its closure in 2005. Emblem of the USSR and its Politburo, it had 12 heights and the main tower reached until the 23 plants with a total of 420,000 m^2. The number of rooms was 3170, with a total area of approximately 35,000 m^2. The State Concert Hall with capacity for 2600 spectators and an anti-nuclear bunker with the possibility of hosting 4000 people (Kudryavtseva 2011) was also housed inside. Everything was demolished between 2006 and 2007, without a blasting procedure controlled disregarding the damages that could appear in other bordering buildings, especially in the Kremlin (Mañueco 2006). This disappearance motivated the presence of a large vacant 13-hectare site in the center of Moscow. New developments were proposed for the place, some by prestigious firms such as Foster and Partners, and always envisaging the construction of a new neighborhood that would restore the historical urban fabric while erasing the scar that the Rossiya demolition had supposed (Posokhin 2012).

We must note how the destruction of a complex and consolidated fabric, both by its urban structure, and by its identity, opened a crisis in the Moscow layout and became an unsolved problem up to the present day. The different proposals that should have organized a new neighborhood, composed by city-buildings, turned out to be a failure in all their versions. The brilliance of those who backed them did not understand the needs of a part of the city massacred by an authoritarian policy and a lack of historical and social perspective. Could an urban form different from the previous ones recover this territory? Could a city-specific structure address the needs of a key location and a post-Soviet society belonging to the XXIst century and facing the West?

In 2013, six years after the demolition of the Rossiya Hotel and the generation of the main Muscovite *terrain vague*, the Moscow City Government promoted the International Competition for the Zaryadye Park with the technical support of Strelka Institute for Media, Architecture and Design. The presentation document was exhaustive and set out in detail all the conditions that should be considered in the proposals submitted: from the historical remains, the topographical and urban conditions or the main relations with the rest of the city. A special attention should be paid to the relationship between the project en the future green areas (Strelka 2013). The intentions that underlie the call are marked by the following paragraph: "The aim of this competition is to develop an architecture and landscaping design concept that will form the basis of the creation of a contemporary park with a high quality infrastructure that will be open for the public all year round" (Archsovet 2013). The city opted for a green choice as a way of regeneration, absolutely new in this environment and with the idea of building a space dedicated, fundamentally, to the whole citizenship: a public space genuinely contemporary (Sease 2015).

That same year, the winning project of the competition was presented under the motto "Wild Urbanism". Diller Scofidio + Renfro (DS + R), New York-based office won the call and for its development had the collaboration of Hargreaves Associates and Citymakers with Buro Happold, Transsolar, Arup and Dimitry Onischenko among others. The jury established as second classified the project proposed by TPO Reserve. The third classified project was the proposal of MVRDV (Archinect 2013). A total of 90 consortiums from 27 countries had submitted their proposals, which gives us a clear idea of the magnitude of the event convened and especially of the interest that arose in a great number of offices of architecture throughout world (Zaryadye 2013). It is currently under construction under the direction of Sergey Kuznetsov, chief architect of Moscow, and with the American team as associate authors (Biaar 2016) (Fig. 5).

The work of DS + R, gathering a sensitivity especially localized in North America (Sease 2015), articulates a park that is built thanks to the union of the four main ecosystems located in the Russian Federation: the steppe, the forest, the swamp and the tundra. Along with these four main areas, several buildings, mostly hidden in the green surface, will close the set of cultural and leisure areas in the place. The authors explain the keys to their intervention:

> "Wild Urbanism" is a proposal to create a surprising set of alliances between the urban and the wild, the local and the national. To begin, a sampling of Russia's national landscapes are transported into the heart of Russia's capital city. Intermixed with this soft structure is a system of gradated hardscapes inspired by neighboring civic squares upon which a constellation of pavers fuse and disperse to provide meandering walking surfaces above while concealing program spaces below. The park becomes a mediator between the local and the distant, the built and the natural, condensing diverse atmospheres into a familiar yet unknown park experience. (Harkema 2013)

The concept photomontages used by the architecture office presented in an attractive way what the previous words pointed out: Muscovite Metro stations

Fig. 5 Diller Scofidio + Renfro (2013) Zaryadye Park Competition, Winner Project by Diller Scofidio + Renfro, entrance from Red Square

emerging along streams, balconies and scenes set in valleys of rocks and fir trees, glass vaults that cover scrub of wide plains, ring-shaped lamps that appear suspended in the middle of forests, etc. A pleiad of images that mark the contrast between the natural and the artificial that is implanted in that context. The whole means an iconography that comes up with a new reality, born of that friction between entities of different origin and qualities (Harkema 2013). An option of similar nature was previously explored by other teams such as the mythical Florentine Superstudio and his proposal for the Monument Continuo.

The intervention, marked by these four landscapes outside the city, is articulated through a system of two inclined central stripes, the forest and the steppe, and the two resulting squares for the marshes, next to the river, and the tundra in the nearer and denser urban part. They all compose a kind of campus where historical buildings are absorbed by this fractionated tapestry and where the new services are hidden, for the most part, taking advantage of the unevenness of the plot in its vertical section. The traffic is located in the perimeter of the project. In the interior there are no paths; people can roam freely and without curbs for the whole extension of the proposal. They can sit, eat and enjoy a new area that can withstand any type of leisure and daily enjoyment throughout the year. The authors seek direct and massive contact between users and this imported nature. At one point, it resonates with the important green areas bordering the Kremlin, both in the main access area and on the banks of the Moskva, but Diller Scofidio + Renfro clearly indicate that this nature is a planned and limited gardening. Their project will contrast, especially in the ideological, with the whole of the Palace. The Red

Square, paved, will become the border between both realities while the Cathedral of San Basilio will receive a renewed strength as an icon or gate for the intervention. The authors projected a wild-looking pole that enhances the environment while enriching and complementing the urbanite's experiences.

Throughout the operation, energy management is capable of conditioning the place to render it comfortable during the periods of harsh weather suffered by the city (Dsrny 2013). Larger roofs act as solar collectors by transferring the generated electricity to radiant floor systems, even to heat water in the swamp area, in order to theoretically create, from renewable sources, thermal differences up to 15 °C (with 0 °C being the possible value of the external ambient temperature, a progressive gradient is constructed, which grows from 8 to 15 °C in the most protected position of, for example, the museum). The biofuel is used in the restaurant in order to raise its indoor temperature, being able to reach 20 °C in comparison with the 0 °C of the outside. In all cases the trees act as passive protectors against the incidence of the wind that usually sweeps this place in the severe days of autumn and winter. This system of planning a place in energetic terms affects the four main patterns/landscapes. By means of different passive systems, mainly forest masses, and of distribution of heating, the respective microclimates are defined through the control of temperature, humidity and sunlight to be true in all their magnitude (Harkema 2013).

There is an interesting debate between how to understand and intervene with the wild in the city (Box and Harrison 1993). In the days of Brezhnev "savages" were those who left the government-sponsored vacation to enjoy camping or fishing outside the official resting places (Premiyak 2015). It is paradoxical that the current authorities are encouraging the construction of this area, typical of the ancient "savages", in the center of the capital. This puts us at a point closer to the phenomenological question: surely the central value of the new park could make us understand that these natural objects give way to the nostalgia that some feel about the times when men lived in a closer relationship with nature (Premiyak 2015).

The reality of the project subtracts values from it and renders it less ambitious but provides it with interesting results. We are facing an intervention that will implant the tundra, the steppe, the forest and the marshes in a place that corresponds only, biologically, with the forest. It is an essentially artificial choice, new in this considered context although easily tolerable by it. The key is the formalization of a piece of rupture and reconfiguration of urban relations using energy as a structuring material with a crucial role (DS + R Consortium 2014) (Fig. 6).

Finally everything will be due to how the energy resources that will feed the park are managed in order to propitiate the life of these four imported ecosystems while facilitating the constant use of the place. The big choice is to materialize a micro-system that develops, in a sense, independently from the natural climatological cycles. It is a commitment to define a new hybrid territory of natural and artificial values: the natural form and the artificial substance that combined give us an attraction that connects us, as humans, with our more atavistic roots (Sease 2015).

Fig. 6 Diller Scofidio + Renfro (2013) Zaryadye Park Competition, Winner Project by Diller Scofidio + Renfro

Table 1 Authors' elaboration with the data provided in Harkema (2013)

Roofed spaces	Description	Area (m^2)
Area of exhibitions and cultural activities	Its access is protected by a grove. The roof has a solar capturing system and at the same time it has air conditioning through underfloor heating	6000
Ice cave	It takes advantage of the entrance of fresh wind from the outside to enhance the creation and maintenance of the frozen place	500
Restaurant	It is protected by the grove and with cover of solar capture. Features a large biofuel fireplace	900
The house of tea	It is located beside the marshes, also has solar cover and a radiant floor capable of heating part of the wetland where it is located	500
Amphitheater	Semi-enclosed building. Future headquarters for the Philharmonic Orchestra	1450
Parking	Two areas	20,000
Park lab	Center of management and learning	2500

Table 2 Authors' elaboration with the data provided in Harkema (2013)

Outdoor spaces	Description	Area (m^2)
Celebration square	Located in the South corner of the intervention and the Philharmonic	10,000
Great meadow	The heart of the intervention	11,500
High forest		8000
Low forest		2500
Tundra		2500
Varvaka walk		2500
Wetland bridge		1800
Partial recovery of the historical layout of the West area	Pre-existent space	300
Footbridge	Pre-existent space. It will provide connection between the park and the river	575

The project presents distributed in 13 Ha the following spaces that we specify in the Tables 1 and 2 with approximate areas that give us an account of its dimensions as well as its main energy characteristics.

This succinct description demonstrates the enormous complexity of the project under the pretext of the recreation of the four main landscapes of Russia. The approach is extremely ambitious in all its central readings, both the exclusively restoratives of the plot, as well as those concerning the functional and, obviously, the natural.

The authors point out that their work "High Line" in New York City was useful to investigate these questions of the natural and unfinished (Millington 2015). In this project they maintained the vegetations that had taken over the abandoned railway structure and incorporated them with all its values into a new urbanization that would enable an effective use by the neighborhood and its future visitors. They built a different linear park where what was traditionally repudiated and eliminated became the main differential value and engine of recovery of a degraded place. It is true that the photographs of Joel Sternfeld (Millington 2015), prior to the work, mythologized the values of the pre-existent urban ruin but finally the project would be the reality who brings together this previous poetics with a new landscape that substantially transforms the place and its interpretation. Architects are conditioned by the local and impregnated values of the existing nature that dwells in the infrastructure. Ricardo Scofidio declared that "they protected the place of architecture" (Millington 2015).

There are notable differences between the two projects, of which the following stand out: the one of NYC is promoted by the residents, Moscow is a governmental question. The first is a part of a large railway infrastructure, the second is derived from of a devastated solar. In the American project nature was there and, in Russian, its presence is forced in the place. High Line generates, eventually, an extraordinary gentrification and, at the same time, a real estate explosion (Millington 2015). Zaryadye, in its foundations, seeks a reconfiguration of the Moscow public space to increase the welfare of its citizens (Zaryadye 2013).

For DS + R the pre-existence of a pseudo-natural context was decisive for the conception of Zaryadye, one of the star operations of the new urban project for Moscow, and will also be one of the maximum exponents of urban design in the present century. In the present time the natural is not a matter of ornamentation, but a vital necessity in the face of the tremendous environmental problems of our XXIst century cities. We must be aware that more than 70% of the world's population will live in a city in 2050 (Ahern 2013).

We still do not have enough time perspective, but Moscow is presenting different operations that begin to construct a paradigmatic case to understand and act in the great metropolises of the present and of the future.

After the collapse of the USSR in the early 1990s of the last century the new Russia emerged but with a socio-economic context that was tremendously complex, difficult and uncertain for all its institutions and citizens. The whole country and its capital had to be redefined. Mayor Yuri Luzhkov managed the city in the main post-communist period, from 1992 to 2010. After being dismissed by the president of the federation, Dimitri Medvedev, e a new phase of development of the city began under the government of Sergei Sobianin who is still in charge of the city council. These minimal reviews account for the importance implied in the government and rehabilitation of Moscow for the whole of the USSR. The support and guidelines from the Central Government are key to understanding the city's urban development in its double aspect: citizen/proximity and representative/global capital. Today there is a strategic plan agreed between the two administrations until 2025 (Moscow 2015).

We have determined a historical sequence of urban plans in which the concern for the green is remarkable. We are in a context of high pollution generated by the industrial concentration of the region, the largest in the country. We can highlight the following plans (Maria Gulieva in Schindhelm et al. 2011; Bulanova 2014, Moscow 2015):

- 1918, "City of Future" by Boris Sakulin. It raises the expansion of Greater Moscow through two rings of eminently residential character around the Kremlin. Between them there would be an additional green ring.
- 1923, "New Moscow" by Ivan Zholtovsky and Alexei Shchusev. It proposed one of the most comprehensive plans for the idea of garden cities where they installed the new working class. The incipient industrialization established measures that meant that no new residential area was more than 600 m away from a green area.
- 1971, M. Posokhin proposed a structuring of the city and its growth in areas organized around important public spaces. Each sector would connect with the others through green areas.

- 2010, "Masterplan of necessities" by Alexander Kuzmin. In this plan there is a choice for the connectivity between sectors through "green wedges", the regeneration of the fluvial riverside of the metropolis focused on the integral rehabilitation of the capital.

We can see how the concern for the construction of a green and permeable context for the whole city in different forms is present in all the contemporary planning. The Western-based hygienist concepts, which are the basis of important present decisions, have been transformed into needs that encourage the improvement of the habitability of the city. Moscow, according to recent studies, is ranked 62, over 64, in terms of quality of life (Filippova 2015).

There is no doubt that the urban design of the city of Moscow, as we have seen, is marked by the Soviet heritage in a profound way. This situation presents a unique work platform that further underpins the last Development Plan. Moscow shows a distortion in comparison with the rest of the western capitals: its percentage of public space in the whole territory is 53% compared to 25% in the cities of the European Union (Anna Trapkova in Schindhelm et al. 2011). This percentage is due to a collective inheritance that defined the city, using open spaces as a reflection of a determined ideology (Argenbright 1999). This requires an extraordinary effort by the administrations that must adapt and qualify it according to the needs of a very different citizenship for which it was formerly thought. It is a challenge but there is an important global consensus that public space is crucial to building a high quality city. The Muscovites possess these areas but they are underused, especially the central ring. The WHO recommends a figure of 9 m^2 of green areas per citizen (World Health Organization 2010), the center of Moscow has a number of around 4.8 m^2 (Maria Gulieva in Schindhelm et al. 2011). The data is overwhelming if we take into account the concern for the planning of green areas in all the urban proposals throughout the XXth century. This reveals that the effort in this direction was concentrated in the periphery of the city, in the *raions*, where the majority of residents are lodged. This harmful situation is aggravated when considering the busy traffic, perimetral and transversal to the center, with the pollution derived from it. In 2006 there was a fleet of 2.6 million cars and, in 2012, the amount of vehicles reached 4.5 million (Horn 2014). It should also be considered that anti-icing treatments applied on public roads eliminate a large number of plant species (Argenbright 1999).

This situation makes the concern, emphasized by external advisors such as the Danish Jan Gehl, for sustainability not optional (House 2015). This is understood by the Plan of 2010 which establishes as one of the main vectors of action the qualification of the city from the public, green space and for the full enjoyment of the Muscovites (Maria Gulieva in Schindhelm et al. 2011). Moscow today has regained this tradition of planning green areas trying to reach at the same time a major impact in its most emblematic areas. In 2011, the old Gorky Park had a deep renovation to adapt it to the needs of the users. It was restructured to remain the great multifunctional reference park in the city (Horn 2014). In 2013 the renovation of the area of the waterfront of Krymskaya was tackled. The proposal of the park

Zaryadye continues to support the plans of the city government. In 2014 the Meganom study wins the international contest for the recovery of the banks of the Moskva (Stott 2014). Around 2017 Gorky and Zaryadye will form an almost continuous extension of about 7 km in the heart of the city. The recovery of the river, with a calendar, initially similar, will mean the rehabilitation of 120 km of riverside. These operations link the Russian capital, in a sense, with similar operations that take place in other large cities such as New York, where its "Designing the Edge" program seeks to recover its waterfronts (Ahern 2013).

Two forums for architectural and urban debate, based in Moscow, emerge between 2010 and 2011: the Strelka Institute, with a strong involvement of OMA/AMO (Schindhelm et al. 2011) and The Moscow Urban Forum. Both teams will work, from an international perspective, in the study of their city and in the new strategies and experiences around the strategies of management and development of cities. Moscow has protected and used them to regain a relevant role in the whole of the world's cities. In 2013, in the Urban Forum, the municipal government distributes the document "Moscow, the city for life" presenting the following priority lines of action: (1) Mobile City; (2) Comfortable urban environment; (3) Healthy city; (4) Well-educated city; (5) Socially protected city; (6) New economics of Moscow; (7) Open Moscow (Transparency) (Horn 2014). All this conceived with the purpose of healing a city conceived, during the most part of the XXth century, as a simple accumulation of work and production while making efforts to redefine itself as a place for citizens (Argenbright 1999).

The project for Zaryadye is not an isolated one: it is part of a larger program with an emblematic role within it. President Vladimir Putin supports this initiative which also beholds those values of new nationalism and Russian grandeur. Perhaps we may consider it, in the future, as a new global icon in the context of world reference parks. Nowadays it is already a leader in the implementation of these natural visions. Meganom's project for the recovery of the Moskva presents common approaches in some of its sections (Stott 2014).

The project is close to landscape urbanism (Sease 2015) but is fully connected with the duality of scenic aesthetics or ecological aesthetics (Jorgensen 2011). It is true that, without considering the impact of the construction of this intervention, the city will have gained an additional green lung and will be able to mitigate the lack of enjoyable green spaces that the city center demands both for its inhabitants and for the requirements of its precarious environment. The main and decisive value of the operation is the ability of this micro-system to recreate a particularly harsh and livable context, a pair of contrasted values that will qualitatively improve and sophisticate the city's wide network of public spaces. This work, based on its aesthetic and phenomenological character, will build a unique territory in Moscow at the same time that it begins to radiate its influence in different projects of notable offices of architecture and outstanding cities. This union between anthropological and ecological conditions is founding a new way of building the contemporary metropolis where the weight of sustainability should be decisive throughout the whole process.

Once defined the main features of the two interventions analyzed in this work, we provide the capacity of their respective green areas to sequester CO_2. Both interventions present figures, estimated on ideal models, which are interesting when analyzing their contribution to the reduction of CO_2 in the atmosphere. It is true that the case of Padua was not defined to combat this polluting situation but today it is playing a modest role in this area. In similar way, it is expected that Zarydaye Park will contribute to the mitigation of the remarkable air pollution problems in Moscow (Moscow Times 2014). In order to obtain a quick approximation based on the theoretical parameters reported by Roehr and Laurenz (2008, 144) regarding the CO_2/m^2 of green surface, we estimate the following quantities, considering the most unfavorable value of the three proposed by Schaefer (2004 cited in Roehr and Laurenz 2008). We make the simplification of assimilating the surfaces of both parks to a theoretical area "grassy" as the predominant vegetable in the two interventions and with a value of capacity of sequestering CO_2 of 4.38 kg/m^2 per year. For Padova with a green surface of 18,635 m^2 the value would be of 81,621 kg of CO_2 per year and for Moscow with a surface 127,232 m^2 the value would be of 557,276 kg of CO_2 per the same period. In any case, these numbers would be more favorable if the different tree species, already existing in the case of Padova and projected in the case of Moscow, were accurately taken into account.

Based on the ideas of the *Isola Memmiana* of Padua and the *Wild Urbanism* of the Zaryadye Muscovite Park, this chapter has presented how cities are enriched due to the increase in the quality of their public spaces thanks to the green experiments they support, regardless of the intentions with which they were conceived. It is noteworthy that the functional constant survives as a fundamental engine that activates these operations. Both in Padua and in Moscow, the improvement of health, either by the improvement of water flows or by the increase

Fig. 7 Diller Scofidio + Renfro (2013) Zaryadye Park Competition, Winner Project by Diller Scofidio + Renfro, tundra landscape

of green areas according to World Health Organization ratios in order to the fight against industrial pollution, are not conceived in isolation from other parameters that will ensure its success over time. Memmo considered the definition of a commercial space suitable for holding major fairs together with the creation of a place of enjoyment and leisure and even for the attraction of more visitors. Diller Scofidio + Renfro attend to this green mass as an improvement and sustainable repair of the center of the city, but also as a necessary step for the construction of a microclimate that makes the natural environment enjoyable throughout the year, helping to define, simultaneously, a new identity for Moscow and its inhabitants (Fig. 7).

The diachronic interpretation establishes guidelines that emphasize the functional reason of both interventions right from their beginning, although Zaryadye will still have to validate it with the passing of the years. Emotional, perceptive, experiential and enjoyment reasons are currently the main value of green interventions in the heart of the cities. If modern and contemporary society is characterized by a remarkable provision of time for solace, among other values, the park reaches a paradigmatic role in this context. Also the constitution of a singular emblem through the materialization of natural spaces makes parks capital issues for the characterization of any city.

References

Ahern, J. (2013). Urban landscape sustainability and resilience: The promise and challenges of integrating ecology with urban planning and design. *Landscape Ecology, 28*, 1203–1212.

Albardonedo Freire, A. (1998). Las trazas y la construcción de la Alameda de Hércules. *Laboratorio de Arte, 11*, 135–165.

Albardonedo Freire, A. (2015). La alameda, un jardín público de árboles y agua. Origen y evolución del concepto. Anuario de Estudios Americanos, 72, 2, Sevilla (España), julio-diciembre, pp. 421–452.

Amrusch, P. (2007). The relation between public services and the demand for sustainable city tourism. In A. Kungolos (Ed.), *Sustainable developement and planning III* (Vol. 2, pp. 1039–1048). Southampton, Boston: WIT Press.

Archinect. (2013). *Diller Scofidio + Renfro wins Zaryadye park competition in Moscow*. http://archinect.com/news/article/86380679/diller-scofidio-renfro-wins-zaryadye-park-competition-in-moscow. Accessed September 12, 2016.

Archsovet. (2013). *Zaryadye park. International competition for architecture and landscaping design concept for Zaryadye Park*. http://archsovet.msk.ru/en/competitions/zaryadye-park. Accessed September 12, 2016.

Argenbright, R. (1999). Remaking Moscow: New places, new selves. *The Geographical Review, 89*, 1–22.

Azzi Visentini, M. (2005). Le Prato della Valle à Padoue: un espace urbain plurifonctionnel à l'époque des Lumières. *La nature citadine au siècle des Lumières: promenades urbaines et villégiature* (Vol. 5, pp. 131–144). Bordeaux: William Blake & Co; Art & Arts.

Biaar. (2016). *Zaryadye park Moscow*. http://biaar.com/realizaciones/zaryadye-park-moscow/. Accessed December 1, 2016.

Bluestone, D. M. (1987). From promenade to park: The gregarious origins of Brooklyn's Park Movement. *American Quarterly, 39*(4), 529–550.

Box, J., & Harrison, C. (1993). Natural spaces in urban places. *Town and Country Planning, 62*(9), 231–235.
Bulanova, E. (2014). *Reconstruction plan of Moscow 1935. As an urbanistic concept*. https://issuu.com/glenka/docs/urbanistic_concept_repor2t. Accessed December 2, 2016.
Citizenarcane. (2005). *Architecture fit for an Emperor*. http://www.citizenarcane.com/index.php/archives/2005/04/23/. Accessed December 10, 2016.
Clark, K. (2011). *Moscow, the Fourth Rome*. Cambridge: Harvard university press.
Diller Scofidio + Renfro. (2013). *Zaryadye park competition, winner project by Diller Scofidio + Renfro*. http://www.e-architect.co.uk/images/jpgs/moscow/zaryadye-park-competition-moscow-s121113-d1.jpg. Accessed January 14, 2017.
Diller Scofidio + Renfro. (2013). *Zaryadye park competition, winner project by Diller Scofidio + Renfro, entrance from red square*. http://aiany.aiany.org/320741.pdf. Accessed January 14, 2017.
Diller Scofidio + Renfro. (2013). *Zaryadye park competition, winner project by Diller Scofidio + Renfro, tundra landscape*. http://www.e-architect.co.uk/images/jpgs/moscow/zaryadye-park-competition-moscow-s121113-d1.jpg. Accessed January 14, 2017.
Diller Scofidio + Renfro. (2013). *Zaryadye Park Competition, Winner Project by Diller Scofidio + Renfro, view from Bolshoy Moskvoretsky Bridge*. http://aiany.aiany.org/320741.pdf. Accessed January 14, 2017.
DS + R Consortium. (2014). *Zaryadye Park. Augmented Atmospheres. Poster from 14. Mostra Internazionale di Architettura. Eventi collaterali, Venezia*. http://www.rieder.cc/fileadmin/USER_FILES/News/Biennale_2014/Zaryadye_Park_Biennale2014.pdf. Accessed October 3, 2016.
Dsrny. (2013). *Zaryadye park*. http://www.dsrny.com/projects/zaryadye-park. Accessed September 12, 2016.
Filippova, A. (2015). *Grand designs: Can a Stalinist propaganda park become an appealing public space?*. http://calvertjournal.com/articles/show/3568/vdnkh-moscow-stalinist-renovation-urbanism-public-space. Accessed October 3, 2016.
García Lorca, A. M. (1989). El parque urbano como espacio multifuncional: origen, evolución y principales funciones. *Paralelo 37,* (13), 105–111.
Gennari, G. (1796). *Informazione Istorica della Città di Padova*. Venezia: Remondini.
Goethe, J. W. (1891). *Viaje a Italia*. Madrid: Imprenta de la Viuda de Hernando y C.ª.
Gordon, C., & Shirley, P. (2002). All things to all people-parks and urban greenspace in 21st century Britan. UK-MAB Urban Forum, May, 2002, pp. 2–21.
Harkema, L. (2013). *Winning proposal, Zaryadye park international design competition*. http://www.lindsayharkema.com/Wild-Urbanism-Zaryadye-Park. Accessed November 15, 2016.
Horn, C. (2014). Moscow the city for life—Russia. http://urbanplanet.info/urbanism/moscow-city-life/. Accessed October 15, 2016.
House, A. (2015). *Capital ideas: Interviews from the Moscow urban Forum*. https://calvertjournal.com/features/show/3578/Interviews-Moscow-Urban-Forum-Kapkov-Revzin-Kuznetsov-Burger Accessed September 10, 2016.
Howard, E. (1902). *Garden cities of tomorrow*. London: Swan Sonneschein & CO. Ltd.
Jorgensen, A. (2011). Beyond the view: Future directions in landscape aesthetics research. *Landscape and Urban Planning, 100*(2011), 353–355.
Jiménez Garnica, A. (2002). El Paseo de El Prado madrileño: Un ejemplo de saludable armonía entre la Naturaleza y el Arte en el espacio urbano. *Anales de la Real Academia Nacional de Farmacia, 68*(1), 158–205.
Kudryavtseva, V. (2011). *Emptiness with a history*. http://fotokonkurs.afisha.ru/en/alumni/kudryavtseva. Accessed on September 15, 2016.
Lawrence, H. W. (1988). Origins of the tree-lined boulevard. *Geographical Review, 78*, (4), 355–374.
Lawrence, H. W. (2008). *City trees: A historical geography from the Renaissance through the nineteenth century*. Charlottesville: University of Virginia Press.

Lopezosa Aparicio, C. (2009). Paseos por la imagen de lo cotidiano: un contexto para múltiples realidades. In *Congreso Internacional imagen y Apariencia. Repositorio Institucional de la Universidad de Murcia*. https://digitum.um.es/xmlui/handle/10201/43609. Accessed December 20, 2016
Mañueco, R. (2006). *Moscú dice adiós a otro mito soviético*. http://www.abc.es/hemeroteca/historico-06–03-2006/abc/Internacional/moscu-dice-adios-a-otro-mito-sovietico_142623631546.html. Accessed September 15, 2016.
Martinelli, R., & Parmini, G. (1996). *A renaissance fortification system: The Walls of Lucca*. UK: Lacey Books Ltd.
Memmo, A. (1786). *Elementi dell'Architettura Lodoliana o sia l'arte del fabbricare con solidità scientifica e con eleganza non capricciosa*. Roma: Stamperia Pagliarini.
Merian, M. (1668). *"Luca". Bella veduta a volo d'uccello della città e delle mura fortificate*. https://www.maremagnum.com/stampe/lucca-luca-bella-veduta-a-volo-d-uccello-della-citta-e-delle/130084055 Accessed December 1, 2016.
Millington, N. (2015). From urban scar to "park in the sky": Terrain vague, urban design, and the remaking of New York City's High Kine Park. In *Environment and Planning A* (Vol. 47, pp. 2324–2338).
Morley, P. (2016). *Friendship park*. http://www.moscow.info/parks/friendship-park.aspx. Accessed October 5, 2016.
Morriset, L. K., & Noppen, L. (2003). De la ville idéelle à la ville idéale: l'invention de la place royale à Québec. *Revue d'histoire de l'Amérique française, 56*(4), 453–479.
Moscow. (2015). *Urban development plan*. http://www.old.mos.ru/en/about/plan/. Accessed December 2, 2016.
Moscow Times. (2014). *Moscow air pollution reaches especially dangerous levels*. https://themoscowtimes.com/articles/moscow-air-pollution-reaches-especially-dangerous-levels-37365. Accessed October 12, 2016.
Mostafavi, M., Hight, C., et al. (2003). *Landscape urbanism: A manual for the machinic landscape*. London: AA Publications.
Mostafavi, M., & Doherty, G. (2015). *Ecological Urbanism*. Cambridge, Massachusetts: Harvard University Press.
Neveu, M. J. (2011). Prato della Valle, Reconsidered. In A. Pèrez, J. Gòmez, & S. Parcell (Eds.), *CHORA: Intervals in the philosophy of architecture* (Vol. 6, pp. 159–180). Montréal: McGill Queen's University Press.
Noro, M., Lazzarin, R., & Busato, F. (2014a). A first experimental survey on the urban heat Island in Padua (Italy). In I. Dincer, A. Midilli, & H. Kucuk (Eds.), *Progress in sustainable energy technologies: Generating renewable energy* (pp. 683–698). London: Springer.
Noro, M., Lazzarin, R., & Busato, F. (2014b). Three years of study and Urban Heat Island in Padua. *Sustainable Cities and Society, 10*(2014), 251–258.
Park-Gorkogo. (2015). *Gorky park*. http://park-gorkogo.com/en/. Accessed September 10, 2016.
Patte, M. (1769). *Mémoirs sur les objets les plus importants de l'architecture*. Paris: Rozet.
Piranesi, F. (1786). *Prospettiva della Nuova Piazza dietro la generale idea già concepita ed in gran parte efettuata dall Eccemo Sig. Andrea Memmo*. http://dioscorides.ucm.es/proyecto_digitalizacion/index.php?b19835875 Accessed December 1, 2016.
Posokhin. (2012). *The project for development of Zaryadye district*. http://www.posokhin.ru/en/activity/?SID=63&ID=370. Accessed December 10, 2016.
Premiyak, L. (2015). *Wild at heart: What a new park in the centre of Moscow means for the city*. http://calvertjournal.com/articles/show/3577/zaryadye-park-wild-urbanism-moscow-protest. Accessed October 3, 2016.
Radicchio, V. (1786). *Descrizione della general idea concepita, ed in gran parte effetuata all' eccellentissimo signore Andrea Memmo [...] sul materiale del Prato che denominavasi della Valle*. Roma: Antonio Fulgoni
Roehr, D., & Laurenz, J. (2008). Greening the urban fabric: Contribution of Green surfaces in reducing CO_2 emissions. In 1st WSEAS International Conference on

LANDSCAPE ARCHITECTURE (LA '08), Algarve, Portugal, June, 11–13, 2008, pp. 140–146.
Rodríguez-Avial Llardent, L. (1980). *El sistema de espacios libres en una ciudad y su planeamiento*. Escuela Técnica Superior de Arquitectura de Madrid, Madrid: Tesis doctoral.
Schindhelm, M., Grigoryan, Y., et al. (2011). *Public space*. Moscow: Strelka Institute for Media, Architecture and Design.
Sease, A. (2015), Landscape (and) urbanism? Engaging Nolli. *Journal of Urbanism, 8*, (4), 352–372.
Stott, R. (2014). *Project Meganom wins contest to transfrom Moscow Riverfront*. http://www.archdaily.com/576688/project-meganom-wins-contest-to-transform-moscow-riverfront. Accessed December 18, 2016.
Stratico, S. (1795). Dell'antico teatro di Padova. Padova: Stamperia del Seminario.
Strelka. (2013). *International landscape and architectural competition for the design of Zaryadye Park*. http://www.dkas.si/files/zaryadye_request_for_expression_of_interest.pdf. Accessed September 12, 2016.
Unknown. (1881). *Map of Zaryadye District*. http://dighist.fas.harvard.edu/courses/2015/HUM54/exhibits/show/moscows-lost-structures/zaryadye-district. Accessed January 14, 2017.
Wilkes, I. (2013). *Wild urbanism' in the middle of putin's Moscow*. https://nextcity.org/daily/entry/wild-urbanism-in-the-middle-of-putins-moscow. Accessed 27 December 2016.
Williamson, R. (1996). *Andrea Memmo's Project for the Prato della Valle*. 84th ACSA Annual Meeting, Regional Papers, pp. 620–623.
World Health Organization. (2010). *Urban planning, environment and health: From evidence to policy action*. http://www.euro.who.int/__data/assets/pdf_file/0004/114448/E93987.pdf?ua=1. Accessed December 20, 2012.
World Health Organization. (2012). Health indicators of sustainable cities. Rio + 20 UN Conference on Sustainable Development.
Zaggia, S. (2010). Isoletta sacra al commercio ed all'arti » . Andrea Memmo, Melchiorre Cesarotti e il Prato della Valle come esperimento di riforma del paesaggio urbano. In: Fabio Finotti (a cura di) Melchiorre Cesarotti e le trasformazioni del paesaggio europeo, EUT Edizioni Università di Trieste, Trieste, pp. 112–128.
Zaryadye. (2013). *Friends of Zaryadye*. http://en.zaryadye.org/competition.html. Accessed September 12, 2016.

Chapter 6
Occupation of the Territory and Sustainability in Transport in Madrid

Alberto Leboreiro Amaro

Abstract The use of private car plays a fundamental role in the GHG. This work analyzes the effect that the changes in the model of land occupation, in the last two decades in the Community of Madrid, have had on mobility. It has produced a high consumption of land in the last two decades, in an increasingly peripheral way, in increasingly distant nuclei and explained in part by the increase of single-family housing. These growths, functionally specialized, have been supported in the network of radial roads, losing density, diversity and the compact model of the Mediterranean city. At the same time, it is verified the relocation of the industry and the set of increasingly peripheral and unconnected employment to the residential space, together with the change in the labor model and the scarce residential mobility, the product of a model of housing in property, it generate an increase of the mobility due to work. The appearance of shopping centers once more peripheral, based on the road network and unrelated to the residence, leads to an increase in private vehicle travel due to the impossibility of providing support with public transport. These changes have resulted in an increase in private vehicle travel, with a clear loss of the percentage in public transport, which contributes significantly to the generation of GHGs and thus to climate change, as well as a high energy consumption. Therefore, the land occupation model has a clear effect on the mobility of the metropolitan area, requiring coordination between the urban model and the transportation model that leads to a more balanced region.

A. Leboreiro Amaro (✉)
DIrección General de Urbanismo, Consejería de Medio Ambiente,
Administración Local y Ordenación del Territorio,
Comunidad de Madrid, Madrid, Spain
e-mail: alberto.leboreiro@madrid.org

R. Álvarez Fernández et al. (eds.), *Carbon Footprint and the Industrial Life Cycle*,
Green Energy and Technology, DOI 10.1007/978-3-319-54984-2_6

1 Introduction

The evolution of greenhouse gases in the Community of Madrid has been parallel to the increase in economic activity and as it can be seen in the high increase of the private vehicle use, with its highest point in 2007, with the ending of the boom of real estate market.

According to data from Special Euro barometer 406, published in 2013, 50% of European citizens use the private vehicle daily, while only 16% use public transport and 12% use the bicycle. Urban transport accounts for about 25% of total CO_2 emissions and about 70% of all emissions in urban areas responsible for climate change. The DG Climate Action: Reducing CO_2 emissions from passenger; shows that cars the cars are responsible for around 12% of total EU emissions of carbon dioxide (CO_2), the main greenhouse gas.

The dependence of car and the congestion of the traffic, suppose for the set of European cities an estimated cost of about 80,000 million per year. Between 15 and 40% of European citizens are being exposed to a level of pollution that exceeds EU quality levels.

Within the study period (1990–2015) we see an increase in GHG emissions over the emissions in Spain, failing to meet the target of the Kyoto Protocol (see Fig. 1). Emissions growth until 2007, with data on GHG emissions to the atmosphere in the Community of Madrid. GHGs in 1990 represented a 4.9% of the national and in 2010, a 6.2% without an equivalent increment of the region economy. They represented a peak in 2007 of 25,665.11 Kt CO_2 equiv.

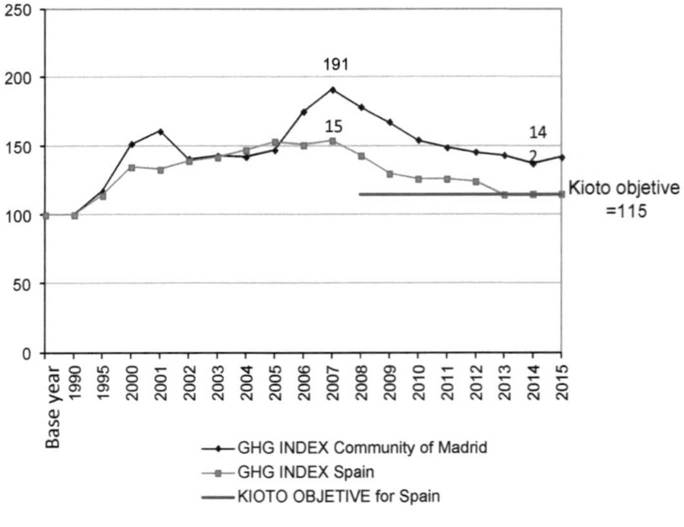

Fig. 1 Index of evolution of emissions of greenhouse gases in the community of Madrid and Spain (base year = 100) on CO_{2equiv}. *Source* D.G. Medio Ambiente, C.M. Inventory of emissions to the atmosphere in the Community of Madrid. Years 1990–2015

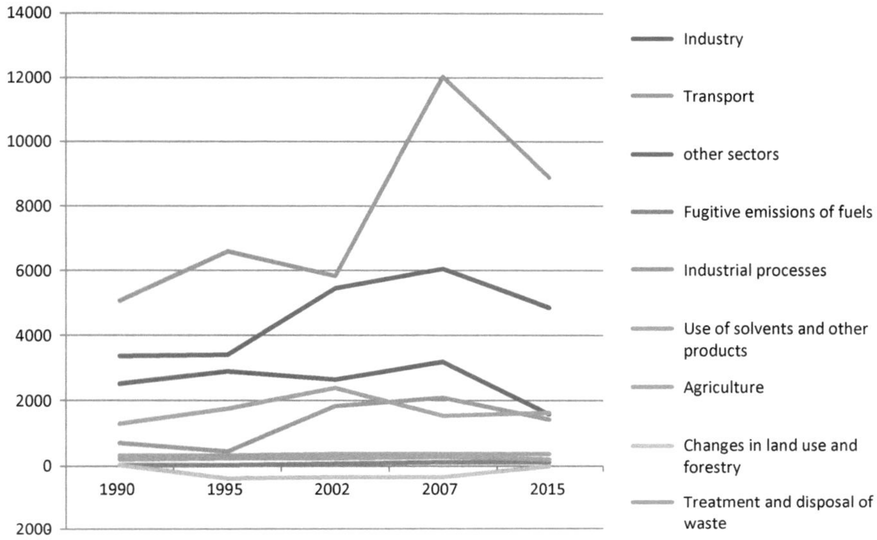

Fig. 2 Evolution of emissions equivalent 1990–2015. *Source* D.G. Medio Ambiente, C.M. Inventory of emissions to the atmosphere in the Community of Madrid. Years 1990–2015

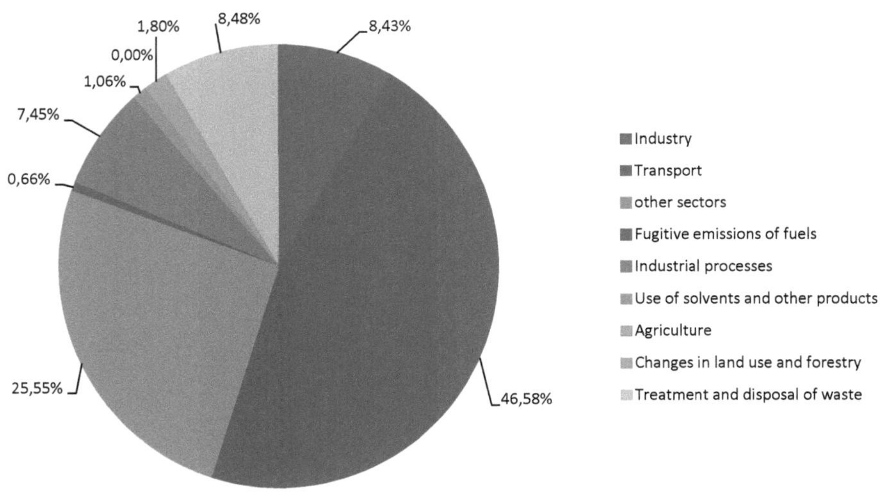

Fig. 3 Emissions equivalent 1990–2015. *Source* D.G. Medio Ambiente, C.M. Inventory of emissions to the atmosphere in the Community of Madrid. Years 1990–2015

If it is taken GHG as the basis for the evolution analysis in the Community of Madrid the CH_4 (t), CO_2 (kt), N_2O (t), SF_6 (kg), HFC, (kg) translated to tones of CO_2 equivalent as the inventory of GHG on CRF (Common Reporting Format) classification of the Community of Madrid, in Fig. 2 we see the strong influence of the transport sector, that pass for a peak of 5,852.68 CO_{2equiv} in 2017 (Fig. 2).

As shown in the Fig. 3, in the year 2015, the sectors that contributed the most to the direct emissions of greenhouse gases was transport sector with a percentage of 46.58%. The total GHG emission of transport sector passes from 5,055.71 Kt CO_{2equiv}. in 1990 to 8,871.86 Kt CO_{2equiv} in 2015.

Being the transport sector the main emitter of GHGs, increasing its share in the total in a continue way until the date. Moving from representing 37.60% of total emissions in 1990 to 46.58% in 2015.

By geographic location, GHG emissions are located in the urban core associated with traffic and the residential sector due to urban expansion, as well as in the surroundings of the main roadways, and in the surroundings of Madrid Airport. See Fig. 4.

In addition, pollutants with direct greenhouse effect are subject to a complementary analysis and if we take the data for the private vehicle, we see already mentioned above ie CO_2 emissions have almost doubled from 3,012 of tn of CO_2 equiv. In 1990 to 5,810 CO2 equiv. in 2015, being the urban driving the unique sector that has grown more when producing a slight reduction in the interurban trips in 2007 (Fig. 5).

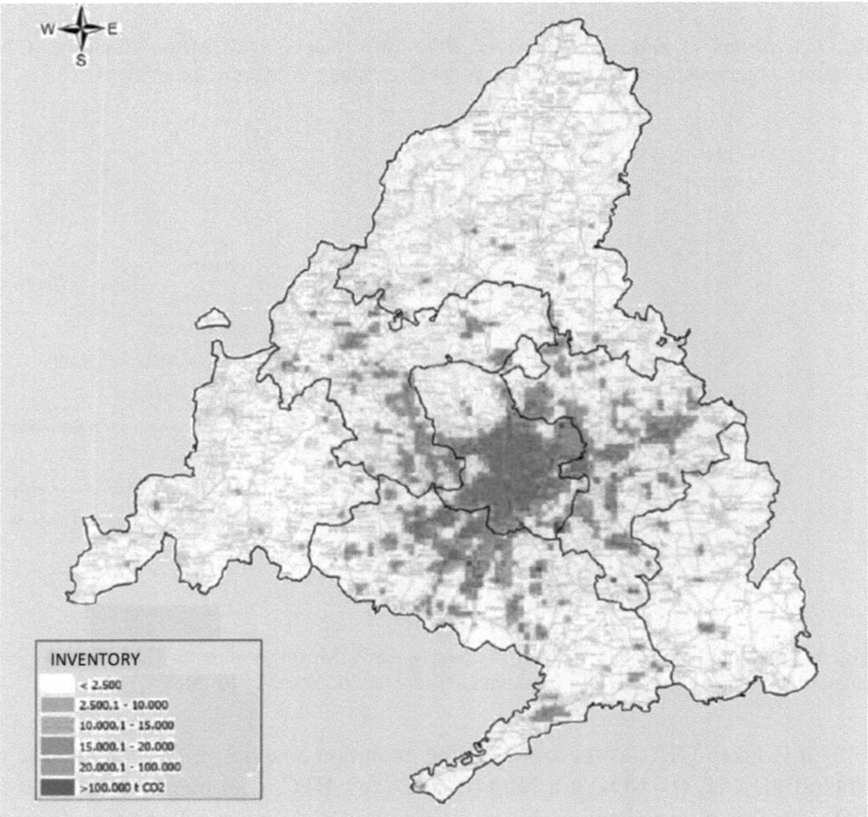

Fig. 4 Inventory of GHG emissions ($CO_{2equiv.}$) 2010. *Source* D.G. Medio Ambiente, C.M. Inventory of emissions to the atmosphere in the Community of Madrid. Years 1990–2015

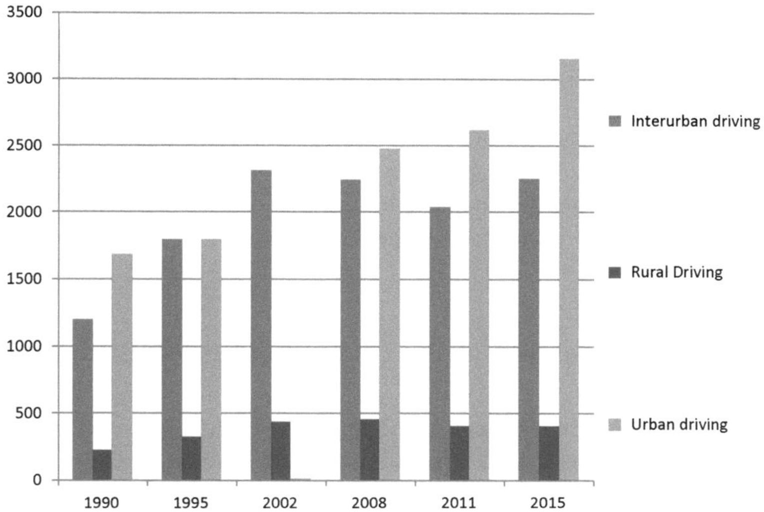

Fig. 5 Evolution of CO_2 emissions. *Source* D.G. Medio Ambiente, C.M. Inventory of emissions to the atmosphere in the Community of Madrid. Years 1990–2015

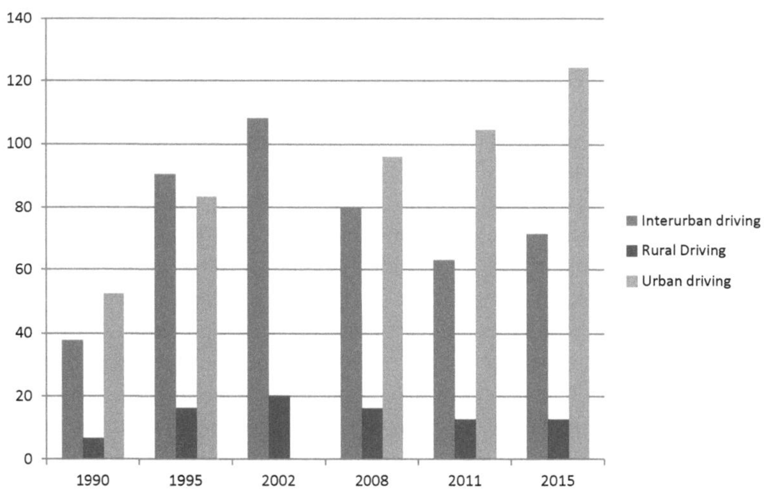

Fig. 6 NO_2 emissions. *Source* D.G. Medio Ambiente, C.M. Inventory of emissions to the atmosphere in the Community of Madrid. Years 1990–2015

The analysis of NO_2 gives a similar result with an increase in urban and decline, from the crisis, in intercity has gone from 97 t in 1990 to 208 t in 2015. In the NO_2 other reason was the increase of the share of diesel vehicles due to the promotion of this type diesel by the governments (Fig. 6).

In the NO_2 other reason was the increase of the share of diesel vehicles due to the promotion of this type diesel by the governments.

An increase of GHG is reflected in this period, including years after economic crisis despite the improvement of the technology, it could be due to the increased length of trips by car. The explanations could resides on the urban development, the sprawl, the relocation of the employment, the emergence of malls centres, in conclusion at the urban development of the last decade, without any regional planning, neither link with a policy of public transport.

2 Sustainable Mobility in the Community of Madrid

In the last two decades, Community of Madrid has experimented a phenomenon of land occupation and use that have contributed to the decline in the use of public transport. On the one hand, Community of Madrid has been suffering a loss of population in the center, causing a reduction in the number of users of public transport in the central area, the better equipped. On the other hand, this population has moved to the periphery towards areas where it exists a lack of public services.

Consolidation of the model based on radial and ring roads by expanding existing and building of new roads, such as toll roads and M-40, M-45, M-50, caused a call effect in the past. This fact led to the colonization of more peripheral spaces, based on the use of private vehicles and amplifying its effect with the help of other factors, such as an increased income level, the increasing number of single-family houses, lower density, and dispersion of activities or the emergence of great mall centers in example.

This urban growth supported on the roads, with monofunctional developments, lacking of diversity generated an increase of the mobility based on the use of private vehicle for business, studies, shopping and leisure in the absence of public transport in the new development territories. This dispersion of the population did not carry out a trend of job generation "in situ", contributing to the residence-employment imbalance and causing an increase of trips between home and work. At the same time the southern spaces have been progressively losing manufacturing employment, while in the center is concentrated much of the employment of the region, thereby causing even greater imbalances than already just mentioned. Besides all this, in the municipalities of the North and West close to the capital (Madrid), a high percentage of financial services is concentrated in an area with little public transport and a with social group that prefers using the private vehicle due to comfort and "status" reasons. This generates more and more imbalance, increasing the number of great shopping centers located on the edge and nodes of the road network, and those that are close to the road have double built area than the most outlying shopping centers. These malls do not have households around, attracting people of distant areas obliged to use of their private cars. Due to the low density of users in their proximity, it makes economically unsustainable the public transport lacking of a sufficient critical mass.

The increase in gross floor area occupied by single-family homes have also generated greater reliance on private vehicles. This phenomenon has occurred mainly in the last decade of the last century, mainly in the West, but its popularization caused a lot of growth in municipalities in the borderland of the Community, searching for better environmental values, absent in the city core.

In short, the absence of strategic master plan, that will expand the new areas or will renew the existing city, linking its developed lands to public transport that would have fostered factors as density, diversity and urban design, led the region to unsustainable mobility.

3 Planning of Land Use and Its Relation to Transport

Land occupation phenomenon has a crucial effect on the sustainability of transport. How this occupation is developed, in terms of density or dispersion of activities, the interstices of non-developed land in the territory, monoculture of single-family homes, the number of mall centers and the economic activity hubs outside the residential areas; the regional imbalances of employment location and incomes led to increase mobility in large metropolitan areas. Besides, transport infrastructures cause negative effects on the territory, colonizing outlying areas of high environmental value, encouraging the use of private vehicles, producing as well strong environmental impacts.

It is clear that the process of densification of the existing city promotes sustainable transport, but it must arise from a clear planning, in order to avoid harmful effects of functional and social substitution, or creating monofunctional homogeneous tissues, or segregated spaces within the city with congestion and loss of urban diversity.

As Giampino (2010) points out, “the projective solution for the territories of the dispersion can not be identified with banal forms of compacting of tissue built, but should move towards a unitary project, made of diverse elements, in which spaces lower density are juxtaposed to compact spaces, in a logic of a vertebrate system by intermodal transport”.

Against the model of “sprawl”, with its corresponding consumption of resources (land, materials, infrastructures, water and energy supply), it should be considered a strategy based on the existing city. Prioritizing high quality life by promoting information technologies and knowledge, reduction of pollution, with more efficient mobility and more quality public spaces, lower energy consumption and more integrated in their territorial environment.

Through regional planning it could be generated a greater territorial balance, but never should be replaced the attractiveness role of the central city for both, activities and high-level technicians. Diffuse city should play a role in the reorganization of the territory. According to Burchell (2002), it should be promoted an smart growth oriented strategy, limiting the construction of new road infrastructure and the provision of urban services by creating models based on more compact settlements, reducing the impact in rural areas or high environmental value. This strategy would promote the location of new developments near urban centers and, at the same time, tends to reduce the settlements in the most remote areas. Burchell (2002) proposed a set of measures aiming to reduce the negative effects of "sprawl", in particular in relation to transport:

- More compact population models, defining the urban-rural boundaries, imposing taxes on low-density growth, in order to internalize the extra costs of transport infrastructure and its maintenance.
- Reduce automobile dependency acting on policies restrictive transport as rising fuel prices, tolls entry to cities, controlled parking, incentive policies of public transport and modal interchanges, and the extension and improvement of the transport network public as well as park and ride in origin.
- Enter planning rules that limit sprawl development, but not only at local levels, but also at the regional level, establishing minimum densities or charging the associated costs to be internalized into the developments.
- Establish metropolitan agencies designed to the control and coordination of the policies of the various municipalities and creating, at the same time, a series of metropolitan institutions or voluntary cooperation organizations, mainly in transport like the Consortium of Transport of the Community of Madrid.
- Development of strategic management plans for large areas to correct the urban deregulation. These plans must be based on those resources to likely to be protected, looking for an urban design that encourages the use of public transport.

In short, and in order to remedy the dispersion of cities due to the impossibility of perpetuating a model based on the car (with strong congestion, pollution and energy consumption effects), solutions must be based on urban policies that promote the greater use of public transport and the development of integrated plans for large areas. As Gibelli (2007) pointed out: "solutions that will require, in order to be formulate coherent and consensual stably, research a long-term impacts of the different population models, the cooperation between local public actors, relevant plans and projects integrated". Instead of straighten, indiscriminately, the triad of Development/Diffusion/Dispersion (3D) it should be considered promoting a polycentric model, a model that discourages territorial dispersion of low density, creating new peripheral centralities with high functional diversification, prioritizing concentration and grouping around transport nodes, infrastructure and major public transport corridors, integrating transport in the urban form.

As part of the anti-sprawl initiatives, some movements or theories that have settled down at international level and that should be emphasized, as the "Transit Oriented Development" (TOD) and the "*New Urbanism*", which are probably the two approaches most universally recognized by the scientific literature. The first concept has been developed with the aim of decreasing the demand for motorized travel, and was coined by Calthorpe (1993). As defined by the American Public Transit Association, the TOD is a "Compact, *mixed-use development near new or existing public transportation infrastructure that serves housing, transportation and neighbourhood goals. Its pedestrian oriented design encourages residents and workers to drive their cars less and ride mass transit more sport*". New urbanism principles can be summarized as: walkable community, multi-functional and diverse, with a mix of residential typologies, increased density, transport "ecological" or "green" environmental friendly and the focusing the use of public transport.

At European level, the closest movement to New Urbanism would be the Urban Villages, as Neal (2003) describes. The formal characteristics that should have an urban village would be: "*Development of adequate size or critical mass, a walkable environment for pedestrians, a good uses mix, good employment opportunities, a system of mixed tenure for both residential use and for offices, enough provision of basic social and sanitary equipment and some degree of self-sufficiency for shopping*". Other reformist designers (Cervero 1997) indicate that the 3Ds, i.e. density, functional diversity in the city and an inclusive urban design that encourages pedestrian and cyclist mobility are key achieving these goals.

Several authors have recognized density as a key factor when considering the demand of travel. However the connection between trip generation and functional diversification and inclusive urban design, has been less reflected in literature. In the case of trips for a cause different than work, more compact settlements, where there exists shops and services nearby along with a pleasant environment, foster intermodality, through collective transport, walking and cycling, and the trips proximity trips of short distances. Multifunctionality and density in areas that have a significant number of pedestrian routes could favor the shared city (sharing resources, goods and services and experiences), more than the scattered and monofunctional city car-oriented. Therefore, the characteristics of the points of departure and arrival, both in residential neighborhoods and business centers, shaped as modal interchanges, represents a great influence in the choice of transport mode.

One of the fundamental objectives of regional planning is the intervention and regulation of land use. Our society is not aware that soil is a scarce resource, like air or water. The soil is the part of the system that sustains life and plays a crucial role in climate change, as an immense natural carbon reservoir. The land is essential to ensure food availability and it is a key factor of landscape and understanding of human relationship with the ecosystem.

Throughout this article, it will be shown that the evolution of metropolitan area in Madrid based on market supply instead of on public demand and lacking planning, has created today an unsustainable mobility. For the analysis, it has been taken the years 1995, 2002, 2008 and 2014, covering the beginning of the housing born until the nowadays crisis.

4 The Occupation of the Territory in the Community of Madrid

During the period 1995–2001 the land occupation in the Community of Madrid was increased in 8893 ha, representing 15% of the existing soil and, in addition, this is the period with the highest growth detached houses. Obviously, the town with the greatest growth was Madrid with 2528 ha and a percentage over the land occupied in 1995 of 12%, representing in whole region 28%, much higher than the increase of its population (23%), in a process of reducing the population density. The areas in which there was a greater increase of land occupation were those located at the West of the Community. Municipalities with a growth exclusively based development of single-family houses, models of low density promoted by public administration in other municipalities as Meco, Arroyomolinos and Rivas Vaciamadrid in example.

In the second period 2001–2008 studied, the overall increment of occupied land in the Community was 7283 ha. The protagonist remains in the West area, but with less intensity. Land occupied is distributed through all the Community increasing the peripheral municipalities in all areas. This process is explained in the real world due to a post-crisis period, due to the increase of the price of housing, in the first place, in those with higher market value, but finally generating a widespread increase in prices. The occupations are more peripheral and in areas of lower environmental value or with lower level endowments. In the last period considered (2008–2014), during the economic crisis, the increase in land occupied is 5522 ha with a 7.47% of increase, approximately half that in the first of the studied periods. Table 1 shows the detailed display of the commented data.

According to Urban Audit, a study developed in the year 2012 by Spanish National Statistics Institute (INE), Pozuelo, Las Rozas and Majadahonda, all municipalities located through the A-6 motorway, were placed at the top level of the ranking of cities per household incomes of Spain, with low unemployment rates, a high percentage of single-family homes and of vehicles per household. Figure 7 presents the income variation figures in the Community of Madrid in those years. It can be seen that, in the West and in the North positive figures in income have been achieved in recent years, which indicates the beginning of processes of economic inequality growth, with concentration of economic capacity as well as an increase in the use of private vehicle.

For the whole Community shows that in the first period (1996–2001) the metropolitan area is the one that achieves a greater growth, highlighting the West and Southeast. In the second period (2001–2008) the South begins to grow more and, in the third period (2008–2014), the growth has been extended to all the periphery with the exception of mountainous areas in the South and North, but extended to the Southeast meadows. In Table 2 it can be observed the increase of population of the West Metropolitan and Northeast and Southwest of the Community, areas of single family houses.

Attending to population, the Community of Madrid grew by 27.82%, less than the land occupied, that was the 37.44%, what reflects the decline in gross densities. In Table 3 is reflected the variation of occupied floor, it can be observed that the model is not due to a land-use planning, or a policy of optimization model of public transport, but a development of the road network and economic interests of developers and local micro-policies.

Doing a trend analysis of how the occupation has varied related to the average of the Community of Madrid, is observed a peripheral growth as leapfrog. The sprawl in the West Metropolitan, that through the A-6 extends to the Escorial with the colonization of surroundings of Guadarrama the Southwest of the Community.

Table 1 Variation of occupied land in the Community of Madrid (per hectares, ordered by municipalities with greater growths by periods)

Municipality	Surface occupied 1995	Occupancy variation 2001–1995	% Occupancy variation 2001–1995	Municipality	Surface occupied 2001	Occupancy variation 2008–2001	% Occupancy variation 2008–2001	Municipaltiy	Surface occupied 2014	Occupancy variation 2014–2008	% Occupancy variation 2014–2008
Total	57,843.84	8,893.81	15.38	Total	66,706.35	7,238.29	10.85	Total	73,930.43	5,522.45	7.47
Madrid	21,152.48	2,528.40	11.95	Madrid	23,680.88	1,367.00	5.77	Madrid	25,047.89	627.72	2.51
Boadilla del Monte	958.96	395.80	41.27	Getafe	1,374.27	298.10	21.69	Alcobendas	1,575.87	369.60	23.45
Alcala de Henares	1,597.20	314.33	19.68	Alcala de Henares	1,911.53	264.87	13.86	Tres cantos	406.04	333.81	82.21
Rivas-Vaciamadrid	296.53	263.99	89.02	Leganes	1,466.53	264.57	18.04	Pozuelo de Alarcon	1,224.31	228.60	18.67
Getafe	1,116.30	257.97	23.11	Boadilla del monte	1,354.76	242.23	17.88	Getafe	1,672.37	219.23	13.11
Fuenlabrada	1,214.87	227.32	18.71	Rivas-Vaciamadrid	560.52	216.69	38.66	Boadilla del Monte	1,596.99	180.83	11.32
Las Rozas de Madrid	1,584.53	227.23	14.34	Meco	191.21	197.93	103.51	Arroyomolinos	355.82	179.90	50.55
Villaviciosa de Odon	695.53	225.81	32.47	Pinto	446.75	189.45	42.41	Leganes	1,731.11	155.24	8.97
Majadahonda	524.47	220.07	41.96	Arroyomolinos	179.34	176.47	98.40	Valdemoro	650.59	130.55	20.07
Villanuevade la Cañada	426.7	208.93	48.96	Las Rozas de Madrid	1,811.75	175.25	9.67	Rivas-Vaciamadrid	777.21	130.44	16.78
Leganes	1,275.88	190.66	14.94	Valdemoro	490.96	159.63	32.51	Torrejon de Ardoz	888.7	114.80	12.92
Mostoles	1,035.55	183.38	17.71	Aranjuez	601.63	159.16	26.45	Parla	556.83	111.24	19.98
San Sebastian de Los Reyes	779.29	172.99	22.20	Majadahonda	744.55	149.74	20.11	Las Rozas de Madrid	1,987.00	106.10	5.34
Pozuelo de Alarcon	1,066.65	135.85	12.74	San Sebastian de Los Reyes	952.28	146.93	15.43	Cie mpozuelos	302.82	92.18	30.44

(continued)

Table 1 (continued)

Municipality	Surface occupied 1995	Occupancy variation 2001–1995	% Occupancy variation 2001–1995	Municipality	Surface occupied 2001	Occupancy variation 2008–2001	% Occupancy variation 2008–2001	Municipaltiy	Surface occupied 2014	Occupancy variation 2014–2008	% Occupancy variation 2014–2008
Alcobendas	1,344.18	123.10	9.16	Alcorcon	913.21	133.55	14.62	San Sebastian de Los Reyes	1,099.21	92.12	8.38
Alcorcon	791.29	121.92	15.41	Fuenlabrada	1,442.18	120.43	8.35	Villalbilla	373.94	87.10	23.29
Pinto	327.09	119.66	36.58	Arganda del Rey	440.32	114.29	25.96	Alcorcon	1,046.76	81.62	7.80
Arroyomolinos	76.03	103.31	135.88	Parla	445.64	111.18	24.95	Pinto	636.2	72.96	11.47
Torrelodones	569.43	100.18	17.59	Navalcarnero	258.38	109.45	42.36	Aranjuez	760.79	70.43	9.26
Meco	96.15	95.06	98.87	Alcobendas	1,467.28	108.58	7.40	Mostoles	1,327.44	68.90	5.19
Parla	358.17	87.48	24.42	Mostoles	1,218.93	108.52	8.90	Fuenlabrada	1,562.61	67.35	4.31

Source Interpretation on orthophoto of the Community of Madrid and developed by author

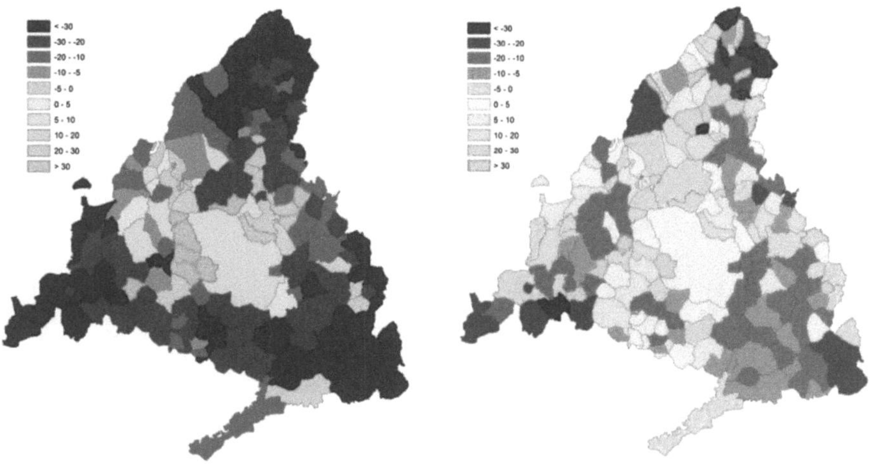

Fig. 7 Income level 1996 (*left*) and income level variation from 1996 to 2010 (*right*). *Source* INE, modified by the author

An extension more compact in the Northeast and finally, the municipalities aforementioned, Arroyomolinos, Rivas Vaciamadrid and Meco with public support, but not responding to a rational criterion of growth (Fig. 8).

In these areas were built new roads in order to continue having growth potential, without recovering any capital gain that could cover the public expense of such infrastructures or the necessary social facilities. But, at the same time, these growths created monofunctional islands, without any relationship neither complementarities between them, due to that as nodes, each one of them contributes with no more than commercial malls of highways, achieving an autonomous process that does not take into account the population of these territories.

4.1 *Evolution of Land Occupied in the Fringe of Radial Roads*

The development of roads has served on the Madrid to open the territory and colonize new land for residential use, in some areas such as the West it can be seen a continued growth in the edge of A-6. Not only radial roads have played this role, also the ring roads and the new highways, had play a role in the location of economic activity and give access to territory of high environmental value lacked public transport. The new roads were developed at the mid-eighties by means of plans, but these were abandoned. The last approved definitely was in 1994, and from that time the roads are designed by previous studies to solve the environmental assessment and road projects. There was no plan for spatial planning, that integrate planning and transport, besides the sectoral plans of transport were forgotten, so the reason for building new roads was exclusively to facilitate the colonization of new lands (Fig. 9).

Table 2 Population variation

Statistics zones	1996	2001	2008	2015	Variation 1996–2001 (%)	Variation 2001–2008 (%)	Variation 2008–2015 (%)	Variation 1996–2015 (%)
Comunity of Madrid	**5,036,172**	**5,372,433**	**6,271,638**	**6,436,996**	**6.68**	**16.74**	**2.64**	**27.82**
Madrid	2,874,775	2,957,058	3,213,271	3,141,991	2.86	8.66	−2.22	**9.30**
North Metropolitan	216,314	247,728	298,367	329,022	14.52	20.44	10.27	**52.10**
East Metropolitan	437,134	480,944	614,280	650,717	10.02	27.72	5.93	**48.86**
South Metropolitan	1,005,810	1,060,920	1,234,877	1,297,089	5.48	16.40	5.04	**28.96**
West Metropolitan	228,766	327,233	439,740	479,723	43.04	34.38	9.09	**109.70**
Northern Sierra	21,025	24,586	37,609	40,401	16.94	52.97	7.42	**92.16**
Northeast Comunity	21,657	29,477	53,063	60,953	36.11	80.01	14.87	**181.45**
Southeast Comunity	57,395	62,983	92,666	102,951	9.74	47.13	11.10	**79.37**
Southwest Comunity	43,041	56,566	98,894	130,862	31.42	74.83	32.33	**204.04**
South Sierra	21,080	22,982	32,227	34,142	9.02	40.23	5.94	**61.96**
Central Sierra	109,176	101,956	156,644	169,145	−6.61	53.64	7.98	**54.93**

Source Instituto de estadistica de la CM

Table 3 Variation of occupied land in Community of Madrid (per hectares, ordered by municipalities that more have grown)

Municipality	Surface occupied in hectares in 1995	Occupancy variations 2014–1995	% Occupancy variations 2014–1995	Occupied area in hectares meters in 2014
Total	57,843.84	21,654.54	37.44	79,420.16
Madrid	21,152.48	4,523.12	21.38	25,675.60
Boadilla del Monte	958.96	818.86	85.39	1,777.82
Getafe	1,116.30	775.30	69.45	1,891.60
Alcala de Henares	1,597.20	643.46	40.29	2,240.66
Rivas-Vaciamadrid	296.53	611.11	206.09	907.64
Leganes	1,275.88	610.47	47.85	1,886.34
Alcobendas	1,344.18	601.29	44.73	1,945.47
Las Rozasde Madrid	1,584.53	508.58	32.10	2,093.11
Arroyomolinos	76.03	459.69	604.60	535.72
Pozuelo de Alarcon	1,066.65	431.77	40.48	1,420.19
Fuenlabrada	1,214.87	415.10	34.17	1,629.96
San Sebastian de los R	779.29	412.04	52.87	1,191.33
Majadahonda	524.47	401.59	76.57	926.07
Pinto	327.09	382.07	116.81	709.16
Tres Cantos	368.43	371.42	100.81	739.85
Valdemoro	416.77	364.37	87.43	781.14
Mostoles	1,035.55	360.79	34.84	1,396.34
Alcorcon	791.29	337.09	42.60	1,128.38
Parla	358.17	309.90	86.52	668.06
Villaviciosa de Odon	695.53	306.99	44.14	1,002.52
Meco	96.15	296.35	308.23	392.50

Source Photo Interpretation on orthophoto of the Community of Madrid and developed by author

Fig. 8 Variation built with respect to the average area of the community of Madrid 1995–2014. *Source* Photo Interpretation on orthophoto of the Community of Madrid and developed by author

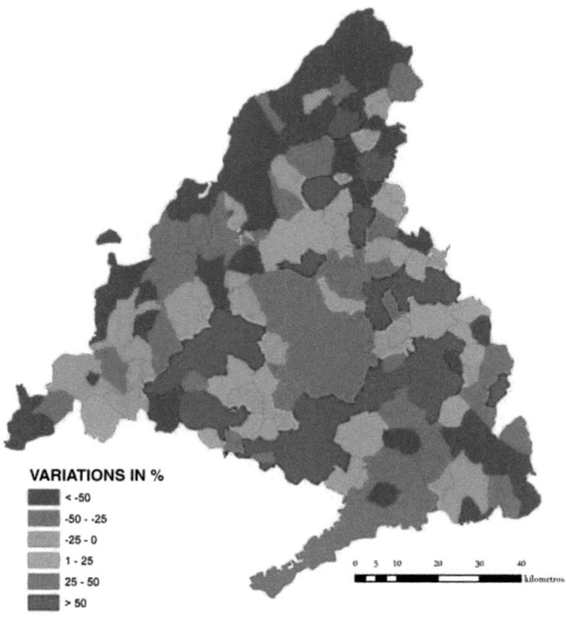

Fig. 9 Highways by date period of construction. *Source* Photo Interpretation on orthophoto of the Community of Madrid and developed by author

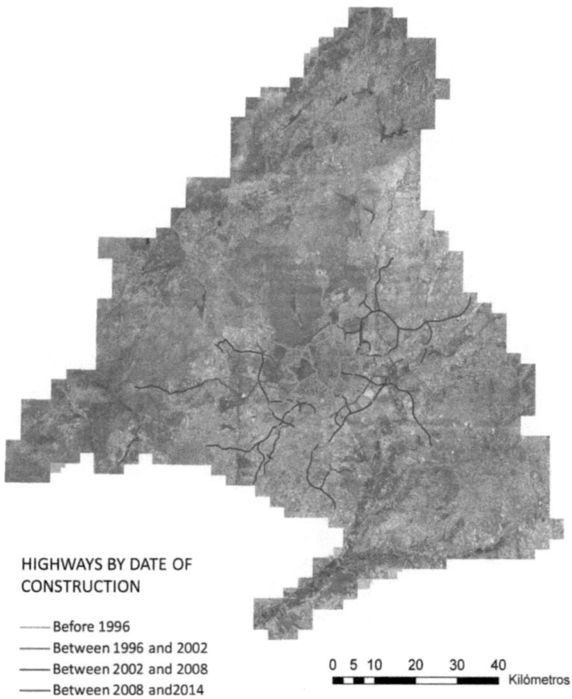

It has been analyzed the evolution of land occupation in the space near the radial roads, a distance of 1200 m on each side, with intervals of 15 km length starting at the M-30. The analysis has taken into account that in the period 1995–2014 a series ring roads were built in some of the cores of the roads analyzed, which can produce some distortions in the results. However, the occupation in many of the cases supposes the silting up of existing tissue, eliminating possible ecological corridors. The results are a percentage over of the ground already occupied what gives an idea of the value of the results. Total occupancy near the radial roads has increased from a low value in the A-2 in its first stretch of 26%, where there was a high occupancy of 44%, to a maximum of 109% in the second section of the A-5, where there was an occupancy rate of 14%. As a whole, most of the values exceeded 37% average occupancy of the Community of Madrid. In absolute terms, the first section of the A-1, where they started from a high occupancy and in the period grew to 609 ha by the growth of PAUs Madrid and developments of Alcobendas and San Sebastián de los Reyes, and the first stretch of A-4 with 530 ha. In the second stretch in the A-2 with 330 ha, the A-4 with 259 ha, the A-6 with 251 ha and A-5 with 244 ha (Table 4). In the third stretch in the A-6 with 154 ha and the A-1 with 110 ha, these two roads are the symbol of the dispersed growth in an important natural area of Madrid Community, and both roads without a good public transport infrastructure.

The roads have led to unbalanced growth, producing a functional specialization, if map of occupation in 2014 is analyzed (Fig. 10), it shows the following:

- The A-1 in its first stage is one of the roads with more variety of uses, including financial services, but not balanced in terms of its location, concentrated, demanding greater use of private vehicles, being a clear example of "zooning", and a high occupancy of single family houses in the three stretches.
- The A-2 is the motorway with more industry around, as reflecting the historical formation of the of the Corridor, the uses are denser as seen in the stretch full of population, with a low percentage single family houses with the 12 and 6% in the first and second tranche
- The A-3 has low occupations, compared with the others, mainly because the two cores, Rivas and Arganda, located only on the left side of the road, have a high percentage of industry in Arganda and a high percentage single-family housing in the third stretch.
- In the A-4 and the A-42 dominates the industry and multifamily houses in all its sectors, without single family-houses.
- The A-5 has a lower occupancy, due to the empty area of Campamento, with a first stretch specializing in multifamily houses in the municipality of Madrid, but in its third stretch specialized in single family houses (Móstoles, Arroyomolinos and Navalcarnero), with a percentage of 94%, and a higher percentage of tertiary in the second stretch (Alcorcón Norte), Hipercor, etc.
- The A-6 road is specialized in single-family homes in the three tranches and without industry, or representation of tertiary sector except in the first stretch. In this one, the use of private vehicles is dominant, therefore the success of its

Table 4 Ocupation in radial roads

Range km	Motorway	Total occupied total 2014 in has	Total edificado 1995–2014 has	% occupied 1995–2014	Occupied by single family home 2014 has	Occupied by single family home 2014
15	A-1	1,543.49	609.07	60.54	368.39	23.87
30	A-1	776.34	205.89	73.48	335.24	43.18
45	A-1	423.28	110.74	73.84	290.34	68.59
15	A-2	1,517.18	314.65	79.26	182.91	12.06
30	A-2	929.18	330.58	64.42	56.35	6.06
15	A-3	966.02	353.48	63.41	93.39	9.67
30	A-3	285.37	124.37	56.42	67.54	23.67
45	A-3	167.12	28.04	83.22	90.25	54.00
15	A-4	1,112.71	529.93	52.37	21.76	1.96
30	A-4	708.12	259.27	63.39	61.96	8.75
15	A-42	1,482.08	315.46	78.71	55.46	3.74
30	A-42	507.07	246.17	51.45	99.78	19.68
15	A-5	691.18	92.38	86.63	22.69	3.28
30	A-5	468.80	244.90	47.76	150.62	32.13
45	A-5	48.94	20.71	57.68	45.98	93.96
15	A-6	1,153.35	326.65	71.68	677.92	58.78
30	A-6	1,037.84	251.24	75.79	762.83	73.50
45	A-6	643.81	154.88	75.94	436.47	67.80
15	M-607	605.11	181.12	70.07	122.73	20.28
30	M-607	214.01	96.97	54.69	82.39	38.50
45	M-607	308.71	67.71	78.07	250.46	81.13

Source photo interpretation of orthophotos Community of Madrid, Municipal register figures and developed by author

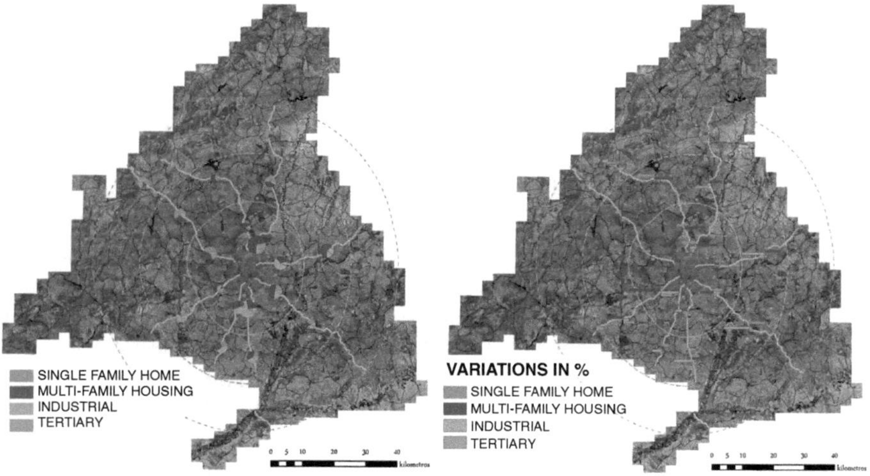

Fig. 10 Area occupied on the roads in a margin of 1200 (*left*) and percentage difference built 1995–2014 and existing meters per ring of 15 km in the period 1995–2014 (*right*). *Source* photo interpretation of orthophotos Community of Madrid, Municipal register figures and developed by author

"bus-HOV lane", but at the same time it has experimented a very high growth in employment in the three stretches, equilibrating both traffic directions.

- An arc seen from the A-5 to A-1 in which the single-family houses are totally dominating, in the areas of more environmental value of the community and with a low level of public transport, by its low density. Although all third stretch has a specialization in single-family houses.

The development of roads during the analyzed period has served on the Madrid to open the territory and colonize new land in some areas, like the west, where it can be seen a continuous growth, not only from radial roads, but in the new highways that give access to areas of high environmental value and lacked public transport. This development of new roads has obeyed sectorial plans, the last of them was approved definitely in 1994, and then the plans were made as previous studies, to solve the environmental assessment, or projects of roads but without being integrated with urban or territorial plans as it would have been logical. On the other hand in those models used for the evaluation of new roads, it could be shown that demand is unlimited in the scope of the metropolitan area. Any highway built begins in service relatively quickly: the proof could be the M-45, that was ahead of all the forecast and where the concessionaire by shadow toll increased its benefits, not the case when the route is a direct toll. The roads were projected by demand and pressure from different local administrations, obeying more to pork-barrel politics, skipping even the environmental impact procedures on some of them.

4.2 Evolution of Land Occupied by Single-Family Houses

Single-family homes showed a minimum at the beginning of the nineties, in 1993 with 5,446 homes, in a low cycle of the economy, see Table 5, but in the last five years of that decade, more single-family homes were built achieving a maximum in the year 1999, with 16,887 single-family homes that accounted for 26.82% of all homes built. Due to the increase of prices in the first decade of 21th century the building of single-family houses has dropped substantially from 26.34% in 2001 to rates around 19%, in 2007 with 5783 houses and already in 2013, immersed a full crisis period, were they were built 793 single-family houses representing 10.51% of the total houses. The percentage of single-family houses over the total has been increased in all municipalities between 1995 and 2014, with the exception of city of Madrid, where they have practically disappeared.

In the first period 1995–2001, see Table 5, and grouping municipalities from largest to smallest area occupied, it can be seen that the evolution of areas occupied by single-family houses were stronger in the municipalities of the first ring of the West. In the second period the intensity turned to the second ring around Guadarrama river and two peculiar municipalities already mentioned Rivas Vaciamadrid and Arroyomolinos, the first one with strong support based on social housing and the second by means promoting of a public company. In the next period 2001–2008 it can be seen a growth in the A-2 road, and again in Rivas Vaciamadrid and Arroyomolinos, the West continues to grow albeit with less force. This is consistent with statistical data of building single-family homes of the Ministry of Development where it is reflected the greatest dispersion process occurred in the last decade of nineties. In the following period 2008–2014, and during the crisis, the growths of single-family houses are more peripheral and smaller dimension.

In the entire period between 1995 and 2014 the West is the area with higher single-familiar houses developed, highlighting the growth in East Rivas Vaciamadrid and South Valdemoro, strange typology for that territory. What it is true is that the occupancy of land for single-family houses has no relation with job or business opportunities nor superior planning, but rather political benefits and local promotions (Valdemoro. Rivas, Boadilla del Monte, Arroyomolinos, Villanueva de la Cañada) with a supply model that generated its own demand.

If the analysis is performed focusing on the percentages of land occupied by single-family houses over the total developed area in the municipalities, those in which smaller area of land for single-family houses in the period 1995–2014 is occupied are located in the South. If the Community is analyzed as a whole, 60.72% of occupied land was for single-family house use. It must be taken into account that the traditional home building in rural areas and the largest area occupied by the single-family house. In Table 6, it can be see the percentage of land occupied by single-family homes in the metropolitan area municipalities in 2014. The percentage of land occupied by this type of houses has decreased in relation with the initial state.

Table 5 Evolution of the land occupied by single family houses in the metropolitan area the period 1995–2015 ordered by municipalities with less average

Municipality	Total surface occupied by houses in hectares 1995	Surface occupied by single family house in hectares in 1995	% Of occupied land by single family house over total residential surface 1995	Total residential surface ocuppied in hectares 1995–2014	Residential occupied surface single family house in hectares 1995–2014	% Of land occupied by single family house 1995–2014	Total residential surface in hectares 2014	Total surface occupied by single family house in hectares 2014	% Of Occupied land by single family house over total residential surface 2014
Fuenlabrada	354.29	21.11	5.96	166.84	2.42	1.45	521.13	23.53	4.25
Leganes	133.55	17.48	13.09	237.98	6.82	2.87	371.53	24.30	6.54
Aranjuez	135.53	44.85	33.09	231.12	14.09	6.10	366.65	58.94	16.08
Tres Cantos	154.94	95.61	61.71	89.47	5.75	6.43	244.41	101.36	41.47
Torrejon de Ardoz	168.34	22.40	13.31	148.19	10.84	7.32	316.53	33.24	10.50
Coslada	168.21	60.67	36.07	40.81	3.54	8.68	209.02	64.21	30.72
Alcorcon	141.90	33.98	23.95	161.71	15.25	9.43	303.61	49.23	16.22
Getafe	418.10	85.94	20.55	279.26	29.00	10.38	697.36	114.94	16.48
Madrid	6,125.58	896.76	14.64	1,348.24	157.75	11.70	7,473.82	1,054.51	14.11
Parla	77.61	0.00	0.00	156.51	28.17	12.00	234.12	28.17	12.03
Alcobendas	685.40	536.36	78.26	220.55	32.60	14.78	905.95	568.96	62.80
Pinto	52.09	9.11	17.49	91.58	14.79	16.15	143.67	23.90	16.64
Alcala de Henares	334.35	46.54	13.92	176.74	42.02	23.77	511.09	88.56	17.33
Ciempozuelos	85.05	11.20	13.17	29.28	7.59	25.94	114.33	18.79	16.43
san Sebastian de Los	439.91	287.09	65.26	358.95	99.09	27.61	798.86	368.18	48.34
Ajalvir	15.73	6.82	43.36	5.33	1.81	34.00	21.06	8.63	40.97
Valdemoro	72.21	10.88	15.07	185.10	85.90	46.41	257.31	96.78	37.61
Majadahonda	275.54	149.50	54.24	142.86	68.12	47.69	418.50	217.62	52.00
Las Rozas de Madrid	733.37	610.94	83.31	341.47	165.03	48.33	1,074.84	775.95	72.19

(continued)

Table 5 (continued)

Municipality	Total surface occupied by houses in hectares 1995	Surface occupied by single family house in hectares in 1995	% Of occupied land by single family house over total residential surface 1995	Total residential surface ocuppied in hectares 1995–2014	Residential occupied surface single family house in hectares 1995–2014	% Of land occupied by single family house 1995–2014	Total residential surface in hectares 2014	Total surface occupied by single family house in hectares 2014	% Of Occupied land by single family house over total residential surface 2014
Mejorada del Campo	54.78	31.27	57.08	34.81	18.53	53.24	89.59	49.80	55.58
Total comunidad	24,213.08	14,970.13	61.83	**10,655.25**	**6,202.97**	**58.22**	**34,868.33**	**21,173.10**	**60.72**

Source photo interpretation of orthophotos Community of Madrid, Municipal register figures and developed by author

Table 6 Percentage of soil surface occupied by single-family housing over the total of residential land in 2014

Municipality	% of land occupied by single-family housing over the total residential	Municipality	% of land occupied by single-family housing over the total residential
Torrejon de Velasco	99.66	Fuenlabrada	4.52
Villalbilla	99.62	Leganes	6.54
Grinon	98.74	Torrejon de Ardoz	10.50
Villaviciosa de Odon	97.06	Parla	12.03
Torrejon de la Calzada	96.92	Madrid	14.11
Villanueva de la Canada	92.17	Alcorcon	16.22
Torrelodones	90.32	Getafe	16.48
Boadilla del Monte	89.10	Pinto	16.64
Paracuellos de Jarama	87.24	Alcala De Henares	17.33
Sanagustin de Guadalix	86.75	Coslada	30.72
Galapagar	86.40	Mostoles	34.93
Meco	85.57	San Fernando Dehe	36.11
Colmenar Viejo	79.77	Valdemoro	37.61
Arroyomolinos	79.64	Trescantos	41.47
Pozuelo Dealarcon	77.62	San Sebastian delos Reyes	48.34
El Molar	75.67	Navalcarnero	50.23
Villanueva del Pardillo	74.59	Humanes Demadrid	51.61
Brunete	74.45	Majadahonda	52.00
Las Rozas de Madrid	72.19	Arganda del Rey	54.76
Rivas-Vaciamadeid	71.88	Alcobendas	62.80

Therefore Alcobendas, with 62.80% or San Sebastián de los Reyes or Las Rozas de Madrid suffered a drop in the percentage of family-houses in this period, increasing density and with a clear evolution towards a new and denser model known as "mini blocks" due to the higher prices of development land. There are municipalities totally specialized in single-family houses. This is the case of Villaviciosa de Odon with 97.06%, Torrelodones with 90.32% or Boadilla del Monte with 89.10% and closer to the city of Madrid, Pozuelo de Alarcon with 73.08%.

The municipalities of the South and Henares Corridor (developed in the 70s and 80s decades in last century), despite the popularization of the single-family house in the nineties, retain a high density. The municipality with less land occupied with single-family house is Fuenlabrada with only 4.52% (Fig. 6). On the map (Fig. 11), is clearly seen that the Henares Corridor and the South of the Community have accomplished the lowest proportion of single-family houses along the Municipality of Madrid. This is one of the advantages for transportation in the Community;

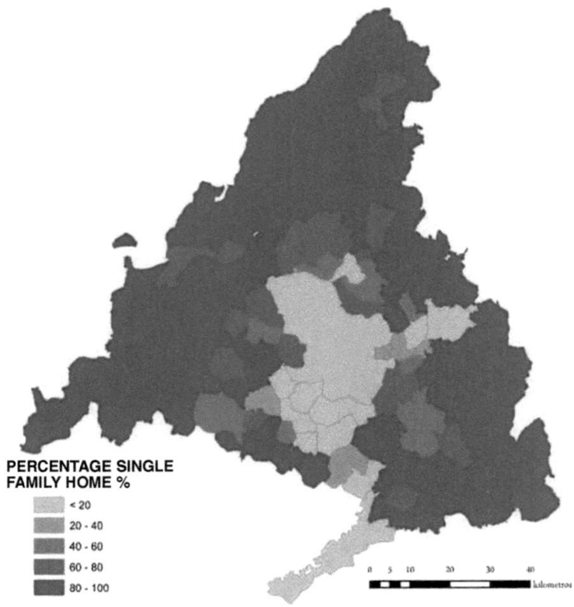

Fig. 11 Proportion of single family houses compared to totals in 2014. *Source* photo interpretation of orthophotos Community of Madrid, Municipal register figures and developed by author

the higher density allows a better sustainable mobility and at the same time guided transport systems of great capacity. In this sense, a concentric model, together with the string of compact cities, allows a suburban network more efficient system. The phenomenon of dispersion appeared mainly in the last two decades, with a stronger effect in last decade of the twentieth century, generating a transfer of public transport trips to private vehicles.

4.3 *Occupation of Industrial Space*

It is seen in Fig. 6 an industrial specialization in the period 1996–2014 in those municipalities that have increased their soil above the average of the Community, which was of 11.24% on that period, the municipalities of the south have lost industrial space, the industry mainly establish in the Southeast and Northeast and in more peripheral municipalities in the second metropolitan crown (Fig. 12).

In the West, with no industrial tissue, in the municipality of Madrid its percentage decreased by 55.12% compared to the average of the Community, the industry abandoned the capital due to the high prices of the land and and the lack of promotion from municipal administration caused the disappearance of industrial sites located in the inner city. Therefore, Industrial area is located in the South and East and the Henares Corridor, as employment does, but mainly attending to logistic activities requirements, the manufacturer industries tend to be located at the edges of the region, increasingly peripheral, lower prized land attractive for those activities.

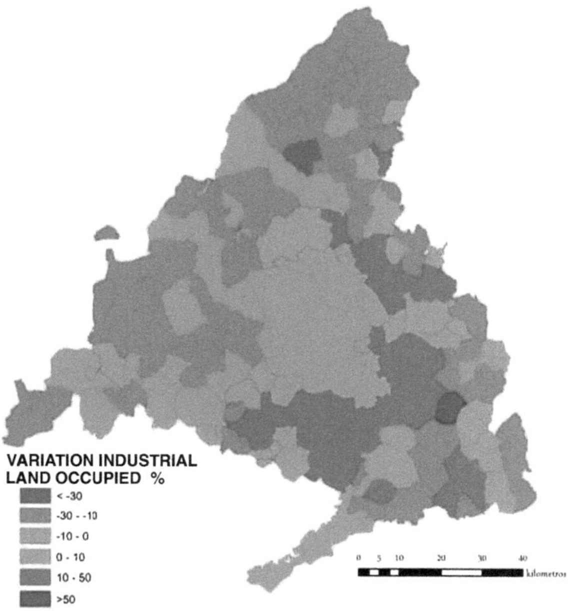

Fig. 12 Variation of the industrial constructed area with respect to the average of the Community of Madrid from 1995 to 2014. *Source* Directory of economic activity units. Institute of Statistics of the Community of Madrid

4.4 The Malls

Their appearance is relatively recent in the community, in the year 1984 there were only three malls in the capital, three store centers of shopping mall chain "El Cortes Inglés" and two malls in the West (Rozas and Majadahonda), two in Alcobendas and one in Alcala de Henares. In 2015 there were 114 shopping centers (Table 7).

Their location has always been made along the radial roads and mainly on their crosses, points of maximum visibility, but complex accessibility. One of the first was promoted from the administration (Hipercor of Alcorcon) that collapsed, in its beginnings, the A-5 road due to the forming long lines of vehicles at the entries (Fig. 13).

From an economic point of view, developers of shopping centers take advantage of the positive externalities created by the highways (large number of vehicles on the road, potential buyers) thus appropriating a public good without contributing to its cost, promoting negative synergies as collapse of the roads, congestion and time lost. At the same time, generating new necessity of public investments without being recovered none of the gains by the administrations.

Shopping malls executed in new residential developments by concentrating all commercial buildable, act as a monopolist occurs a loss of the diversity in the city and a loss of local commerce, resulting in an increase in the use of private vehicles for daily shopping.

Table 7 Large shopping centers created by periods

Statistics zones	Before 1984	Between 1984–1990	Between 1990–1996	Between 1996 and 2002	Between 2002–2008	Between 2008–2014	Total
Municipality of Madrid	9	5	8	7	8		37
North Metropolitan	2	1	5	3	5		16
East Metropolitan	1	4	5	3	5	1	19
South Metropolitan			6	7	4	3	20
West Metropolitan	2	3	4	9	1	2	21
Southwest Comunity					1		1
Total	**14**	**13**	**28**	**29**	**24**	**6**	**114**

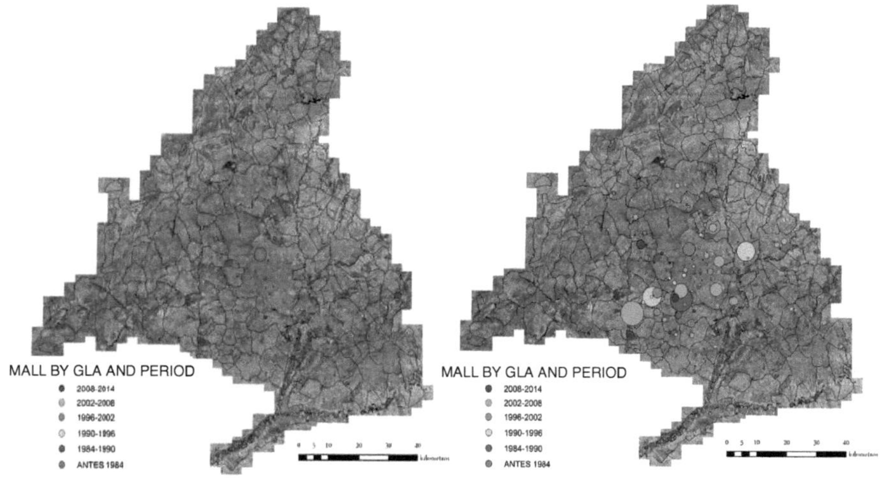

Fig. 13 Mall centers 1984 (*left*) Mall centers 2014 (*right*). *Source* Directory of Malls 2014. Spanish Association of Shopping Centres and Malls, and developed by author

Today nobody doubts that malls are the real nodes of activity during the weekends versus traditional urban centers, congested and lacking of accessibility (Table 8).

What is possible to be deduced is that mobility generated by these centers is based mostly on the use of private vehicle and that therefore the sustainability of the transport system is questioned. Fig. 6 stands out the increase in Madrid capital, during the period 2002–2008 with eight shopping centers and a gross leasable area of 462,367 m^2, with the malls of Hipercor in Sanchinarro, the Gavia in Vallecas and “Isla Azul” in Carabanchel. These malls contributed to the financing of the urbanization of PAUs but at the cost of absorbing all the commercial capacity of these monofunctional areas, creating spaces surrounded by highways, as Indovina notes “urban archipelagos” car-dependent lacking of retails of proximity (Indovina 2007). More sustainable policies would have been possible as evidenced the opening of the new tunnel Atocha-Chamartin and the construction of transport interchange of “Puerta del Sol”, which substantially changed the chances of a historical center in decline, allowing a 20 min trip to the million inhabitants in South and achieving its revitalization.

If it is analyzed the existing inhabitants in a radius of 600 m around each mall to see if it is feasible to go walking or by bicycle, it is observed that nobody lives in the vicinity ratio of 600 m in five of them. In the picture it can be observed that one of the oldest, Pinar de la Rozas, there are no appreciable population in their environment and looking to data in Table 9, they have high gross surface per capita, so the lack of critical mass cannot be covered by public transport.

A total of 26 shopping centers with around 600 m them more than 10 m^2 of surface per capita, 33 from 2 and 10 m^2, and 12 between 0.70 and 1 m^2, only the

Table 8 Surface by periods

Zones	Before 1984	Between 1984–1990	Between 1990–1996	Between 1996–2002	Between 2002–2008	Between 2008–2014	Total
Municipality of Madrid	306,157	51,496	203,019	153,175	462,367		1,176,214
East Metropolitan	14,738	254,758	237,819	56,783	192,575	7,850	764,523
North Metropolitan	3,4378	14,390	121,758	134,240	169,660		474,426
West Metropolitan	37,963	31,091	146,948	161,522	33,700	73,050	484,274
South Metropolitan			210,435	240,906	70,218	77,684	599,243
Southwest Comunity					152,887		152,887
Total	393,236	351,735	919,979	746,626	1081,407	158,584	3,651,567

Source Directory of Malls 2014. Spanish Association of Shopping Centres and Malls, and developed by author rest (30%) of shopping centers may urban considered (Fig. 14). Analyzing the centers in relation to the roads (Fig. 15) it can be seen that 55 malls, which are a distance less than 500 m from the radial road, have an average gross leasable area of 40,391 m^2 and the other 43, that are outside this range, have an average area of 24,863 m^2. The major road network creates the possibility of larger sizes of malls.

In the road network, it is shown that the A-5 has the largest area of shopping centers, followed by the A-1, A-2 and A-6 and the lower the A-3 and A-4 respectively. By ring road, it would have to be considered the proximity of the capital and the opening date being the order M-30, M-40 and M-50.

The road network shows that shopping centers are larger as much further they are of the capital, since the greater distance needs to be compensating with larger size and greater attractiveness. All these centers are newer and therefore integrate the last generation.

Shopping centers are an example of new elements, product of globalization, and changes in lifestyle, which are located in the territory, outside any transparent process of public governance, producing an appropriation of public goods in metropolitan areas and encouraging urban sprawl throughout the region, acting as a urban islands, supported by the use of private vehicles to public transport front.

As for gross leasable area can be seen that Madrid capital had 0.26 m^2 per inhabitant while Arroyomolinos 6.68 m^2 per inhabitant, San Sebastián de los Reyes 3.19 m^2 per inhabitant, Rivas VaciaMadrid 2.35 m^2 per inhabitant and Torrelodones 1.60 m^2 per inhabitant. The location of shopping centers is unbalanced in relation to the resident population since its location is based fundamentally on the road network (Fig. 15).

Table 9 Population around mall center in a radius of 600 m

Named	Surface	Population	Surface per inhabitant
Plenilunio	70,000	2612	26.8
Hipercor Pozuelo	64,429	2377	27.11
Parquesur	151,200	5247	28.82
H20	50,000	1490	33.56
Factory Las Rozas	9550	242	39.46
Moraleja Green	27,708	638	43.43
Megapark Madrid	98,140	2245	43.71
Ionorte Parque Comercial	24,850	427	58.2
Islazul	90,700	1047	86.63
Madrid Xanadu	152,887	1151	132.83
Tres Aguas	64,310	379	169.68
Gran Plaza 2	57,500	184	312.5
Parque Oeste Alcorcon	125,000	130	961.54
Carrefour Pinar de Las Rozas	29,725	10	2972.5

Source Directory of Malls 2014. Spanish Association of Shopping Centres and Malls, and developed by author

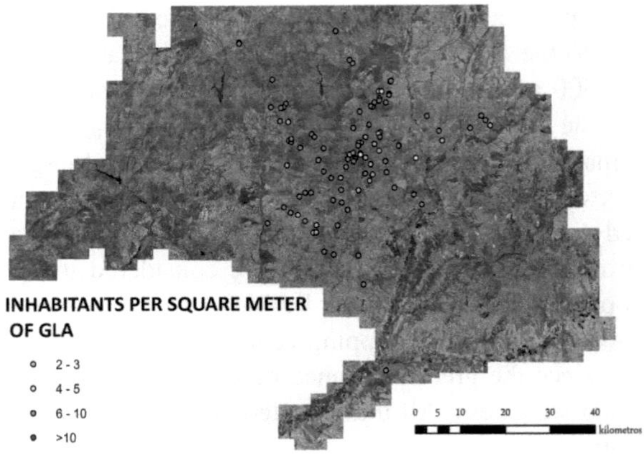

Fig. 14 Population around mall center in a radius of 600 m. *Source* Directory of Malls 2014. Spanish Association of Shopping Centers and Malls, and developed by author

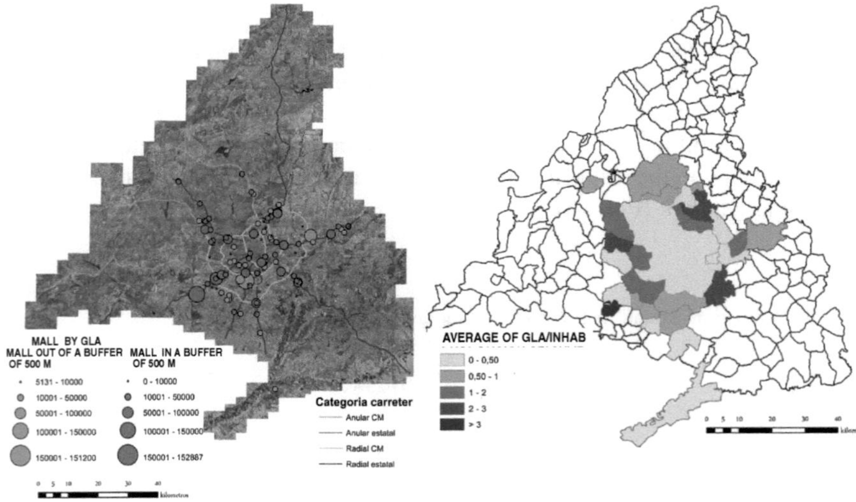

Fig. 15 Nearby commercial centers las radiales (*left*); ratio of gross leasable area per capita 2014 (*right*). *Source* Directory of Malls 2014. Spanish Association of Shopping Centers and Malls, and developed by author

5 Evolution of Densities in the Community of Madrid

In a first step, it has been taken exclusively population density on the residential surface or net residential density and, subsequently, the occupied density over the entire surface of any use, which it is called gross urban density. In 1995 the lowest density, within the functional metropolitan area, were located at the municipalities belonging to the West, throughout the first and second period, from 1995 to 2008,

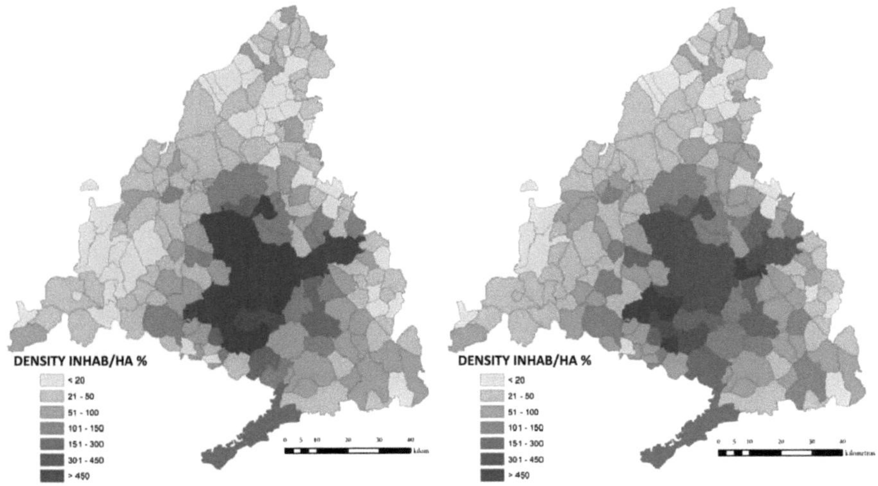

Fig. 16 Net residential density of population 1995 (*left*) net residential density of population 2014 (*right*). *Source* photo interpretation of orthophotos Community of Madrid, Municipal register figures and developed by author

all municipalities with low densities, without exception slightly increase its density. It must be taken into account the processes due to the occupancy of secondary seasonal and vacant dwellings in the variation in densities, but everything seems to indicate the emergence of a model of denser housing, maintaining the detached house typology.

The municipalities of higher densities are those who had their maximum development processes in the late sixties and early seventies, the South municipalities and Henares Corridor, where a continued decreased of densities is observed over the period considered. The territory would function as communicating vessels in which the densities tend to balance albeit very slowly, with an equilibrium point in Arganda del Rey which remains at 240 inhabitants per hectare (Figs. 10 and 16).

Analyzing the gross densities, (Fig. 17 and Table 10) depending on the entire surface occupied, obviously are lower, and even more in those municipalities that have a greater variety of uses. This occurs when cities have a higher population rate, increasing the facilities and endowments demand, and have certain density and purchasing power. Looking at these gross densities it is seen that the South is still the highest, but with a sharp decline over the three periods. The municipality of Madrid (with 135.64 ha per inhabitant in 1995) decreases to 124.91 in 2014.

If it is maintained throughout all periods the limits of existing census sections in 1995, and analyzed the variation of population, it is observed that the municipalities of the South and Henares Corridor, denser and built with multi-family housing, decreased its density over all periods, in reverse the municipalities the North and the West, with lower density, have increased its density, as it was stated in the analysis of net densities. The analysis of the residential net density of population in rings,

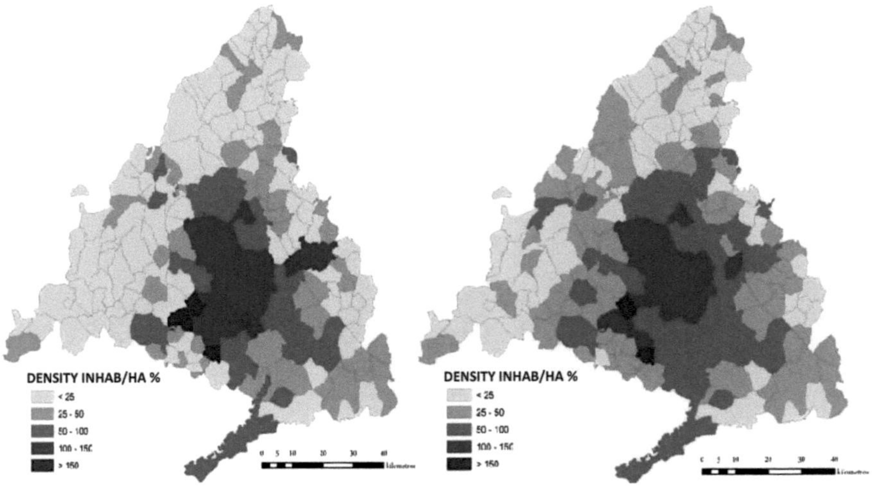

Fig. 17 Gross urban density 1996 (*left*); gross urban density 2014 (*right*). *Source* photo interpretation of orthophotos Community of Madrid, Municipal register figures and developed by author

Table 10 Net and gross densitity 2014 higher and lower

Municipality	Gross	Net	Relation
San Fernando de Henares	84.98	597.52	7.03
Human de Madrid	48.03	263.34	5.48
Leganes	99.13	503.31	5.08
Pinto	66.09	326.23	4.94
Alcala de Henares	91.41	400.76	4.38
Coslada	111.2	437.39	3.93
Alcorcon	150.46	559.18	3.72
Mejorada del Campo	71.63	257.25	3.59
Daganzo de Arriba	51.54	178.88	3.47
Ciempozuelos	60.63	209.48	3.45
Mostoles	147.85	508.94	3.44
Madrid	124.91	429.13	3.44
Meco	32.93	108.02	3.28
Somosierra	12.57	41.18	3.28
Torrejon de Ardoz	123.33	391	3.17
Fuenlabrada	121.18	379.02	3.13
Valdemoro	91.63	278.18	3.04
Galapagar	34.89	35.18	1.01
Pozuelo de Alarcon	52.17	68.17	1.31
Villalbilla	25.25	35.44	1.4

(continued)

Table 10 (continued)

Municipality	Gross	Net	Relation
Colmenar Viejo	91.5	135.79	1.48
San Sebastian de Los Reyes	68.91	102.76	1.49
Torrelodones	29.83	48.99	1.64
Brunete	67.12	111.05	1.65
Rivas-Vaciamadrid	86.08	145.12	1.69
Villanueva de la Cañada	26.41	45.39	1.72
Villanueva del Pardillo	77.25	136.7	1.77
Villaviciosa de Odon	26.91	49.58	1.84
Arroyomolinos	45.38	86.18	1.9
San agustin de Guadalix	50.4	95.85	1.9
Boadilla del Monte	26.77	51.93	1.94
Las rozas de Madrid	43.86	85.41	1.95
Torrejon de Velasco	39.92	81.55	2.04
Velilla de san Antonio	73.29	218.04	2.98

Source Municipal register figures and developed by author

starting in the "kilometer zero" point located in Community of Madrid and establishing an incremental radius of 10 km, turn to corroborate a part of what it was stated. Clearly, in the shortest distances (from 0 to 10 km), the densities are higher in the year 1995, with 468.46 inhabitants per hectare and in the ring of 40–50 km 40.01 inhabitants per hectare, nearly 10% of the first. In the year 2014 (Fig. 11) density has decreased in the first ring, being 439.58 inhabitants per hectare, it is observed a decrease of the 6.57% in the third ring (40–50 km) which represents a 18.38%. However, and what is most peculiar, the second ring (10–20 km) with densities of 256.19 inhabitants per hectare in 1995 has decreased until 204.59 inhabitants per hectare in 2014. This represents a decrease of 25.22%, while observing the rest of the rings is possible to see that in the third ring and in the fourth one, density increased in 12.55 and 16.38%, respectively. This is because in the first crown density decreases slightly at processes of dispersion of the population, expulsion of low incomes people due to the increase of housing prices in the core of the capital, the decreased of household size and the reusing of houses for business activities, but the city center keeps its attractiveness. In the second crown, where there was a high density of population living in multi-family and smaller houses, with larger families (theory), the pressure is higher and the decline is therefore greater. In the third, a settlement of new population occurs, through occupation of secondary seasonal and vacant dwellings with new smaller families (Fig. 18).

In order to analyze the influence of roads (Fig. 19) in the colonization of the territory and its densities, the territory has been subdivided in segments composed by the bisectors between radials roads, in an inner circle of 45 km. Over these areas, it has been disposed the population of the census. The result is interesting in order

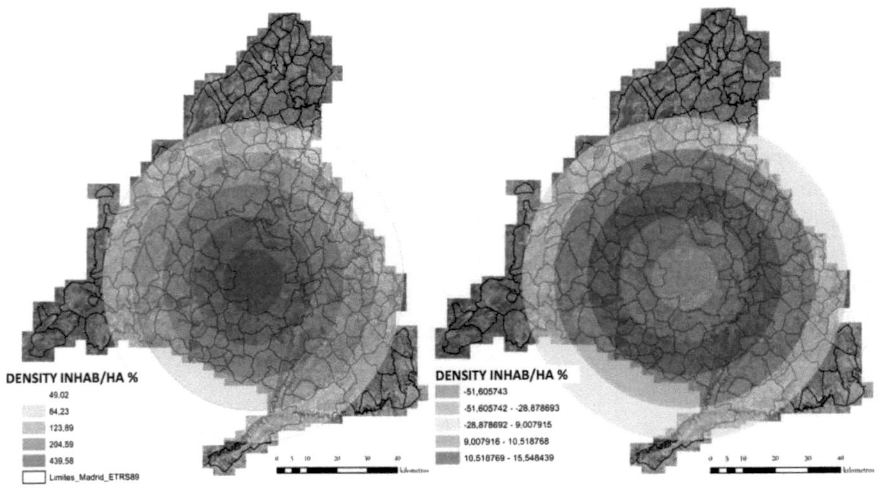

Fig. 18 Density for rings 2014 (*left*) and average of density for rings between 1995 and 2014 (*right*). *Source* photo interpretation of orthophotos Community of Madrid, Municipal register figures and developed by author

Fig. 19 Population density near the radial roads. *Source* photo interpretation of orthophotos Community of Madrid, Municipal register figures and developed by author

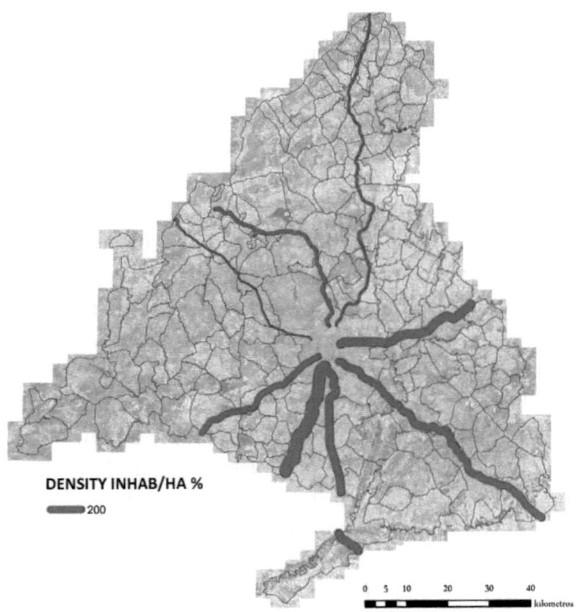

to see the functional specialization of the road. The highest density occurs in the South (road A-42) with 247.69 inhabitants for hectare, obviously as it is concentrated there all the cities of Southern with certain entity. The same pattern is observed in the A-4 motorway. Besides density is radically lower in the two roads of the territories specializing in single family houses (A-1 and A-6). This last motorway A-6, with the lower density of 63.80 inhabitants for hectare, in this value would also have to take into account the influence of protected environmental areas such as the Monte del Pardo.

6 The Territorial Imbalance, Location of Employment

The Directory of Units of Economic Activity of Statistics Institute of Autonomous Community of Madrid shows a series from the years 1998 to 2010 that allow an analysis during period of the housing bubble, but it is not possible to deepen in the effect of the economic crisis. The mobility survey developed in 2014 easy shows that during the economic crisis the capital of the Community was the more employment destroyed. In 1998, the 63% of the total employment in the region were located in the Capital and within the center of it the 36%, in the metropolitan area the 25%, where the metropolitan South accounted for 12%. Nevertheless, in 2008 the weight of the Capital had fallen to 58% but in the center of the Capital the employment had increased to 43.5%, therefore the area that loses more employment was the periphery of the Capital, meanwhile the metropolitan area increase to 29%.

Employment in the capital, represented 58% in 1998, increasing to rise 63% in 2008, while in the crown remained at 29%, i.e. increase more the premises in the crown but not the employment, where the premises are smaller, with more small and medium-sized enterprises. Comparatively the periphery loses weight in employment rates compared to the center. This caused an imbalance increased to the weight of housing and population in the periphery. That means a dispersion of activities towards the crown and periphery, but a concentration of employment in the municipality of Madrid takes place.

In the year 1998, the number of employment in the Community was 1,474,682 and the number of the premises 208,326, which represents an average of 7.08 employments by premises. Low figures if compared with other metropolitan areas in Europe. The poor figures are a consequence of the lack of big companies, a characteristic of the dynamics of the economy and the closure of the scant industry that there exists, due to the economic crises from the year 1973 to 1991, therefore the activity was highly fragmented. Analyzing the territories, the smaller values are located at the southern outskirts of the capital, with 3.53 employments, and the non-metropolitan municipalities.

During the period 1998–2008 the number of premises increased in 64,834 and the number of employees in 1,164,540, representing in the whole of the community increments of 31 and 78% respectively, resulting in a slight increase in the size of premises until 9.42 jobs per premise.

By areas, the largest increase in size of premises occurred in the northwest outskirts of the municipality with 14.89 jobs by premise; the same behavior is seen in the eastern periphery, around Campo de las Naciones. Increases in the metropolitan periphery were higher, pointing to a dispersion of activities with an

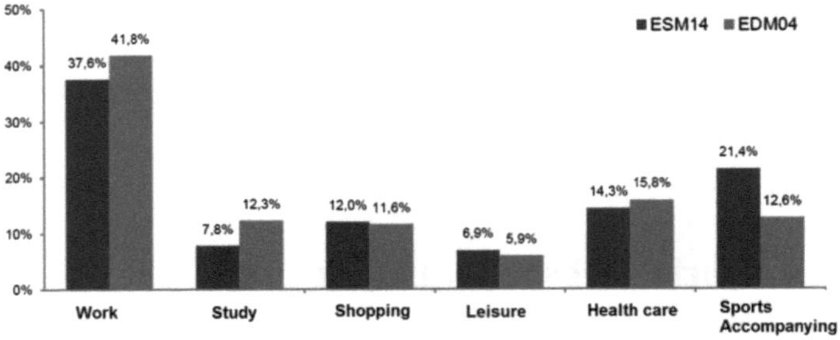

Fig. 20 Distribution of trips for priority motives (esm 14). *Source* CRTM

increase of 51% in premises and 135% of employment in absolute value. It would be the South crown with 9747 premises more and with an increase of 139,542 jobs, but in relative terms, the largest increase occurred in the metropolitan west with increases of 98% in local, 236% in employment and 102,622 jobs in absolute values respectively.

Trips for work are the most important because of its number (Fig. 20). Everything points to a problem for public transport, due to the dispersion of activities to the crown but with less employment by premises than in the center of the capital, and at the same time, due to the smaller size of the premises, occurs a lower concentration of trips. Like other metropolitan areas of capital city of developed countries, the industrial activity in the Community is disappearing, becoming to more peripheral. By sectors, the employment focus first in the financial sector, then services to companies and subsequently administration and trade, though the latter moves with the population and its purchasing power towards the metropolitan crowns mainly to the West. It has been observed, during the analyzed period, a dispersion trend of activities in the territory of the metropolitan area, moving to the periphery, albeit with a lower job creation in relation with the premises.

This model was broken with the crisis, as already pointed. The center of the capital loses a significant number of jobs, and that will be an influence in peripheral trips for work purposes by public transport. It has been analyzed the evolution of employment and the employed population located in the territory through data available in the Transport Consortium Survey Mobility of years 88, 96 and 2004. The trips for work purposes are the most important in the Community and although its importance, due to the crisis and unemployment, declined slightly as the main reason to trip.

In the case of the municipality of Madrid for those years, the ratios between employment and working population are 1.12, 1.19 and 1.20 respectively i.e. surplus in employment and a positive development. However, the imbalance between neighborhoods is remarkable between the center and the east with peaks around 2.5 employment versus working population ratio and at the other end, the Southeast, around 0.40. Evolution is negative in the center and in the district of

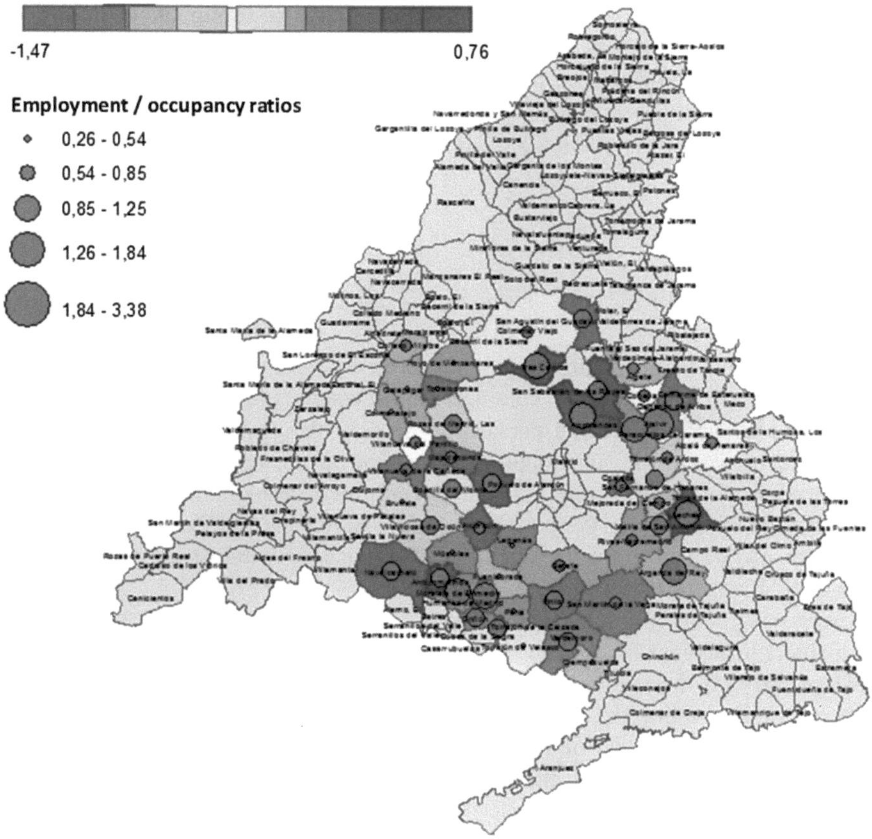

Fig. 21 Employment/occupancy ratios and evolution of the ratio 1988–2004. *Source* CRTM

Salamanca, but the values are always above 1, and for Chamartin, Chamberí and Barajas the percentage increases as the employment did.

In the Community there were only values for the first crown available (Fig. 21), in the inquire of 2004 (EDM04), highlighting Ajalvir with 3.39 jobs/employed population ratio, Humanes with 1.75, Moraleja de Enmedio 1.78, Tres Cantos 1.84, the West Arroyomolinos with 1.20 and Pozuelo 1.28, being also positive in Alcobendas and San Sebastián de los Reyes.

Looking the map (Fig. 21) in a period with widespread employment growth it is remarked in blue circles the ratio of employment and in uniform color the index variation. Employment remains in the capital and increases in the closest municipalities, but especially in the municipalities where employment is concentrated in the service sector (such as Alcobendas, San Sebastián de los Reyes or Pozuelo de Alarcón), but to a lesser extent in the industrial towns of the South and the Corridor, which reaches lower ratios due to the loss of jobs in the industrial sector.

Increases in employment in the central zone had a strong effect during the economic boom. As it was already noted, then occurred a strong loss of employment (ESM14), the dispersion of activities to the periphery and the loss of jobs in the center caused a loss of the relations among radials of periphery/center, and increased the cross relations with less, endowed with poor public transport infrastructure (Table 11).

7 Private Vehicles

Increase in the use of private vehicles has been a consequence of multiple factors: the creation of new roads to colonize new lands in the metropolitan area, the increasing of income level, dispersion of the activities towards the periphery, the lack of density and the functional specialization of the territory. In short, a new territorial model. A clear effect that it is confirmed throughout this article is the increase in population in the periphery as it was reflected in the synthetic survey Mobility EMS/14 of 2014, the increase of population within the regional crown 2004–2014 was a 31.59% (Fig. 22).

Improving economic capacity and more dispersed residential new model has led to a sharpen increase in cars ownership rate, from 0.93 by household in 1996 to 1.41 in 2014. If the analysis is performed by crowns, it is observed the sharpen increase of them along the time and, moving away from the center, being the index in the metropolitan crown of 1.55 vehicle/household. The increase versus time is higher in the urban periphery and this is consistent with the change in the territorial model (Fig. 23).

The ESM/14 confirms, as indicated throughout the article, a shift of activities and hence employment to the periphery, as shown in the Table 12, it is noteworthy the loss of employment in the central zone from the year 2008, where had increased until that year according to the existing data.

The crisis deteriorate the employment in the core of the capital city (Fig. 24) and increased slightly in the periphery, even despite the crisis this evolution of employment to the periphery leads to greater mobility, and the absence of nodes of concentration of employment, therefore increase the transversal mobility.

It must be taken into account other factors, such as the home ownership, which represented 96% in 2007 in the Community of Madrid, and went down to 93% with the crisis in 2015. However, compared with any European metropolis is extremely high. Keep in mind that changing in housing has a high tax and administrative cost, increasingly more contracts are temporary or part time, this entails an increase in transverse mobility (Fig. 25).

As shown in the Fig. 26 the number of trips on all modes grew until the beginning of the crisis coming in 2004 to 2.25 trips per capita (14–68 years).

The motorized trips per capita/day in 1974 represented 1.09 with a significantly increasing until 2004 with 1.63. It is observed a sharp decline of motorized travel in the year 2014, being 1.39 trips per capita. The decreasing the number of trips

Table 11 Comparison between the ratio employment in the municipality and labor force resident 2014 (ESM14)

Corridor	Employment	Ratio employment/work force	Corridor	Employment	Ratio employment/work force
Alcobendas	94,752	3.5	Villanueva de	2,484	0.1
Getafe	64,306	2.4	Villanueva	2,606	0.1
Pozuelo de	61,493	2.3	Torrelodones	3,109	0.1
Alcalá de	61,308	2.3	Navalcarnero	3,217	0.1
Tres Cantos	45,665	1.7	Colmenarejo	3,284	0.1
Alcorcón	44,678	1.6	San Agustin	3,381	0.1
Rozas de	39,357	1.4	Velilla de San	3,401	0.1
Fuenlabrada	39,220	1.4	Villaviciosa	4,886	0.2
Leganés	38,341	1.4	Arroyomolino	5,265	0.2
Torrejón de	37,854	1.4	Galapagar	5,288	0.2
San Sebastián	37,161	1.4	Ciempozuelo	5,999	0.2
Mostoles	30,836	1.1	Paracuellos	6,062	0.2
Arganda del	27,997	1.0	Pinto	7,385	0.3
Rivas-	27,366	1.0	Algete	7,641	0.3
Coslada	26,907	1.0	San Martin de	7,902	0.3
Boadilla del	24,040	0.9	Parla	8,645	0.3
Valdemoro	21,349	0.8	Humanes de	9,873	0.4
San	16,783	0.6	Colmenar	10,569	0.4
Majadahonda	14,752	0.5	Collado	10,979	0.4

Source Regional Transport Consortium of Madrid

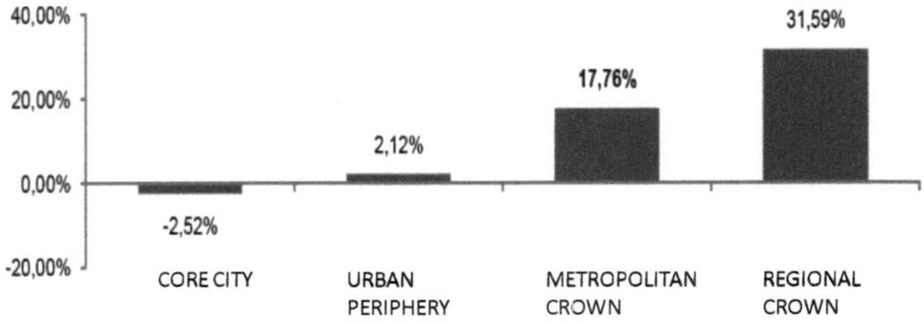

Fig. 22 Variation of population between 14 and 80 years old in the transport crowns 2004–2014 (ESM14) of the Community of Madrid. *Source* Regional Transport Consortium Madrid

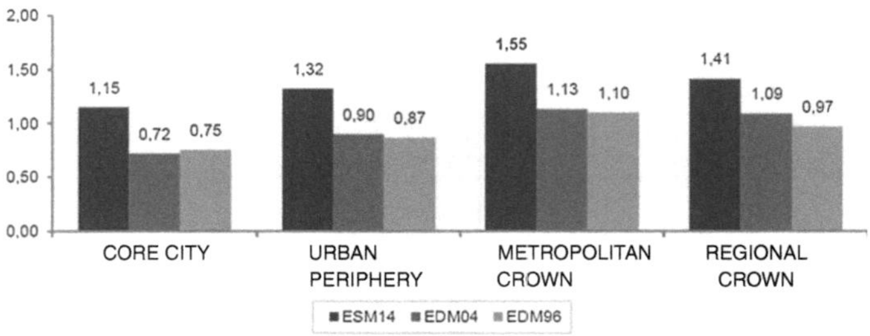

Fig. 23 Comparison between the family motorization index 1996–2014 per zones. *Source* Regional Transport Consortium Madrid

Table 12 Comparisons by crowns of the employment in the community of Madrid 1996–2014 (ESM14)

	ESM14	%	EDM04	%	EDM96	%	% Variation 14/04
Core city	450,220	16.20	959,877	34.70	686,020	38.70	−53.10
Urban periphery	993,485	35.80	815,648	29.50	508,844	28.70	21.80
Metropolitan crown	1,148,805	41.40	893,982	32.40	516,384	28.70	28.50
Regional crown	184,112	6.60	93,561	3.40	63,237	3.50	96.80
Total	**2,776,623**	**100.00**	**2,763,067**	**100.00**	**1,774,494**	**100.00**	**0.50**

Source Regional Transport Consortium Madrid

affected in a more important way to the motorized trips for the reasons stated before. Non-mechanized trips lesser extent decrease, going from 0.62 to 0.60 trips per capita. The overall travel figures in the Community of Madrid has increased significantly until 2004 moving from 12,873,643 trips to 12,925,051 trips in 2014, a slight increase of 0.4%. But decreased motorized trips in 2.82% and increased non-motorized (Table 13).

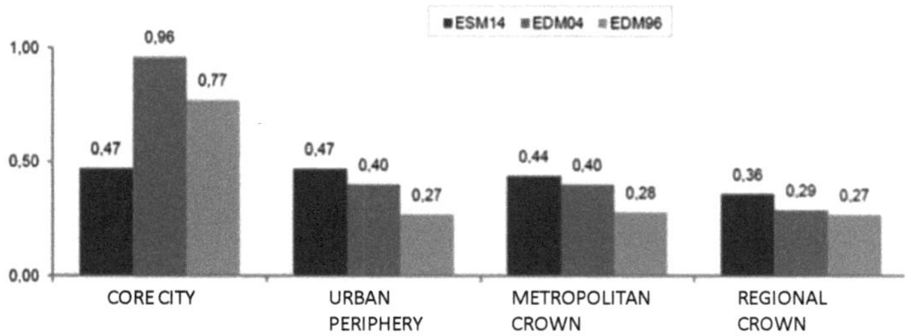

Fig. 24 Comparison of the employment-to-population ratio 1996 to 2014 (ESM 14). *Source* Regional Transport Consortium Madrid

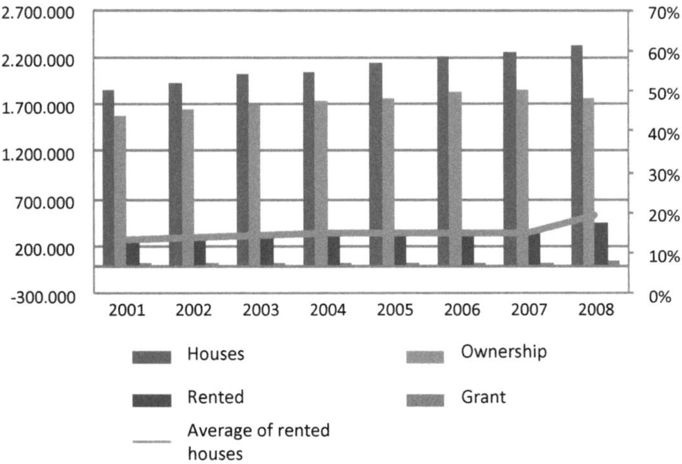

Fig. 25 Evolution of housing ownership. *Source* Civil Works Ministry

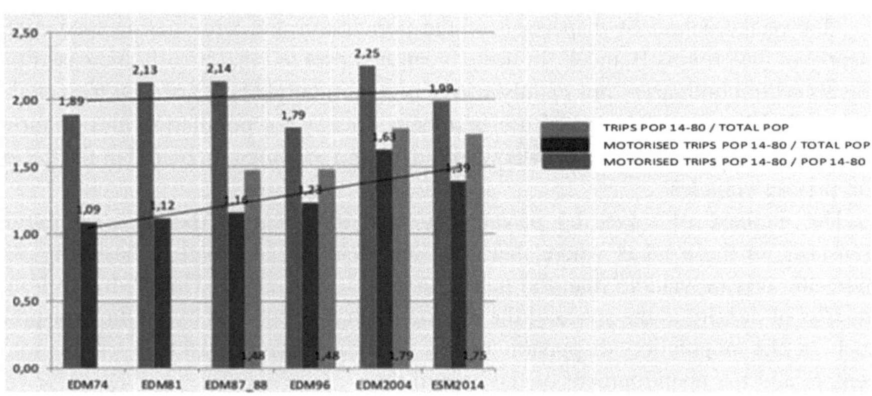

Fig. 26 Evolution of mobility indicators. *Source* Regional Transport Consortium Madrid

Table 13 Comparison between the total

	ESM14	%	EDM04	%	Variation 14/04
Motorised	9,058,931	70.10	9,321,884	72.40	−2.82
Non motorised	3,866,120	29.90	3,551,759	27.60	8.85
	12,925,051	100.00	12,873,643	100.00	0.40

Table 14 Evolution of motorized mobility per inhabitant

Enquire	Total population	Functional area	Motorised trips	Motorised trips/per inhabitant	% Variation
EDM74	3,755,500	Metropolitan area	4,080,787	1.09	–
EDM81	4,196,265	Metropolitan area	4,688,800	1.12	2.8
EDM87	4,517,697	Crown A and B	5,617,247	1.16	3.8
EDM96	5,022,289	Total Comunity	6,197,829	1.23	6.3
EDM04	5,718,942	Total Comunity	9,321,884	1.63	32.5
ESM14	**6,495,551**	**Total Comunity**	**9,058,931**	**1.39**	**−14.7**

Source Regional Transport Consortium Madrid

In the total non motorized trips, an increase of 8.85% that fails to be compensated by the increase in population, a decrease of the number of trips per capita of 4.16% is observed. Trips per person decrease; this can be attributed to the sharp increase in unemployment and reduced income levels (Table 14).

In the period 2004–2014 as it is shown in Table 15, motorized trips in private vehicles decreased in the inner of the A crown, where there is more density of population and better public transport. Between the crown A, see Fig. 27 while the rest of the crowns increased. This was a result of the dispersion of residential growth and activity, Rising the trips in the whole crown A in a 4.16%, mainly due to trips to the crown B but also to the C and proportionally greater when going away from the center.

This phenomenon of dispersion of the population and activities is seen more clearly in the crowns B and C. In these it can be observed that greatly increased the travels within both areas but mainly with a high number 411,523 of new trips in the crown B. The consolidation of the periphery, increasing population and employment, led to a growth in the crowns within them and between them and in the use of the private vehicles.

The increase of travel by private cars had its effect in the trips by public transport, in those areas where increased the mechanized travel, it has increased more the private vehicle than the public transport (Table 16). This is due to both, the loss of density in the space of the metropolitan area covered by public transport, such in the population dispersion and/or activities, the appearance of the mall centers and the abandonment of the central city.

Table 15 Comparison between trips by crowns in private car 2004–2014 (ESM14)

	ESM14	%	EDM04	%	Difference	% variation	% Variation
Mobility private vehicle	5,211,840	100.00	4,527,159	100.00	684,681	15.12	15.12
Crown A Madrid—Crown A Madri d	1,458,534	27.99	1,554,291	34.33	−95,757	−2.12	−6.16
Crown A Madrid—Crown B Metropolitan	1,152,035	22.10	989,739	21.86	162,296	3.58	16.40
Crown A Madrid—Crown C Regional	107,744	2.07	79,478	1.76	28.266	0.62	35.56
Crown A Madrid—outside Community of Madrid	33,790	0.65	18,612	0.41	15,178	0.34	81.55
Total	2,752,103	52.80	2,642,120	58.36	109,983	2.43	4.16
Crown B Metropolitan—Crown B Metropolitan	1,901,909	36.49	1,490,386	32.92	411,523	9.09	27.61
Crown B Metropolitan—Crown C Regional	229,863	4.41	151,512	3.35	78,351	1.73	51.71
Crown B Metropolitan—outside Community of Madrid	65,632	1.26	37,914	0.84	27,718	0.61	73.11
Total	2,197,404	42.16	1,679,312	37.11	517,592	11.43	30.31
Crown C Re gional—Crown C Regional	241,362	4.63	191,159	4.22	50,203	1.11	26.26
Crown CRegional—outside Community of Madrid	16,583	0.32	12,394	0.27	4,189	0.09	33.80
Total	257,945	4.95	203,553	4.50	54,392	1.20	26.72
Outside Community of Madrid—outside Community of Madrid	**4,337**	**0.08**	**1,674**	**0.04**	**2,713**	**0.06**	**162.07**

Source Regional Transport Consortium Madrid

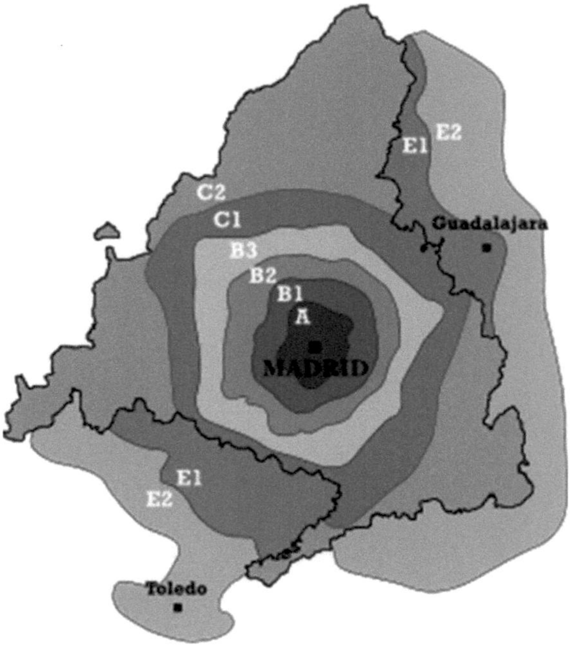

Fig. 27 Transportation areas. *Source* Regional Transport Consortium Madrid

Table 16 Comparison of the share public/private in the relationship between crowns 2004–2014

	ESM14		EDM04	
Mobility motorised	Public (%)	Private (%)	Public (%)	Private (%)
Crown A Madrid—Crown A Madrid	59.27	40.73	61.85	38.15
Crown A Madrid—Crown B Metropolitan	42.28	57.72	46.98	53.02
Crown A Madrid—Crown C Regional	47.07	52.93	52.44	47.56
Crown A Madrid—outside Community of Madrid	31.03	68.97	24.21	75.95
Total	52.83	47.17	57.05	42.95
Crown B Metropol itan—Crown B Metropolitan	16.88	83.12	25.02	74.98
Crown B Metropol itan—Crown C Regional	8.31	91.69	19.79	80.21
Crown B Metropol itan—outside Community of Madrid	23.05	76.95	13.50	86.50
Total	16.28	83.72	24.34	75.66
Crown C Regional—Crown C Regional	4.08	95.92	6.47	93.53
Crown C Regional—outside Community of Madrid	8.78	91.22	3.23	96.77
Total	4.40	95.60	6.28	93.72
Outside Community of Madrid—outside Community of Madrid	13.62	83.38	0.00	100.00

Source Regional Transport Consortium Madrid

In the synthetic mobility survey (2014) a loss in the whole Community is observed (Table 17) in trips in public transport in favor of private vehicle (39 vs. 28% of public transport) This variation of trips in public transport and its proportion reflected more clearly that it is determined by the territorial model formed during the last two decades.

And if it is compared the percentage of motorized trips doing in public transport against private car, it had declined since 1986 from 54.08 to 40.47%, despite growing the number of trips in public transport in more than 20% (Table 18).

The total number of trips of the survey 2014 is 12,925,051 trips/day, equivalent to 1.99 trips per person. It is noteworthy that due to the high density of the cities that make up the metropolitan area, the number of trips by foot is more significant than in other European cities. Although the model changes cause a decrease, although in the period crisis walking trips have increased in the Madrid central area, and in the 2014 survey in this central area still represent 36% while in the crown B 39%. Since 1986, the percentage of total foot travel drop from 47.6 to 29.9% in 2014 according to synthetic survey (Table 17). That represents a remarkable decrease of the total number of trips on foot, reflecting the change of territorial model and the way of life. It has a harmful effect in the health, in the way of sharing the public space of the city as a place of social relations, intergenerational relationship and increasing factors of social inequality in the region. If it is made an analysis by radial corridors, total trips in each radial, it is seen that the number of mechanized trips in the 2004 survey increased without exception, but observed that the corridor that more increases is the A-6, as a response to the observed dispersion of population, occupying single-family houses, but decreasing trips due to the economic crisis, like in the A-4 and in A-42 (Fig. 28).

If the percentage of travel in public transport is analyzed against private vehicle it shows that the number of trips is clearly favorable to the public transport in the A-5, A-4, and Henares Corridor although with a strong decline. Moreover it is noted that except for the M-607, public transport decreases constantly over time (Fig. 29).

The lowest percentages of trips by public transport are located in the A-1, due to lack of efficient public transport. That has not seen a significant change with the new to the subway and railway, along the A-6 and the A-3 with a scattered population and housing model house versus the dense pattern of Southern cities. In the A-3, the appearance of the subway is observed in the mobility survey 2004, but this trend did not have continuity. Figure 30 clearly reflects the territorial model in terms of dispersion and densities in each of the radial corridors.

Table 17 Comparison of modal split between EDM 88, EDM 96, EDM 2004 and ESM14

Mode	EDM88		EDM96		EDM2004		EESM14	
	TRIPS	%	TRIPS	%	TRIPS	%	TRIPS	%
Walk	5,283,880	47.6	3,892,556	37.2	4,511,080	31.1	3,866,120	29.9
Public transport	3,037,979	27.4	3,311,533	31.6	4,581,874	31.6	3,666,419	28.4
Private transport	2,270,184	20.5	2,954,462	28.2	4,952,645	34.1	5,022,890	38.9
Others	499,649	4.5	313,549	3.0	465,799	3.2	369,623	2.9
Total	**11,091,692**	100.0	**10,472,101**	100.0	**14,511,398**	**100.0**	**12,925,051**	**100.0**

Source Regional Transport Consortium Madrid

Table 18 Mechanized trips in the community of Madrid

Mode	EDM88		EDM96		EDM2004		EESM14	
	TRIPS	%	TRIPS	%	TRIPS	%	TRIPS	%
Mobility motorised	5,617,247	50.64	6,197,829	59.18	9,321,884	64.24	9,058,931	70.09
Public transport	3,037,979	54.08	3,311,533	53.43	4,581,874	49.15	3,666,41 9	40.47
Private transport	2,270,184	40.41	2,954,462	47.67	4,952,645	53.13	5,022,890	55.45

Fig. 28 Trips to Madrid. *Source* Regional Transport Consortium Madrid

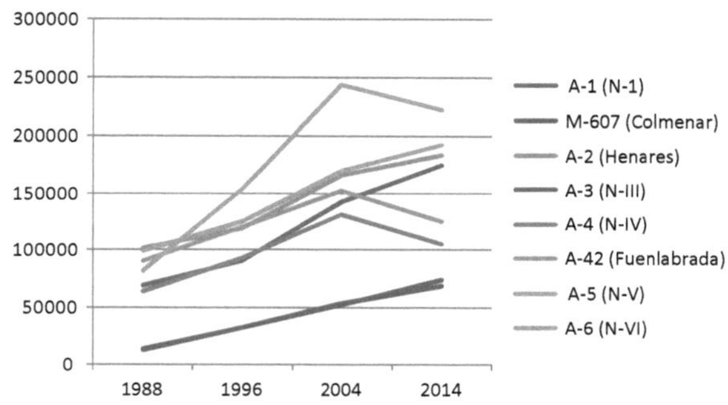

Fig. 29 Percentage of trips by public transport to the core of the capital city by corridors. *Source* Regional Transport Consortium Madrid

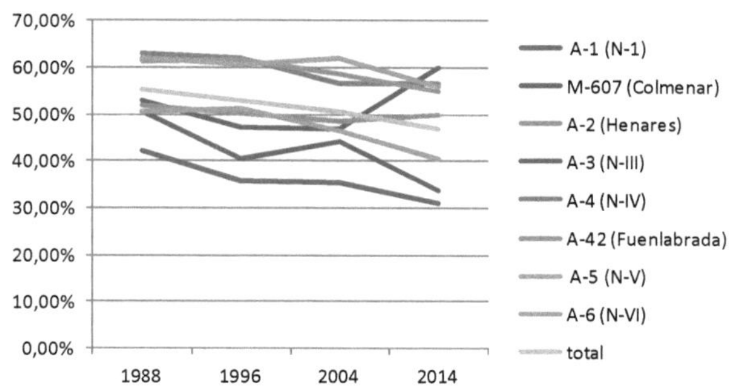

Fig. 30 Percentage of total trips attracted to the center of Madrid by corridors. *Source* Regional Transport Consortium Madrid

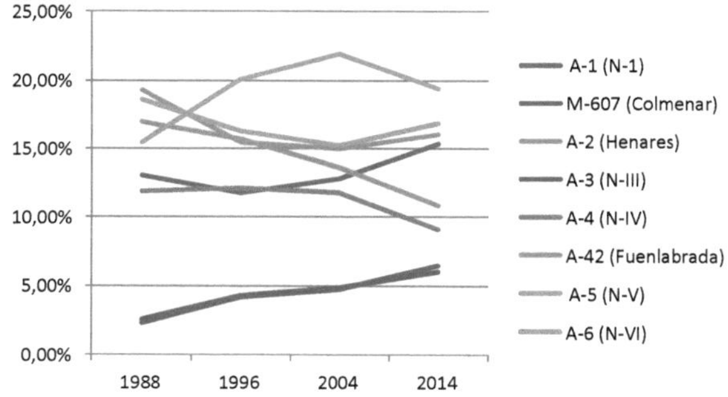

8 Conclusions

Throughout this text has it has been analyzed the sharp rise in land occupied in Community of Madrid, increasingly in more peripheral municipalities. Land consumption not only for homes, but also for infrastructures to support that growth, close to the roads, in areas of high environmental value, as the West or North, areas lacking an efficient public transport network.

It has been reflected a widespread decrease in density population in the central areas, such as capital or large cities of the South and the Henares Corridor, spaces historically with public transport of high capacity, by displacement of the population to more remote areas, which has accompanied the decline the use of public transport.

The growth of land occupied, in a model of low density and single-family home, was especially remarkable in the last decade of the previous century, in remote areas or on the outskirts of existing cities, far from endowments and services urban and without a policy for your generation in order to serve the population of arrival. With a growing occupation in the West and North of the region, as it has been noted, but with a transferred model due to its popularization to a more peripheral areas, depending on the price of land towards the south and east on the edge of the regional.

The dispersion of activities throughout the territory, product of high land prices in central areas, has led to less diversity, while the appearance of the malls has caused a drop in trade in the historic city and what it means to livability loss of these spaces.

The occupation of large areas of land of housing, or the appearance of peripheral nodes of businesses or industrial, isolated nodes and unrelated, rather than through the road network and outside the urban tissue, has led to zoning of the territory with the effect of weight loss of public transport.

The increase in the number of private vehicles, reflected in its use. It has a clear effect in the more peripheral metropolitan crowns, product of urban model. It the loss of the radial model has been produced towards a oil stain model, of islands in the territory, generating a greater need for new ways to connect, due to increasing the transversal mobility, which causes an increase of the private car trips.

The model developed in recent decades, has favored the sprawl of cities, with a peripheral growth depending on the prices of land. This model has increased the dependency in the use car and had its support in the development of the road network.

In this process a greater imbalance employment versus residence occurs, due to the specialization of the territory, by means of the location of activity nodes peripheral outside the existing city or of the new residential growth. The loss of industrial employment in some cities in the South and its offshoring partly to the border of the region and, at the same time, the concentration of the financial sector on North and West periphery created more imbalances and therefore contributed to increase travel and travel for work purposes.

Although it is difficult, it would be necessary to balance the relationship between residence and employment through a set of public policies. En the last two decades the dispersion of activities supported by the road network, the relocation of industry outside residential areas, further away on the edge of the region due to: the land prices, the scarce urban regulations and lower salaries, along with the emergence of malls centers has generated a loss of employment close to neighborhood and all this has increased the territorial unbalance.

It is necessary to link both, regional and urban planning with an integrated planning of the transportation system, promoting density in general and especially around transport interchange, avoiding urban expansion by encouraging the regeneration of the existing city. It is key promoting the diversity of uses within residential spaces and the introduction of designs that favor those modes of transport more sustainability such as pedestrian or bicycle. These measures will promote a better quality of the air, less consumption of resources and time in transport, the recovery of spaces of relation in our cities, in addition a better quality of life for all citizens.

References

Burchell, R., Downs, A., McCann, B., & Mukherji, S. (2005). *Sprawl costs: Economic impacts of unchecked development*. USA: Island Press.

Burchell, R. W., Lowenstein, G., Dolphin, W. R., Galley, C. C., Downs, A., Seskin, S., ... & Moore, T. (2002). *Costs of sprawl—2000*.

Calthorpe, P. (1993). *The next American metropolis: Ecology, community, and the American dream*. Princeton architectural press.

Cervero, R., & Kockelman, K. (1997). Travel demand and the 3Ds: Density, diversity, and design. *Transportation Research Part D: Transport and Environment, 2*(3), 199–219.

Consorcio Regional de Transportes de Madrid. (2016). Encuesta Sintética de Movilidad 2014. http://crtm.gfk.es. Accessed January 27, 2017.

Europea, A. A. (2006). *Urban sprawl in Europe, the ignored challenge*.

Giampino, A. (2010). ¿ Metrópolis dispersas? Cuadernos de investigación urbanística, (72).

Gibelli, M. C. (2007). Los costes económicos y sociales de la ciudad de baja densidad. In La ciudad de baja densidad: lógicas, gestión y contención (pp. 277–306). Diputació Provincial de Barcelona.

Indovina, F. (2007). *La ciudad de baja densidad: lógicas, gestión y contención*. Barcelona: Diputación de Barcelona.

Neal, P. (Ed.). (2003). *Urban villages and the making of communities*. Taylor & Francis.

Chapter 7
Private Vehicle and Greenhouse Gas Emissions in Spain: A War Without Winner

Francesc Clar and Roberto Álvarez

Abstract The region of Madrid, which includes the capital and surrounding towns, has increased the number of private vehicles up to almost 4 million cars in 2016, plus about a million trucks, motorcycles and vans. Current situation in Barcelona is similar, with up to 2.5 million cars. Both countries (two of the most urbanized cities in Spain) have begun thinking about taking drastic steps to reduce car exhaust pollution. In fact, in 2016 some actions (insufficient and unpopular) have already been carried out. It can be said that a war against pollution has begun. Will there be winners? And how have they come to this limit situation? This work tries to show the evolution of greenhouse gas emissions in these representative countries, including factors as urbanism and economic crisis.

1 Introduction

Every day, citizens of urban cities breathe an atmosphere whose concentration on harmful particles has grown dramatically and uncontrollably throughout the Twentieth century. Internal Combustion Engines (ICE) used in cars, trucks, ships, trains and planes mostly burn petroleum based fossil fuels and strongly contribute to global warming and air pollution. ICE emits different gases and particles included in the well-known. Greenhouse Gases (GHG). In order to use a standard unit that express the impact of each different particle emissions in terms of the amount of CO_2 that would create the same effects, an emitting source that includes many different greenhouse gases is commonly expressed as a single number (CO_{2eq}). Conversion of the various gases into equivalent amounts of CO_2 is based in the Global Warming Potential (GWP), which describes the impact of each

F. Clar (✉)
Electric Vehicle (Z.E), Renault Group, Madrid, Spain
e-mail: ingenieroxclar@gmail.com

R. Álvarez
Department of Engineering, Universidad Nebrija, Madrid, Spain
e-mail: ralvarez@nebrija.es

R. Álvarez Fernández et al. (eds.), *Carbon Footprint and the Industrial Life Cycle*, Green Energy and Technology, DOI 10.1007/978-3-319-54984-2_7

greenhouse gas relative to CO_2. The weighting factors currently used are the following: carbon dioxide = 1, methane = 25, nitrous oxide = 298, and sulphur hexafluoride = 22,800.

Figure 1 shows the global distribution of direct emissions by economic sectors, showing that the transport sector produced 7.0 $GtCO_{2eq}$ of direct GHG emissions (including non-CO_2 gases) in 2010 and hence was responsible for approximately 23% of total energy-related CO_2 emissions (IPCC 2015) (Figs. 2 and 3).

Looking at European Union, in 2014, greenhouse gas emissions in the EU-28 were down by 22.9% compared with 1990 levels, near to surpass its 2020 Kyoto Protocol target, which is to reduce GHG emissions by 20% by 2020 and by 40% by 2030 compared with 1990 (Eurostat 2016).

The Kyoto Protocol represented the most important global environmental international agreement ever. In fact, the Kyoto Protocol (adopted in Kyoto on December 11th, 1997 and went into force on February 16th, 2005) was a mandatory call for states to work together on a various environmental standards focusing in the reduction of pollution. In order to achieve Kyoto protocol targets, the EU agreed that average of emissions from new passenger cars should not exceed 120 g of CO_2 per km by 2012 (Directive 93/116/EC). Nevertheless, in 2005, the European parliament replaced the voluntary commitments for mandatory since major automakers figures were unsatisfactory. Later, in 2008, new schedule to phase in the CO_2 emission standard with increasing percentage levels from 2012 to 2015 for the 130 g/km. By 2021 the fleet average to be achieved by all new cars is 95 g of CO_2 per kilometre. This means a fuel consumption of 4.1 l/100 km of petrol or 3.6 l/100 km of diesel. The 2015 and 2021 targets represent reductions of 18 and 40% respectively compared with the 2007 fleet average of 158.7 g/km.

Fig. 1 Global greenhouse gas emissions

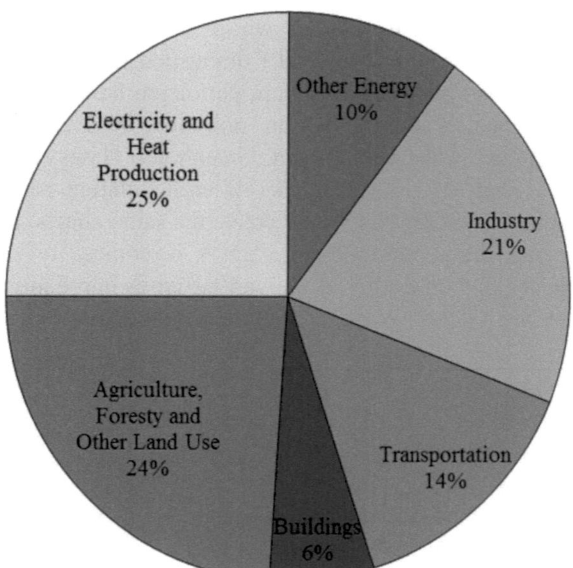

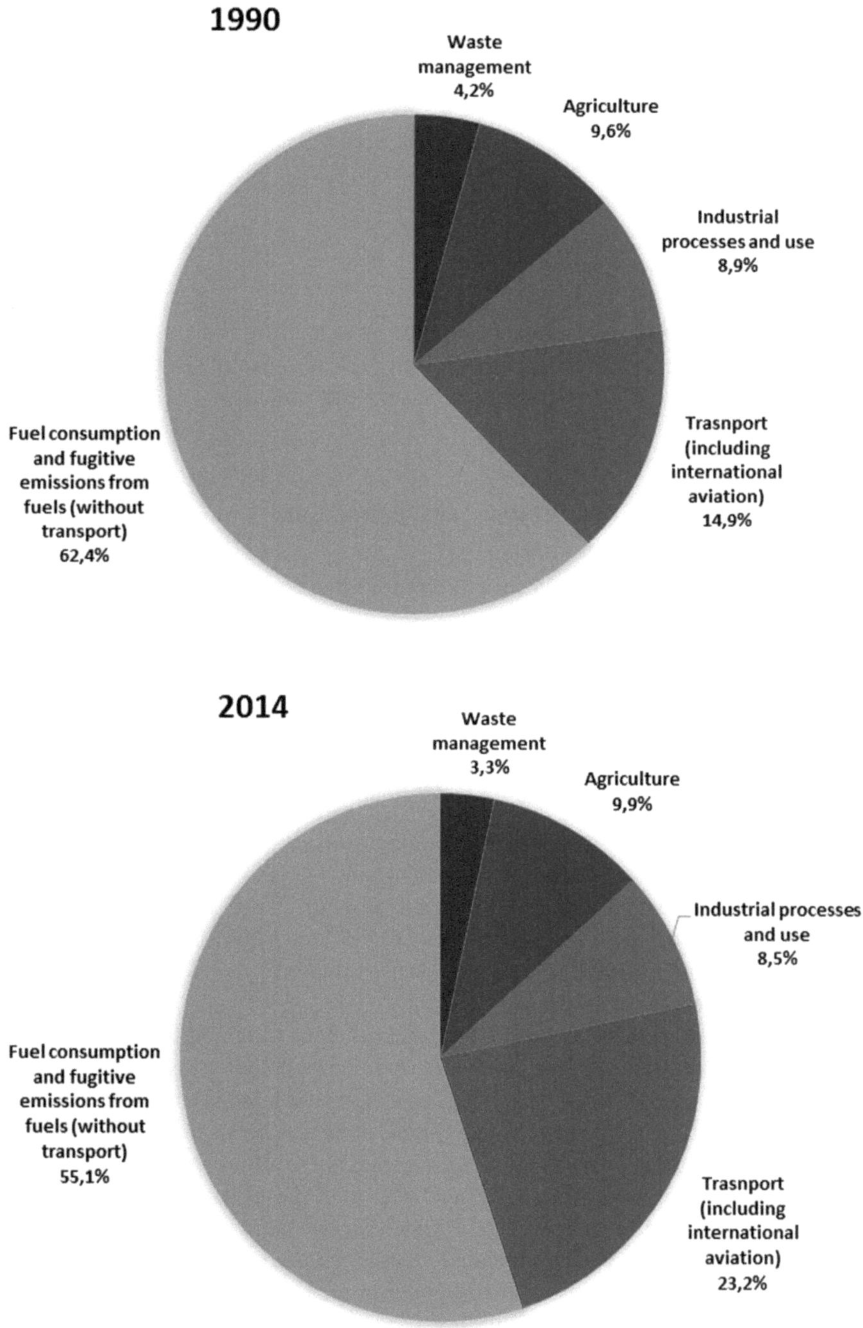

Fig. 2 Greenhouse gas emissions comparison 1990 and 2014 in the EU-28

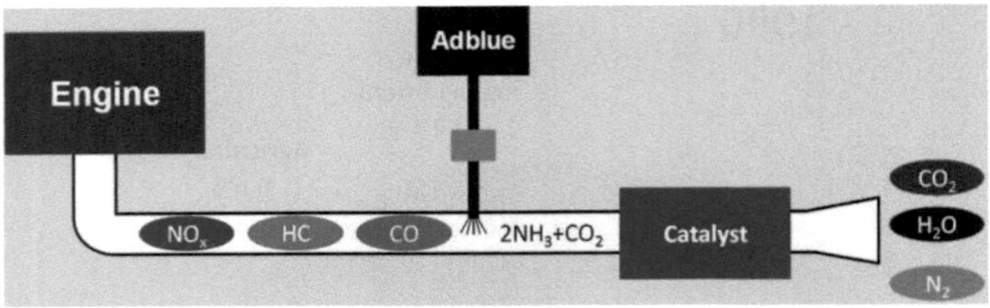

Fig. 3 Catalyst scheme

In the present study, the author's ambition is to analyze the current situation of Spanish efforts in order to achieve a progressive decarbonization, focusing in two of the most urbanized cities in Spain: Madrid and Barcelona.

2 Vehicles and Greenhouse Gas Emissions

Different types of energy sources (or fuels) for vehicles are currently used to fuel road transport vehicles:

- Petrol and diesel.
- Biofuels: ethanol and biodiesel.
- Liquefied Petroleum Gas (LPG).
- Natural gas.
- Electricity (produced from various energy sources).

Petrol and diesel vehicles tailpipe emissions are mainly composed by: (CO) carbon monoxide, (HC) Hydro–Carbons, (NO_x) Oxides of nitrogen, (PM) Particulate matter and (CO_2) Carbon dioxide. There has been substantial progress in reducing the emissions (NO_x, CO, HC, and PM) linked to local photochemical air pollution and it seems likely that this progress will continue in the future (Wallington et al. 2008).

Emissions from petrol cars have been reduced thanks to the introduction of new technologies as catalytic converters, which oxidize pollutants such as CO to CO_2. Comparisons in CO, HC and NO_x emissions of a petrol catalyst vehicle are much lower than in a non-catalyst one, but CO_2 emissions are higher due to the oxidation of carbon monoxide to carbon dioxide. Despite these improvements, petrol cars with catalysts still produce more CO and HC than diesel cars, although exhaust emissions of NO_x are much lower than diesel cars. When compared to petrol cars with a catalyst, diesel vehicles achieve higher emissions of particulate matter, perhaps the most characteristic of diesel emissions, responsible for the black smoke associated with diesel powered vehicles.

Ethanol and biodiesel are considered green fuels because they come from renewable resources. Today, most of the biofuels used in vehicles are additives to petrol and diesel fuel. The use of these combustibles contributes to a reduction of some emission components vehicles, but not to the elimination. Even sometimes, as it occurs in the NO_x case, it can increase them (McCormick et al. 2006).

Liquefied Petroleum Gas (LPG) is a mixture of propane and butane. It is considered a less pollutant substitute than petrol and it is widely used every day for cabs, buses and trucks as it contributes to a considerable reduction in consumption and emissions. It does not means that LPG does not emit GHGs (Won et al. 2007; Ristovski et al. 2005). Table 1 represents an example of the difference in emissions with LPG, petrol and diesel vehicles.

Natural Gas Vehicles (NGVs) are considered a climate-friendly alternative to petrol and diesel that use Compressed Natural Gas (CNG) or Liquefied Natural Gas (LNG) as a cleaner alternative. CNG and LNG are considered alternative fuels under the Energy Policy Act of 1992. Natural gas powers about 150,000 vehicles in the United States and roughly 15.2 million vehicles worldwide (Alternative Data Center 2017). It is important to remark that natural gas vehicles are not the same as vehicles powered by LPG, which is a fuel with a different composition. Natural gas vehicles reduce emissions of CO_2 by 20–30%, CO by 70–90% and NO_x by 75–95% (Thigpen Energy 2017). Finally, the concept of "electric car" has been growing during the last ten years, including Battery Electric Vehicle (BEV), Hybrid Electric Vehicle (HEV), Plug-in Hybrid Electric Vehicle (PHEV), Extended Range Electric Vehicle (EREV) and Fuel Cell Electric Vehicle (FCEV). In the sense, hybrid powertrains include an ICE, and they produce greenhouse gas emissions. So the only electric vehicles that do not directly emit GHG are the battery electric vehicles and those fueled by hydrogen (FCEV market penetration figures are very poor nowadays). As far as battery electric vehicles are the substitutes of the traditional internal combustion vehicles, the reduction of greenhouse gas emissions through the progressive use of private electric vehicles will depend mostly on factors as the power generation mix (Alvarez et al. 2015). Because BEVs do not burn fossil fuel and therefore, there are no tailpipe emissions when driving, but production of the electricity necessary to charge the batteries does generate global

Table 1 Example of emissions for patrol, diesel and LPG fueled vehicles (Hendriksen et al. 2003)

	Petrol	Diesel	LPG
CO g/km	1.48	0.10	1.39
HC g/km	0.13	0.02	0.10
NO_x g/km	0.10	0.80	0.07
HC + NO_x g/km	0.24	0.83	0.18
NO_2 Mg/km	0.002	0.37	0.01
PM g/km	0.006	0.046	0.005
CO_2 g/km	208.1	180.5	189.3
Fuel consumption l/100 km	8.86	6.78	11.74
NH_3 mg/km	17.3	0.9	50.6
SO_2 mg/km	8.9	3.7	2.8
N_2O mg/km	3	7	3
OC/EC mg/km	1.1/0.6	11.5/26.1	0.4/0.2

warming emissions, that has to be taken into account. Only BEVs charged entirely from renewable sources like wind and solar power produce virtually no emissions.

3 European Strategy for Low-Emission Mobility

Carbone dioxide emissions limits generated by vehicles were subject to a voluntary agreement between the EU and car manufacturers (ACEA agreement). Ultimately, the objective of the European Union with voluntary agreements was to contribute to achieving an average CO_2 emission of 120 g/km for all new tourism vehicles for the year 2012 (Directive 93/116/EC). Seeing that manufacturers do not voluntarily reduce emissions, the European Commission decided in 2009 to force a progressive emission reduction, which aims to achieve the average target of 95 g/km per car manufactured. The percentage of vehicles of each manufacturer below average will increase progressively: 65% in 2012, 75% in 2013, 80% in 2014 and 100% in 2015. If the average emissions of the fleet manufactured by a company increases with respect to 2012, it must pay a penalty. In 2020, the target for CO_2 emissions is 95 g/km.

Oxides of nitrogen (NO_x) and particulate matter (PM) have been reduced through a series of directives, each amendments to the 1970 Directive 70/220/EEC, that constitute the legal EU framework for controlling transportation emissions. The following is a summary list of the EURO standards, when they come into force and which EU directives provided the definition of the standard:

- Euro 1 (1993) for passenger cars—91/441/EEC. Also for passenger cars and light trucks 93/59/EEC.
- Euro 2 (1996) for passenger cars 94/12/EC (& 96/69/EC). For motorcycle 2002/51/EC 2006/120/EC.
- Euro 3 (2000) for any vehicle 98/69/EC. For motorcycle 2002/51/EC 2006/120/EC.
- Euro 4 (2005) for any vehicle. 98/69/EC (& 2002/80/EC).
- Euro 5 (2009/9) for light passenger and commercial vehicles. 715/2007/EC.
- Euro 6 (2014) for light passenger and commercial vehicles. 459/2012/EC.

Nowadays Euro 6 is the European Union directive to reduce harmful pollutants from vehicle exhausts. All mass-produced cars sold from this date need to meet these emissions requirements. The aim of Euro 6 is to reduce levels of harmful car and van exhaust emissions, both in petrol and diesel cars. Euro regulations (Regulation (Ec) No 715/2007; Regulation (Ec) No 692/2008; Regulation (Ec) No 595/2009; Regulation (EU) No 566/2011; Regulation (Ec) No 459/2012) have historically set different emissions standards for petrol and diesel cars, as a reflection on the different pollutants that the most used fuels produce.

Euro directives have been highly restrictive for diesels, which has achieved that the levels of NO_x emitted dramatically dropped from 780 mg/km (Euro 1) down to a maximum of 80 mg/km (Euro 6), a great step if compared to the 180 mg/km level that was required for cars that met the previous Euro 5 emissions standard (Table 2).

Table 2 Euro standards emission limits (Bishop et al. 2017)

Euro standard	Entry into force		Emission limits		
	New approvals	All new registrations	Petrol No_x (mg/km)	Diesel No_x (mg/km)	Diesel PM (mg/km)
Euro 0	1 Oct 1991	1 Oct 1993	100	1600	(no limit)
Euro 1	1 Jul 1992	31 Dec 1992	490	180	140
Euro 2	1 Jan 1996	1 Jan 1997	250	730	100
Euro 3	1 Jan 2000	1 Jan 2001	150	500	50
Euro 4	1 Jan 2005	1 Jan 2006	80	250	25
Euro 5	1 Sep 2009	1 Jan 2011	60	180	5
Euro 6	1 Sep 2014	1 Sep 2015	60	80	5

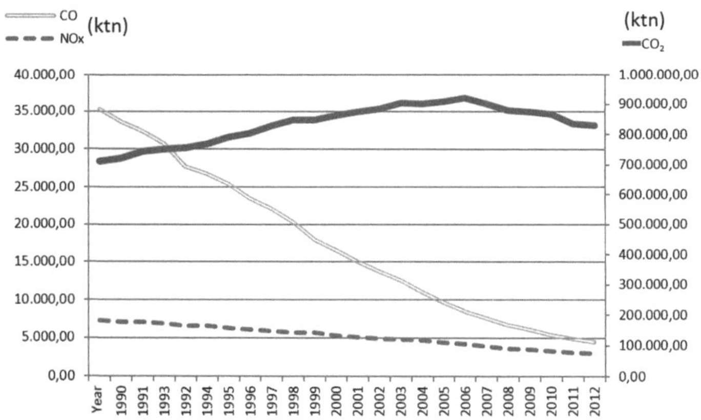

Fig. 4 Evolution of EU-28 gas emissions by road transport (Eurostat 2016)

In contrast, the NO_x limit for petrol cars remained unchanged from Euro 5, as it was already low at 60 mg/km, before the great change form 490 mg/km (Euro 1) to 60 mg/km (Euro 5 and Euro 6).

Thanks to these restrictions, Fig. 4 shows a great decrease of CO, undoubtedly due to the advances and the mandatory use since 1992 of the use of catalysts where the applied that allowed achieving lower values of pollutant gas emissions. This reduction did not affect in the same way to NO_x (with a decreasing trend) and CO_2 emissions that have increased figures until year 2006 (when the economic crisis began), as the petrol price grew up and higher unemployment rates reduce work-related trips.

The European strategy for low-emission mobility directly points to BEV, as it is the shortest way to reduce private vehicle transport emissions. BEV has been presented as the magic remedy for the decarbonization of large urban areas. But reduction of greenhouse gas emissions through the progressive use of private electric vehicles will depend mostly on the power generation mix of each country. In this sense, emissions linked to the charge of a battery electric vehicle will depend

Table 3 Power sources to electricity generation in Europe 2014

Power source	Quantity (%)
Thermal	47.70
Nuclear	27.60
Renewable	21.50
Others	3.20

Table 4 Emission factor in Spain from 2008 to 2015 (kg CO_{2eq} per kWh)

2008	2009	2010	2011	2012	2013	2014	2015
0.310	0.297	0.206	0.267	0.300	0.248	0.267	0.302

directly on the power generation mix (Table 3), which is defined as the percentage of thermal, nuclear and renewable energy, such as hydropower, used to produce electricity.

So, the emission factor EF_{mix} specific value is an average of the mix between the years 2008 and 2015, based on the data of electric power generation published for each country by Eurostat (Vestreng 2009) and emission factors for each source published by the Intergovernmental Panel on Climate Change (IPCC). In this way, the concrete expression to obtain the emission factor of the generation mix is based in the sum of the different proportions that each source represents on the overall energy generation and then multiply it by the corresponding emission factor EF_i. This is reflected in Eq. 1 as follows:

$$EF_{mix} = \sum_{i=1}^{i=n} EF_i \left(C_i / \sum_{i=1}^{i=n} C_i \right) \quad (1)$$

The values for Ci coefficients could be deduced from those measured at power generation central stations terminals, without considering the losses when the electricity is transported to the end user by electric conductors. As an example Table 4 displays the emission factor figures in Spain (kg CO_{2eq} per kWh) from years 2008 to 2015.

The main target from 2020 to 2050 is the decarbonisation of Europe. This is to delete completely the use of fossil fuels and open the door to the use of alternative mobility based in renewable energy as main power source. International Energy Agency (AIE) has presented several studies, analysing hypothetical scenarios, about the possibility of a 80% reduction of greenhouse gases and an 85% of CO_2.

4 GHG Transport Emissions in Spain, 15 Years Evolution

Pollution gases figures in Spain have been changing, the same as across Europe, during the years, parallel to the implementation of EURO standards to limit the production of gases from the ICE vehicles. The levels of emissions were rising

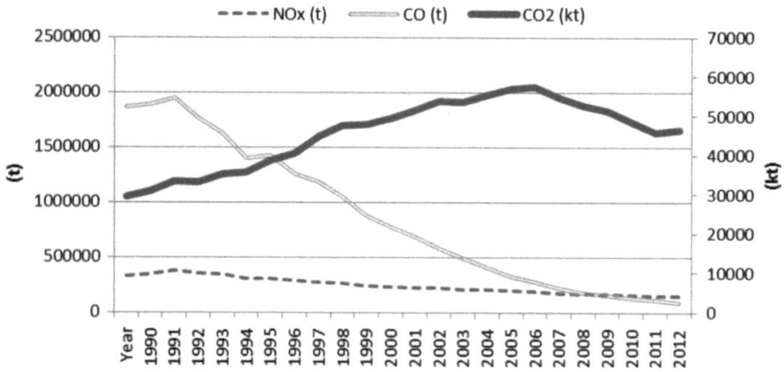

Fig. 5 Evolution of GHG emissions by road transport in Spain

continuously caused by the daily increase in the number of vehicles on the road. Between 1990 and 1993, there is a reflexive point where the emissions of NO_x and CO were decreasing (see Fig. 5); however, there was a kiloton of these gases emitted directly to the atmosphere. Alongside this, there are the CO_2 emissions, which did not reduce and instead continued to increase until 2007, when the new Euro Standard 5 was implemented, which was more restrictive than the previous one.

Also, the new Euro Standards for contamination, the increase of fossil fuel sales price and the crisis which arose, contributed to the decreasing of the power sources and consequently the emissions. Even though the contamination levels are still increasing by thousands of kilotons per year, in 2012 there were 150,678.80 tonnes of NO_x, 87,785.23 tonnes of CO and 46,399.18 kt of CO_2, forcing the application of new rules to fight against the effects.

All comments can be checked in Fig. 5, this information was obtained from Instituto Nacional de Estadística (INE 2016a, b) during period 1990–2012.

Another important factor to take into account is newly manufactured vehicles, which replace previous models. The new automobile generation emits lower quantities of polluting gases as a result of the European restrictions; however, the new technologies cannot correctly determine the levels of CO_2 emissions. Within Spain between 2008 and 2012 the 24.50% of greenhouse gases were emitted, exceeding the 15% imposed by Kyoto Protocol (Santamarta and Higueras 2013), this protocol was introduced to reduce the quantities of pollution in the atmosphere, however it does not state the recommended values in tonnes. In Fig. 6 it is shown that in spite of the decrease in new automobile sales in 2015, the levels of CO_2 kilotons emitted to the atmosphere are not directly related which was unexpected.

The vehicle sales decreased caused by the crisis effect and their devastating consequences to the population. A big quantity of people lost their jobs and those who kept them, focused their economic efforts on other life subjects, not buying a new car. Besides the increasing in the petroleum costs caused a reduction in vehicle use with the intention of reducing the individual economy.

It is necessary to take into account the driving behavior which, in general, it is aggressive, with accelerations and brakes, fast starts and stops, similar to other big cities or places where the traffic is higher. With increased unemployment, there was

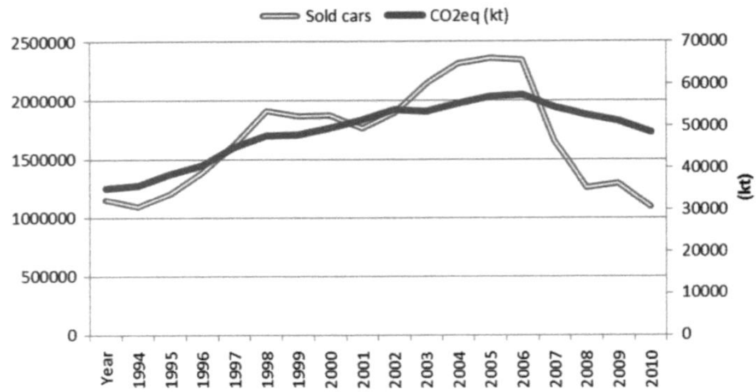

Fig. 6 Sold cars against emitted CO_{2eq} by road transport in Spain

a decrease in the daily automobile movement which in turn caused a reduction of CO_{2eq} emittance. The population continues to use their old vehicles, which are standardised by old Euro standards and therefore less restrictive in the pollutants emissions. This will, therefore, cause dangerous levels of toxic gases in the environment.

Nevertheless, the daily movement of vehicles to travel to and from work was reduced significantly, with the unemployment that rose from 8.23% in 2006 to 21.39% in 2010 and continued to rise to 24.44% in 2014 (Instituto Nacional de Estadística 2016). The decline in the active population brings a reduction in the number of vehicles circulating every day, this point also provides a significant reduction of the greenhouse gases emitted, and that is why the CO_{2eq} level was reduced by tonnes; it was not caused by the technology advances or social behaviour but by the decrease in automobile use. It leads on to suggest that when unemployment begins to reduce and the active population rises, the greenhouse gases will also rise because more private cars will be used again, and with the fear of a crisis people will not consider buying new less pollutant car models.

Due to this, the excess levels of CO_2 will come back in larger quantities if the necessary measures are not taken in order to find a plausible solution. As per the information obtained from Instituto Nacional de Estadística (which provides details of the emitted CO_{2eq} levels until 2011) and from Ministerio de Agricultura, Alimentación y Medio Ambiente de España, these hypotheses are correct. Although it is necessary to add the new vehicle's emissions, with more restrictive Euro standards, they still emit GHG gases.

A high rate of pollution creates consequences for the environment and the health of human beings. Clearly, the emission limits imposed by European Air Quality Direction were exceeded (Jiménez-Herrero 2008). Thanks to the information obtained from Environment Minister of Spain (Ministerio de Agricultura, Alimentación y Medio Ambiente 2015), during the period of 23 years, it can be observed that the emission levels of CO_{2eq} have changed and not in a positive way. The levels have changed so much so that the lowest numbers were in 1990 when the automotive industry was not as advanced as today. The levels of CO_2 emitted during the year 2007 were in great excess, 441,720 kt, which is 150,984 kt more than 1990. Spain was a country with a lot of pollution and it was necessary to urgently reduce gas emissions.

The recorded measurements of emissions in 2013, showed that there was a reduction of 119,716 kt, but this was not enough to meet the expected reductions. The Spanish communities causing the most pollution within previous years are; Andalucía, followed respectively by Cataluña and Castilla-y-León, Galicia, Community of Valencia and Madrid. However, the last available information (2013) shows that Andalucía (49,892 kt) and Cataluña (42,768 kt) still are the communities which emit the most CO_2 in Iberian Peninsula. However, it is also important to consider that these communities have more factories than the others, caused by the quantity of industries based in these regions. However, Andalucía and Cataluña, Galicia and Castilla-y-León also create some of the highest pollution levels in Spain, this is due to the vehicle companies, petroleum sector, chemicals, power generation and importantly the agriculture sector, where companies use a really big quantity of machinery, causing 12,110.04 kt of CO_2 in Spain (2013). In addition, the majority of the CO_2 concentration is caused by the road transportation, other companies, houses etc. It is observed daily in the big regions like the Communities of Madrid and Cataluña, which emitted 21,494 and 42,768 kt respectively during 2013 (Subdirección General de Energía y Cambio Climático de Madrid 2015).

Those polluting levels are reflected in environmental security measures taken in Madrid and Barcelona with anti-pollution protocols caused by the volume of emitted gases into the atmosphere.

5 Electricity Generation in Spain

In Spain, power generation has increased by 0.4% in 2015 compared with 2014 (Instituto Nacional de Estadística 2016). However there is not the same quantity of GWh obtained from each power source, the percentages have changed from 2014 to 2015. The power generation in Spain is distributed in the following way (Table 5).

The power generation by nuclear and carbon sources are the methods used most. This is a negative point or not, taking into account the environmental and hazardous consequences of this technology. Referring to 2014, the use of nuclear is 1.80% lower but it is still the most used power source along with carbon which is 21.90% higher. This increase in electricity production, using carbon sources, contributes directly to the CO_2 emissions, creating approximately 50 million tonnes. The total CO_2 provided by power generation was almost 75 million tonnes during 2015 and the use of renewable energy sources has decreased by 44.30% since the end of 2014 and the end of 2015; this is bad taking into account the necessity to increase these power sources (hydraulic, wind power, solar and renewable thermal) that they

Table 5 Electric power source generation distribution in Spain

Power source	Total GWh	% 15/14
Nuclear	56,796	−1.00
Carbon	56,672	21.90
Fuel/gas	6891	3.40
Combined cycle	30,217	16.60
Cogeneration and others	27,183	5.00
Renewable energies	93,040	−44.30

represent just the 34.36% of total energy production in the country at the present time (Red Eléctrica de España 2015).

This is very bad news and confirms the need for change, by finding alternative sources of energy which create less pollution. Due to this being an important issue for the planet's health, there are measures which have been put into place and should be achieved by 2050 with total decarbonisation of all European countries, this would reduce the greenhouse gas emissions by 80% (Regulation (EU) No 566/2011).

Measurements will be useful but without total implication, it will be not possible to obtain the targets, like increasing the use of renewable power sources by 55% of the gross final energy consumption, lower the use of nuclear energy and obtain the improvement energy efficiency of the country and Europe.

6 GHG Emissions in the Community of Madrid. 15 Years Evolution

In big cities, there is a lot of daily movement carried out by the vast population such as going to work, social events or completing domestic tasks. This kind of movement requires a lot of different types of transportation to be used as can be seen in cities such as Madrid.

The city of Madrid, during last few years, suffered an urban sprawl which is considered the reason for most of the long commutes, and the increasing use of vehicles. The increasing in the use of vehicles is stronger and therefore, there are more emission of greenhouse gases (Lumbreras et al. 2008), even higher than expected. The limits of different greenhouse gases within the atmosphere to maintain healthy air quality was imposed by the Real Decreto (102/2011) and also was updated in the Real Decreto (678/2014). The Community of Madrid is the one that represents many municipalities that surpass or are near the limits arranged according to the information obtained by the year 2015.

The daily concentration limit of PM10 particles is set at 50 $\mu g/m^3$, being exceeded by Colmenar Viejo and Villa del Padro at 100 $\mu g/m^3$; Alcalá de Henares, Rivas Vaciamadrid, Móstoles, Majdahonda, Guadalix de la Sierra, El Atazar and Orusco de Tajuña by two times (150 $\mu g/m^3$); Alcobendas, Arganda del Rey, Coslada, Torrejón de Ardoz, Aranjuez and Fuenlabrada by three times (200 $\mu g/m^3$); Leganés and Getafe by four times (250 $\mu g/m^3$) (Vestreng et al. 2009). This is not the case of PM2.5 particles where the annual limit set is 25 μg/m^3 and this was not exceeded (Consejería de Medio Ambiente, Administración local y Ordenación de Territorio de la Comunidad de Madrid 2016). The ozone emissions (O_3) are also regularised with the objective limit at 120 μg/m^3, information limit at 180 μg/m^3 and alert limit at 240 μg/m^3. Although these limits have not been exceeded, it is necessary to keep them under control.

The time limit of nitrogen oxide (NO_2) is 200 μg/m^3 and the annual limit to protect the health is 40 μg/m^3. In spite of not exceeding the time limit, there were times when the emissions came close to it. In the case of annual limit, it was exceeded by Cosalda and Alacalá de Henares with 46 and 42 μg/m^3 respectively and Getafe was on the boarder (Consejería de Medio Ambiente, Administración Local y Ordenación de Territorio de la Comunidad de Madrid 2016).

The time limit of sulfuric dioxide (SO_2) is set at 350 μg/m^3 and annual limit for the health protection is 125 μg/m^3. Alcalá de Henares had the major pollution value (time limit) at 6 μg/m^3 (Consejería de Medio Ambiente, Administración Local y Ordenación de Territorio de la Comunidad de Madrid 2016) in spite of there is no infraction of SO_2.

In carbon monoxide (CO) case, there is an established maximum 8-h value of 10 mg/m^3; it was not exceeded. However, these are required to improve in order to reach the 2050 target.

This study put the focus on Municipality of Madrid, the most populous city in the autonomous community of Madrid, which in 2013 emitted 10,498 kt of CO_{2eq} and 21% of them come from road transportation (Subdirección General de Energía y Cambio Climático de Madrid 2015).

Clearly, CO_{2eq} emission levels dispñayed in Fig. 7 shows that the transport sector contribution represents an important part of the total emissions within Madrid. At the same time, it is necessary to include the increasing in private transport use; which most of the time, is a vehicle per person. It contributes to the increase in polluting gases and automobile density on the highways and roads of Madrid, especially on those which are directly an access point to the city.

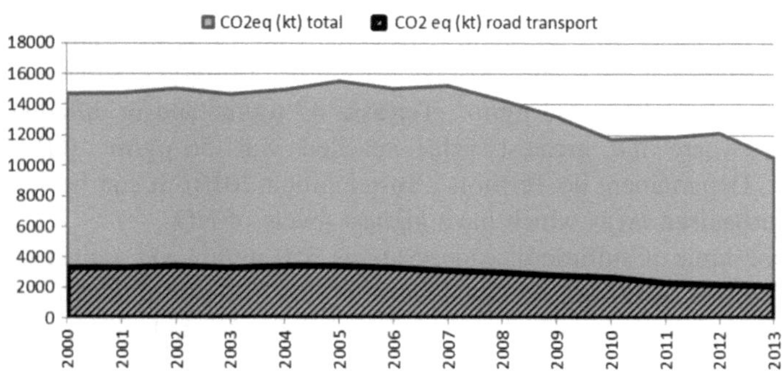

Fig. 7 CO_{2eq} (kt) direct CO_2 emissions Municipality of Madrid by year (2000–2013)

7 GHG Emissions in the Community of Cataluña. 15 Years Evolution

As well as looking at this study of Madrid it is important to look at a similar study for the second largest city in Spain Barcelona. This is a huge urban nucleon, which according to the Barcelona town hall in 2015 (Ajuntament de Barcelona 2016) had a population up to 1,064,555 people.

According to the report of Barcelona´s air quality (Selvas i león 2014), no place within the city exceeded the limit for the levels of CO (8-h), SO_2 (time limit), annual PM2.5 or PM10, but the levels were close in case of suspended particulates. However, referring to the annual media of NO_2 at different points in the city, the limit was exceeded in 2011, 2012, 2013 and 2014 during the seasons with a lot of road traffic (Selvas i León 2014); because of this, there was forced improvement plan of air quality for the Barcelona city implemented; like that which was implemented in Madrid. This was caused by the need to reduce the polluting gases which contribute to the greenhouse effect.

As indicated in point 6, the Real Decreto 102/2011 and the later update in the Real Decreto 678/2014, also set the allowed limits within Barcelona. Cataluña presents a number of municipalities that exceeded or are close to the limits according to information from Generalitat de Catalunya in 2015. The concentration of daily limit of PM10 is 50 $\mu g/m^3$. In 2015 the PM10 limit was exceeded in four places, the zones of Vallès-BaixLlobregat, Plana de Vic and Terres de l'Ebre (Generalitat de Catalunya, Departament deTerritori i Sostenibilitat 2016). This is not the case with the levels of PM2.5 (limit 25 $\mu g/m^3$), which were not exceeded.

The ozone emissions (O_3) are also regularised with an objective limit at 120 $\mu g/m^3$, information limit at 180 $\mu g/m^3$and alert limit at 240 $\mu g/m^3$ (Real Decreto 102/2011). In 2015 there are 115 h of excess of the limit of information (Generalitat de Catalunya, Departament de Territori i Sostenibilitat 2016) this happened in different places, Barcelona, Camp de Tarragona, Plana de Vic, Maresme, Comarcas de Girona, Alt Llobregat, Pirineu Oriental, the Prepirineo and Terres de Ponent. In spite of this, the alert limit was not exceeded.

The limit of nitrogen oxide (NO_2) is 200 $\mu g/m^3$ and the annual limit to protect the health is 40 $\mu g/m^3$ (Real Decreto 102/2011). The annual limit was exceeded in Sant Adrià del Besòs with 42 $\mu g/m^3$, Badalona with 41 $\mu g/m^3$, Barberà del Vallès 41 $\mu g/m^3$, Martorell 41 $\mu g/m^3$, Mollet del Vallès 46 $\mu g/m^3$, Sabadell 42 $\mu g/m^3$, Sant Andreu de la Barca 43 $\mu g/m^3$, Terrassa 47 $\mu g/m^3$ and in different points of Barcelona, where the greatest value reached was 56 $\mu g/m^3$ (Generalitat de Catalunya, Departament de Territori i Sostenibilitat 2016). It can be seen that it is the most urbanised areas which have highest levels of NO_2.

The time limit of sulfuric dioxide (SO_2) is 350 $\mu g/m^3$ and annual limit for the health protection is 125 $\mu g/m^3$ (Real Decreto 102/2011). In this case, there are no zones, within Barcelona, which have exceeded the limit. In the case of carbon monoxide (CO), it is the same as sulfuric dioxide, in that it did not exceed the limits

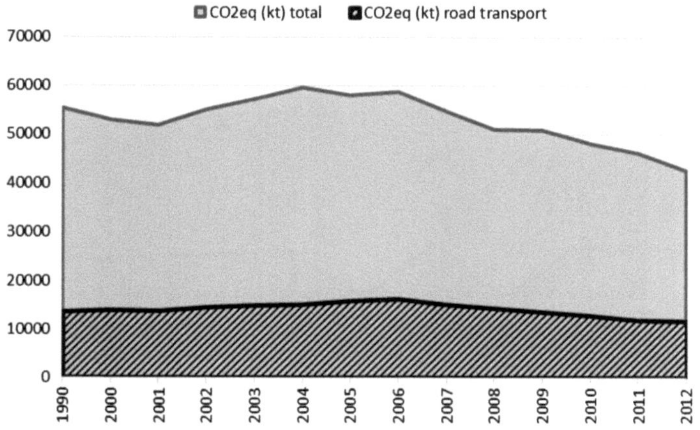

Fig. 8 CO_{2eq} Cataluña emissions

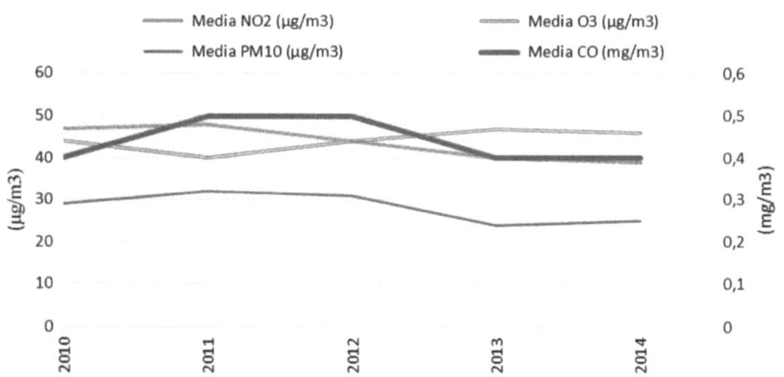

Fig. 9 GHG gas emissions evolution of Barcelona

set. It is, however, necessary to check the evolution of CO_{2eq} in Cataluña from 1990 to 2013 (Fig. 8).

According to this information from the Generalitat de Catalunya, it is clear that environmental standards have not always been met. The level of CO_{2eq} kt which came from road transportation is between 24.52 and 27.82% of the overall pollution.

A change was required and it forced a plan for the Cataluña air quality improvement, and specifically in Barcelona (Ajuntament de Barcelona 2015) exactly as it occurs in Madrid. In some zones of Cataluña the PM10, O_3, NO_2 and CO limits were also exceeded. Being that, this study was focused in Barcelona, the bigger city of Cataluña, the main focus was the levels of emitted pollutants (Fig. 9).

Thanks to information obtained from Barcelona city website (Ajuntament de Barcelona 2016) it can be seen that the levels of the gases have changed. The NO_2 decreased gradually after the implementation of more restrictive standards which has forced it to decrease (from 47 $\mu g/m^3$ in 2010, there is media 39 $\mu g/m^3$ in 2014);

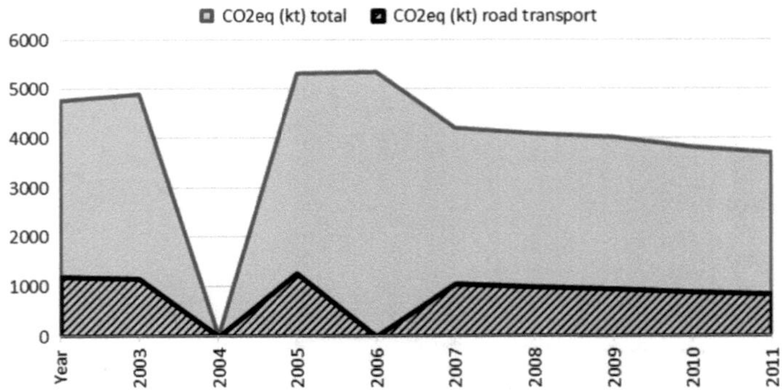

Fig. 10 CO_{2eq} emissions of Barcelona

similar to this is the case of PM10, which decreased until 2013 but then in 2014 this type of particles began to rise again reaching 25 μg/m^3.

In the case of recorded levels of O_3, there is a point where it increased from 2011, even though it remains under the critic limits for the people health at 120 μg/m^3 (Real Decreto 102/2011). The situation of CO level is slightly different because their measurements are recorded in mg/m^3. The evolution of these levels until 2012 is not favourable, but the number are still improving changing from 0.5 to 0.4 mg/m^3 in 2013 and this level was maintained during 2013 and 2014. It is necessary to reduce CO_{2eq} emission levels in the city of Barcelona (Fig. 10).

Looking at the information provided by Barcelona Townhall (Ajuntament de Barcelona 2016), we can see the total level of CO_{2eq} kt and the level caused by road transport which changes between 22.84 and 25%. There was no available data for the year 2005 and for the CO_{2eq} road transportation 2007. However, with the figures which are available the values are really high, if you take into account the evolution of automobile technology and given the mentioned necessity of electric vehicle revolution; nowadays this is the way forward but the population is not too confident with this type of vehicle because of the autonomy, the recharging, the security and the reduction of pollutant gases. In 2015 the sales of alternatively powered vehicles (hybrid and electric) were just the 2.5% of total new vehicles (ANFAC 2016). Through the Barcelona Townhall website (Ajuntament de Barcelona 2016) it is possible to check the air quality index of the city, and the level of polluting gases are practically correct in all the city.

8 Comparison Between Madrid and Barcelona

The two most important cities in Spain and where this study is focused are Madrid and Barcelona. As indicated, the daily movements within both cities and levels of pollution have reached excessive points taking into account technological advances.

The Community of Madrid has 8022 km^2 of space and of this, the city of Madrid occupies 605.77 km^2 of them (Consejería de Medio Ambiente y Ordenación del Territorio 2004). Madrid had a population of 3,141,991 people in 2015. Cataluña, on the other hand, has 32,091 km^2 of land, of which 102.15 km^2 is Barcelona (Consejería de Medio Ambiente y Ordenación del Territorio 2004). The population in Barcelona was 1,604,555 people in 2015. There are also territorial differences which also have an effect, for example, Madrid is located in the middle of the country and Barcelona is located by the sea.

A lot of companies has their headquarters in Madrid and Barcelona. However, the number of them are not the same in both cities. In Madrid, there are 415,976 companies and in Barcelona 244,259 (Einforma 2016a, b). Madrid is surrounded by flat land, providing the possibility of the creation of rural towns on the outskirts of the city for mainly housing. Most of the companies have chosen this city to have their offices, where the workers have to commute daily. The apartment prices in Madrid have increased exponentially and therefore many people live in other cities close to Madrid, where the apartments are cheaper.

Barcelona is different, where most of the companies are installed in the surroundings and the workers live in the main city. Several companies are in the industrial estates which it tends to be located in Sabadell, Terrassa or Tarragona. Even though the presence of the sea in Barcelona creates the possibility to invest in the port areas, this is not the case in Madrid. However, this positive industrial point causes many dangerous gases to be emitted by industrial machinery and boats. Within the year 2013, the Catalan community emitted 42,768 kt of CO_{2eq} and the Madrid community 21,494 kt CO_{2eq}. The difference between them is remarkable, but it is important not to forget the types of industries in each city.

Madrid and their surroundings have not got the same agricultural companies as Barcelona does. It can be seen that this increase in CO_2 and greenhouse gases emissions is a direct result of an increase in the use of agricultural machines. The Catalan province registered 2450 new machines within the year of 2015, which was 0.74% more than 2014; while the province of Madrid registered 129 machines in 2015, which is 9.79% less than previous year (Ministerio de Agricultura y Pesca, Alimentación y Medio Ambiente 2016). Also, it is necessary to add into the mix the old agricultural machines and in Madrid maybe there are not as many agricultural companies as Barcelona, but there are many companies who have factories which treat the products produced; these also cause excessive pollution. Both cities have a large number of companies which cause the population to commute daily. This, in turn, results in a lot of the CO_2 emissions in the atmosphere, traffic saturation and noise pollution; the most used transport is private vehicles and specifically vehicles which use fossil fuels.

9 Conclusions

The current transport system is practically the same as 15 years, although it has developed, this is not enough in reference to emissions of polluting gases, which is reflected in this document.

As mentioned above (quantities of gaseous pollutants, non-compliance with air quality standards, improvement of the transport system), it is correct to say that pollution information is alarming, in that it is exceeding the limits set and not finding efficient solutions. This pollution is contributing to the increase in gases related to the greenhouse effect and damaging the planet. If efforts are not invested in improving the current situation, future pollution and resource scarcity will cause an irreversible and irreparable damage. However, the need is too great to avoid environmental sanctions for large companies and countries that focus on hiding their actual emission levels and very few are seeking solutions.

This paper have shown the true problems caused by the use of road transport with fossil fuels, which includes the increase in the greenhouse effect and the climatic deterioration. The main objective is to use renewable energy such as the electric vehicles, motorcycles or bicycles, which emits 0% of direct pollutants, 0% of noise pollution and are much more efficient than ICE vehicles. Through the use of electric vehicles, it will contribute to the most notable reduction of pollution in many cities.

References

Ajuntament de Barcelona. (2015). Plan de mejora de la calidad del aire en Barcelona 2015–2018. Abril 2015.

Ajuntament de Barcelona. (2016). Abril 2016. http://www.bcn.cat/estadistica/castella/indicadors/index.htm

Alternative Data Center. (2017). Natural gas vehicles. http://www.afdc.energy.gov/vehicles/natural_gas.html

Álvarez, R., Zubelzu, S., Díaz, G., & López, A. (2015). Analysis of low carbon super credit policy efficiency in European Union greenhouse gas emissions. *Energy, 82,* 996–1010.

ANFAC, Asociación Española de Fabricantes de Automóviles y Camiones. (2016). Nota de prensa –Estadísticas - Turismos. Junio 2016. http://www.anfac.es/estadisticas.action?accion=estad_turismos

Bishop, J. D. K., Molden, N., & Boies, A. M. (2017). Real-world environmental impacts from modern passenger vehicles operating in urban settings. *International Journal of Transport Development and Integration, 1*(2), 203–211.

Consejería de Medio ambiente y Ordenación del Territorio. (2004). Mapas Políticos. http://www.madrid.org/cs/Satellite?blobcol=urldata&blobheader=application%2Fpdf&blobheadername1=ContentDisposition&blobheadervalue1=filename%3DCapitulo+1+mapas+politicos+de+Atlas_Med_Amb4.pdf&blobkey=id&blobTable=MungoBlobs&blobwhere=1158609635785&ssbinary=true.

Consejería de Medio Ambiente. (2016). *Administración local y Ordenación de Territorio de la Comunidad de Madrid.* Marzo: Red de calidad del aire de la Comunidad de Madrid.

Einforma. (2016a). Directorio de empresas Geográfico. https://www.einforma.com/informes-empresas/BARCELONA/EMPRESAS-BARCELONA.html
Einforma. (2016b). Directorio de empresas Geográfico. https://www.einforma.com/informes-empresas/MADRID/EMPRESAS-MADRID-4.html
Eurostat Comisión Europea. (2016). Mayo 2016. http://ec.europa.eu/eurostat/statistics-explained/index.php/Electricity_production_and_supply_statistics
Generalitat de Catalunya, Departament deTerritori i Sostenibilitat. (2016). La Qualitat de l'aire a Catalunya, Anuari 2015—Resum. Secció d'Immissions Direcció General de Qualitat Ambiental, Abril 2016.
Hendriksen, P., Vermeulen, R., Rijkeboer, R., Bremmers, D., Smokers, R., & Winkel, R. (2003). Evaluation of the environmental performance of modern passenger cars running on petrol, diesel, automotive LPG and CNG, TNO-report 03.OR.VM.055.1/PHE.
INE. Instituto Nacional de Estadística. (2016a). Tasa de paro de la población. DOI http://www.ine.es/consul/serie.do?s=EPA136022
INE. Instituto Nacional de Estadística. (2016b). Estadística de Transporte de Viajeros (TV). Nota de prensa.
IPCC, Intergovernmental Panel on Climate Change. (2015). *Climate change 2014: Mitigation of climate change* (Vol. 3). Cambridge: Cambridge University Press.
Jiménez-Herrero, L. M. (2008). Calidad del aire y sostenibilidad urbana. *Ambienta: la revista del Ministerio de Medio Ambiente, 76*, 18–27.
Lumbreras, J., Valdés, M., Borge, R., & Rodríguez, M. E. (2008). Assessment of vehicle emissions projections in Madrid (Spain) from 2004 to 2012 considering several control strategies. *Transportation Research Part A: Policy and Practice, 42*(4), 646–658.
McCormick, R. L., Williams, A., Ireland, J., & Hayes, R. R. (2006). Effects of biodiesel blends on vehicle emissions: Fiscal year 2006 annual operating plan milestone 10.4 (No. NREL/MP-540–40554). National Renewable Energy Laboratory (NREL), Golden, CO.
Ministerio de Agricultura y pesca, alimentación y medio ambiente. (2015). Avance de la inscripción de maquinaria agrícola nueva, en los registros provinciales 2015 http://www.magrama.gob.es/es/agricultura/estadisticas/avance_2015.aspx
Ministerio de Agricultura, Alimentación y Medio Ambiente. (2015). Emisiones de GEI por Comunidades Autónomas a partir del inventario español—serie 1990–2013. Noviembre 2015.
Ministerio de Medio Ambiente. (2016). http://www.mambiente.munimadrid.es/sica/scripts/index.php
Real Decreto. (102/2011). Boletín Oficial del Estado, Enero 2011.
Real Decreto. (678/2014). Boletín Oficial del Estado, Agosto 2014.
Regulation (EC) No 715/2007 of the European Parliament and of the Council of 20 June 2007.
Regulation (EC) No 692/2008 of the European Parliament and of the Council of 18 July 2007.
Regulation (EC) No 595/2009 of the European Parliament and of the Council of 18 June 2009.
Regulation (EU) No 566/2011 of 8 June 2011.
Regulation (EC) No 459/2012 of the European Parliament and of the Council de 29 May 2012.
Ristovski, Z. D., Jayaratne, E. R., Morawska, L., Ayoko, G. A., & Lim, M. (2005). Particle and carbon dioxide emissions from passenger vehicles operating on unleaded petrol and LPG fuel. *Science of the Total Environment, 345*(1), 93–98.
Santamarta, J., & Higueras, M. A. (2013). *Informe de emisiones de gases de efecto invernadero en España 1990–2012*. Spain: World Wildlife Fund.
Selvas i León. (2014). La calidad del aire de Barcelona. CONAMA 2014.
Subdirección General de Energía y Cambio Climático de Madrid. (2015). Inventario de emisiones de gases de efecto invernadero del municipio de Madrid en el año 2013. Ayuntamiento de Madrid.
Thigpen Energy. (2017). Natural gas fueling. http://www.thigpenenergy.com/natural-gas-fueling/transportation/cng/. Accessed September 10, 2016.
Vestreng, V., Ntziachristos, L., Semb, A., Reis, S., Isaksen, I. S., & Tarrasón, L. (2009). Evolution of NO_x emissions in Europe with focus on road transport control measures. *Atmospheric Chemistry and Physics, 9*(4), 1503–1520.

Wallington, T. J., Sullivan, J. L., & Hurley, M. D. (2008). Emissions of CO_2, CO, NO_x, HC, PM, HFC-134a, N_2O and CH_4 from the global light duty vehicle fleet. *Meteorologische Zeitschrift, 17*(2), 109–116.
Won, J. H., Park, J. Y., & Lee, T. G. (2007). Mercury emissions from automobiles using gasoline, diesel, and LPG. *Atmospheric Environment, 41*(35), 7547–7552.

Part II
Urban Planning Policy Implications

Chapter 8
Implementing COP 21

Jan-Erik Lane

Abstract The global policy of halting climate change has only been decided upon in Paris 2015, as the implementation process has yet to start. Many more meetings and conferences are held by The United Nations Framework Convention on Climate Change almost monthly, but the real putting into effect of means and instruments that promote the COP21 goals is still lacking, at least from a global point of view. I will point out the problematic with the COP21 Agreement, employing two well-known social economics models and illustrating it with a few examples from the real world. The findings include that several countries face an energy-emissions conundrum that the climate change project must face.

1 Introduction

Mankind has through the United Nations chosen the COP21 Agreement as the most hopeful response to the climate change challenge, involving decentralized policies in favour of decarbonisation, given some form of international governance oversight. All the hurrahs in and around the Paris accords now face the grim realities of policy implementation—Wildavsky's hiatus—and the strong links of emissions with economic development through the energy connections—Kaya's model. This paper spells out the implications of these two major models in political science and economics for the probability of success of the COP21 project. First I state some theoretical notes on implementation and the links: GDP-energy-emissions. And then I provide a few empirical examples of the difficulties the models entail.

J.-E. Lane (✉)
Public Policy Institute, Belgrade, Serbia
e-mail: janeklane@googlemail.com

R. Álvarez Fernández et al. (eds.), *Carbon Footprint and the Industrial Life Cycle*,
Green Energy and Technology, DOI 10.1007/978-3-319-54984-2_8

2 Policy Making Versus Implementation

In the wake if the enlargement of the welfare state in advanced economies, a lot of research was conducted into policy outcomes in order to establish whether policy ambition or policy objectives were fulfilled in reality. The theory of policy implementation entailed a profound hiatus between goals and outcomes (Pressman and Wildavsky 1973, 1984). The recognition of this gap between high level policy-making and legislation on the one hand and low level policy execution had not been fully theorized in the discipline of public administration. This hiatus applies of course with a vengeance to the COP21 project, which truly large, involving the governments of the countries in the world in an implementation that is to last some 50 years or more. Can the enormous goal of global decarbonisation really be achieved? And if so, with what means or instrument by whom?

2.1 The Wildavsky Gap

Two rounds of implementation studies may be identified with different lessons for policy-making, both relevant to the concerns of global climate policies. First, we have the Wildavsky (1979, 1987) gap between policy and implementation:

- Divergence between primary goals and later goal developments.
- Resistance to change from local groups.
- Goal reinterpretation to handle non-conformity of outcomes.
- Dynamic evolution of policies to adapt to implementation difficulties.
- Own goals with the implementors deviating from officially stated ones.
- Long drawn out implementation processes reduce the likelihood of successful goal fulfillment.
- The occurrence of policy-implementation paradoxes, as goals are pursued but resulting in opposite outcomes.
- Goals may find no relevant outcomes, or goals may be conducive to dysfunctional outcomes.

These results were supported by empirical finds, examining huge welfare state programs, and made understandable in terms of decision theory—incrementalism—and game theory—asymmetric information and opportunism.

An interesting example of policy dys-functionality is the recent global decision on the so-called HFCs or hydro fluorocarbons, which is a very harmful greenhouse gas. It was put in refrigerators and air-conditioners as a result of the Ottawa protocol to try to heal the ozone layer hole. But it proved to be noxious for global warming, while at the same the ozone layer has hardly healed very much.

2.2 Strategy of Implementation

In the second round of implementation enquiry, P. Sabatier and associates took the initiative to theorize how policy implementation may succeed, if at all (Mazmanian and Sabatier 1989). His answer was the "policy coalition" framework or "policy advocates" approach (Sabatier 1985, 1993, 1998). This innovative model emphasized a number of important elements in policy implementation with a chance of successful goal fulfillment:

- Private-public partnerships.
- Team work or networks.
- Dynamic goal evolution.
- Relevance of incentives.
- Bottom-up instead of top down approach.

Now, the COP21 objective of decarbonisation must take the lessons from the two implementation rounds of study into account when it now mover forwards, from the stage of talking to the stage of performing, delivering outcomes.

Three goals are essential in the COP21 project, namely:

- Goal I: halting the upward trend in CO_2 emissions.
- Goal II: reduction of CO_2 emissions by 40% by 2030.
- Goal III: almost complete decarbonisation by 2075.

How are these goals to be promoted? They require a gigantic management effort by the single governments and societies as well as assistance from international governance. The implementation problematic is especially difficult, not only because it is a matter of global policy-making in a decentralized approach, putting the chief responsibility with the countries themselves, but because there is vital restriction upon policy implementation, viz economic development must be maintained.

The COP21 Agreement contains little about the implementation of these three objectives above. It only states:

- Implementation is to be decentralized to the governments of countries.
- Assistance is to be provided by means of a Super Fund of Stern (2007) type.
- Some form of overview and evaluation will be put in place by the UN Convention on Climate Change.

The COP21 project will now hopefully become more concrete through the COP22 meeting. The Third World is eager to hear how the Super Fund will operate: budgeting and funding. "We need help from them because they created the problem in the first place"—say Third World governments. And they certainly need massive assistance, as we will show below. This brings us to the Kaya model. It models the CO_2 emissions from anthropogenic sources, which is also what the CO21 projects targets.

3 Emissions, GDP and Energy

A major effort to model the CO_2 is the deterministic Kaya equation. The Kaya identity runs with these factors: environmental (I)mpact against the (P)opulation, (A)ffluence and (T)echnology. Technology covers energy use per unit of GDP as well as carbon emissions per unit of energy consumed (Kaya and Yokoburi 1997).

3.1 Theory

In theories of climate change, the focus is upon so-called anthropogenic causes of global warming through the release of greenhouse gases (GHG). To halt the growth of the GHG:s, of which CO_2:s make up about 70%, one must theorize the increase in CO_2:s over time (longitudinally) and its variation among countries (cross-sectionally). As a matter of fact, CO_2:s has very strong mundane conditions in human needs and social system prerequisites. Besides the breading of living species, like Homo sapiens for instance, energy consumption plays a major role. As energy is the capacity to do work, it is absolutely vital for the economy in a wide sense, covering both the official and the unofficial sides of the economic system of a country. The best model of carbon emissions to this day is the so-called Kaya model. It reads as follows in its standard equation version—Kaya's identity:

(E1) Kaya's identity projects future carbon emissions on changes in Population (in billions), economic activity as GDP per capita (in thousands of $US (1990)/ person year), energy intensity in Watt years/dollar, and carbon intensity of energy.

Concerning the equation (E1), it may seem premature to speak of a law or identity that explains carbon emissions completely, as if the Kaya identity is a deterministic natural law. It will not explain all the variation, as there is bound to be other factors that impact, at least to some extent. Thus, it is more proper to formulate it as a stochastic law-like proposition, where coefficients will be estimate using various data sets, without any assumption about to stable universal parameters. Thus, we have this equation format for the Kaya probabilistic law-like proposition, as follows:

$$\text{(E2) Multiple Regression}: \mathrm{Y} = \mathrm{a} + \mathrm{b1X1} + \mathrm{b2X2} + \mathrm{b3X3} + \cdots + \mathrm{btXt} + \mathrm{u}$$

where

Y the variable that you are trying to predict (dependent variable);
X the variable that you are using to predict Y (independent variable);
a the intercept;
b the slope;
u the regression residual.

Thus, using the Kaya model for empirical research on global warming, the following anthropogenic conditions would affect positively carbon emissions:

(E3) CO_2:s = F(GDP/capita, Population, Energy intensity, Carbon intensity), in a stochastic form with a residual variance, all to be estimated on most recently available data from some 59 countries.

3.2 *Empirical Findings*

I make two empirical estimations of this probabilistic Kaya model, one longitudinal for 1990–2014 as well as one cross-sectional for 2014.

3.2.1 Longitudinal Analysis

I make an empirical estimation of this probabilistic Kaya model—the longitudinal test for 1990–2014, World data 1990–2015 (Fig. 1):

$$\text{(E4) } \ln CO_2 = 0.62 * \ln(\text{Population}) + 1.28 * \ln(\text{GDP/Capita}) + 0.96 * \ln(\text{Energy/GDP}); \quad R2 = 0.90$$

The Kaya model findings show that total GHG:s go with larger total GDP. To make the dilemma of energy versus emissions even worse, we show in Fig. 1 that GDP increase with the augmentation of energy per capita. This makes the turn to a sustainable economy (Sachs 2015a, b) unlikely, as nations plan for much more energy in the coming decades.

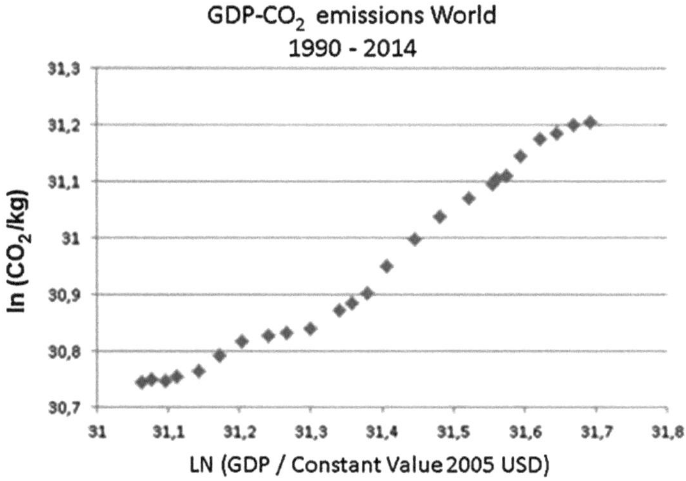

Fig. 1 Global GDP-CO_2 link: y = 0.80x + 5.96; R^2 = 0.97 (N = 59)

Decarbonisation is the promise to undo these dismal links by making GDP and energy consumption rely upon carbon neutral energy resources, like modern renewables and atomic energy.

We need to model this energy-emission dilemma for the countries of the COP21 project. To understand the predicament of Third World countries, we need to know whether GHC:s or CO_2:s are still increasing (Goal I) and what the basic structure of the energy mix is (Goal II). Thus, I suggest:

$$\text{GDP} - \text{GHG}(\text{CO}_2)\ \text{link},\ \text{energy mix}$$

as a model of the decarbonisation feasibility in some Third World countries, to be analysed below, following the so-called "Kaya" model. The first concept taps the feasibility of Goal I: halting the growth of GHG:s or CO_2:s, whereas the other concepts target the role of fossil fuels and wood coal like charcoal.

3.2.2 Cross-Sectional Analysis

In a stochastic form with a residual variance, all to be estimated on data from some 59 countries, I make an empirical estimation of this probabilistic Kaya model—the cross-sectional test for 2014:

$$\begin{aligned}(\text{E5})\ \text{k1} = 0.68;\ \text{k2} = 0.85;\ \text{k3} = 0.95;\ \text{k4} = 0.25\\ \text{R}^2 = 0.80\end{aligned}$$

Note that:

$$\begin{aligned}\ln(\text{CO}_2) = \text{k1} * \ln(\text{GDP/Capita}) + \text{k2} * (\text{dummy for Energy Intensity})\\ + \text{k3} * \ln(\text{population}) + \text{k4} * (\text{dummy for Fossil Fuels/all})\end{aligned}$$

where:

Dummy for fossils = 1 if more than 80% fossil fuels; k4 not significantly proven to be non-zero, all others are (N = 59).

4 The Emission-Energy Conundrum

The findings here entail that economic growth and population increases are the key determinants of carbon emissions. This creates a formidable challenge for the signatories of the COP21 Agreement. The government face the management task of a complete overhaul of national energy systems, from fossil fuel, especially coal, to renewables like solar, wind, geo-thermal and atomic power.

One hand hoped that the energy part of GDP would go down when economies mature. Figure 2 shows that this is doubtful on a global scale. Figure 3 shows not

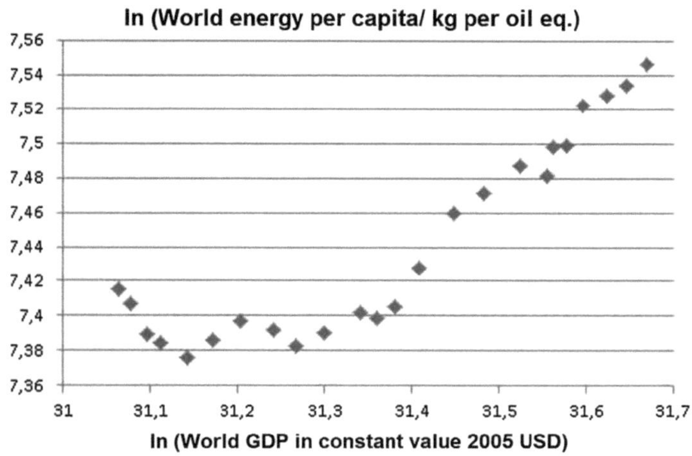

Fig. 2 GDP against energy per person (N = 59)

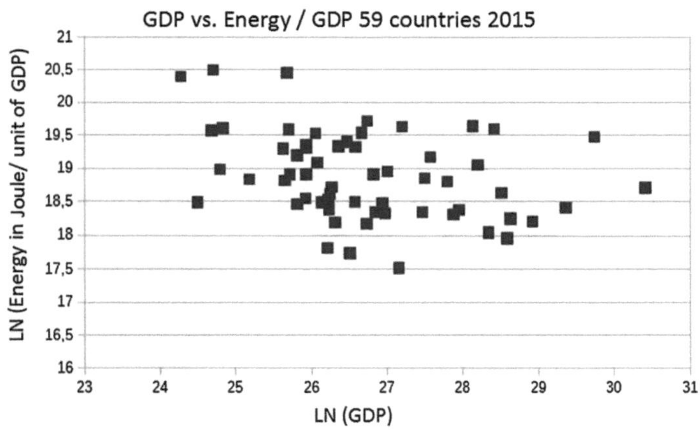

Fig. 3 Global energy efficiency ($y = -0.17x$; $R^2 = 0.11$; N = 59)

tendency to minimize energy/GDP as countries grow richer. A few countries use an incredible amount of energy person to deliver electricity for very high standards of living, viz, the Gulf States. As global warming proceeds, more and more people need air-conditioning all the time.

Now, we can also show that the ratio fossil fuels/energy resources are not displaying any economy of scale with GDP (Fig. 4). Had this been the case, it would have been an example of a Kutnez' environmental curve.

The difference between global warming concern and general environmentalism appears clearly in the evaluation of atomic power. For reducing climate change, nuclear power is vital, but for environmentalism atomic power remains a threat. From a short-term perspective, the global warming concerns should trump the fear

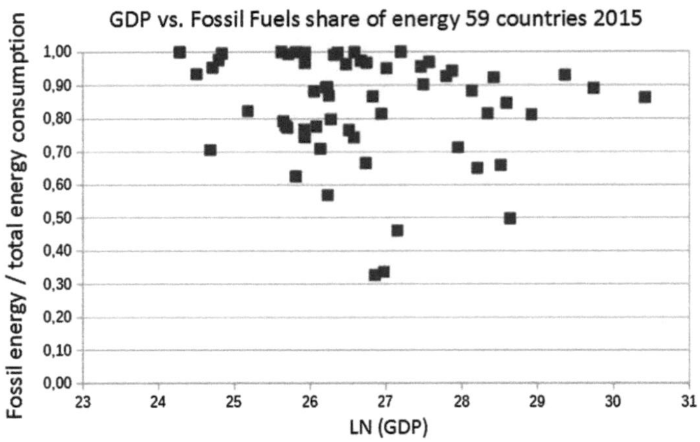

Fig. 4 Carbon efficiency of energy (y = −0.018x; R^2 = 0.0222; N = 59)

of radioactive dissemination, as global warming will hit mankind much sooner. In the Third World, nuclear power plants are increasing in number, whereas in the mature economies their number is being reduced. New nuclear technology is much safer, why also advanced countries should use this option, like for instance the UK.

5 Comparative Country Inquiries: Energy-Emission Conundrums

It cannot be enough underlined that the COP21 Agreement resects a key principle in public international law (PIL), namely state sovereignty. This entails that the chief responsibility for taking action against global warming rests with the country governments. Some minimum oversight from international governance is foreseen, but the main activity of the international bodies, active in climate programs, is to set up and run a giant global fund for helping countries make the crucial energy transition involved in decarbonisation—the Super Fund with 100 billion dollars per year in the coming decade.

Now, what has been completely lacking from the COP21 project discussion is the recognition that the fulfillment of its objectives requires massive energy management, monitored by international governance. The costs of a tremendous energy transformation, from fossil fuels and wood coal, to renewables in a short period of time of about 1–3 decades will be astronomical. The talk about a Super Fund along the ideas of Stern (2007) is most appropriate, as many countries are in dire need of financial assistance.

The energy-emission problematic is much aggravated by the ambition of almost all nations to increase their energy consumption up to 2050. Thus, not only have old energy sources to be replaced but also new ones erected—what a burden for investments in solar, wind and geo-thermal energy besides the controversial atomic power plants (British Petroleum 2016a, b; Outlook 1998; Wildavsky 1979, 1987).

Below, we give a few telling examples of what the management of the COP21 project involves at the nation level. Governments have accepted he responsibility to promote the COP21 goals. In each of these countries, governments will have to work with the markets and civil society. But the probability of implementation success is low, unless the Super Fund can make a huge difference, which I doubt very much, given the financial situation in Western nations.

5.1 Huge Poor Countries in Need of Assistance

5.1.1 India

India will certainly appeal to the same problematic, namely per capita or aggregate emissions. The country is more negative than China to cut GHG emissions, as it is in an earlier stage of industrialization and urbanization. Figure 5 shows the close connection between emissions and GDP for this giant nation.

India needs cheap energy for its industries, transportation and heating (Fig. 6) as well as electrification. From where will it come? India has water power and nuclear energy, but relies most upon coal, oil and gas as power source. It has strong ambitions for the future expansion of energy, but how is it to be generated, the world asks. India actually has one of the smallest numbers for energy per capita, although it produces much energy totally. Figure 6 shows its energy mix where renewables play a bigger role than in for instance China.

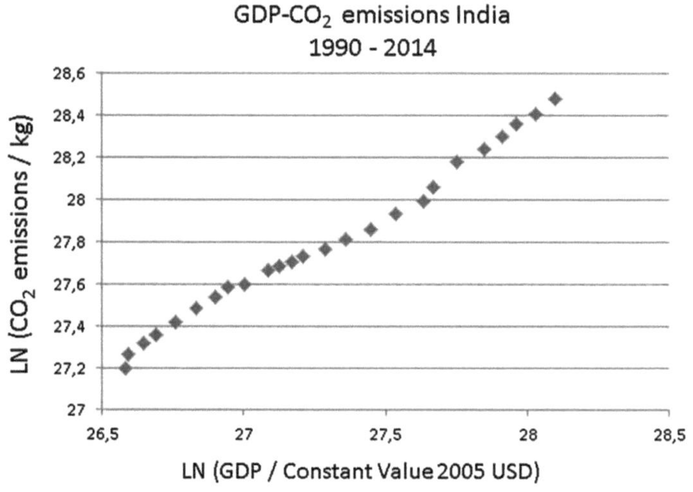

Fig. 5 India's link GDP-CO_2: $y = 0.77x + 6.79$; $R^2 = 0.99$

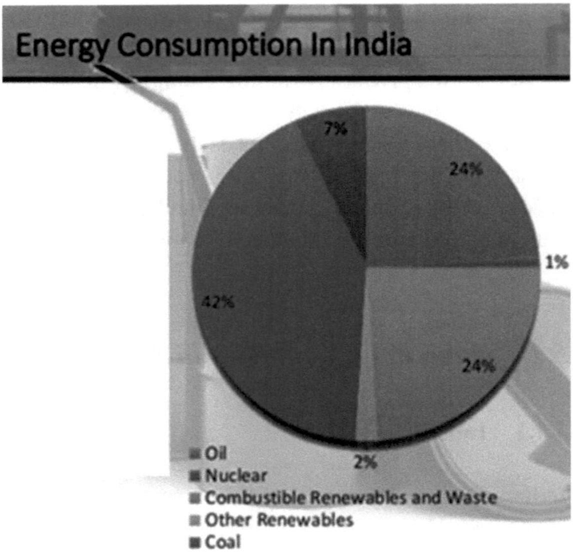

Fig. 6 India energy mix. *Source* IEA (International Energy Agency)

India needs especially electricity, as 300 million inhabitants lack access to it. The country is heavily dependent upon fossil fuels (70%), although to a less extent than China. Electricity can be generated by hydro power and nuclear power, both of which India employs. Yet, global warming reduces the capacity of hydro power and nuclear power meets with political resistance. Interestingly, India uses much biomass and waste for electricity production, which does not always reduce GHG emissions. India's energy policy will be closely watched by other governments and NGO:s after 2018.

5.1.2 Indonesia

One may guess correctly that countries that try hard to "catch-up" will have increasing emissions. This was true of China and India. Let us look at three more examples, like e.g. giant Indonesia—now the fourth largest emitter of CO_2:s in the world (Fig. 7).

Indonesia is a coming giant, both economically and sadly in terms of pollution. Figure 7 reminds of the upward trend for China and India. However, matters are even worse for Indonesia, as the burning of the rain forest on Kalimantan and Sumatra augments the GHG emissions very much. Figure 8 presents the energy mix for this huge country in terms of population and territory.

Only 4% comes from hydro power with 70% from fossil fuels and the remaining 27% from biomass, which also pollutes.

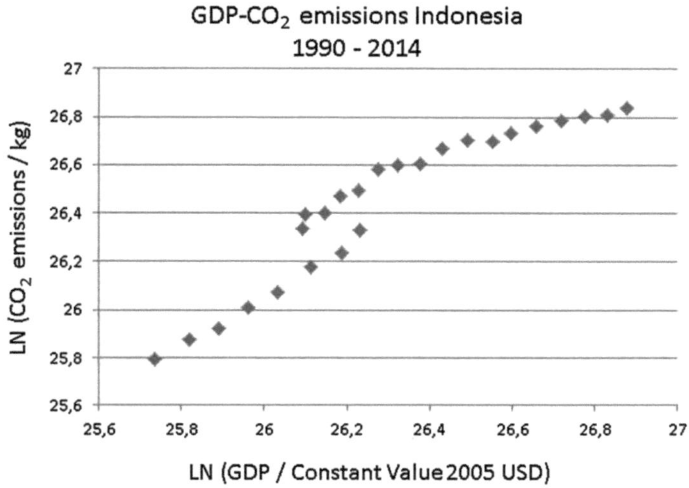

Fig. 7 Indonesia's link GDP-CO_2: y = 0.95x + 1.58; R^2 = 0.89

Fig. 8 Indonesian energy in 2009. *Source* http://missrifka.com/energy-issue/recent-energy-status-in-indonesia.html

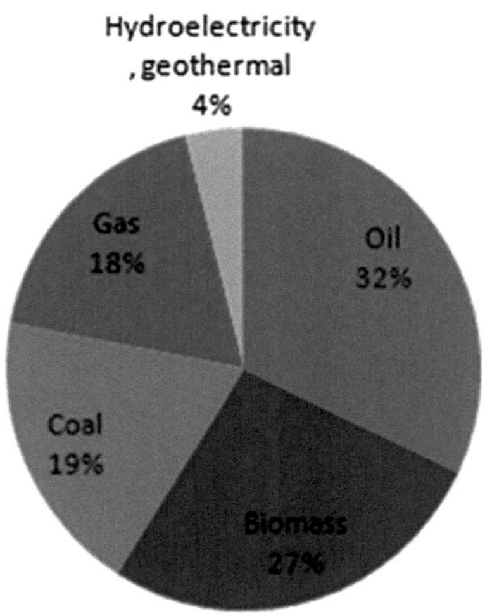

5.1.3 Brazil

Let us look again at a BRIC nation, namely the ethanol country par preference: Brazil. Figure 9 shows a considerable drop in total emissions, but it is followed by huge increases that tend to flatten out.

Brazil employs the most biomass in the world, but the emissions stay at a high level, which is a reminder that renewables may also have GHG:s. One advantage for Brazil is the large component of hydro power, but the overall picture for the largest Latin American country is not wholly promising when it comes to reduction

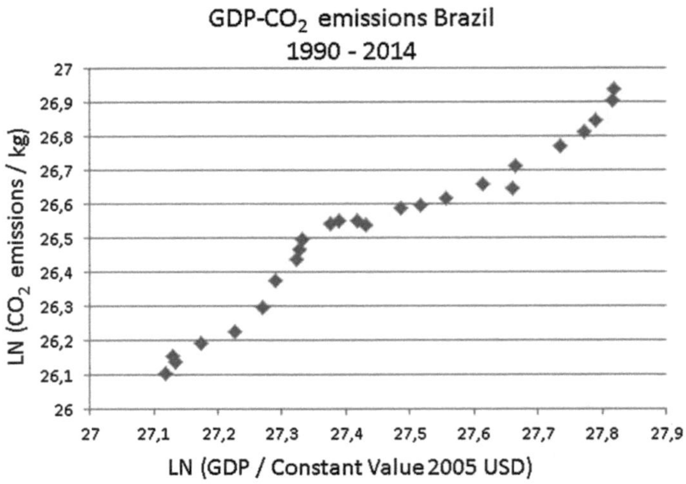

Fig. 9 Brazil's link GDP-CO_2: y = 1.03x − 1.72; R^2 = 0.95

Fig. 10 Brazil's energy consumption 2013. *Source* Energy Matters

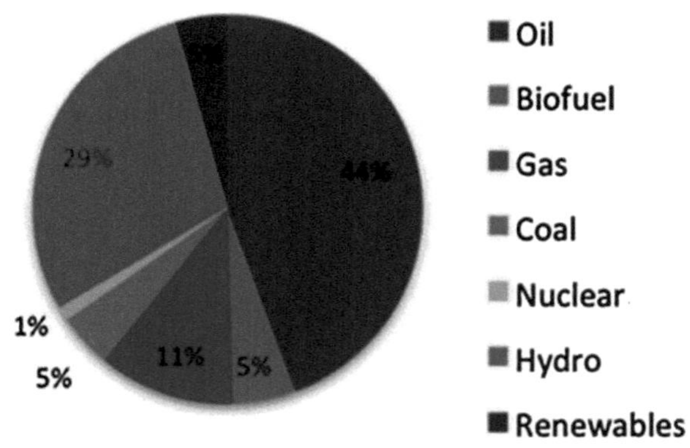

of emissions. Global warming reduces the potential of hydro power, and Brazil has very little nuclear power (Fig. 10).

5.2 *Big Advanced Economies in Need of Fundamental Energy Transformation*

A few nations do not depend upon any foreign assistance, because they are highly developed technologically and can draw upon own substantial financial resources. One may find that the emissions of CO_2 follows economic development closely in many countries. The basic explanation is population growth and GDP growth—more people and higher life style demands

5.2.1 China

Take the case of China, whose emissions are the largest in the world, totally speaking (Fig. 11). China was a Third World country up until yesterday.

The sharp increase in CO_2:s in China reflects not only the immensely rapid industrialization and urbanization of the last 30 years, but also its problematic energy mix (Fig. 12).

Almost 70% of the energy consumption comes from the burning of coal with an additional 20% from other fossil fuels. The role of nuclear, hydro and other renewable energy sources is small indeed, despite new investments. This makes China very vulnerable to demands for cutting GHG emissions: other energy sources or massive installation of highly improved filters? Relying upon market incentives (Hayek 1991), China wants to maintain high economic growth in the decades to come (de Bruyn 2012; Erikson 2013).

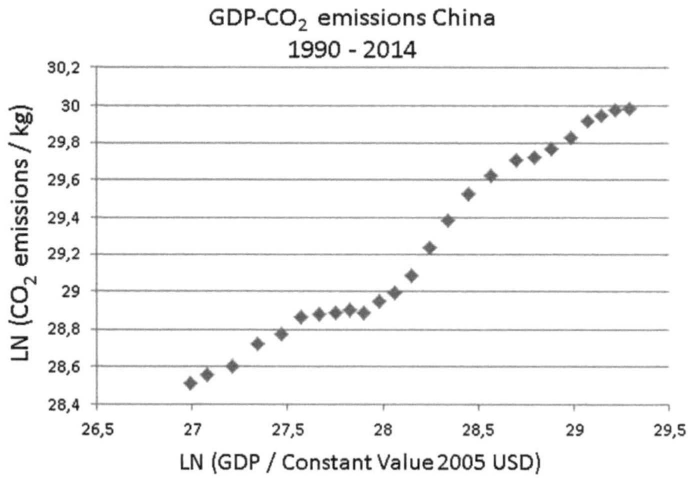

Fig. 11 China's link GDP-CO_2: y = 0.703x; R^2 = 0.97

Fig. 12 China's energy consumption. *Source* Energy Matters

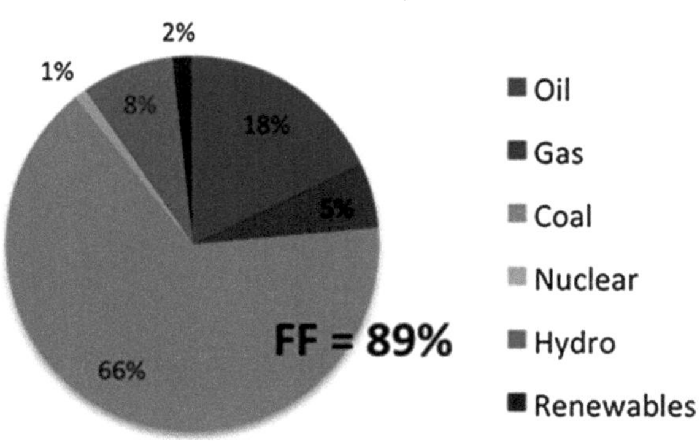

It should be pointed out that several small countries have much higher emissions per capita than China. This raises the enormously difficult problematic of fair cuts of emissions. Should the largest polluters per capita cut most or the biggest aggregate polluters? At COP21 this issue was resolved by the creation of a Super Fund to assist energy transition and environment protection in developing counties, as proposed by economist Stern (2007). But China can hardly ask for this form of foreign assistance. It is true that China energy consumption is changing with much more of renewables ad atomic plants. But so is also demand increasing with new and bigger cars all the time plus increased air traffic on huge new airports. Can China really cut CO_2:s with 40% while supply almost 50% more energy power, according to plan?

5.2.2 South Korea

Industrial giant South Korea is very interesting from the perspective of the COP21 Agreement, because the basic trend violates both Goal I and Goal II. An entirely different trend than that of other mature economies is to be found in South Korea (Fig. 13), which has 'caught up' in a stunning speed but with enormous GHG emissions.

Lacking much hydro power, South Korea has turned to fossil fuels for energy purposes, almost up to 90% (Fig. 14). Now, it builds nuclear plants, but South Korea needs to move aggressively into solar power to reverse trends.

It differs from China only in the reliance upon nuclear power, where the country is a world leader in plant constructions. Reducing its GHG emissions, South Korea

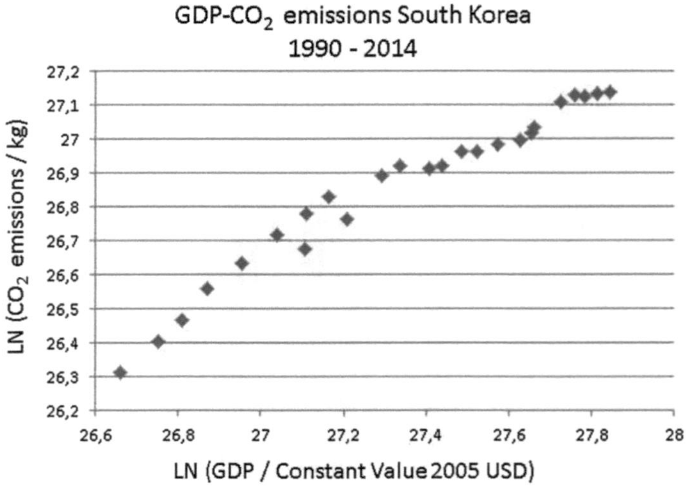

Fig. 13 South Korea's link GDP-CO_2: $y = 0.65x + 9.19$; $R^2 = 0.96$

Fig. 14 Energy in South Korea. *Source* EIA

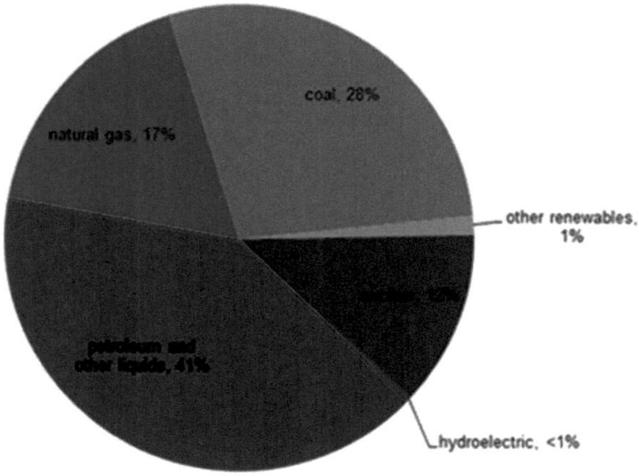

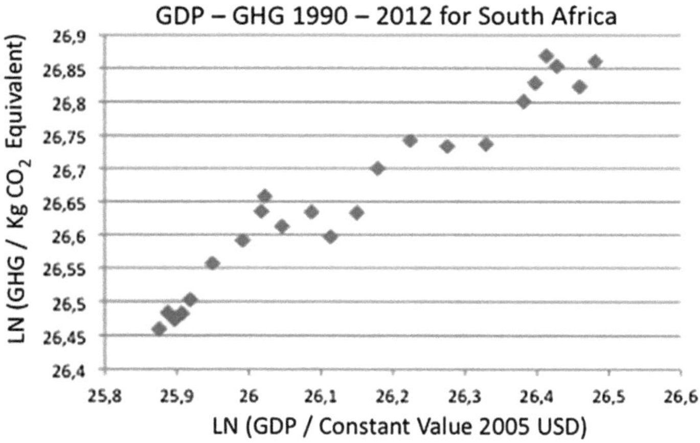

Fig. 15 South Africa: ln(GHG/kg CO_2eq) and ln(GDP/Constant Value 2005 USD)

will have to rely much more upon renewable energy sources, as well as reducing coal and oil for imported gas or LNGs.

6 Fairness in Emission Reduction

Since the emissions of GHG:s or CO_2:s per capita are much higher in rich countries than poor nations, the United Nations Conference on Climate Change will have to discuss "fair" cuts in emissions, given the implications for cheap energy and economic development. Consider for instance a BRIC country like the RSA (Fig. 15).

Emissions are high in the RSA, because South Africa uses a lot of coal to generate electricity (Fig. 16). Decarbonisation will be difficult and costly.

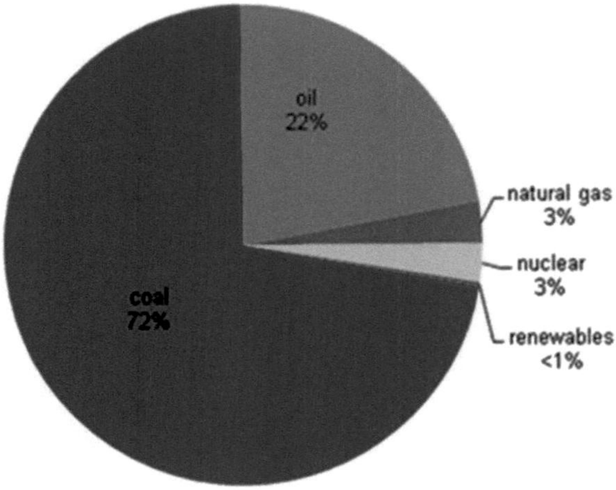

Fig. 16 RSA energy mix. *Source* EIA

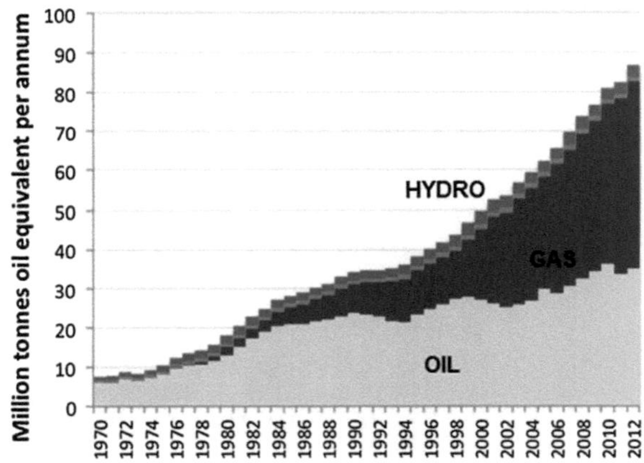

Fig. 17 Egypt's energy mix. *Source* Energy Matters

The reliance upon coal in this large economy in Africa is stunningly high (Fig. 16). No wonder that the RSA has started to look for shale oil and gas.

Or consider huge and poor Egypt that has neither much hydro power nor oil assets, but huge natural gas deposits. The emission trend is clear. It has a huge population with high unemployment and mass poverty besides a certain level of political instability, resulting from religious conflicts. But surely it has electricity from inta giant Assuam dam and the Nile? No, it does not count for very much, where most people live in the Nile delta (Fig. 17).

The share of hydro power is stunning low for a country with one of largets rivers in the world. Actuallu, the water of the Nile is the source of interstate confrontation between Egypt, Sudan and Ethiopia.

As Egypt relies upon fossil fuels, it has massive CO_2 emissions, the trend of which follows its GDP (Fig. 18).

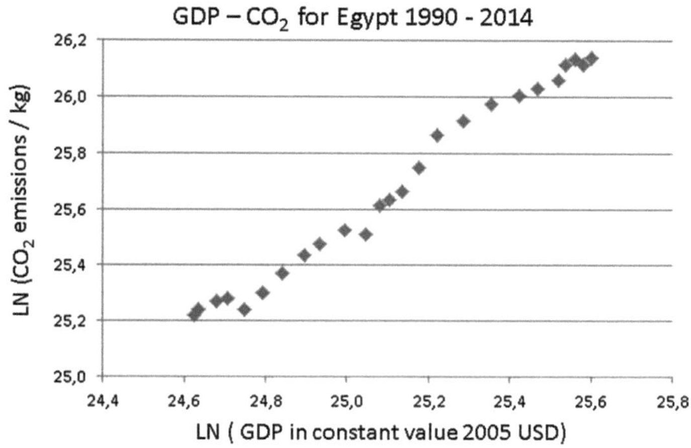

Fig. 18 GDP-CO_2 for Egypt: y = 1.02x; R^2 = 0.99

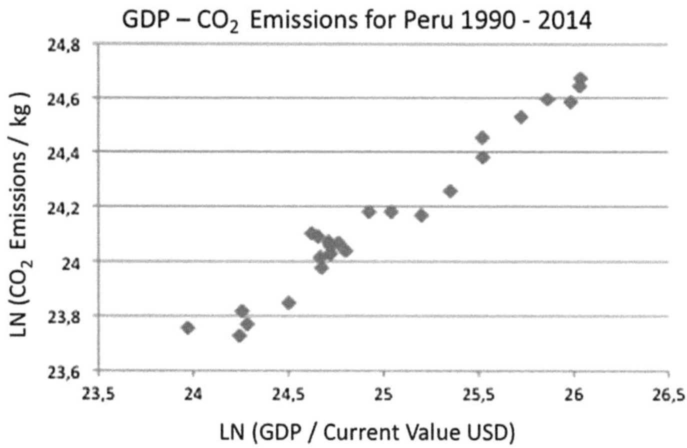

Fig. 19 Peru's link GDP-CO_2: y = 0.47x; R^2 = 0.97

It will be very difficult for Egypt to make the COP21 transformation, at least without massive external support. But where to build huge solar power plants in a country with terrorism, threat or actual?

6.1 Peru

In Latin America, there are several countries that envy the energy and emissions per capita in rich countries, especially the Gulf States with their incredible electricity

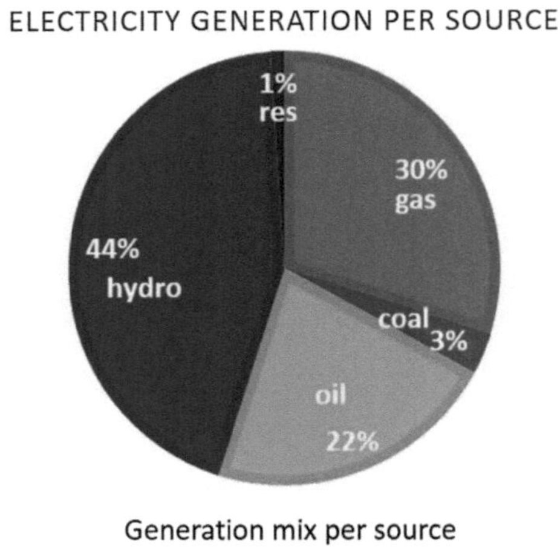

Fig. 20 Energy in Peru: Electricity. *Source* https://windhop.wordpress.com/2015/09/11/electricity-situation-in-peru/

consumption per person. Take Peru for instance. How can it break its upward trend for GDP-CO_2:s (Fig. 19)?

Peru must cut back on CO_2:s while maintaining socio-economic development. How? The country uses hydro power, but is still dependent upon fossil fuels, like for electricity production (Fig. 20). If transportation had been added, Peru's dependency would be larger and not in conformity with COP goals.

Peru like all Third World countries need assistance from the First World and the UN.

7 The Wood Coal Problematic

Renewables should be preferred over non-renewables in the COP21 project. Yet, this statement must be strictly modified, as there are two fundamentally different renewables.

Traditional renewables: wood, charcoal and dung. They are not carbon neutral. On the contrary, employing these renewables results in severe pollution, not only outside but also inside a household. New renewables: solar, wind, geo-thermal and wave energy that are indeed carbon neutral, at least at the stage of functioning. In the poor African countries with about half the population in agriculture and small villages, traditional renewables constitute the major source of energy. Let us look at a giant nation in the centre of the African continent (Fig. 21).

One notes how little of hydro power has been turned into electricity in Kongo, but economic development and political instability, civil war and anarchy do not go together normally. At the same, one may argue that an extensive build-up of hydro power stations would pose a severe challenge to the fragile environment in the centre of Africa. Kongo can now move directly to modern renewables like solar power. This enormous reliance upon traditional renewables is to be found also in

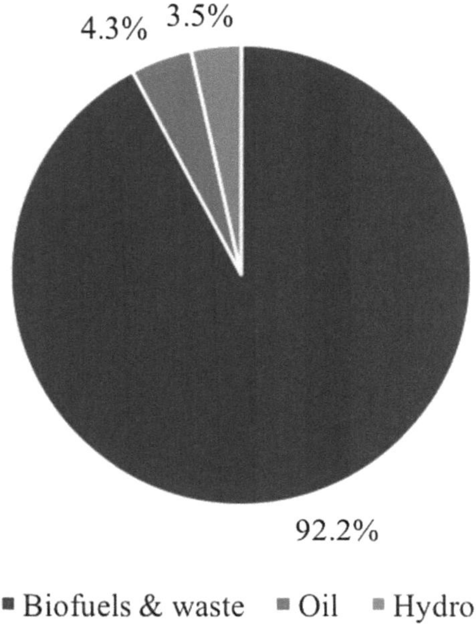

Fig. 21 Democratic Republic of Congo. *Source* Energy Outlook, Kungliga Tekniska Hoegskolan

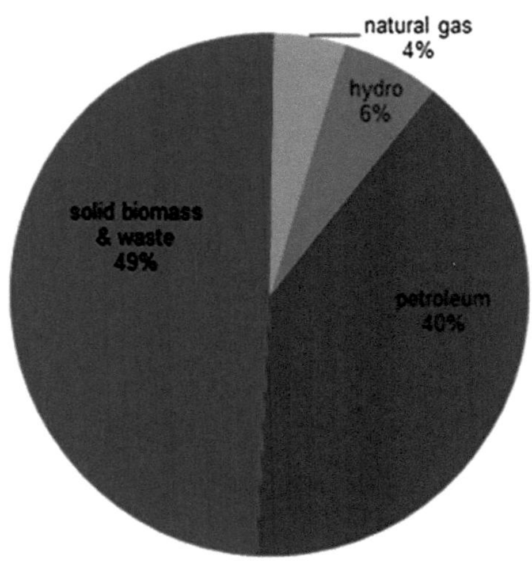

Fig. 22 Angola's primary energy consumption, 2012. *Source* EIA

Angola and Nigeria, although both have access to both hydro power and fossil fuels. Figure 22 describes the energy mix for Angola.

Angola like Kongo has suffered from long and terrible civil war. In the mass of poor villages, energy comes from wood, charcoal and dung—all with negative environmental consequences. Angola has immense fossil fuels—oil and gas, but the political elite family may prefer to export these resources instead of using them for electricity generation. Giant Nigeria has a resembling energy mix—see Fig. 23.

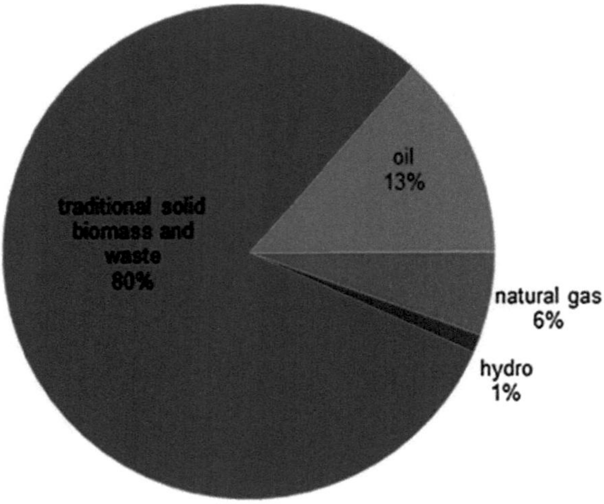

Fig. 23 Nigeria's energy mix. *Source* EIA

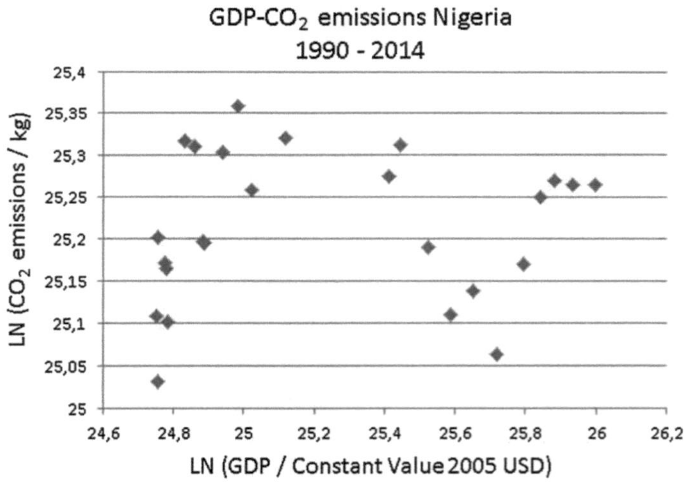

Fig. 24 Nigeria's link GDP-CO_2: y = 0.016x + 24.8; R^2 = 0.006

Nigeria would have to diminish the use of traditional renewables in order to meet the COP21 goals. The very same policy recommendation applies to two countries in the Nile valley, namely Sudan and Ethiopia—extremely poor countries relying mainly upon traditional renewables (Fig. 24).

The countries that rely upon traditional renewables to an extent up to 50% or more have to reflect upon how to bring these figures down considerably with modern renewables. The massive employment of wood coal leads to deforestation and desertification.

8 The Option of Reneging

The United Nations Framework Convention on Climate Change has just held its 22nd meeting, but there is little in terms of real achievements so far. The amount of CO_2 in the atmosphere is still augmenting and the acidification of the oceans is up. No plan how to build up the Super Fund has been delivered by the UNFCCC—how to finance such a huge fund with what kinds of charges or taxes?

It is true that some countries reduce coal energy, which is a positive. But at the same time the number of cars and their size increases. Nothing has been done to reduce emissions from air or sea transportation that is simply speaking just enormous. Instead, many countries plan for extensions of their airports and the airlines launch huge plain orders. Soon China can produce its own aircraft for civilian purposes at its immense airports. It is also true that renewable energy is rapidly expanding—another positive. But the management tasks of supply solar and wind energy on a gigantic scale are mind-boggling. Atomic energy is cut back in some countries, which is negative, but expanded in a few others. Nobody knows what to do with the reliance upon wood coal in poor nations or how to stop the cutting down of forests, not only the rain forests, for having wood coal or for agricultural produce or conspicuous consumption like Gabon. In reality, the coordination needs are daunting. How to make sure that governments simply sign the COP21 Agreement and then renege upon the implementation of its goals, citing poverty, lack of funding and no technological expertise?

8.1 Japan

Governments make plans, but they may not hold for unforeseen developments. Take the case of Japan (Fig. 25).

Japan is today more dependent upon fossil fuels than earlier due to the debacle with its nuclear energy program. Could it not happen elsewhere too? When forced, governments will turn back to the fossil fuels, as for them economic growth trumps the environment. After all, nations are brutally egoistic, at least according to standard teachings in international relations.

8.2 Iran

Countries may rely upon petroleum and gas mainly—see Iran (Fig. 26). CO_2 emissions have generally followed economic development in this giant country, although there seems to be a planning out recently, perhaps due to the international sanctions against its economy.

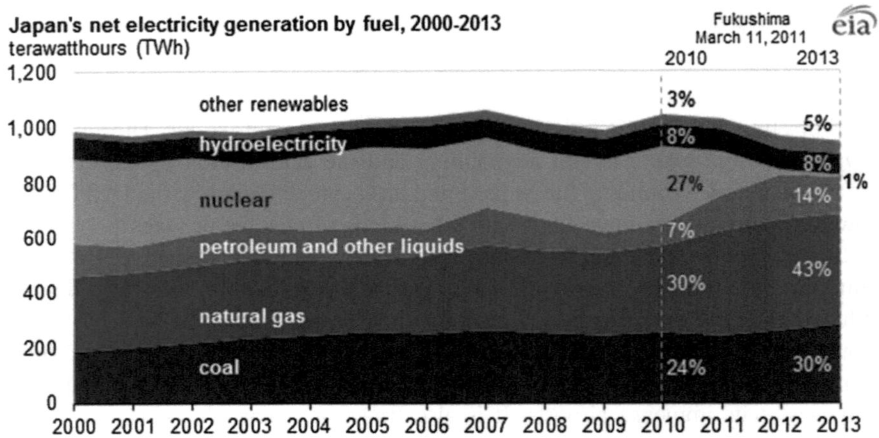

Fig. 25 Energy for electricity in Japan. *Source* EIA

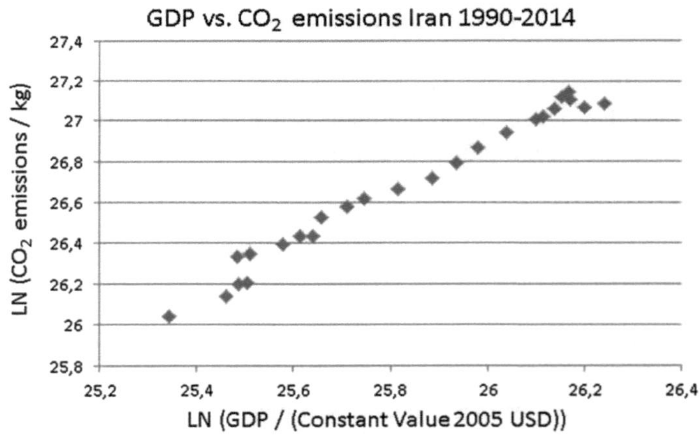

Fig. 26 Iran GDP-CO_2 link $y = 1.2229x - 4.91$; $R^2 = 0.98$

Iran is together with Russia and Qatar the largest owner of natural gas deposits. But despite using coal in very small amounts, its CO_2 emissions are high. Natural gas pollute less than oil and coal, but if released unburned it is very dangerous as a greenhouse gas. Iran relies upon its enormous resources of gas and oil (Fig. 27).

Iran needs foreign exchange to pay for all its imports of goods and services. Using nuclear power at home and exporting more oil and gas would no doubt be profitable for the country. And it would also help Iran with the COP21 goals achievement, or decarbonisation. Oil producing countries with economic difficulties like e.g. Venezuela and Saudi Arabia may renege upon the radical decarbonisation plan of the UN.

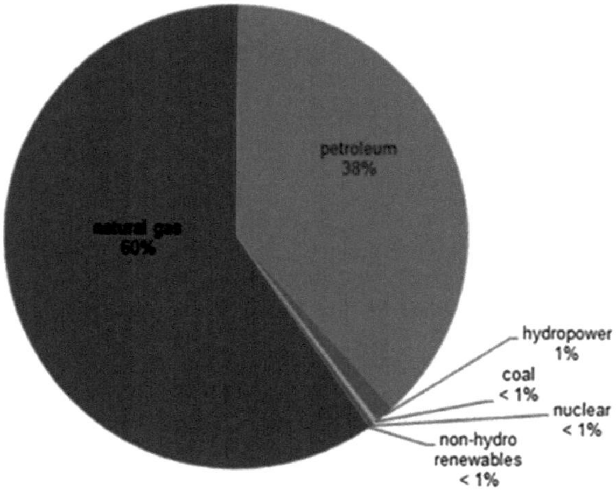

Fig. 27 Iran energy mix. *Source* EIA

9 Conclusions

Although awareness is growing about the risks with climate change for mankind, and despite many new initiatives to develop renewable energy sources, the basic parameters of the human conditions have not changed (Stern 2007, 2016).

What is completely lacking is the application of social science models that teach how difficult state coordination is and how transaction cost heavy international governance tends to be. Implementation of policy goals constitutes and art for the management science. It is problematic to arrive at successful policy accomplishment, especially when mundane incentives play a major role, as in the GDP-energy-emissions conundrum.

In a recent book, Stern (2016) underestimates the hindrances to the fulfilment of the COP21 objectives. The COP21 or COP22 project may derail due to:

- Lack of funds for huge poor nations.
- Economic growth trumps the COP21 goals in catch-up nations.
- Countries are forced to renege due to unforeseen developments.
- Countries with traditional renewables lack alternatives.
- Oil producing countries find oil and natural gas cheaper than solar, wind or atomic power.
- Nations with hydro power are faced with water shortages and droughts.
- The risk that any of the big polluters simply reneges entirely upon the whole CO_2 1 project, like the USA.

The UNFCCC offers more utopian objectives than down to earth policy implementation. Its promises of real outcomes—almost complete decarbonisation—are open to the most powerful strategic response of sovereign states, namely government reneging. Countries aiming at quick socio-economic development, having experiences their "take-off stage" (Rustow 1960) and pursuing a "catch-up" strategy (Barro 1991, 1992, 1995) will only implement decarbonisation goals, if compensated from the Super Fund for the giant costs.

References

Barro, R. J. (1991). Economic growth in a cross section of countries. *The Quarterly Journal of Economics, 106*(2), 407–443.

Barro, R. J., & Sala-i-Martin, X. X. (1992). Convergence. *Journal of Political Economy, 100*(2), 223–251.

Barro, R. J., & Sala-i-Martin, X. X. (1995). *Economic growth*. New York: McGraw Hill.

British Petroleum. (2016a). *Statistical review of world energy*. London: BP.

British Petroleum. (2016b). *Energy outlook*.

de Bruyn, S. M. (2012). *Economic growth and the environment: An empirical analysis*. Berlin: Springer.

Eriksson, C. (2013). *Economic growth and the environment: An introduction to the theory*. Oxford: OUP.

EU Emissions Database for Global Research EDGAR. http://edgar.jrc.ec.europa.eu/

EU Joint Research Centre Emission Database for Global Atmospheric Research. http://edgar.jrc.ec.europa.eu/overview.php

Hayek, F. (1991). *Fatal conceit. The errors of socialism*. Chicago: University of Chicago Press.

International Energy Agency (IEA). http://www.iea.org/statistics/ieaenergyatlas/

Kaya, Y., & Yokoburi, K. (1997). *Environment, energy, and economy: Strategies for sustainability*. Tokyo: United Nations University Press.

Mazmanian, D. A., & Sabatier, P. A. (1989). *Implementation and public policy Paperback*. Lanham: UPA.

Outlook, E. (1998). *Energy Information Administration (EIA)*: Washington DC.

OECD National Accounts data files. http://www.oecd.org/std/na/

Pressman, J., & Wildavsky, A. (1973, 1984). *Implementation*. Berkeley: University of California Press.

Rostow, W. W. (1960). *The stages of economic growth: A non-communist manifesto*. Cambridge: Cambridge University Press.

Sabatier, P. A. (1985). Top-down and bottom-up. Approaches to implementation research: A critical analysis and suggested synthesis. *Journal of Public Policy, 6*(1), 21–48.

Sabatier, P. A. (1993). *Policy change and learning: An advocacy coalition approach*. Westview: Boulder.

Sabatier, P. A. (1998). The advocacy coalition framework: Revisions and relevance for Europe. *Journal of European Public Policy, 5*(1), 98–130.

Sachs, J. D. (2015a). *The age of sustainable development*. New York: Columbia University Press.

Sachs, J. (2015b, August 10). Sustainable development for humanity's future. (http://jeffsachs.org/2015/08/sustainable-development-for-humanitys-future/)

Stern, N. (2007). *The economics of climate change*. Oxford: OUP.

Stern, N. (2016). *Why are we waiting? The logic, urgency, and promise of tackling climate change*. Cambridge: MIT Press.

UN Framework Convention on Climate Change. http://unfccc.int/ghg_data/ghg_data_unfccc/time_series_annex_i/items/3814.php

World Bank Data Indicators. data.worldbank.org

Wildavsky, A. (1979, 1987). *Speaking truth to power*. Piscataway: Transaction.

World Bank national accounts data. data.worldbank.org

World Resources Institute CAIT Climate Data Explorer. cait.wri.org

Chapter 9
Do Municipalities Have the Right Tools to Become Zero Carbon Emissions Cities? *ACCENT*, a Pan-European Decision-Support Tool to Take Refurbishment Decisions at City Scale Based on Buildings Energy Performance

Carolina Mateo-Cecilia, Vera Valero-Escribano and Miriam Navarro-Escudero

Abstract Buildings are responsible for almost 40% of energy consumption and over a third of CO_2 emissions in the European Union. Most of the buildings that will exist in 2050 are already built. Renovation of the existing building stock is therefore crucial to meet long term energy and climate goals. The public sector is an important driver in supporting market transformation towards more efficient energy systems and buildings. To succeed in the energy transition, it is important not only to mobilize local administrations but also to engage other local stakeholders: citizens and service providers. But the development, financing and implementation of ambitious sustainable energy plans and measures should be based on reliable data, and here we find the big challenge. The aim of this paper is to present the main insights of ACCENT (Accompany Cities in Energy Strategy), a new decision-support tool for local administrations developed under a pan-European project upheld by Climate KIC. ACCENT is a GIS web-based platform that supports local administrations to monitor building energy performance and plan actions on the building stocks of the city. ACCENT faces global pan-European challenges, such as the need to share data regarding buildings energy consumption or the reluctance of some energy suppliers to offer information in this regard. Additionally, in some countries like Spain, specific local barriers are addressed such as the lack of connection between public bodies, the dispersion of available data,

C. Mateo-Cecilia (✉) · V. Valero-Escribano · M. Navarro-Escudero
Valencia Institute of Building, Valencia, Spain
e-mail: cmateo@five.es

V. Valero-Escribano
e-mail: vvalero@five.es

M. Navarro-Escudero
e-mail: mnavarro@five.es

R. Álvarez Fernández et al. (eds.), *Carbon Footprint and the Industrial Life Cycle*, Green Energy and Technology, DOI 10.1007/978-3-319-54984-2_9

the need to make citizens aware of energy renovation, or the inaccessibility of citizens and technicians to energy data. ACCENT four main functionalities—mapping, planning setting up scenarios and connecting—allows the Local Administrations to take refurbishment decisions at city scale based on buildings energy performance.

1 Introduction

Cities consume three quarters of the world's natural resources, and part of their responsibility falls on the building sector. The residential building stock was responsible for about one quarter of the final energy consumption in 2014. This consumption, together with non-residential buildings' consumption, reached about 38% of the final energy consumption, representing the largest energy consuming sector in Europe (Eurostat 2016).

Buildings sector worldwide, comprising both the residential and services sub-sectors, generates up to about 17% of total direct energy-related CO_2 emissions from final energy consumers. In addition, if indirect upstream emissions attributable to electricity and heat consumption are considered, the sector contributes about one-third of global CO_2 emissions (International Energy Agency 2013).

Previous research has studied whether the construction of new buildings is preferable to the refurbishment of existing ones. A high number of studies remark that the demolition and replacement with highly efficient new buildings have less operational (in-use) carbon emissions. On the contrary, the development of new buildings consumes huge quantities of energy that could be saved by reusing the existing ones (Ireland 2008). Moreover, considering that the average lifespan of a building is over 50 years and a complete renewal of the building stock would take 100 years, investment in building renovation should be a priority (THINK 2012).

European cities are completely aware of this and energy policies focused on energy efficient cities are being continuously launched specially since 2011 (European Commission 2011). The Covenant of Mayors (CoM), a major European network involving local and regional authorities towards sustainability and resilience to climate change, has gathered more than 6600 European signatory cities as of April 2016 (Covenant of Mayors for Climate & Energy 2016). These signatories voluntarily agree to reduce their CO_2 emissions by at least 20% by 2020 with respect to a Baseline Emission Inventory (BEI), a quantification of the amount of CO_2 emitted due to energy consumption in the territory of the signatory within a given period of time, being 1990 the recommended base year. After establishing the BEI, they have to submit a Sustainable Energy Action Plan (SEAP) outlining how they intend to reach this CO_2 reduction target by 2020, identifying the activities and measures planned, together with time frames and assigned responsibilities.

It is well known the crucial importance of data collection when planning public policies (Genre et al. 2000). A successful SEAP needs to rely on a relevant and detailed diagnosis of their territory in order to identify potentials o energy savings and renewable resources and building appropriate mid and long-term strategies. However, urban renovations are usually characterized by high levels of uncertainty and risk, and local governments need to get a real overview on the energy performance of the city, to overcome the lack of information and complexity of tasks (Ferreira et al. 2013). The collection of reliable data with an acceptable level of accuracy is one of the most difficult tasks in designing, implementing and monitoring a SEAP (Bloem et al. 2015).

In order to achieve long term urban renovation processes, monitoring and managing decision-support tools are needed (Ascione et al. 2012; Stylianidis et al. 2012; Hailu 2012). The two most important aspects of these decision-support tools are (1) the manageability (Bhanot and Jha 2012) and (2) the integration of real data from different city stakeholders as service providers, citizens, policy makers and local authorities (Müller and Siebenhüner 2005).

In this context, ACCENT (ACcompany Cities in ENergy sTragegy) tries to fill the gap by providing a decision support tool within the building sector easy to manage. It has been developed as an interactive GIS-web-based tool aiming at making energy data accessible to a wide variety of stakeholders (local authorities, service providers and citizens). ACCENT aims to be a decision support tool to overcome the stakeholders' barriers by tackling their main needs when implementing energy efficiency building measures at any scale. Within the next pages, it will be explained (1) the co-creation design process of the ACCENT tool and (2) its main functionalities.

2 Co-creation Design Process

2.1 Why Don't Existing Tools Answer to Municipalities Needs?

There are several city planning tools existing already in the market, however, why aren't they more implemented in a daily based management at municipalities? As first step, it was needed to do a state of the art analysis in order to avoid mistakes of current tools in the market.

The competitive analysis was done in 2014 and 2015, and it covered nearly 40 initiatives. The main factors to analyze the initiatives were:

- **Type of offer proposed**: The different kinds of offers depend on the financial investment to run the tool. The categories evaluated were: Software, Study/Report, On-line Platform, and spreadsheet.
- **Inputs**: The information that the different tools use to calculate the energy model is essential to determine how the software is conceived and how it works. On which data the maps and estimations are based (statistical, real data, etc.)? The categories itemized were: Mainly real consumption data, real thermic data, Statistical and injection real data, Statistical data.

- **Outputs:** It answers the question: what type of results does the tool provide (map, graphs, etc.)? The categories spitted were: Follow-up instrument, Map, Scenario/Forecast, Tables and Graphs.
- **Scale of inputs/outputs**: The last factor analyzed was the scale: municipality in a whole, district and building.

The results obtained were synthesized in 4 graphs (Fig. 1). They represent the scale of results, the type of offer proposed by the tool, the inputs, and outputs of the tool. It's needed to remark that the categories analyzed under each factor are not always exclusionary, as the different tools can represent several of the categories analyzed.

In Fig. 1 it is shown that practically 50% of the tools analyzed display the results on a building scale. It is also displayed that a third of the tools analyzed were scientific research without specific tool developed so far, about 60% correspond to computer tools, software or on-line applications; while the number of tools developed based on excel sheets were not very representative. On the other hand, tools based on real thermic data inputs and offering a follow-up instrument, weren't very common.

To sum up, the main weak points identified were:

- Planning tools were set at a local framework, and it was difficult to interact or to make modifications of the existing tools.
- There was a lack of tools focused on the building stock management as a whole.

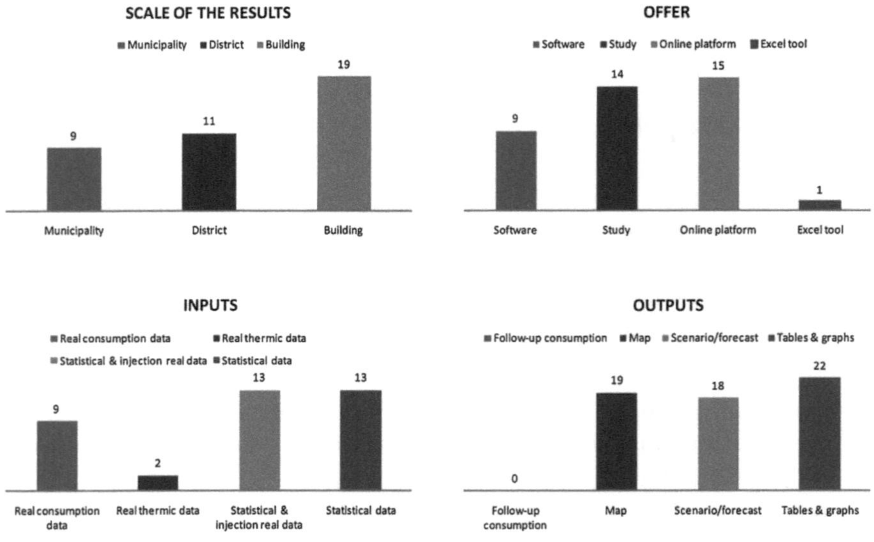

Fig. 1 Synthesis of the tools and factors analyzed

- There was a lack of tools providing urban energy planning assistance and gathering on the same platform the main actors involved in energy efficiency: citizens, municipalities and service providers.
- At the moment of the analysis, none of the online platforms, providing global estimations or maps of the energy consumption, offered a networking option to connect with other cities.

2.2 *Identifying Needs and Barriers at European Wide Scale, with Local and Empirical Information*

The competitive analysis was completed with a market research consisting of a set of 103 semi-structured interviews with potential users of the tool in 5 countries: France, Germany, Italy, Spain and Switzerland, that was later extended to United Kingdom, Belgium, the Netherlands and Denmark. Within the 103 interviews, 54 were made to local authorities, mainly from the Covenant of Mayors (CoM) network. The remaining 49 were made to service providers such as network operators, Energy Service Companies (ESCOs); construction and refurbishment companies, renewable energy developers, utilities, energy consultancy companies, urban planners or building construction materials' suppliers.

The market research was focused on collecting needs and barriers from local authorities and service providers when defining and implementing their energy strategy and also, on gathering their opinion regarding a tool of such characteristics. Besides, it allowed us to detect if local authorities and service providers were using tools such as those identified in the competitive analysis and how these met their specific needs.

To have a clearer overview of the municipalities needs, five structured workshops and interviews were held in France, Italy and Spain, involving local authorities, service providers and citizen representatives along the co-design process. With the information collected in these meetings, the ACCENT mock-up was adapted and refined to the specific needs of stakeholders in the selected countries, to then be used in four cities as a pilot: Paris, Valencia, Reggio-Emilia and Ferrara. In Table 1 we can see the main needs identified by the different actors:

Table 1 Synthesis of needs related to the different stakeholders

Needs	Local administrations	Service providers	Citizens
Visualization of energy data	✗	✗	✗
Follow-up instrument for Action Plan	✗		
Decision-support tool for energy planning (based on the construction of different scenarios)	✗	✗	
Learn from other similar experiences	✗		
Establishing links with other actors	✗	✗	✗

Energy Data Visualization Needs

It was identified that local authorities have a very strong need to have an understanding of the building stock energy consumption in order to be able to draw a strategy of energy efficiency and reduction, set targets and evaluate the impact of actions on energy consumption and CO_2 emission reduction.

Interviews undertaken showed that normally municipalities get yearly real energy consumptions of the building stock aggregated at the city scale. Most of the time there is a rough segmentation of buildings in three categories: residential, tertiary and industry. Furthermore, data are few years behind the current year of exploitation, depending on the type of data and the source. They also count on information regarding buildings and households provided by Statistics organizations (NSIs). Such organizations get data through population census and provide them to actors that evaluate the energy consumptions and CO_2 emissions. But most of the local administrations interviewed pointed that to go further they need more detailed data at all levels. The scale of data needed depends on the actions to be carried out. Generally, lower scale is needed by technical actors, and upper scale information is targeting policy and strategy actors. The interviews have revealed that district and block scale is generally sufficient for municipalities because they are unable to target one precise building and they organize strategies at district level. But some municipalities would prefer information at building level, because they would have a more precise and reliable evaluation of the consumption. On the other side, going down to the level of the dwelling could enable the public authorities to address issues of occupants, but for most actors it is consider as an intrusion into the privacy of citizens.

On the other hand, service suppliers have different needs concerning the scale of information needed. Interviews revealed that they would like data at all types of scales, depending on their core business and their specific projects. Utilities' big investments would require data at regional level. For energy network companies, the needs are at building and district levels, in order to have a more global overview. To develop district heating network, operators need to have at least the information at district level and preferably at building level too. For construction companies doing refurbishment works, data at building level is sufficient but they could need data at dwelling level. Some appliances companies would like the information at dwelling level or even more detailed inside the dwelling in order to have a precise analysis of the consumer behavior and offer them a dedicated service.

Finally, within the citizen's needs, interviewees agreed that the appropriate scale of data for citizens is dwelling/housing scale. Furthermore, citizens are interested to monitor their energy consumption and get precise advices on energy saving measures and potential refurbishment works to improve energy efficiency of their homes. Additionally, it was also considered very interesting to provide them information on energy consumption at city scale, as a necessary exercise of local government transparency, but also to include them in a collective effort, to allow citizens to understand how their actions influence city energy consumptions.

Needs of Follow-Up Tools for SEAPs Implementation

Mostly all the municipalities belonged to the Covenant of Mayors network, and had already developed their SEAP. SEAP defines the activities and measures set up to achieve the targets (reduction of energy consumption and CO_2 emissions), together with time frames and assigned responsibilities. Every two years the municipality has to report on its SEAP implementation, to check the compliance of the interim results with the foreseen objectives in terms of measures implemented and CO_2 emission reductions. But tools and/or follow-up instruments for monitoring are lacking.

Need of Decision-Support Tool for Energy Planning

It was fully agreed that cities and service providers need a tool that allows them to design and evaluate different scenarios and alternatives based on a defined set of attributes. This tool would be very useful to prioritize group of actions per different criteria as: set a feasible level of investment maximizing energy savings, given an emission saving target find the cost-optimal set of actions or define sets of actions with a balance between short- and long-term payback.

Need of Communication Tools

Today there few meeting points between public authorities, service providers and the public. This meeting point is strongly needed. Local administrations have the duty to orchestrate the development of the energy transition on their territory. They do so by directing, closely, the services and products supplied by providers, and by motivating the citizens to have virtuous behaviors regarding their energy consumption. Almost all cities interviewed have information and awareness rising programs through different communications means. On the other hand, service providers want to communicate their activities and reach the market with their services or products. Finally, the citizens need to have clear information about the challenges of their territory and their building regarding the energy consumption. Awareness raising and consciousness of environmental and economic issues is essential to success.

On the other hand, sharing information with other cities implementing similar actions to the ones set out in their action plan could anticipate success or failures. This way the results that a city is willing to share would be displayed on the platform, enabling other municipalities to compare the results achieved with their own policies.

Main Barriers Identified

The main obstacle related to energy data visualization is where to get the information from. Some countries—among them Spain—have an energy system dominated by few major utilities that are reluctant to share their data with local governments. On the other hand, data confidentiality is an issue that must be considered. Although interviews reflected that stakeholders are willing to share their information, they are not always willing to share it for free. Lastly, most part of the information is not digitalized, what means it should be processed first in order to use as an input in the GIS tool.

On the other hand, barriers to develop follow-up instrument for Action Plans were that Local administrations sometimes lack resources, competences or technicians able to introduce their Action Plans in the interactive map. Also, bureaucracy works slow in some countries, where most part of the information cannot be easily reached or updated. Sometimes there is a mismatch between energy efficiency policies and political decisions. People planning energy efficiency actions may not be the same implementing these actions due to changes in the political scheme.

The possibility to learn from other similar experiences is dependent on the Communication in between public administrations, a communication that could be improved. Sustainable Energy Action Plans (SEAPs) are not aligned nor related by any instrument or tool. SEAPs from municipalities at the same territory should be aligned to regional policies and vice versa.

Finally, the main communication barrier with citizens is that there are a significant proportion of citizens who are not learned in the use of energy terms and internet applications. A GIS web-oriented tool would not reach this part of the population, so other complementary means to address citizens would be also positive and welcomed: face-to-face, school campaigns, neighborhoods associations, etc.

3 ACCENT Tool Functionalities Which Answer to Real Challenges

Once the barriers were identified and the needs stated, the ACCENT challenge was how to deal with all of them in a singular tool. A first version of the ACCENT tool is already working in 4 pilot cities including Valencia, as it is shown in the project website (ACCENT project partnership, s.f.). The tool, based on a web-platform, provides different and specific user interfaces depending on users' role: local administrator, service provider or citizen. That enables to later deliver functionalities specifically developed to accomplish each user profile needs. The main features and functionalities of the tool have been built around four main functionalities: mapping, action sets evaluation, scenario and connection with other cities, responding in that way to the needs and barriers identified, as it is shown in Table 2.

The project is now finishing but some meetings, interviews and workshops are still ongoing, involving all actors, public administrations, service providers and citizens, so that to obtain the final validation of the functionalities and services developed so far.

Table 2 Synthesis of needs, barriers and functionalities regarding the different actors involved in the city energy transition

Actors	Needs	Barriers	Functions
Visualization of energy data			
LA	Realistic monitoring of the current situation Updated data More detailed data at all scales	Difficulty to access/obtain data from other stakeholders, such as some energy providers Data confidentiality Inaccessibility to existing information on the state of the buildings Lack of energy geo-located information	Mapping
SP	Easy access to municipality's energy plans More detailed data at all scales		Action sets evaluation Mapping
C	Get to know energy consumption at city scale		Mapping
Follow-up instrument for action plan			
LA	Monitor SEAP actions Interactive updated maps Easy access information	Lack of specific training and resources in local administrations (specially in smallest ones) Mismatch between the planning of energy efficiency measures and the implementation timeframe	Mapping Scenarios Action sets evaluation
Decision-support tool for energy planning			
LA	Design and evaluate different scenarios Prioritize group of actions Set a feasible level of investment maximizing energy savings Find the cost-optimal set of actions	Lack of financial capacity Lack of employees with strong competences and tools	Mapping Scenarios
Learning from other similar experiences			
LA	Learn from other cities experiences Share and compare results with other cities	Lack of tools allowing comparison and sharing of results	Connection with other cities
Establishing links with other actors			
LA	Get to know the action realised by service providers on their territory Have a stronger communication with citizens	Not all citizens have internet access	Action sets evaluation Mapping
SP	Communicate their activities and reach the market Know the energy strategy of their operation territory		Action sets evaluation
C	Have clear information about their territory and energy consumption		Mapping/Action sets evaluation

3.1 Mapping

The interface is based on a GIS map that enables to visualize geo-located energy-related information regarding the buildings and the city including estimated energy consumption or greenhouse gas emissions by energy use (heating, cooling, etc.) and building characteristics. The information is shown at different scales: building, sub-district, district, and city level, as long as it is available.

As you can see in Fig. 2, a dedicated button splits the map in two windows (vertical, horizontal or scope) to let the user compare two different indicators in each one. An info pop-up window is shown on the map when user clicks either on a building, sector or district. Buildings/Sector/Districts are scale dependent and are never shown at the same time. This info pop-up card contains data regarding the identification code, building use, surface (m^2), year of construction, the figure/s of the indicator chosen for the building, sector or district.

A legend tab shows the color code of the indicators displayed on the map. In addition, the mapping functionality allows setting the Basemap between Streets, Satellite, Topography and OpenStreetMap.

The selection tool is used to select groups of buildings. The user can define the selection under geographical criteria and/or under attribute criteria. Available geographical criteria are: all the buildings of the city, all the buildings of a district or a sector, the buildings included inside a polygon defined by user, or all buildings visible on the map. The selection can be also created based on attribute criteria. Users can add as many attribute criteria and conditions as they want. The selections can be saved, and shared with other users or keep them private, or deleted. Finally,

Fig. 2 Mapping functionality screen

a statistics panel allows comparing indicators and data of two different selections side-by-side, what enables to carry out an urban diagnosis.

3.2 Action Sets Evaluation

Local administrations can use the ACCENT tool to support them in the construction of an energy strategy (SEAP) by means of mapping and scenarios interfaces. In case the SEAP is already defined, it can be directly incorporated in ACCENT.

Through the Action sets evaluation, users can check, modify and manage their previously defined building selections, using the same tool previously explained in the mapping functionality section. The tool allows comparing set of actions chosen by the user from a predefined catalogue, and applied to the previously defined selections. This new set of actions are evaluated, and the engine returns the estimated gains of final energy consumption, of GHG emissions, the total investment required, savings on energy costs, etc. A comparison function enables to assess the benefits of each set of actions as it's shown in Fig. 3.

3.3 Scenarios

As buildings are changing over time, data must be stored for the different periods. Distinct building evolutions are possible over time, which represent different

Fig. 3 Action sets evaluation functionality screen

timelines from a beginning year. The mapping functionality shows the actual state of buildings of the city. Scenarios are intended to evaluate and to plan actions over time. It can so be seen as one potential evolution of the buildings from the city based on one reference year of Building stock.

The most interesting actions set can be selected and transferred to one specific scenario, while the actions can then be distributed over it. So when a scenario has to be evaluated, the system has to consider in one hand the buildings in the scenario timeline for those impacted by one or several actions, and in the other hand the buildings in the building stock timeline, at the scenario reference date, for all others.

Through the construction, analysis and comparison of different scenarios, the tool allows taking decisions and defining strategies based on objective and well-founded criteria.

As it's presented in Fig. 4, analysis tab permits to assess the gains (Energy consumption, GHG emissions, and savings), data investment (payback period and rate effort) and energy consumption savings evolution along the implementation of the actions included in the scenario.

3.4 Connection with Other Cities

This functionality, specifically designed for cities, enables them to share and benefit from other cities feedback on specific actions. The city manager of the tool can decide to publish information on the ACCENT network. The information could be regarding a specific action, scenario and/or strategy. Different types of sharing

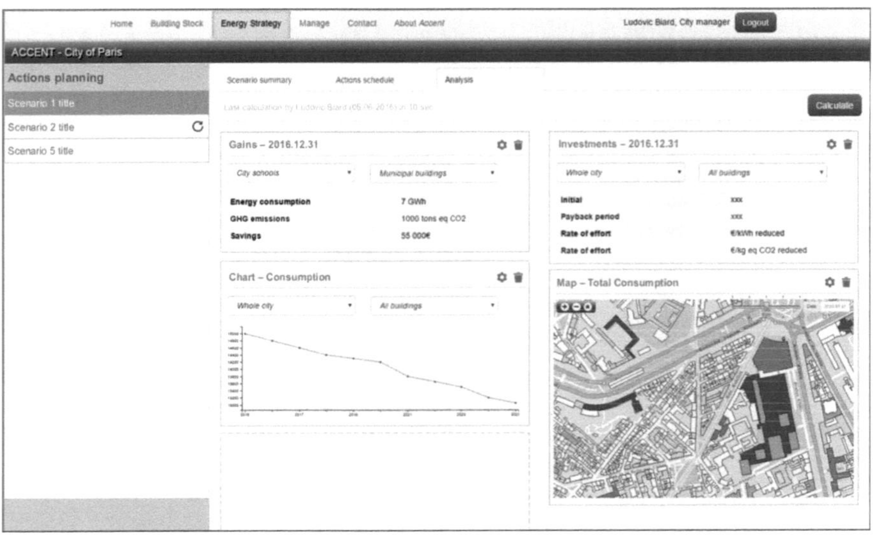

Fig. 4 Scenarios functionality screen

settings have been defined (with a group of other cities, on public website). Comments highlighting successful practices and systems, or difficulties arisen during the implementation can be linked to the published actions in order to work for the common good of energy efficiency along with other cities.

4 Making It Happens: How Does ACCENT Tool Internally Work?

ACCENT is a GIS web-based platform that covers all the functionalities described in the previous section, but how does Accent tool internally work to provide the final users with these functionalities? In order to make it happens, the configuration of ACCENT tool has been built up through the integration of different technical modules. ACCENT works according to the software architecture defined in the Fig. 5.

In the first level we can find the web carto HTML5/API js for AGS is the application used to offer an interface. It represents the real connection between the tool and the final user. Through the use of this application, the user is able to have a private access to the city data and all the functionalities included in ACCENT tool.

Behind this application, the cartographic server Esri ArcGIS for Server is the web service that permits the creation of dynamic and sophisticated maps that meet stakeholders' needs regarding the visualization of energy data associated to a specific geographic location at different scales: building, district, city, etc.

As third layer, we can find the database server PostGreSQL + ArcSDE is the technical block in charge of managing and storing all the data needed to make ACCENT tool works. PostGreSQL is the database management system (DBMS), which means this is the application that permits the definition, creation, update and administration of databases. The other part of this technical module is the application server ArcSDE, whose purpose is to facilitate the storage and management of

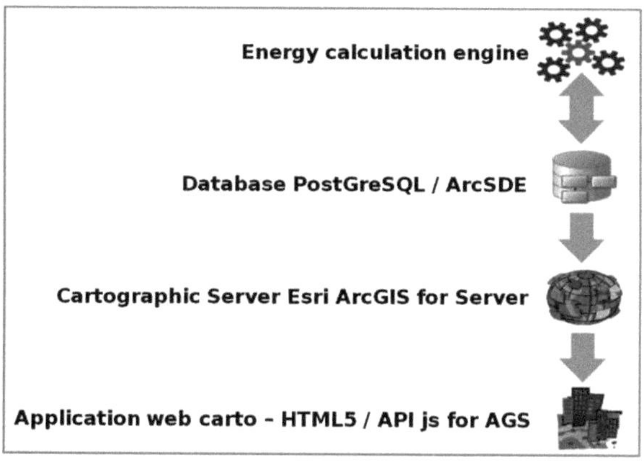

Fig. 5 ACCENT Architecture—technical modules

the spatial data in the database management system PostGreSQL and makes the data available (buildings data, renovation actions, scenarios, users, etc.)

Finally, the energy module is the responsible for the calculation of energy efficiency based on a specific engine used to evaluate energy and GHG ratios of a building based on more than 30 known or estimated building characteristics (construction date, surface, type of use, heating and/or cooling system, building heating system installation year, resistance of insulation and windows, etc.), and on 50 city parameters (architecture, solar irradiation, mean average temperature per month; etc.) (Fig. 6).

The energy module calculation implemented in the ACCENT tool has been developed based on both, the European standard EN ISO 13790 "Energy performance of buildings—Calculation of energy use for space heating and cooling", and the energy model calculation proposed in the European project TABULA "Typology Approach for Building Stock Energy Assessment", that also derives from the standard EN 13790.

Considering both sources, energy consumption ratios disaggregated by energy use: heating, cooling, domestic hot water and specific electricity; are calculated according to the following methodology:

- Domestic hot water (DHW): the building DHW demand is characterized through the definition of an annual ratio per square meter based on the type of the building use and the occupancy data. The type of DHW system, the type of energy used and the year of installation allow to define a global efficiency of the DWH installation that enable to calculate the energy consumption associated.
- Specific electricity: an annual ratio per square meter that depends on the type of building is used to characterize the specific electricity consumption of the building.
- Heating and cooling: the calculation is made for the whole building considering the following criteria:

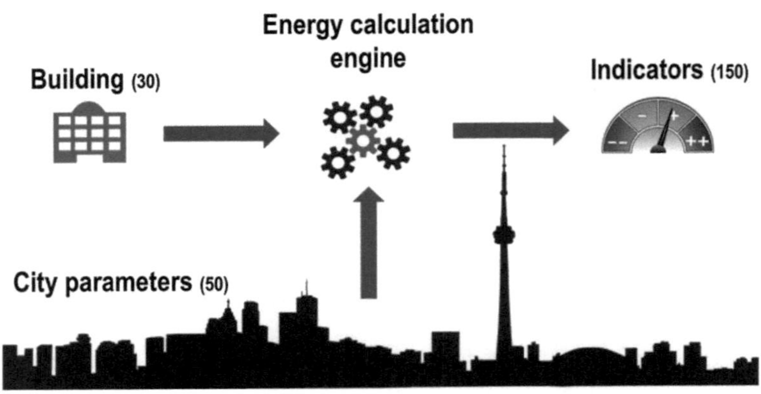

Fig. 6 ACCENT energy module calculation

- Energy losses by transmission are calculated considering both, losses through envelope elements depending on their surfaces and thermal resistances; and thermal bridging characterized by a global coefficient relative to the external dimension of the building.
- Energy losses by air change include the air change by ventilation which depends on the type of building use; and air change by infiltration depending on each specific building.
- Monthly heat gains are calculated considering internal heat gains that depend on the type of building use, the occupancy and the specific electricity consumption in the building; and solar heat gains, considering that the solar irradiation is calculated for each building surface in function of its orientation and inclination. This calculation only considers solar gains through windows following the calculation procedure used in TABULA.
- Monthly energy needs for heating and cooling depends on the climatic conditions, the temperature set point in the building, the losses coefficients, the heat gains, the specific building characteristics (thermal mass and inertia, and default values according to TABULA method) and the occupancy ratios.
- The global efficiency of the heating installation is defined by the type of heating system, the energy used and the year of installation enabling to calculate the heating consumption ratio.
- In case a building is cooled, the global efficiency of the cooling installation is defined by the energy used and the year of installation of the cooling system enabling to calculate the cooling consumption ratio.

5 A Replicable but Completely Adjustable Tool. The Customisation Made to the City of Valence (Spain)

ACCENT tool has been designed to be easily replicable. That means that a harmonized data model and a common calculation method have been defined for all the cities that decide to be part of ACCENT. However, the tool is also user-adjustable and adaptable to meet the data available on each territory. ACCENT tool provides output results at different scales depending on the granularity of data each territory or city is willing to provide. In addition, it is also susceptible to be improved by means of database updates, corrections made by the specific users or other improvement' measures, such as the intelligent feeding of data through the use of smart meters.

To initialize calculations, it is required to provide the tool with the input data necessary to make ACCENT tool engine works. For each city, different levels of spatial data are needed: building, sub-district and district data. In addition, other input parameters such as CO_2 emission factors or primary energy factors are considered at national scale. The collection of input data (parameters, scale and sources) carried out by the team in order to launch the calculation for the City of Valencia (Spain) is shown in Table 3.

Table 3 Data collection for the city of Valencia (Spain)

Input data		
Parameter	Scale	Source
Building characteristics (year of construction, height, use, property, etc.)	Building	Cadastre/City of Valencia/Urban Atlas
Building footprints	Building	Cadastre/City of Valencia
District boundaries	District	City of Valencia
Type of heating system (collective central heating, individual central heating, no heating installation, no heating installation but individual room heaters)	District/Subdistrict	National Statistics Institute (INE)
% of heating energy source in residential buildings climatic zone B3 (oil, natural gas, heat pump, gas in bottles or in tanks, electricity)	City	From TABULA-EPISCOPE project
% of residential buildings cooled	City	Sech-Spahousec Project (IDEA)
Performance of energy systems	National	Spanish official software of energy certification TABULA-EPISCOPE project
CO_2 emission factors and primary energy factor	National	Defined by the Ministry of Industry, Energy, Tourism and Ministry of Public Works
Climatic data	City	Guide of climatic conditions in the Valencian Community
Energy costs	National	IDAE

6 Conclusions

The ACCENT functionalities developed so far are the result of an iterative co-design process involving public administrations, energy-related service providers and citizens, extended with the results gathered from project partners' research on the new trends and possibilities already available in the sector. We can sum up three main ACCENT benefits:

First, it supports local authorities in identifying where and how to act in priority, monitoring the city progress towards sustainable development, creating future scenarios and relating to other cities which may implement pilot energy policies.

Second, service providers need energy data or building characteristics in order to develop their business strategy. For them it is also important to be aware of the energy strategy of their operation territory, something delivered by the tool. In addition, specific service providers as building managers need to prioritize the actions and investment on their building stock. ACCENT provides them geographical information and action plan evaluation.

Finally, citizens need to know where to go when looking for regulations, products and services for energy efficiency as well as information on service providers.

However, as a bottom up market answer tool, it has also its limitations: the speediness of the ITC market and the rising needs of municipalities in order to reach their low carbon goals, has opened a very fast changing business niche. It means that as long as we write this paper, other planning tools are popping up into the market creating confusion and also new solutions.

In conclusion, ACCENT can be currently part of successful policy-making and policy-changes, but it would need to be continuously updated in order to answer to real-time needs.

Acknowledgements The research presented in this paper is partially supported by funds from Climate-KIC, the Europe's largest public-private innovation partnership focused on climate change. The authors gratefully acknowledge the support provided by LGI consulting, together with the consortium partners: ENGIE, NUMTECH, ESRI, ARXIT, EIVP, IVE, INNDEA, ASTER and SINERGIS, especially by: Solenne Tesseron, Amira Bentahar and Romain Petinot from ENGIE. The authors also acknowledge the support of Silvia Navarro during her stage at IVE.

References

ACCENT project partnership. (n.d.). Retrieved from www.accentproject.com

Ascione, F., De Masi, R., de' Rossi, F., Fistola, R., & Vanoli, G. (2012). Energy assessment in town planning: Urban energy maps. In M. Pacetti & C. Brebbia (Eds.), *The sustainable city VII* (pp. 205–216).

Bhanot, J., & Jha, V. (2012). Moving towards tangible decision making tools for policy makers: Measuring and monitoring energy access provision. *Energy Policy, 47,* 64–70.

Bloem, H., Boguslawski, R., Borzacchiello, M., Piergiorgio, C., Kona, A., Martirano, G., et al. (2015). *Location data for buildings related energy efficiency policies*. Luxembourg: European Commission—Joint Research Centre.

Covenant of Mayors for Climate & Energy. (2016). Retrieved from http://www.covenantofmayors.eu/index_en.html

European. (2011). *Communication from the commission to the European parliament, The council, The European economic and social committee and the committee of the regions. A roadmap for moving to a competitive low carbon economy in 2050—COM (2011) 112 final*. Brussels: European Comission.

Eurostat. (2016). *Energy, transport and environment indicators*. Publications Office of the European Union.

Ferreira, J., Duarte Pinheiro, M., & de Brito, J. (2013). Refurbishment decision support tools: A review from a Portuguese user's perspective. *Construction and Building Materials, 49,* 425–447.

Genre, J.-L., Flourentzos, F., & Stockli, T. (2000). Building refurbishment: Habitat upgrading. *Energy and Buildings, 31,* 155–157.

Hailu, Y. G. (2012). Measuring and monitoring energy access: Decision-support tools for policymakers in Africa. *Energy Policy, 47,* 56–63.

International Energy Agency. (2013). *Transition to sustainable buildings—Strategies and opportunities to 2050*. International Energy Agency.

Ireland, D. (2008). *New tricks with old bricks*. London: The Empty Homes Agency Ltd.

Müller, M., & Siebenhüner, B. (2005). Policy instruments for sustainability-oriented organizational learning. *Business Strategy and the Environment, 16*(3), 233.

Stylianidis, E., Karanikolas, N., & Kaimaris, D. (2012). A Gis for urban sustainability indicators in spatial planning. *International Journal of Sustainable Development and Planning, 7*(1), 1–13.

THINK. (2012). *How to refurbish all buildings by 2050*. Firenze: Publications Office.

Chapter 10
Towards a Climate-Resilient City: Collaborative Innovation for a 'Green Shift' in Oslo

Hege Hofstad and Jacob Torfing

Abstract The starting point of this chapter is climate change as a wicked and unruly problem that requires collaborative innovation to create local climate solutions. We pay special attention to the role of institutional design and public leadership and management in facilitating collaboration and spurring innovation. The chapter provides an analytical framework that aims to combine a process perspective on networked collaboration and creative problem solving with an institutional and management perspective to enable and sustain processes of collaborative innovation. The city of Oslo, with its highly ambitious climate goals and its dependence on innovation in governance systems to spur new solutions contributing to goal attainment, forms the empirical basis of the chapter. Our analysis of the Oslo case shows that the city's strategy for reaching these ambitious goals tends to cohere with ideas and principles of collaborative innovation. The city of Oslo is currently making a huge effort to design and lead collaborative arenas that may spur the development of innovative solutions. However, our analysis reveals that, at this early point in the process, there is still a long way to go before the city government can begin to reap the fruits of cross-sector collaboration. So far, the city government's in-house focus has overshadowed attempts to build society-wide arenas for collaborative innovation that can mobilize the knowledge, resources and energies of all the relevant and affected actors in the pursuit of innovative climate solutions. Hence, future research should concentrate on formulating and testing hypotheses about the conditions, drivers, barriers and impact of collaborative innovation as a promising new approach to public policy-making.

H. Hofstad (✉)
Norwegian Institute for Urban and Regional Research, Oslo and Akershus University College of Applied Sciences, Oslo, Norway
e-mail: Hege.hofstad@nibr.hioa.no

J. Torfing
Roskilde University, Roskilde, Denmark
e-mail: jtor@ruc.dk

R. Álvarez Fernández et al. (eds.), *Carbon Footprint and the Industrial Life Cycle*, Green Energy and Technology, DOI 10.1007/978-3-319-54984-2_10

1 Introduction

Big cities around the world are facing severe and urgent challenges in mitigating and adapting to the impacts of climate change (Bulkeley 2013). On the one hand, cities continue to grow and become mega-hubs for consumption, production and transport, and are thus a significant source of carbon emissions and global warming. On the other hand, the same cities are vulnerable to the impacts of climate change in terms of heat waves, prolonged spells of drought, problems with water supply, rising sea levels, and increasing risk of cloudburst rain (IPCC 2007). For such reasons, cities are important strategic arenas and actors for generating a proactive response to climate change that seeks to both reduce carbon emissions and adapt to climate change by developing innovative solutions that disrupt common wisdom and established practices. While both international organizations and national governments have been caught up in political conflicts over climate measures and, until recently, have been hesitant to launch new, ambitious climate solutions, municipal governments have taken on the formidable challenge of reducing emissions and adapting urban infrastructures to the new climate conditions. Recognizing the limits of their own knowledge, resources and capacity, local governments tend to involve a variety of public and private actors in urban climate governance through the creation of networks, partnerships and green compacts that cut across organizations, sectors and scales.

This chapter claims that climate change is a wicked and unruly problem that calls for collaborative innovation of local climate solutions. The empirical case is the Norwegian city of Oslo, which, with its new and highly ambitious climate and energy plan, aims to position itself at the forefront of low-carbon development and climate adaptation. Since collaborative innovation is unlikely to arise spontaneously and faces numerous challenges in terms of collective action problems, destructive conflicts, tunnel vision and weak implementation capacity (Torfing 2016), we will pay special attention to the role of institutional design and public leadership and management in facilitating collaboration and spurring innovation. As such, we will identify drivers and barriers of collaborative innovation and explore the strategic attempts of the municipal government to enhance the drivers while overcoming the barriers in order to spur the development of innovative climate solutions that will transform Oslo into a low-carbon city.

The structure of the chapter is the following. The first section defines the concept of wicked and unruly problems, and the second section demonstrates that mitigation and adaptation to climate change in the city of Oslo qualifies as a wicked and unruly problem. The third section defines what innovation means in a public sector context, explains why multi-actor collaboration is a preferred strategy for enhancing innovation, and discusses the potential drivers and barriers for collaborative innovation and the need for supportive institutional designs and new forms of public leadership and management. The fourth section introduces the empirical case, accounts for the collected data and describes how the city of Oslo intends to enhance climate change mitigation and adaptation by stimulating processes of collaborative innovation. The fifth section discusses the role of institutional design and public leadership in enhancing collaborative innovation of climate change policies in the city of Oslo. The conclusion summarizes the arguments, draws some theoretical and practical lessons and points out avenues for further research.

2 Wicked and Unruly Problems: The Ultimate Challenge for Local Governments

The research on collaborative governance and public innovation has paid considerable attention to the challenges that local governments face when trying to solve complex societal problems that require equally complex solutions (Koppenjan and Klijn 2004; Ansell and Torfing 2014). Complex problems are frequently referred to as wicked problems due to their uncertain and paradoxical nature. In this section we argue that many societal problems are not only 'wicked' due to cognitive limitations of the decision makers, but also 'unruly' due to political constraints.

The term 'wicked problems' was coined by Rittel and Webber (1973). Unlike 'tame problems' which are relatively simple and well-described and enjoy widespread consensus on both their definition and their solution (Roberts 2000), wicked problems are hard to define and even harder to solve. Wicked problems are unique, complex and tangled and, therefore, difficult to pin down. We can see that something is not working, but it is hard to tell what the real problem is because it is difficult to know exactly what distinguishes an observed condition from a desired condition. Even if we agree on a tentative problem definition, it is often impossible to locate the root of the problem in the complex web of causalities that characterizes open societal systems. Moreover, since there are no clear, objective and undisputed goals in public policy-making, but rather incomplete, changing and contradictory objectives, there is no true or optimal solution to wicked problems, only solutions that are considered 'acceptable' or 'good enough'. When searching for an appropriate solution, public authorities may also find it difficult to rely on processes of trial and error since the consequences of an error might do irreversible damage to users, citizens or private firms. The risks are high and there is often no agreed system for risk sharing. Finally, there is no ultimate test of solutions to a wicked problem because undesirable future repercussions might outweigh the present advantages of a particular solution.

Since the initial definition of the nature of wicked problems, there has been an ever-growing scholarly interest in the phenomenon and in the limitations of the traditional linear methods of problem solving that it reveals (Degrace and Stahl 1990; Roberts 2000; Campbell 2003). At the same time, there has been a growing awareness of the surge of wicked problems in our increasingly fragmented, complex and dynamic societies (Koppenjan and Klijn 2004). However, something seems to be missing, both from the original definition of wicked problems as well as from many of its subsequent elaborations. The definition of wicked problems tends to be defined exclusively in *cognitive terms* and, thus, as a question about the lack of precise and adequate knowledge and information about problems, causes, objectives and solutions. As such, one can consider wicked problems to be rooted in an epistemic deficiency that challenges the linear scientific approach to policy-making and public governance. However, there is also an important *political aspect* to societal problems, which means that they are often both wicked and unruly (Ansell 2013). This is evident from a recent discussion of global climate change as a 'super wicked problem' (Lazarus 2009) that adds the following characteristics to the definition of wickedness: (1) time is running out; (2) the problem is misaligned with capacity or authority; (3) those seeking to solve the problem are causing it; and (4) the future is heavily discounted. Other political constraints that may hamper our ability to solve important societal problems include political disagreement about the nature of the problem; political problem framing that produces symbolic or ineffective solutions; problems are a solution to other problems or a symptom of deeper problems; and solutions are either unavailable, overly expensive, or objectionable (Ansell 2013).

Instead of subsuming all these political constraints under the notion of wicked problems, we propose to retain the original notion of wicked problems, but add the term 'unruly' and thus speak of 'wicked and unruly' problems in order to signify the concurrence of cognitive and political constraints on public problem-solving. Wicked problems are often unruly because cognitive limitations open the way for political disputes over the nature of these problem and their possible solutions, or because political struggles may destabilize the cognitive foundation of societal problem solving. Hence, although problems might be wicked without necessarily being unruly, and vice versa, it makes good sense to talk about wicked and unruly problems in public policy-making.

Koppenjan and Klijn (2004) capture our conceptual expansion of wicked problems into wicked and unruly problems in their discussion of the interlacing of cognitive, strategic and institutional uncertainties. Cognitive uncertainty problematizes the scientific approach to public decision making that was predominant in the 1960s and 1970s. Strategic and institutional uncertainties question the alleged decision-making monopoly of formal government institutions by drawing our attention to the role of competing stakeholders and decision-making arenas and by urging us to recognize the impact of informal rules, norms and procedures in public governance.

3 The Wickedness and Unruliness of Climate Change Mitigation and Adaptation

Climate change mitigation and adaptation are wicked and unruly problems. The twin challenge is to mitigate climate change by cutting CO_2 emissions through the development of low-carbon solutions and to find ways of adapting to the disruptive impact of climate change. However, adaptation and mitigation differ in their degree of wickedness and unruliness. Climate change *adaptation* is about finding physical solutions to rougher and more extreme climate conditions: for example, handling increasing amounts of surface water from big storms and sudden cloud bursts by expanding green capture areas, digging new channels and opening rivers that were previously contained in subterranean tubes. Climate change *mitigation*, on the other hand, requires a fundamental shift in the organization and functioning of society and the economy (Bulkeley 2013, p. 4). Global warming and climate change are not external to society; they are caused by our way of living. In our advanced industrial society, production, transport and consumption are highly dependent on the use of fossil fuels. Through the development of new technologies and the creation of new behavioral patterns, we can contribute to reducing and eliminating CO_2 emissions. However, the pace of technological innovation is highly unpredictable and the effects of technology are uncertain. In addition, it is extremely difficult to change institutionally sedimented behaviors through a combination of explicit economic incentives and more subtle forms of nudging. We have very little knowledge of what works and what does not. When occasionally we succeed in changing people's behavior, it might solve one problem while creating another one. Hence, although economic incentives might induce people to buy and drive electric cars, this might move people from public to private transport systems, thus enhancing traffic congestion and making public transport more costly. Underlying the strategic efforts to transform technological, economic and social systems is the cognitive problem that climate change is relatively invisible and that political awareness and understanding of causes and effects depends on expert knowledge ridden with uncertainty.

While traditional environmental problems, such as smog, deforestation and traffic noise, are tangible in the sense that they are visible, and can be smelled and heard, climate changes are not as palpable (Beck 1992). Rather, they manifest themselves in and through natural incidents such as landslides, heat waves, persistent drought, bush fires, heavy rain, hurricanes, avalanches, glacier melting and so forth. The empirical frequency of these incidents is growing, but the variation might fall within a natural range. Hence, natural science experts play a crucial role in identifying and interpreting the empirical patterns through analysis of historical time series data. This is exactly the function of the IPCC, which is responsible for collecting and analyzing a huge amount of data and communicating the results to relevant decision makers and the larger public (Naustdalslid 2011). The nature of the problem, and the cognitive constraints on the efforts to deal with it, makes climate change mitigation a clear example of a wicked problem.

Climate change mitigation is also an unruly problem. The silo organization of the political and administrative system does not match the crosscutting efforts to reduce emissions. Since climate gas emissions emanate from a diverse set of sources, there is a pressing need for a whole-of-government approach that cuts across organizational and sectoral boundaries. Moreover, despite the concentration of production plants, transport hubs and cars in big cities, city governments will not be able to control the parameters influencing emission levels. Regional planning and national and transnational levels of public governance may hamper, or spur, the climate change mitigation strategies of city governments. The presence of a multi-level governance system further exacerbates the unruliness of climate change mitigation. Last but not least, the political and distributional conflicts over the shift from fossil fuels to renewable energy sources makes it clear that climate change mitigation is a tough challenge that might be blocked by political veto players.

Despite the widespread recognition that climate change is a major societal challenge, there is a lack of strong political climate action (Naustdalslid 2011). It seems to be difficult to find political solutions that are feasible, affordable and enjoy widespread support (Sattertwhite 2014). Controversies concerning the severity of the climate problem and the applicability of the available solutions are found in both the political and scientific communities and seem to amplify each other (Bulkeley 2013). Together, the political conflicts and the dispersion of power along the horizontal and vertical axes make climate change mitigation an unruly problem.

In sum, we can conclude that, although climate change mitigation and adaptation both present severe economic, social and political challenges, their level of wickedness and unruliness differs. Compared to climate adaptation, climate change mitigation calls for more fundamental economic, technological and behavioral changes, which are difficult to bring about due to the fragmented and multilevel character of public governance and the intensity of political and economic conflicts. In the following, we will focus on climate change mitigation, as it provides the strongest example of wicked and unruly problems and thus generates an urgent need for collaborative innovation.

4 Collaborative Innovation: The Proper Response to Wicked and Unruly Problems

In this section we first look at why and how wicked and unruly problems can be solved through public innovation and then consider how multi-actor collaboration can help to tackle wicked and unruly problems by spurring public innovation.

4.1 Public Innovation

For more than a century, innovation has been considered a key driver of growth and prosperity in private business firms (Schumpeter 1934). By contrast, public innovation is often seen as an oxymoron due to the lack of competition and the prevalence of centralized control, red tape and institutional inertia in public bureaucracies. More recently, however, public innovation has moved to the top of the public sector agenda (Kattel et al. 2014). This has happened partly in response to the pressures emanating from the combination of increasing expectations of citizens for the quality and availability of public services despite dire fiscal constraints, and partly in response to the pervasiveness of wicked and unruly problems. Such problems can be solved neither using the available standard solutions nor by allocating more money, staff and administrative resources (Sørensen and Torfing 2011). Hence, when faced with ill-defined and hard-to-solve problems in areas with many stakeholders and a high risk of conflicts, we need to find new ways of framing these problems and creative ways of solving them. We must develop and implement new solutions that break with the conventional wisdom and habitual solutions in a particular context. In short, we need to foster innovation, defined as the development and realization of new ideas that work (Hartley 2005).

Innovation is more than a continuous improvement of existing designs and less than a 'radical transformation' of the entire policy field. Rather, innovation can be seen as a step change that combines old and new elements in the construction of creative solutions that somehow disrupt the established practices and the underlying thoughts and ideas in a particular context (Hartley 2006). Public innovation may aim to transform public discourse, actual policies, organizational designs, public services, or the overall role perception of the public sector (Hartley 2005). It can be more or less radical depending on the size of the steps taken. Further, it can either be a result of the invention of something new or of the adoption and adaptation of innovative ideas from other organizations, sectors or countries (Hartley et al. 2013). Still, the key driver of public innovation, regardless of its forms and sources, seems to be the expectation that innovative solutions will outperform existing solutions and offer new and better ways of doing things at the same or lower costs than before. However, there is no guarantee that public innovation leads to improvement. Innovative solutions may not produce the desired output, and the final judgment of the outcomes of innovation relies on the subjective evaluations of politicians, public managers and employees, private stakeholders and different user groups.

4.2 Collaborative Innovation

Wicked and unruly problems require the crafting of innovative solutions, but what is the best strategy for enhancing innovation? Roberts (2000) compares authoritative, competitive and collaborative strategies and concludes that collaboration is superior to competition and authority when it comes to creative problem solving. Authoritative strategies authorize a particular group of decision makers to define a problem on the basis of their formal position or expertise and urge them to come up with a matching solution; the experts might be wrong, however, because they fail to benefit from knowledge sharing and mutual learning with relevant and affected actors. This weakness of authoritative strategies is especially salient in relation to wicked and unruly problems that are fraught with cognitive and political uncertainties. Competitive strategies engage relevant stakeholders in a zero-sum game in which the winner takes all and eventually gets to define the problem and its solution. While the advantage of these strategies is that competition prompts the search for innovative solutions and challenges institutionalized powers, a lot of resources are nevertheless wasted on rivalry and conflicts and the exchange of ideas among competitors is prohibited by competition. By contrast, collaborative strategies permit the exchange of knowledge, competences and ideas between relevant and affected actors and thus facilitate mutual learning that helps improve the understanding of the problem and identify a range of feasible options (Roberts 2000; Weber and Khademian 2008). In fact, collaboration does not only facilitate knowledge sharing, problem definition and idea generation. It also enables the integration of ideas, the selection of the most promising ones, the assessment and sharing of risks and benefits, the reduction of implementation resistance through the construction of joint ownership, and the diffusion of innovative ideas through the recruitment of a large number of ambassadors (Sørensen and Torfing 2011; Hartley et al. 2013).

As such, it has been argued, collaboration is the only strategy in which it is not the presence of institutional and organizational boundaries that decides who will be involved in the production of innovation, but rather the possession of relevant innovation assets in terms of experience, creativity, financial means, courage, implementation capacity, and so on (Bommert 2010). Hence, collaboration is not only a favorable strategy for dealing with wicked and unruly problems, but also provides a promising method for developing and implementing innovative solutions.

Collaboration does not necessarily involve time-consuming attempts to forge unanimous agreement. In fact, a total consensus that eliminates all forms of dissent is often predicated on the actors agreeing on the least common denominator, which seldom leads to an innovative solution. Rather, collaboration should be defined as a process through which multiple actors aim to establish a common ground for solving multiparty problems through the constructive management of difference (Gray 1989). Hence, we collaborate with other actors because they have resources, knowledge and ideas that are different from our own, and rather than seeking to eliminate these differences, collaboration aims to manage and exploit them in ways that facilitate the development of innovative solutions to common problems.

4.2.1 Conditions for Successful Processes of Collaborative Innovation

The notion of collaborative innovation aims to capture the ability of multiactor collaboration to spur innovation. However, it is often difficult to bring together the relevant and affected actors that are needed to solve complex problems. It is equally difficult to ensure their collaboration, and even if they do collaborate, there is no guarantee that collaboration will lead to innovation. Two factors may help in fostering successful processes of collaborative innovation: (1) institutional design of proper arenas in terms of networks, partnerships and compacts that allow interdependent actors to exchange resources while respecting their operational autonomy; and (2) the exercise of public leadership and management that encourage collaboration and spur innovation.

Let us first consider the role of *institutional design*. Institutional design theory shows how the modeling of institutional arenas for multi-actor collaboration, such as public hearings and consultations, deliberative forums, mini-publics, public-private partnerships and governance networks, influences the motivation of public and private actors to participate, the quality of their deliberation, the policy outcomes that are generated and the ability to hold decision-makers to account for the outcomes (Hysing and Lundberg 2015; Hendriks 2016). Institutional design theory is an integral part of democratic theory and its search for democratic innovation (Smith 2009; Fung 2003, 2007), but there is also an interesting new 'design thinking' that provides important insight into how institutional designs can accommodate co-created policy innovation (Bason 2010; Ansell and Torfing 2014). Design of collaborative arenas is also a key focus point in theories of integrative public leadership (Crosby and Bryson 2010). These theories claim that the success of boundary-spanning collaboration in solving complex problems depends on the creation of procedures that ensure an early agreement on the nature of the problem, help to overcome power imbalances and enable a joint tracking of inputs, outputs and outcomes. Analysis of the institutional design of collaborative arenas typically focuses on the basic questions of who participates, where, how and when and aims to explore how the remit, or other official documents, seeks to define and frame the task, determine the leverage of the collaborative arena, and set up procedures for ensuring accountability.

When it comes to *public leadership and management*, we propose that the drivers of collaborative innovation can be enhanced and the barriers partially overcome if public leaders and managers assume the role of 'conveners,' 'facilitators' and 'catalysts' (Straus 2002; Crosby and Bryson 2010; Morse 2010; Page 2010; Ansell and Gash 2012).

The role of the *convener* is to bring together the relevant actors and spur interaction and the exchange of information, views and ideas. Hence, the convener must:

- Select the team by identifying actors with relevant innovation assets and incite and motivate them to participate in the innovation process
- Clarify the role of the different actors and draw up a process map that delineates who participates when and how in the different phases of the innovation process
- Encourage interaction and exchange between the participating actors by stimulating the recognition of their mutual dependence on each other's resources
- Secure political support for the search for innovative solutions and protect the integrity of the collaborative
- Give direction to the joint search for innovative solutions and align the goals and expectations of the actors.

The role of the *facilitator* is to get the actors to collaborate by constructively managing their differences and engaging in processes of mutual learning that bring them beyond the least common denominator. Hence, the facilitator must:

- Lower the transaction costs of collaborating by arranging good and effective meetings, ensuring smooth communication and selectively activating those actors who are not contributing as much as they could
- Enhance and sustain trust between the actors by creating venues for informal social interaction, encouraging the development of common rules and procedures for interaction and triggering a virtuous cycle of trust-creation through a unilateral display of trust in the other actors
- Develop a common frame of understanding by creating a common knowledge base through knowledge exchange and joint fact-finding missions and developing a common language based on jointly accepted definitions of key terms and ideas
- Resolve or mediate conflicts so that they become constructive rather than destructive and ensure that irresolvable conflicts are depersonalized and conceived as joint puzzles rather than road blocks
- Remove obstacles to collaboration by securing support from the executive leaders of the participating organizations and negotiating how costs and gains of innovative solutions are distributed among the actors.

The role of the *catalyst* is to create appropriate disturbances and stimulate the actors to think out of the box and develop and implement new and bold solutions. Thus, the catalyst must:

- Construct a sense of urgency either by invoking a ‘burning ship’ or demonstrating the presence of a ‘window of opportunity’
- Prevent tunnel vision by encouraging the actors to change their perspective, including new and different actors in the team, or bringing new and inspiring knowledge into play
- Create open and creative search processes by changing the venue and the way that the actors interact and collaborate
- Facilitate the management and negotiation of the risks associated with innovative solutions and coordinate the implementation process to enhance synergy and avoid overlap
- Ensure that the participating actors assume the role of ambassadors and use their strong and weak ties to disseminate knowledge about the innovative solution.

Since processes of collaborative innovation are complex and full of jumps and feedback loops, public leaders and managers must be capable of skillfully combing the their roles as conveners, facilitators and catalysts so as to ensure that the right kind of actors are brought together, encouraged to collaborate and jointly explore and implement innovative solutions. Some of the leadership tasks can be distributed to other actors, such as professional facilitators or private actors partaking in the collaborative innovation processes, but the overall responsibility for driving the processes of collaborative innovation to their conclusion and getting results lies with local government officials. The challenge that they are facing is enormous since, after three decades of New Public Management, most public leaders and managers are used to focusing on how their own organization and employees are delivering a preordained set of results rather than on how a plethora of public and private actors are aiming to create new and disruptive solutions (Torfing 2016).

5 Mitigating and Adapting to Climate Change Problems in the City of Oslo

City governments do not only respond to policy expectations from national government and needs articulated by regional authorities. They also address global problems and participate in global arenas to get new ideas for innovative local solutions. Thus, they take part in global networks, projects, seminars, agreements, etc. in order to learn from and inspire each other to test new ways of tackling environmental problems and climate challenges (Kern and Bulkeley 2009). Paradoxically, participation in city climate networks has become an important tool in the global competition for recognition and branding as a leading green city that is attractive for global businesses and competent urban professionals to invest and settle in (Andersen and Røe 2016; Kern and Bulkeley 2009).

The city of Oslo is an active player in the global arena of climate change mitigation (City of Oslo 2016a). The city government belongs to several global networks such as the C40 network, which promotes sharing knowledge about how the world's megacities may contribute to low- and zero-carbon solutions. Oslo does not really qualify as a megacity, but it is part of the network because it 'has shown distinct environmental and climate leadership' (City of Oslo 2016a), both of which have been gradually more pronounced in the last couple of years.

5.1 Oslo's 'Green Shift'

In the spring of 2015, the center-right city government of Oslo presented its proposal for a new climate and energy strategy, the so-called *Green Shift* (City of Oslo 2015). The aim of the strategy is 'to institute substantial changes in how we as a city function, and we call this the green shift' (City of Oslo 2015, p. 3). The green shift is about spurring a transition to a sustainable society based on renewable energy (City of Oslo 2015). However, soon after the formal adoption of the green shift strategy in June 2015, the local election in September 2015 resulted in a political shift with considerable consequences for the implementation of the climate and energy strategy. The right-center city government was replaced by a new left-green city government. They had an expressed goal of accelerating the achievement of

Oslo's ambitious climate goals. Thus, the new left-green city government revised the green shift strategy in ways that maintained the original goals but shortened the timeframe for their realization and developed some new measures and instruments (City of Oslo 2016b). The new climate and energy strategy was formally adopted in June 2016 (City of Oslo 2016c).

Starting from the left, Fig. 1 illustrates the dynamic between the shift in city government and the acceleration of the climate goals. The original goals of the right-center city government were to: (1) cut CO_2 emissions by 50% in 2030 (compared to the 1991 emission level) and (2) reduce emissions to zero by 2050 (City of Oslo 2015). In the new and revised version of the green shift strategy developed by the new government, the left-green government accelerates the fulfillment of the climate goals by: (1) cutting CO_2 emissions by 50% already in 2020 (compared to the 1990 emission level) and (2) cutting CO_2 emissions by 95% before 2030 (compared to the 1990 emission level) (City of Oslo 2016b). To reach these new and ambitious goals, the new city government must have a strong focus on climate change mitigation and take swift action. Core politicians in Oslo acknowledge the size of the task: "*The city government's goal is to make Oslo into an environmental capital. Oslo shall function as a good example as a climate smart, inclusive, and diverse city,*" says councilor for city development Hanna E. Marcussen (City of Oslo 2016d). Councilor for environment and transport Lan Marie Nguyen Berg adds: "*Climate measures are not going to be implemented somewhere else, in another place, or by someone else. Oslo must take responsibility and use all available municipal means to act here and now*" (City of Oslo 2016d).

To reach its ambitious goals, the city government will speed up the implementation of plans, specify priorities, tasks and actions in the transport sector, enhance organizational capacities and make climate change mitigation the

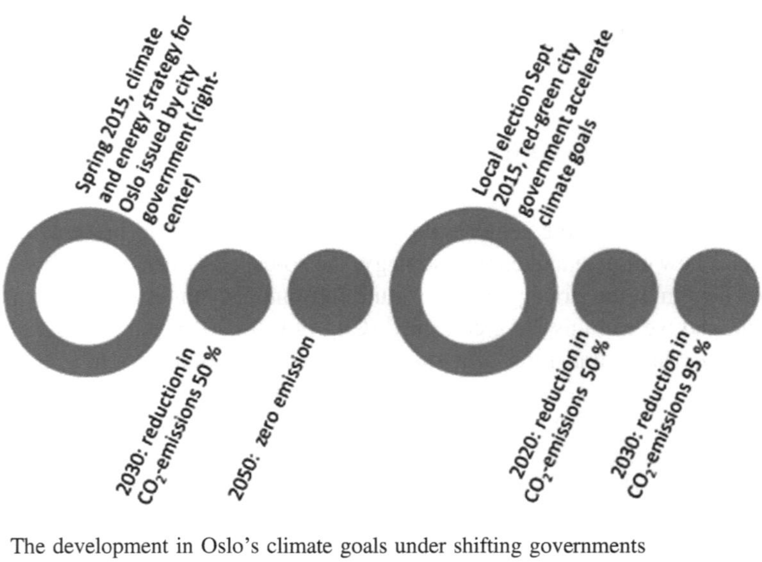

Fig. 1 The development in Oslo's climate goals under shifting governments

Table 1 Oslo's main goals with supporting goals

50% cut in CO_2-emissions by 2020	95% cut in CO_2-emissions by 2030
• 20% reduction in all car traffic • Share of everyday travel by bicycle 16% • 60 km of new bicycle roads, lanes and car-free streets • All new car transport of persons and goods to use renewable fuel or chargeable hybrids • Stop use of fossil energy for heating and exchange it with alternative sources of energy • Public transport on renewable fuels by 2020 • Full scale capture of CO_2 at Klemetsrud in collaboration with the state	• 33% reduction in car traffic transport of persons and goods is emission-free

Table 2 Direct and indirect measures in support of the green shift

Direct measures	Indirect measures
Annual climate budget Municipal acquisition Value management (management through ownership) Available financing/subsidiaries arrangements Use of formal authority (management of different laws and regulations where climate demands can be made Oslo's economic support for citizens and local businesses	Communication Competence building Involvement of citizens Collaboration with business and academic communities

responsibility of the municipality as a whole (City of Oslo 2016b). Table 1 lists concrete goals that must be reached in order to fulfill the two overall goals (City of Oslo 2016e, p. 2).

Implementation of the measures described in Table 1 calls for substantial changes involving innovation in mindsets, institutions and organizational systems, as well as in technology, production systems and urban infrastructures. The city wants to be a "test bed for innovation and cooperation" (Monsen 2016a). Thus, Oslo's 'green shift' is about stimulating innovation within the public sector and society at large.

5.2 *Collaborative Innovation in Oslo*

How will Oslo stimulate innovation? According to the city government, it will deploy a combination of direct and indirect measures when implementing the climate and energy strategy, summarized in Table 2 (City of Oslo 2016e, p. 5).

Direct measures are a type of measure where the city government of Oslo has considerable control and is capable making its own priorities and taking action in ways that have a direct impact on emissions (City of Oslo 2016e, p. 5).

Oslo's annual budget, which sets priorities and allocates resources, is a good example of a direct measure. *Indirect measures* have a more indirect effect on emissions, but are still important as they affect the development and implementation of different tools and measures (City of Oslo 2016e, p. 5). Stakeholder collaboration is an example of an indirect measure that carries the potential of either enhancing the effect of existing measures or creating new and innovative measures that rely on co-production, co-creation and co-governance.

In fact, most of the efforts to reduce emissions rely on interaction with core stakeholders such as citizens, business, private organizations, other municipalities, regional authorities and the state. Knowledge sharing is critical for making well-informed decisions; coordination is important in order to prevent overlaps, gaps and conflicts and to create synergy; and collaboration will help to develop new and disruptive climate solutions and secure ownership of their implementation.

5.3 Identification of Core Stakeholders

The city government acknowledges its dependence on active support and involvement of different stakeholders to reach the ambitious climate goals:

> The transition (…) requires a change in governance structures. Oslo will therefore increase the focus on identifying and implementing appropriate organizational and governance structures that allow for sustainable urban development, especially through the involvement of different stakeholders and citizens to reach acceptable solutions that promote the paradigm shift (City of Oslo, year unknown).

The municipality's dependency on a diverse set of actors permeates the note describing the content and course of the climate and energy strategy (City of Oslo 2016b). The note clearly describes how different public and private actors can contribute to reaching the stated climate goals. *The state* plays a crucial role as facilitator of the green shift by developing judicial instruments, regulatory measures and behavioral incentives that support and transform the transport and energy sector in accordance with the ambition to create a low-carbon society. This is not least evident in the effort to develop, and financially support, infrastructures for sustainable mobility (public transport, bicycling and walking) and sustainable energy production (e.g., through capture and storage of CO_2). Additionally, the state can widen the scope of investment in innovative low-carbon solutions such as car-free city centers, fossil-free city areas, introduction of car pool lanes and rush-hour dependent fees on car-based admission to the city.

The business community plays a pivotal role as partner and investor in the development of infrastructure for renewable fuel (hydrogen and biogas) and energy (solar panels and wind turbines), and it provides a crucial source of technological development. In addition, professional property owners and urban developers drive sustainable city development by initiating new smart city projects. Hence, in the Norwegian context, where private partners can initiate and sponsor local development plans, they are the key to realizing the goal of concentrating urban development around transport nodes meant to boost public transport use.

Neighboring municipalities are a key to unleashing the potential of cutting emissions from the transport sector, which is the single most important source of CO_2-emissions in Oslo (Monsen 2016a). The Oslo metropolitan area consists of Oslo municipality and 22 neighboring municipalities surrounding the city. Together, the municipalities have made a joint plan that aims to curb CO_2-emissions from transport by coordinating land use planning and transport development (Oslo and Akershus 2015). Development will be restricted to areas with good public transport, most notably train stations. Through prolonged negotiations, they have agreed on a map indicating which areas can grow and which areas will allow only limited and sustainable growth in order to maintain the current level of inhabitants and the services they need.

Citizens finally decide the future of Oslo's climate aims and measures. They are important for the uptake and daily functioning of specific policies and measures and, at the same time, they can withdraw their support of the current city government in the next election and potentially disrupt or reverse the current climate projects.

5.4 *Innovation in City Governance*

In order to unleash the cognitive, ideational and political potential that each of the above-mentioned actors has for realizing Oslo's climate ambitions, change in governance structures is important. Neither the traditional bureaucratic forms of top-down governance with specialized administrative silos nor the New Public Management-inspired attempts to spur competition between public service organizations and private service providers will do. In order to facilitate crosscutting collaboration, spur innovation and mobilize resources towards goal achievement, the climate and energy strategy describes a new set of measures aiming to change city governance (City of Oslo 2016b, e):

- *Climate department*—an administrative unit that serves as a crosscutting implementation hub for the climate and energy strategy
- *Climate budgets*—a new governing tool from 2017 aiming to ensure goal attainment through the creation of climate targeted city budgets that shows the scale of the planned reductions in CO_2-emissions, the results that have been achieved so far, and the dates for expected future results
- *Annual input conferences*—an attempt to mobilize active citizens, business actors, private organizations and public authorities from other jurisdictions in order to generate feedback and new ideas that can help to develop new climate initiatives
- *Climate-friendly architecture policy*—a new effort to preserve and develop urban qualities and promote the development of environment- and human-friendly urban spaces

- *Oslo-standards*—a new set of norms for climate-friendly urban development
- *City ecological innovation center*—an initiative that seeks to promote sustainable lifestyles, city nature, and climate adaptation

All these measures are examples of public innovation. Through development of *new organizational structures* (the climate department), *new arenas* (annual input conferences and city ecological innovation center), *tools* (climate budgets) and *policies* (climate-friendly architecture policy and Oslo standards), the green shift strategy aims to create a new way of framing the urban climate challenge and to facilitate the involvement of a plethora of public and private actors in the development of new and innovative solutions. Although there is a risk that the new measures might not be implemented in the way they are planned, or might fail to produce the desired results, there is a strong political will to challenge conventional knowledge and traditional forms of governance and thus create new pathways to climate resilience. Whether the new governance arenas will succeed in creating a collaborative interface between public and private actors that will spur the development and implementation of disruptive climate solution depends to a large degree on the creation of appropriate institutional designs and the exercise of public leadership and management.

6 The Role of Institutional Design and Public Leadership in Spurring Collaborative Innovation

Oslo is one of many global cities with high ambitions with regard to climate mitigation, and with its new climate goals, it aspires to join the league of the most ambitious climate cities in the world: *Vancouver* aims to reduce greenhouse gas emissions by at least 80% below 2007 levels before 2050; *Barcelona* aims to reduce its CO_2 emissions per capita by 40% compared to 2005 figures; *Copenhagen* aims to be the first climate neutral city by 2025—a slightly more ambitious goal than that of her little sister Oslo (City of Vancouver 2016; City of Barcelona 2015; City of Copenhagen 2009). Hence, the problem that Oslo shares with other climate cities is how to find ways of achieving its ambitious goals. This is a huge challenge, and the city government's approach has been to develop a variety of collaborative initiatives, internal as well as external. Multiactor networks and collaborative arenas have been established across the city administration, as well as with the neighboring county and its municipalities, the business community, environmental organizations of various sorts, and the citizens of Oslo. Of all these initiatives, the city government has recently directed most of its attention and energy to activities aiming to improve in-house collaboration between a large number of administrative departments and agencies.

6.1 Oslo's Governance Model

Through institutional design and leadership the city government has developed a new collaborative governance model that, according to our informants, will ensure a 'tight steering' of Oslo's green shift strategy across sectors and organizational boundaries. The core of the new governance model is a mainstreaming of the entire municipal administration. The climate department demands that the central climate goals of the city government guide and direct decisions and activities of all administrative departments and agencies. In order to reduce complexity, the overall goals and tasks have been disaggregated into more manageable parts and the administrative sectors and agencies are encouraged to identify core problems and to apply a variety of direct and indirect measures to cope with those problems. The tracking of problems to be solved has resulted in 76 new initiatives in 16 administrative sectors. The results of the decentered efforts of administrative departments and agencies to fulfill the disaggregated climate goals are carefully monitored on an annual basis as a part of the city budget (City of Oslo 2016e). In addition to all of this, each department regularly receives letters with assignments and expectations from the climate department in order to prompt and inspire decentralized actions, ideally through inter-departmental and inter-agency collaboration.

Now, with its effort to ensure that all administrative departments and agencies contribute to reaching the overall climate goals, the mainstreaming exercise appears to be rather top-down. However, the mainstreaming strategy is deeply concerned with ensuring horizontal coordination and collaboration in order to prevent gaps, overlaps and conflicts between the decentralized climate initiatives and foster complementarity and synergy between them. To this end, the climate department has formed five 'climate groups' with participants from 40 different administrative units. These crosscutting climate groups are arenas for knowledge sharing, exchange of experience, competences and resources, dialogue about problems and solutions, mutual inspiration and learning, identification of common barriers and drivers, experimentation and concerted or joint action (Monsen 2016b). The climate groups are thematic, each of them dealing with a core challenge for climate mitigation: mobility, building, resources, energy, and cross-sector linkages. Each of the five groups meets for three days in a row on a biannual basis to discuss an ongoing joint project involving several administrative units; for example, the energy group will meet to discuss compact urban development in one particular city district (Monsen 2016b). The creation of the crosscutting climate groups adds a horizontal and collaborative dimension to the vertical and rather hierarchical imposition of climate goals on all the administrative departments and agencies. The climate groups provide arenas for coordination and collaboration and may over time develop into arenas for collaborative innovation. The goal of these combined efforts to ensure vertical and horizontal integration is to ensure that the city of Oslo acts as a strategic whole thereby laying the foundation for a consistent, transparent and predictable climate leadership that can direct, design and manage the mobilization of internal as well as external resources and actors. So far, internal capacity building has taken up most of the time and energy of the political and administrative leaders, but new efforts to build collaborative relations with external actors are under way.

6.2 Emerging Collaborative Innovation Efforts

One example of the endeavor to build collaborative relations with external actors is that the city increasingly uses their purchasing power to foster constructive and problem-focused dialogue with urban developers. The municipality uses this dialogue to broadcast and discuss its request for climate-neutral solutions. According to our informants, the request helps to spur new and innovative solutions in the building sector. Hence, strategic leadership and dialogue influence the content of the solutions provided by for-profit actors.

Another example is the Business for Climate network, which is an arena for climate collaboration between the city of Oslo and a broad range of private firms from the Oslo region. About 70 firms are members of the network, and as members, they commit themselves to contribute to attain Oslo's climate goals through the exchange of information and experiences on climate projects, the development of new measures for making emissions cuts, and implementation of specific climate measures in their own firms and to report on these measures and their climate effects. Business for Climate organized a climate summit in 2014 where they challenged each other on green innovation and explored the prospect of further collaboration to reach Oslo's ambitious climate goals, thus making it a leading green city. The summit also aimed to develop concrete business models capable of supporting the development of new green solutions.

A third example is the way that established procedures for consultation and public hearings are used to generate input and support for the development of new climate solutions among local associations and citizens. Norway has well-developed mandatory procedures for the consultation of local actors in urban planning that facilitates dialogue about local problems and challenges, and these procedures are frequently used in relation to issues pertaining to climate change mitigation.

We do not have sufficient data to see how the city government has designed the collaborative arenas where non-governmental actors are involved in developing and implementing innovative climate solutions. However, interviews with key actors in the climate department reveal a strong commitment to the three leadership roles that we presented above. As for the convener role, the city of Oslo seems to have a clear understanding of the importance of spurring interaction between public and private actors in order to exchange information, ideas and resources and perhaps stimulate learning and innovation. They are very conscious of the need to build networks, partnerships and collaborative arenas and carefully reflect on whom to engage, how, where and when. The aim is to involve public as well as for-profit and non-profit private actors. As for the facilitator role, focus is on organizing interaction around concrete problems in order to emphasize the mutual dependence of the actors, which is key to spurring collaboration. This is particularly clear in the case of the climate groups and the Business for Climate network. Finally, the catalyzing role is visible in the way the city administration stimulates development of new technologies in the transport and energy sectors and new forms of smart living in the compact city. The use of purchasing power to stimulate the creation of new planning solutions is a case in point.

7 Conclusion and Future Research

In this chapter, we have aimed to provide a framework for studying how wicked and unruly problems can be addressed through the design and management of arenas for collaborative innovation. We have shown that complex problems often have both a cognitive and a political component and have argued that the proper answer to wicked and unruly problems in the public sector is to search for innovative public solutions that disrupt established ways of thinking and acting. Comparing different strategies for creating innovative solutions, we have shown that multiactor collaboration is a potent driver of innovation. Since collaboration between public and private actors with different ideas and interests is difficult to bring about and will not always lead to innovative outcomes, there is an urgent need for institutional design and leadership in order to enable and sustain processes of collaborative innovation. Therefore, our analytical framework aims to combine a process perspective on networked collaboration and creative problem solving with institutional and management perspectives.

The empirical analysis demonstrates that climate change mitigation is a wicked and unruly problem, probably more so than climate change adaption which tends to be more straightforward. To see how the problem of climate change mitigation is dealt with we introduced the case of the Green Shift strategy in the city government of Oslo that recently set some highly ambitious goals for climate change mitigation. The analysis shows that the strategy for reaching these ambitious goals, both in theory and in practice, tends to cohere with ideas and principles of collaborative innovation. In terms of implementation of the strategy, the climate department is currently making a huge effort to design and lead collaborative arenas that may spur the development of innovative solutions. However, the analysis reveals that, at this early point in the process, there is still a long way to go before the city government can begin reap the fruits of cross-sector collaboration. So far, the in-house focus of the city government has tended to overshadow attempts to build society-wide arenas for collaborative innovation that can mobilize the knowledge, resources and energies of all the relevant and affected actors in the pursuit of innovative solutions to perhaps the most crucial problem of our time.

Further research is needed along three main avenues. First, we need to know much more about drivers and barriers so that relevant and affected actors can collaborate and jointly pursue innovative solutions to climate problems. Second, we need to compare different types of collaborative arenas in order to understand the role and impact of institutional design for collaboration and innovation. Last, we need to study the context dependent efficiency of different leadership and management tools in order to be able to guide public leaders and managers aiming to convene actors, facilitate collaboration and catalyze innovation. Public innovation has only recently come to the fore in academic studies of public policy and governance, and the research on collaborative innovation and the role of institutional design and leadership herein is entirely new. Hence, the explorative character of the research that is called for here is hardly a surprise. Nevertheless, we should insist that the goal of such explorative studies is to formulate and test hypotheses about the conditions, drivers, barriers and impact of collaborative innovation, which seems to offer a new approach to public policy making.

References

Andersen, B., & Røe, P. G. (2016). *The social context and politics of large scale urban architecture: Investigating the design of Barcode*. Oslo: European Urban and Regional Studies. doi:10.1177/0969776416643751

Ansell, C., & Gash, A. (2012). Stewards, mediators, and catalysts: Towards a model of collaborative leadership. *The Innovation Journal, 17*(1), Article 7.

Ansell, C., & Torfing, J. (Eds.). (2014). *Public innovation through collaboration and design*. London: Routledge.

Ansell, C. (2013). *Wicked and unruly problems*. Paper presented at Roskilde University, Denmark, September, 2013, personal communication.

Bason, C. (2010). *Leading public sector innovation*. Bristol: The Policy Press.

Bech, U. (1992). *Risk society—towards a new modernity*. London: Sage Publications.

Bommert, B. (2010). Collaborative innovation in the public sector. *International Public Management Review, 11*(1), 15–33.

Bulkeley, H. (2013). *Cities and climate change*. London: Routledge.

City of Barcelona. (2015). Barcelona's commitment to the climate. In *Document for the 21st Conference of Parties to the United nations Framework Convention on Climate Change (COP-21) in Paris*.

City of Copenhagen. (2009). *Copenhagen carbon neutral by 2025: Copenhagen climate plan*. Copenhagen: The city of Copenhagen.

City of Oslo. (2015). Det grønne skiftet—klima- og energistrategi for Oslo. Oslo: The City of Oslo. [The green shift—climate and energy strategy for Oslo].

City of Oslo. (2016a). Internasjonalt samarbeid på miljø. https://www.oslo.kommune.no/politikk-og-administrasjon/miljo-og-klima/internasjonalt-samarbeid-pa-miljo/ [International environmental cooperation]. Accessed September 20, 2016.

City of Oslo. (2016b). Tilleggssak til klima- og energistrategien. Vedtatt i bystyret 22.06.2016 på bakgrunn av byrådssak 33 av 18.02.2016 [adoption of additional formulations to the climate and energy strategy].

City of Oslo. (2016c). Vedtak av Klima- og energistrategi for Oslo. Vedtatt i bystyret 22.6.2016 på bakgrunn av byrådssak 156 av 26.06.2015 [adoption of climate and energy strategy].

City of Oslo. (2016d). http://www.oslo.kommune.no. Accessed September 29, 2016.

City of Oslo. (2016e). Byrådets budsjettforslag 2017 og økonomiplan 2017–2020, Klimabudsjett 2017 [Budget 2017].

City of Oslo. (year unknown). Green transition in Oslo—a holistic view on sustainable urban development.

City of Vancouver. (2016). Goals and targets. http://vancouver.ca/green-vancouver/goals-and-target.aspx. Accessed October 20, 2016.
Crosby, B. C., & Bryson, J. M. (2010). Integrative leadership and the creation and maintenance of cross-sector collaborations. *The Leadership Quarterly, 21*(2), 211–230.
Campbell, M. C. (2003). Intractability in environmental disputes: Exploring a complex construct. *Journal of Planning Literature, 17*(3), 360–371.
DeGrace, P., & Stahl, L. H. (1990). *Wicked problems, righteous solutions*. Englewood Cliffs: Yourdon Press.
Fung, A. (2003). Recipes for public spheres: Eight institutional design choices and their consequences. *Journal of Political Philosophy, 11*(3), 338–367.
Fung, A. (2007). Mini-publics: Deliberative designs and their consequences. In S. Rosenberg (Ed.), *Deliberation, participation, and democracy: Can the people govern?* (pp. 159–183). New York: Palgrave MacMillan.
Gray, B. (1989). *Collaborating: Finding common ground for multiparty problems*. San Francisco: Jossey-Bass.
Hartley, J. (2005). Innovation in governance and public service: Past and present. *Public Money and Management, 25*(1), 27–34.
Hartley, J. (2006). *Innovation and its contribution to improvement*. London: Department for Communities and Local Government.
Hartley, J., Sørensen, E., & Torfing, J. (2013). Collaborative innovation: A viable alternative to market competition and organizational entrepreneurship. *Public Administration Review, 73*(6), 821–830.
Hendriks, C. M. (2016). Coupling citizens and elites in deliberative systems: The role of institutional design. *European Journal of Political Research, 55*(1), 43–60.
Hysing, E., & Lundberg, E. (2015). Making governance networks more democratic: Lessons from the Swedish governmental commissions. *Critical Policy Studies, 10*(1), 21–38.
IPCC (Intergovernmental Panel on Climate change). (2007). IPCC fourth assessment report: Climate change 2001. http://www.ipcc.ch/publications_and_data/publications_ipcc_fourth_assessment_report_synthesisreport.htm. Accessed September 15, 2016.
Kattel. R., Cepilovs, A., Drechsler, W., Kalvet, T., Lember, V., & Tonurist, P. (2014). 'Can we measure public sector innovation? A literature review'. LIPSE Working Paper Series, 3.
Kern, K., & Bulkeley, H. (2009). Cities, Europeanization and multi-level governance: Governing climate change through transnational municipal networks. *Journal of Common Market Studies, 47*(2), 309–332.
Koppenjan, J., & Klijn, E.-H. (2004). *Managing uncertainties in networks*. London: Routledge.
Lazarus, R. J. (2009). Super wicked problems and climate change: Restraining the present to liberate the future. *Cornell Law Review, 94*(5), 1153–1233.
Monsen, R. D. (2016a). *Turning Oslo into a zero-emission city—climate and energy strategy for Oslo*. Presentation for African and Norwegian researchers June 6 2016, Norwegian institute for Urban and regional research.
Monsen, R. D. (2016b). *Klimagruppemøte August 2016*. Presentation for climate group members August 24 2016, City of Oslo.
Morse, R. S. (2010). Integrative public leadership: Catalyzing collaboration to create public value. *The Leadership Quarterly, 21*(2), 231–245.
Naustdalslid, J. (2011). Climate change—the challenge of translating scientific knowledge into action. *International Journal of Sustainable Development and World Ecology, 18*(3), 243–252.
Oslo & Akershus. (2015). Regional plan for areal og transport i Oslo og Akershus, vedtatt i Oslo kommune og Akershus fylkeskommune desember 2015 [land use and transport plan for the Oslo and Akershus region].
Page, S. (2010). Integrative leadership for collaborative governance: Civic engagement in seattle. *The Leadership Quarterly, 21*(2), 246–263.
Rittel, H. W. J., & Webber, M. M. (1973). Dilemmas in a general theory of planning. *Policy Sciences, 4*(2), 155–169.

Roberts, N. C. (2000). Wicked problems and network approaches to resolution. *International Public Management Review, 1*(1), 1–19.

Sattertwaite, D. (2014). Editorial: Getting local governments, residents and enterprises to respond to the new IPCC assessment. *Environment & Urbanization, 26*(1), 3–10.

Schumpeter, J. (1934). *The theory of economic development*. Cambridge: Harvard University Press.

Smith, G. (2009). *Democratic innovations designing institutions for citizen participation*. New York: Cambridge University Press.

Straus, D. (2002). *How to make collaboration work*. San Francisco: Berrett Koehler Publishers.

Sørensen, E., & Torfing, J. (2011). Enhancing collaborative innovation in the public sector. *Administration and Society, 43*(8), 842–868.

Torfing, J. (2016). *Collaborative innovation in the public sector*. Washington, DC: Georgetown University Press.

Weber, E. P., & Khademian, A. M. (2008). Wicked problems, knowledge challenges, and collaborative capacity builders in network settings. *Public Administration Review, 68*(2), 334–349.

Chapter 11
Energy Innovation in the Environmental Kuznets Curve (EKC): A Theoretical Approach

Daniel Balsalobre-Lorente, Muhammad Shahbaz, José Luis Ponz-Tienda and José María Cantos-Cantos

Abstract This chapter presents the relationship between economic growth and environmental pollution through the theoretical hypothesis of the Environmental Kuznets Curve (EKC). Moreover, it attempts to illustrate the impact of renewable energy sources and energy Research, Development and Demonstration (RDD) on environmental degradation in countries around the world. Many studies have confirmed the existence of an inverted N-shaped EKC pattern in the relationship between income level and the environmental degradation process. These results also indicate that energy regulation processes delay technological obsolescence once economies have reached the early stages of the decontamination process, which, in the long-run, means that an increase in income threshold is required before there can be a return to rising pollution levels. Furthermore, this chapter explains the environmental pollution process through an analysis of low-carbon technologies. It also introduces how income levels affect energy consumption and explains how higher energy demand leads to a larger share of fossil sources in energy mix and, thus, an increase in greenhouse gas (GHG) emission levels. Finally, this chapter offers an empirical approach to the positive impact that energy innovation policies exert over the replacement of polluting sources with renewable ones and explains how these measures help to control environmental pollution levels. In addition, Administrations' regulatory policies help to delay technical obsolescence and also control the scale effect that causes economies to return to

D. Balsalobre-Lorente (✉) · J.M. Cantos-Cantos
Department of Political Economy and Public Finance, Economic and Business Statistics and Economic Policy, University of Castilla-La Mancha, Albacete, Spain
e-mail: Daniel.Balsalobre@uclm.es

J.M. Cantos-Cantos
e-mail: josemaria.cantos@uclm.es

M. Shahbaz
Energy and Sustainable Development, Montpellier Business School, Montpellier, France
e-mail: m.shahbaz@montpellier-bs.com

J.L. Ponz-Tienda
Civil and Environmental Department, Universidad de Los Andes, Bogotá, Colombia
e-mail: jl.ponz@uniandes.edu.co

R. Álvarez Fernández et al. (eds.), *Carbon Footprint and the Industrial Life Cycle*,
Green Energy and Technology, DOI 10.1007/978-3-319-54984-2_11

increasing pollution levels. Although the promotion of energy innovation actions has a direct impact on the reduction of GHG emissions, this chapter concludes that, in the long-term, it is necessary to continue implementing energy innovation measures to delay technical obsolescence and, thus, delay the return to a stage of increasing GHGpc emissions.

1 Introduction

In recent decades, environmental issues have assumed an increasingly prominent role in the global debate about the Earth's future. Greenhouse gas emissions (GHG) are increasing despite joint efforts to implement international agreements. In this context, the energy sector is being recognized for its role in economic development and sectoral transformation, where it is the key to competitiveness among countries. This recognition has been incorporated into environmental analysis in order to address the over-exploitation of non-renewable energy sources, which is exacerbating air pollution.

In the late eighties, a critical view began to emerge among both developed and developing countries worried about the extreme use of natural resources and the lack of consideration of the support capacity of ecosystems. Global average temperature has already increased over the last three decades, and the global surface temperature is likely to rise an additional 1.1–6.4 °C (2.0–11.5 °F) in the 21st century (IPCC 2007a). The linear warming trend of the last 50 years is nearly twice that of the last 100 years. Global increases in concentrations of CO_2 emissions are primarily due to the burning of fossil fuels, which accounts for 56.6% of total emissions (IPCC 2007b).

In this regard, reducing air pollution levels is considered a sign of environmental sustainability. Kuznets (1998) claims that economic growth represents the long-term increase in the capacity of a country to provide its population with increasingly diversified economic goods and that this increased capacity is based on the advancement of technology and on structural and ideological changes required by technical progress. This chapter re-examines various studies that have analyzed the effects of both energy innovation processes and economic growth on environmental correction processes. In other words, empirical studies demonstrate that the environmental process of degradation is conditioned by the energy innovation process and the economic cycle (Aghion et al. 2014). In this regard, some studies have analyzed the relationship between economic growth and environmental pollution (e.g., Grossmann and Krueger 1991, 1995; Panayotou 1993; Selden and Song 1994). Their primary findings have demonstrated the existence of an inverted U-shaped relationship between economic growth and environmental degradation.

This behavior assumes that at the early stages of economic development, environmental degradation increases with per capita income. The next step, after a society reaches a threshold or turning point, is a decline in environmental degradation, which occurs once economic growth causes income levels to rise. In other words, the empirical literature describes how societies undergo an environmental transition from the first stage of economic growth, at the early stages of industrialization (with ascending pollution rates), to a developed stage of economic growth where increases in income levels reduce pollution levels. This theory supports the statement about a development stage during which pollution increases rapidly as a result of the higher priority placed on material output, during which there is an interest in improving productivity rather than environmental quality. On the other hand, once a society enters its developed stage, pollution levels decrease. This relationship, which is defined as the Environmental Kuznets Curve (EKC), reveals that economic growth can be compatible with environmental improvements. According to Stern (2004), p. 1419, "the EKC proposes that indicators of environmental degradation first rise, and then fall with increasing income per capita".

The EKC model also provides the opportunity to consider technical advances in the energy sector—during the developed stage of economic growth—as a necessary mechanism for reducing environmental pollution. The application of energy policy actions seems to be justified by the need both to correct environmental externalities and to reduce dependency on non-renewable resources. Therefore, this chapter attempts to illustrate, through the assessment of selected empirical studies, the positive effects that energy innovation policy actions exert over environmental correction.

Taking this into account, our aim is to present a summary of sustainable development and how this concept is linked with economic growth and environmental pollution.

2 Sustainable Development, Economic Growth and Global Warming

2.1 A Theoretical Approach to the Sustainable Development Concept: The Cost of Climate Change

It was in the 1970s that the Club of Rome raised awareness of the need for economic systems to preserve environmental sustainability (Meadows et al. 1972) and reflect the existence of a trade-off between economic growth and environmental degradation. Recent approaches have argued that it is possible to reach a state of "*sustainable development*" in which economic growth is compatible with environmental quality improvement (WCED 1987; Grossman and Krueger 1991; Munasinghe 1999). For an environmental agency such as the United Nations Environmental Programme (UNEP), environment is an obvious entry point from which to pursue sustainable development of an economy. This requires beginning from environmental impacts and tracing their causes into the economy. Environmental measures and policies serve to demonstrate both the economic and social benefits of environmental policy rather than the comprehensive sustainability of the economy.

Table 1 Selected impacts of climate change (by the end of the XXI century)

2013	2100
Global warming: 0.78 °C (1850/1900–2003/2012)	Global warming: 2.6 °C (0.3–4.8 °C)
Sea level rise: 19 cm (1901–2010)	Sea level rise: 54 cm (26–82 cm)
Arctic sea ice: 3.8% shrinkage per decade (1979–2012)	Arctic sea ice, year-round reduction: 43–94%
Tropical cyclones: increasing no. ("likely" in some regions, since 1970)	Tropical cyclones: increasing no. ("likely" increase in intensity)

Source IPCC (2013)

The 1992 Earth Summit in Rio de Janeiro promoted sustainable development for countries at all stages of development (UNEP 1992). Since 1995, the gatherings of the United Nations Framework Convention on Climate Change (UNFCCC) have used annual meetings to determine what measures can be taken to control global warming and greenhouse gas emissions (GHGs). The first remarkable outcome of the UNFCCC was the Kyoto protocol (1997), which introduced binding obligations for developed countries to decrease their GHGs. Nevertheless, the Kyoto protocol never became a real global agreement because its first and second amendments were binding for only the European Union and several industrialized countries; sadly, the world's largest emitters—USA, Canada and India, among others—did not ratify the protocol.

The 2012 Rio + 20 conference presented the greening of the economy as a tool for sustainable development, and new global development goals are in making (UNEP 2012). One way of achieving a certain degree of integration—or at least coordination—of social objectives with the other dimensions of sustainable development is to set goals and targets for the different dimensions.

Some environmental studies and environmental organizations have warned about disasters if global warming continues at its current pace (UNEP 2012; IPCC 2013). The widely recognized report of the Intergovernmental Panel on Climate Change (IPCC) presents a scientific analysis of climate change (IPCC 2013) that, through the use of biophysical measures, points to potential disaster if no action is taken (Table 1).

It is estimated that global warming will increase by approximately 2–6 °C by the end of this century, resulting in serious problems with sea levels and a reduction in artic sea ice (IPCC 2013).[1] Stern (2006) predicts that global warming will bring about major crop failures in developing countries, with a 25–60% increase in the number of people at risk of hunger; a sea-level rise that threatens London, Shanghai, New York, Tokyo and Hong Kong; the possible collapse of the Amazon

[1]IPCC (2013) shows that during the period of 1750–2011, total CO_2 emissions generated from human activities amounted to 555 billion tons, among which 240 billion tons have been accumulated in the atmosphere. Many scientists have addressed this and related environmental issues in past studies.

rain forest; the extinction of many species; and increasingly intense natural disasters. Following Stern (2006), Lin and Jiang (2009) noted that the global temperature could rise by 2–3 °C, which could reduce global GDP by 5–10% and by more than 10% in poor countries. Data from the IEA (2016) show that the atmospheric CO_2 concentration has increased by 40% compared to the mid-18th century, and there has been a significant increase during the last century. These increasing concentrations have gradually influenced the stability of earth's climate, ecosystems, and socioeconomic system. Even if the concentration of CO_2 emissions were to remain unchanged, climate system changes such as global warming and sea-level rise would continue irreversibly for centuries.

One advance in the evaluation of potential environmental damage has been to relate the impacts of climate change to economic activities and their values. Therefore, several environmental economists point to *cost-benefit analysis* as a method of pricing the social costs and benefits of scarce environmental services (Table 2).

Table 2 shows the application of various techniques of environmental damage evaluation, including market simulation, travel costs to nature reserves, and surveys of people's willingness to pay for environmental benefits and to accept compensation for environmental damages. The IPCC (2007b) estimated the average cost of damages from 4 °C of global warming (presumably in the second half of the twenty-first century) at 1–5% of GDP.

The IPCC (2014) does not provide any damage estimates. Therefore, Table 2 indicates only the difference in mitigation costs between the best and worst

Table 2 Global cost of climate change (% of world GDP)

	Past cost	Expected cost	
	2006	2100	2200
IPCC (2007)		1–5[a]	
IPCC (2014)		1.9[b]	
Stern (2006)		2.9[c]	13.8[c]
Nordhaus (2008)		2.5[d]	
World Bank (2010)	0.4[e]		

Notes: This Table shows illustrative estimates of past and future global costs of inaction on climate change. The different costs are hardly comparable. The reasons are differences in coverage, uncertain projections of carbon emissions and world GDP, and different time frames and valuation methods

[a]Global mean loss (macro-economic cost) for 4 °C global warming in the second half of the 21st century (IPCC 2007)

[b]Difference in mitigation costs in terms of consumption losses between best and worst scenarios (430 ppm CO_2-eq vs. 580–650 ppm), converted to % of GDP (IPCC 2014)

[c]"High-climate" scenario (mean: 4.3 °C global warming). (Stern 2006)

[d]"Best guess" of economic damage in 2100 with no intervention regarding emission control (Nordhaus 2008)

[e]Social cost in 2006 in adjusted net savings accounts at marginal damage cost of \$5.4/t of CO_2 World Bank (2010)

scenarios of greenhouse gas concentrations as a proxy for the possible 'maintenance cost of reducing the impact of climate change'.

The mitigation costs might be lower than the damage costs, at least with regard to their mid-range, which provides strong support for taking climate action now. Stern's (2006) estimate for the year 2100 is within the cost range of the IPCC but is expected to increase dramatically by the year 2200. In contrast, Nordhaus (2008) rejects Stern's low discounting, which ignores the real and more realistic interest rates for climate-related investments, dramatizes climate change events and justifies unnecessarily costly action now. Nordhaus's (2008) model of economic growth and climate change would set a global eco-tax at the level of the optimal carbon prices. Considering both the rise in the concentration of CO_2 emissions in the atmosphere and advances in emissions control and adaptive production and consumption patterns, the optimal tax rate should increase gradually from U$9.2 per ton CO_2-eq in 2010 to U$54 by the end of the century. Meanwhile, the World Bank (2010) applies a damage estimate in its "adjusted net savings" calculations and uses a particularly low marginal damage value of U$5.4 per ton of CO_2 emissions (albeit expected to increase to U$28 by 2030).

On the other hand, Weitzman (2009) presents his dismal "theorem" which disagrees with the statement that the gravity of potential disaster resulting from a global warming of 10–20 °C might "outweigh" the discounting of the low-probability impacts. This study ignores extreme effects in the economic valuation, which could be worse than "not presenting a cost-benefit estimate". Uncertainty about future greenhouse gas emissions and their effects calls for greater caution, but economics appears to be unable to determine how much awareness is needed. Rather than looking at the uncertain future, an assessment of the past might provide better clues for policy making. Weitzman (2009) criticizes the economic models of climate change, and this study considers whether looking back into the past may provide a more objective measurement of the effects of climate change.

2.2 *Energy Use and Evolution of Air Pollution Levels Within OECD Countries*

Thus, initiatives from the Kyoto Protocol to the European Councils[2] have exposed the future implications and consequences of global warming; they have also addressed the design and realization of program strategies and actions, as well as public and private decisions about the energy sector (IPCC 2007a, 2014; EU 2001, 2008, among others).

[2]Energy Efficiency Directive 2012/27/EU of the European Parliament and of the Council of 25 October 2012 on energy efficiency, amending Directives 2009/125/EC and 2010/30/EU and repealing Directives 2004/8/EC and 2006/32/EC.

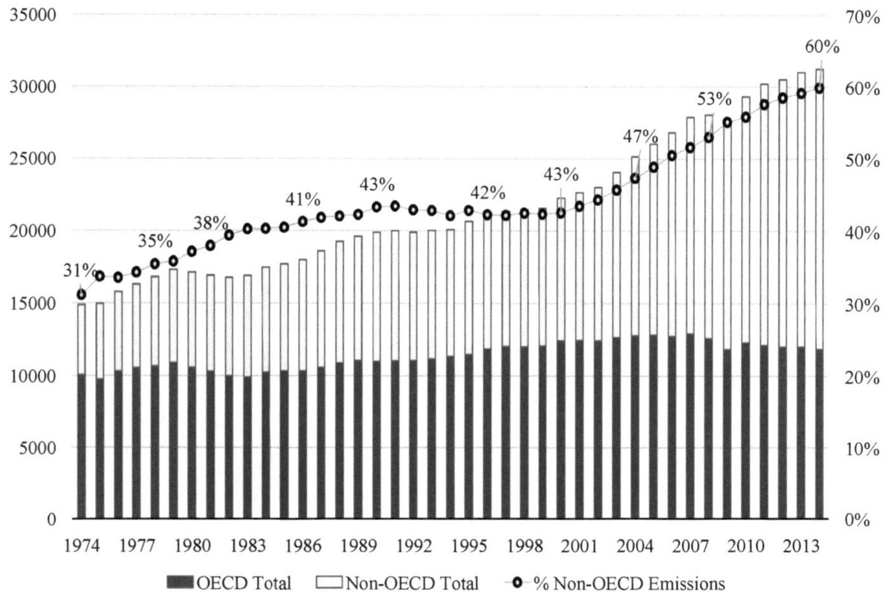

Fig. 1 Total CO_2 Emission-Fuel Combustion (Mt of CO_2) in OECD & Non-OECD countries (1974–2014). *Notes Main axis* OECD and Non-OECD CO_2 Emission-Fuel Combustion (Mt of CO_2). *Secondary axis* Non-OECD Share (%) CO_2 emissions (*Source* IEA 2016)

Several studies have highlighted the crucial role that energy sector currently plays in societies. This sector is becoming an essential issue for the welfare of countries and, by extension, their citizens, either on consumer side or as producers. Below this framework, there has been increasing concern about sustainability, including efforts to promote regulatory initiatives that may reduce air pollution levels. It is recognized that greenhouse gas (GHG) emissions and their evolution are still under study. Analysis focuses on climate change and the adverse changes that are taking place in the environment.

Figure 1 relates the evolution of CO_2 emissions in OECD countries (developed countries) and Non-OECD countries (developing countries). Between 1974 and 2014, both OECD and Non-OECD countries increased their shares of CO_2 emissions. In OECD countries, CO_2 emissions levels have increased from 10077.41 Mtod of CO_2 to 11856.55 Mtod of CO_2 (IEA 2016). Non-OECD countries increased from 4800.76 Mtod of CO_2 to 19395.04 Mtod of CO_2. During the last decade, developing countries' share of CO_2 emissions overcame that of developed countries (Fig. 1). When we analyze the evolution of CO_2 emissions, in OECD countries we can observe a link with the evolution of air pollution levels. In this chapter, we will attempt to show how both economic growth and energy innovation measures are linked with this fact. On the other hand, Fig. 1 also shows the share (%) of CO_2 emissions of OECD and Non-OECD countries. Figure 1 reveals that the share of CO_2 emissions in OECD countries has changed during the last two

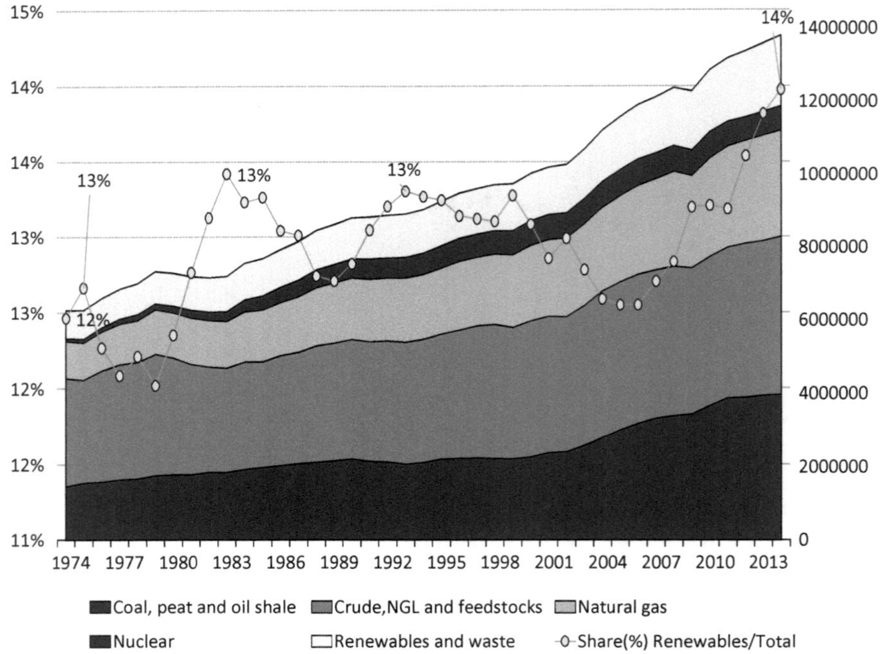

Fig. 2 World Total Energy Production by Sources (ktoe) (1974–2014). *Notes* Share (%) of Renewables and waste/Total Energy Production (*Main axis*). World Total Energy Production by Sources (ktoe) (*Secondary axis*) (*Source* IEA 2016)

decades. In the seventies and eighties, OECD countries had a higher share of CO_2 emissions. Currently, this share has declined and Non-OECD countries represent the main contributors to global CO_2 emissions (60% in 2014).

In 2014, world carbon emissions amounted to 32381.04 Mtod CO_2. This indicates an increase in emission levels. It implies an increase of 110% from 1974 to 2014, although during the last decade the OECD countries have appeared to reduce their CO_2 emissions (IEA 2016).

In recent decades, worldwide economic development has generated a large demand for energy, such that the global demand for energy tripled in 2014 compared to its 1974 level: fossil fuel energy accounted for more than 80% of that increase, and increasing energy demand was a key driver of rising CO_2 emissions (IEA 2013a).

Figure 2 represents a compilation of diverse energy sources that can be used to analyze the advance of these variables during recent decades. First, Fig. 2 shows the evolution of energy production by source (ktoe) between 1974 and 2014. From 1974 to 2014, both OECD and Non-OECD countries increased their non-renewable outputs. On other hand, the production of renewable sources increased from 12 to 14% between 1974 and 2014. By 2014, renewable energy sources had increased to 1,929,079 ktoe of energy production and 1,188,401 ktoe of energy consumption (IEA 2016).

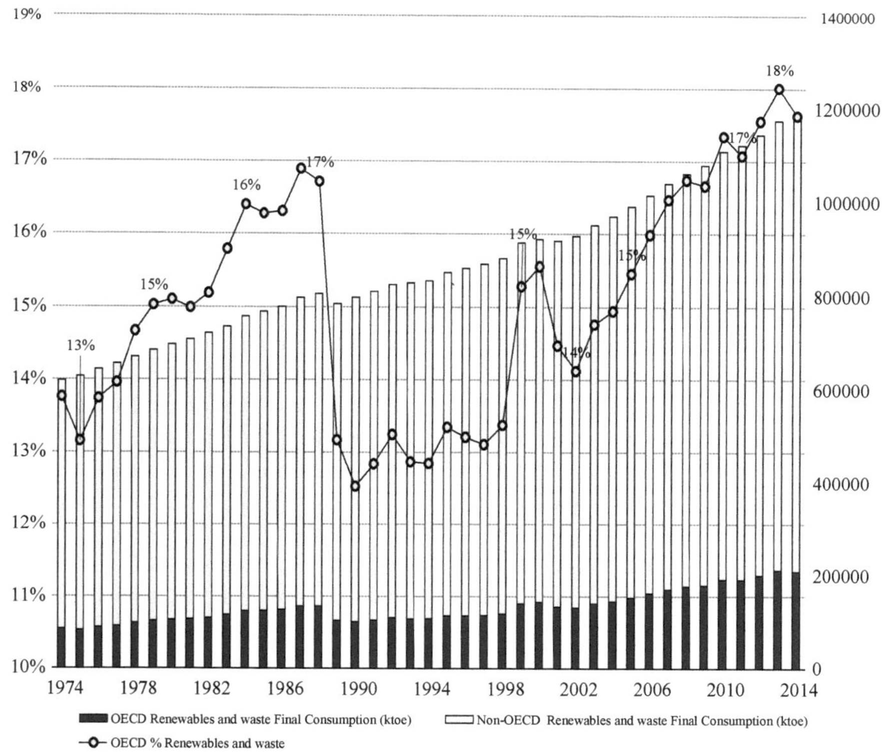

Fig. 3 OECD and Non-OECD countries' Renewables and Waste Consumption (ktoe) (1974–2014). *Notes* Share (%) of Renewables and waste Final Consumption in OECD countries (*Main axis*). OECD and Non-OECD countries Renewables and Waste Consumption (ktoe) (*secondary axis*) (*Source* IEA 2016)

Figure 3 illustrates the global evolution of renewable consumption and waste (ktoe) in OECD and Non-OECD countries. It reveals a net increase in consumption (from 620629.42 ktoe in 1974 to 1075423.01 ktoe in 2014). In 1974, renewable production in OECD countries was 182635.67 ktoe, and consumption was 85435.38 ktoe. In 2014, energy production by renewable sources was 512300.02 ktoe in OECD countries, and consumption of renewable sources accounted for 209453.01 ktoe (IEA 2016).

Climate change has been aggravated in recent years as CO_2 emissions have continued to grow in both developed and developing countries (IPCC 2014). Although OECD countries have a larger responsibility for climate change, non-OECD countries have accounted for almost 50% of global carbon emissions since the last decade. These emissions were linked to the energy sector (OECD 2013).

On other hand, some studies argue that the increase in the share of renewable energy sources is generated by innovation in the energy sector. These measures are connected with the reduction in CO_2 emissions (Foxon and Pearson 2007;

Jefferson 2008; Balsalobre et al. 2015; Dogan and Seker 2016). In addition, to complement these activities, it is also necessary to increase overall efficiency in the energy sector (Eichhammer et al. 2013; Marques et al. 2010; Schleich 2009).

2.3 The Public Budget in Energy RD&D Within OECD Countries: Some Policy Recommendations

Having observed both emissions and energy consumption variables, we will analyze the evolution of expenditures in the public budget on energy research, development and demonstration (RD&D) undertaken by the OECD economy (Fig. 4). This variable complements the influence of other variables that influence the evolution of intensity energy in these countries. So, it is important to recognize that a public budget for energy RD&D strengthens the technological capacity of the company as well as its capacity for innovation and learning, thus enhancing value-added strategies (Balsalobre et al. 2015).

Figure 4 presents the evolution of public budgets for energy RD&D since 1974 for different countries and sources of the OECD. Figure 4 shows a positive evolution of this variable; however, during recent years these efforts have been reduced,

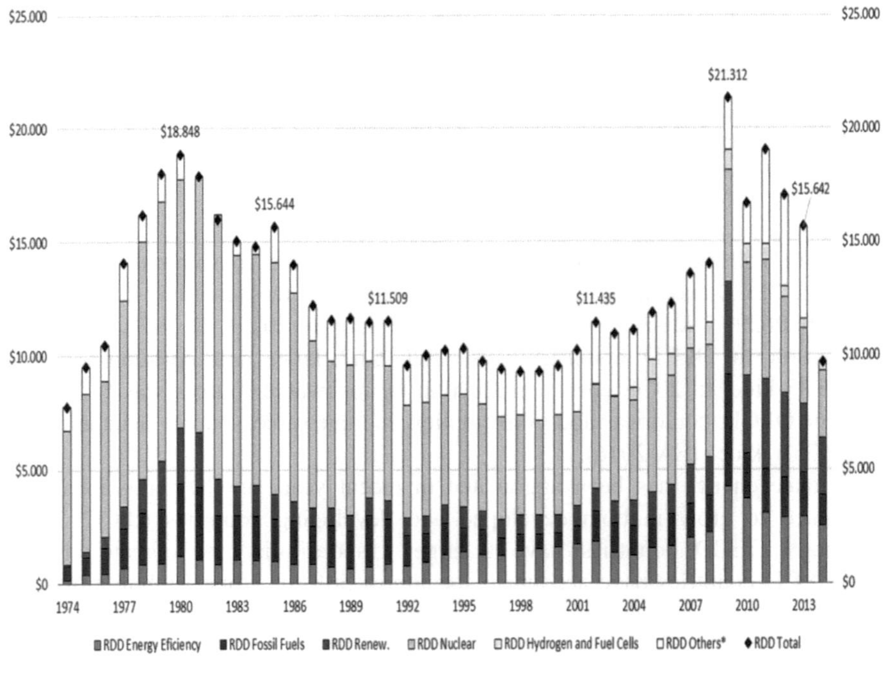

Fig. 4 Public Budget in RD&D (in Million USD 2014 prices and PPP) in OECD Countries (1974–2014). *Notes* energy RDD by sources (*Main axis*); Total energy RDD in OECD countries (*secondary axis*) (*Source* OECD 2016)

probably for reasons linked to the economic cycle. Regarding the nexus between technological innovation and energy efficiency, technological innovation increases the quality of production by augmenting energy efficiency (Brock and Taylor 2005). In fact, OECD countries experience greater energy efficiency gains due to the magnitude of their technological innovation compared to other developing countries (Wong et al. 2013). Technology may reduce energy efficiency marginally; however, in absolute terms, it may produce a rebound in overall energy use (Greening et al. 2000).

These investments in energy RD&D are hindered by different factors, such as high upfront costs, risks, and uncertainty regarding long-term viability of the technology, long payback periods, high regulatory and infrastructural dependency as well as public acceptance (Müller et al. 2011). Therefore, policy makers have to promote public policies that help to reduce these risks, by, for example, reducing the capital costs of creating more efficient technologies; this should be linked with reduction in demand for fossil sources that have been heavily subsidized in the past (Szabó and Jäger-Waldau 2008). However, based on the results of some studies (Friebe et al. 2014; Lüthi and Prässler 2011; Lüthi and Wüstenhagen 2012), relevant decision criteria have been identified, including the assumption of regulatory measures linked with streamlining of the administrative process. Chassot et al. (2014) confirm these statements by highlighting the perceived risk of policies as the main determinant of investment decisions. Furthermore, Szabó and Jäger-Waldau (2008) propose reducing the costs of capital for sustainable energy projects to promote the progress of more efficient capital markets and to promote competitiveness in the energy sector.

In other words, a blend of supportive financial regulation and energy regulation policies would help to positively extend energy improvements (Lüthi and Wüstenhagen 2012; Wüstenhagen and Menichetti 2012; Bergek et al. 2013). Otherwise, whether policy adoption determinants are similar in industrialized countries is questionable, given the differences in political systems, international commitments to mitigating climate change, and economic development. For example, rapid income growth rates during periods of high oil prices may have encouraged the search for alternative energy sources. Furthermore, less affluent countries may need international financial and capacity building support to implement energy technologies with high investment costs. Consequently, we can expect that international climate costs will help to establish new policies. However, we consider it necessary to promote national-level policy development in support of climate change mitigation.

Finally, sustainability measures are even more relevant when the focus is on particular impacts, such as climate change, that are brought about by economic activity. Rather than pretending that the inclusion of climate costs generates overall economic optimality, the sustainability of capital maintenance provides a quantifiable benchmark for tackling climate change: the idea is to reduce greenhouse gas emission to maintain the absorptive capacity of carbon sinks.

3 Economic Growth and Environmental Pollution: The Role of Innovation Within the Environmental Kuznets Curve

In this section, we propose an extension of the stylized fact of the EKC[3] model. We present a review of the economic growth literature and discuss how this variable assumes a fundamental role in the evolution of pollution and the promotion of innovation measures. The EKC concept thrived in the early nineties, as it described the time trajectory that a country's pollution would follow as a result of its economic growth. The EKC is a widely applied empirical model that is used to assess the effect of economies' income levels on their environmental pollution levels. In this section, we also present a review of the literature on economic growth and discuss how this variable assumes a fundamental role in the evolution of environmental quality improvements. When economic growth appears in a poor country, pollution emissions grow because the increase in output generates pollutants and because the country places a low priority on the control of environmental degradation. Once a country obtains a sufficient level of affluence, its priorities shift to controlling environmental quality.

3.1 A Theoretical Approach to the EKC Model

It was not until the early 1990s that environmental pollution problems began to be more frequently described in exiting economics literature (Groosman and Krueger 1991; Shafik and Bandyopadhyay 1992; Panayotou 1993; Selden and Song 1994). Since Grossman and Krueger (1991), many studies have analyzed the connection between economic growth and environmental damage. The underlying argument behind the EKC theoretical model is that environmental pollution is an increasing function of the level of economic activity, until a critical income level is reached. When an economic system reaches a certain income level, a better level of environmental quality appears, at which point environmental destruction diminishes; this pattern usually takes an inverted U-shape (Fig. 5) (Grossman and Krueger 1991; Panayotou 1993; Selden and Song 1995).

Originally, Grossman and Krueger (1991) proposed an inverted U-shaped relationship between environmental degradation and income level (Fig. 5).

Figure 5 shows an inverted U-shaped relationship between income level and environmental pollution. This behavior implies that in the early stages of economic growth, environmental pollution levels rise until a certain turning point is reached,

[3]The Environmental Kuznets Curve (EKC) was initially developed by Grossman and Krueger (1991). It takes its name from the inverted U relationship between economic growth and environmental destruction, whose pattern of conduct resembles that studied by Kuznets (1955) in his analysis of inequality and economic growth (Panayotou 1993).

Fig. 5 Inverted U-shaped Environmental Kuznets Curve (*Source* Prepared by the author)

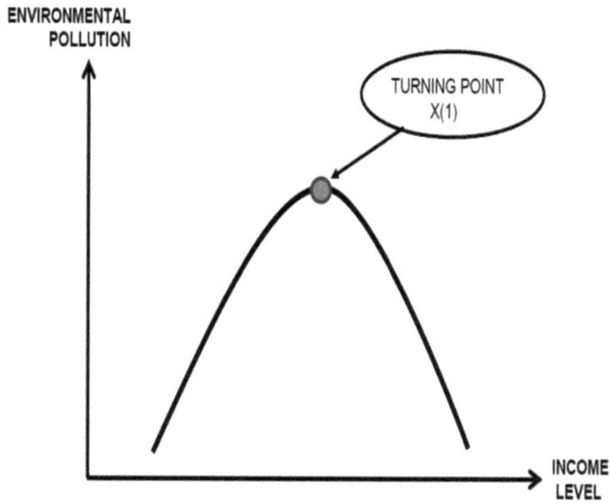

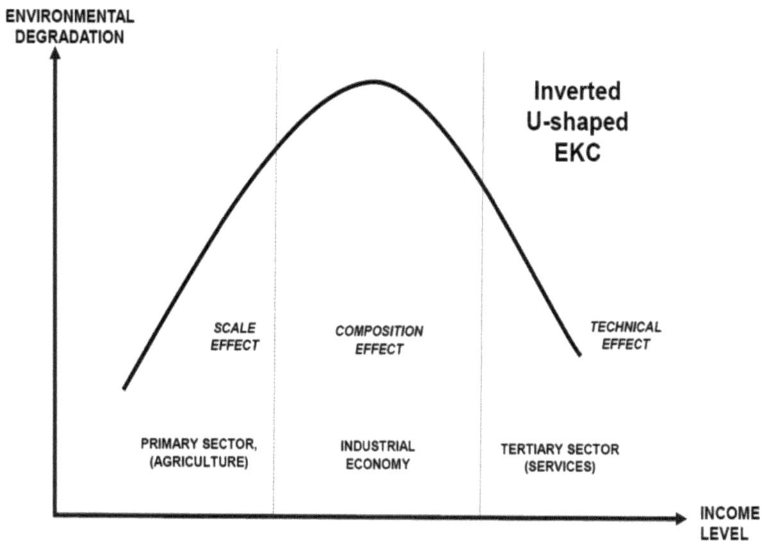

Fig. 6 EKC: Inverted U-shaped relationship between environmental pollution and income level (*Source* Prepared by the author)

beyond which economies experience a reduction in pollution levels. This behavior also implies that economic growth affects environmental quality via three channels (Grossman and Krueger 1991): scale, composition, and technical effects (Fig. 6). This progression also supposes a dynamic structural change connected with economic growth (Dinda 2004).

Figure 6 illustrates how the economic cycle influences environmental quality through three channels: scale, composition and technical effects (Grossman and

Krueger 1995). When the economic system is in a developing stage, the scale effect overcomes the net effect. The *scale effect* asserts that even if the structure of the economy and the technology of the countries do not change, an increase in production will result in decreased environmental quality; therefore, economic growth initially has a negative impact on the environment. This effect suggests that even if the structure of the economy and the technology of the countries do not change, an increase in production will result in decreased environmental quality. In other words, when shares of inputs, and by extension energy demands, are employed to obtain higher output levels, this process involves an increase in environmental pollution levels (Torras and Boyce 1998).

At high levels of economic growth, the composition effect positively affects environmental quality. This effect suggests a positive impact on environmental quality because the economic structure changes from agriculture and heavy manufacturing industries to cleaner industries and the service sector. During the earlier stages of economic development, pollution levels increase, but, during the second stage of development, when the economic structure changes from agriculture to more intensive heavy manufacturing industries, pollution levels decrease as a consequence of shifts in the economic structure towards services and light manufacturing industries. This effect represents the transition from a developing economy, with highly polluting production processes, to a developed system, with a production pattern involving less-polluting activities (Hettige et al. 1998). In other words, the composition effect refers to developing economies' transitions from capital-intensive industrial sectors to service sectors (Fig. 6).

Finally, when economies are in a developed stage (high-income level), the technical effect captures their advances in productivity and their adaptation to cleaner technologies. This process will lead to an increase in environmental quality.

Since Grossman and Krueger (1991), other studies have also examined the EKC pattern (Shafik 1994; Holtz-Eakin and Selden 1995). To analyze the long-term relationship between economic growth and environmental pollution, it is necessary to explore an extension of the EKC's theoretical framework; this allows us to identify different scenarios of the relationship between environmental quality and economic growth (Fig. 7).

Behind the EKC theoretical scheme outlined here, which links a relationship between environmental pollution and income levels to an N-shape, the EKC's cubic scheme is expressed as follows (Eq. 1):

$$EP_t = \alpha_{it} + \beta_1 Y_{it} + \beta_2 Y_{it}^2 + \beta_3 Y_{it}^3 + \beta_4 Z_{it} + \varepsilon_{it} \tag{1}$$

Where EP_{it} is the environmental pollution in the country or region "i", in time "t". Y_{it} is the per capita income level (we also incorporate the quadratic and cubic polynomial), and Z_{it} is the other influential, or auxiliary, variable over environmental quality. The α_{it} coefficient includes the environmental quality average when income has no special relevance to environmental pressure, while β coefficients represent the relative importance of exogenous variables; variable ε_{it} is the error term. Depending on the value allocated to coefficients β (Eq. 1), the EKC can adopt

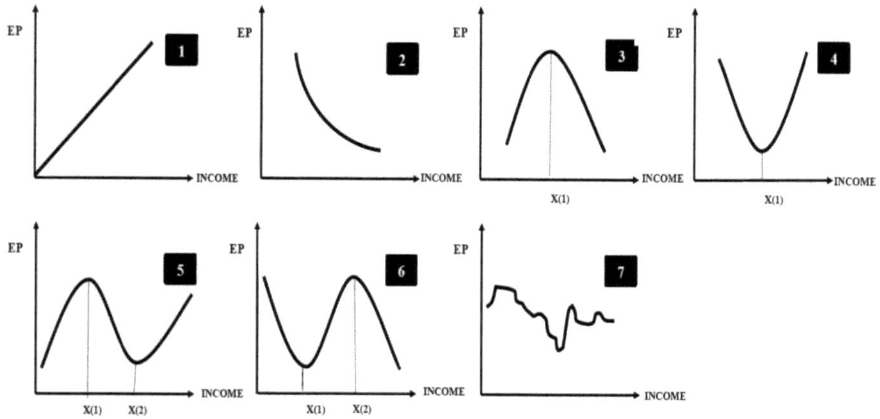

Fig. 7 Potential behaviors between environmental pressure (GHGpc) and GDP per capita. *Notes* *1* If β1 > 0, β2 = β3 = 0, increasing monotone relationship, where high levels of income are associated with high levels of pollution. *2* If β1 < 0, β2 = β3 = 0, decreasing monotone relationship, where high levels of income are associated with decreasing levels of pollution. *3* If β1 > 0, β2 < 0, β3 = 0, quadratic relationship in inverted U pattern, representing the EKC and indicating that high levels of income are associated with decreasing levels of pollution once a certain level of income is reached. *4* If β1 < 0, β2 > 0, β3 = 0, quadratic relationship in U pattern, contrary to the EKC. *5* If β1 > 0, β2 < 0, β3 > 0, cubic polynomial, representing the N shape, where the inverted U hypothesis occurs up to a certain point, from which pollution increases again. *6* If β1 < 0, β2 > 0, β3 < 0, cubic polynomial, inverted N shape. *7* If β1 = β2 = β3 = 0, flat behavior, indicating that emissions are not influenced by the level of income

different forms other than the typical one (Fig. 8). We must pay special attention to the N-shaped pattern (see Fig. 6: pattern 5), following the contributions of Grossman and Krueger (1995), among others. The N-shaped EKC suggests a behavior that amplifies the income-environmental pollution relationship in the long term (Shafik and Bandyopadhyay 1992; Selden and Song 1994; Grossman and Krueger 1995; Moomaw and Unruh 1997; Torras and Boyce 1998). The N-shaped pattern of the EKC makes it possible to discuss aspects related to the scale effect and the long-term effects of energy innovation. This, in turn, makes it possible to analyze the potential return to rising emission levels once economies have achieved negative pollution rates, and environmental technological obsolescence becomes possible (Balsalobre and Álvarez 2016).

Although most of the EKC's theoretical contributions assume an inverted U-shaped pattern (Selden), it is necessary to consider the N-shaped behavior (Fig. 8) to analyze the long-term evolution of the relationship between income level and the environmental pollution process (Shafik and Bandyopadhyay 1992; Grossman and Krueger 1995; Torras and Boyce 1998, among others). The N-shape checks whether environmental degradation returns at very high levels of economic growth.

This N-shaped relationship occurs when the relationship between the environmental degradation of the economy and income level is initially positive,

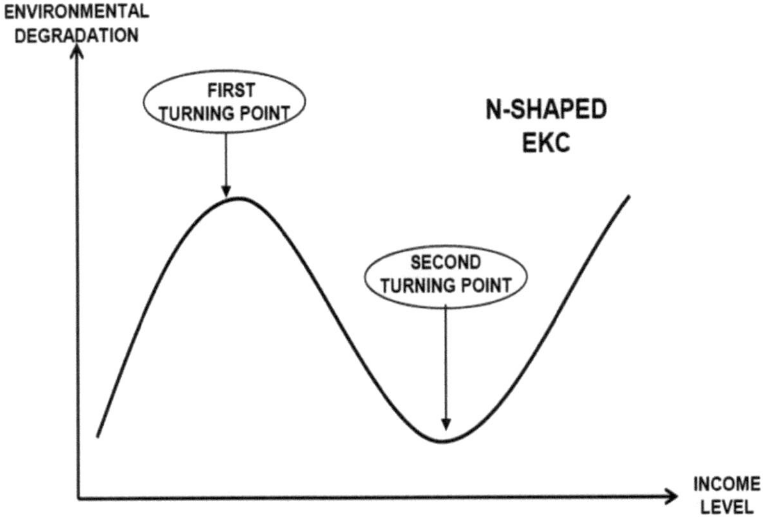

Fig. 8 EKC N-shaped (*Source* Prepared by the authors)

but it becomes negative once a given income threshold is reached, until environmental degradation becomes positive again. In other words, this EKC's behavior means that environmental degradation increases during a developing stage, but once it reaches an initial turning point, environmental degradation decreases (Fig. 8). This process continues until a second turning point, where environmental pollution once again begins to experiment with an increase. This third stage is characterized by high income levels with low economic growth rates and inefficient and insufficient environmental innovation measures. Torras and Boyce (1998) consider that the return to an upward pollution path appears when the margin for successive improvements in the distribution is exhausted; in other words, when there are diminishing returns in terms of technological change reducing pollution because of "obsolescence". At the end of this section, we will more exhaustively develop the technical obsolescence-within-EKC model.

Next, we discuss the theoretical relationship between economic growth and the innovation process.

3.2 Economic Growth and the Energy Innovation Process

The fluctuating energy sector has directly affected societies, changing organizational and social behavior patterns. As presented above, rising energy demands have accelerated the collapse of fossil sources. This process, along with the volatility and pressure surrounding energy prices, has increased environmental degradation. Energy regulation policies have considered the need for savings and energy efficiency. Furthermore, the elasticity to legitimize a response to climate change in society has accelerated technological improvements in the sphere of innovative energy sources, which are being added as a fundamental input to the socioeconomic discussion about positive effects on social, economic and sustainable development.

According to this hypothesis, several studies analyze the positive effects of innovation and structural changes on the evolution of emission levels (Torras and Boyce 1998; Andreoni and Levinson 2001; Balsalobre and Álvarez 2016, among others). Prior to these studies linking technological progress with the EKC model, Romer's theory (1990) proposed that technological progress is the key element in long-term economic growth. This theory also supports the idea of the existence of external effects in the economic system, which are able to generate innovations and to disseminate the knowledge acquired. Romer (1990) also notices that externalities produced by innovation may be able to reduce the amount of environmental degradation.

It is possible to validate that the returns on private investment in energy innovation are lower than the social benefits, which, without public intervention, would be difficult to achieve (Griliches 1992). If it is assumed that environmental pollution is a negative externality, then the empirical evidence suggests a relationship between low-carbon technological innovations and the reduction in environmental pollution (Heyes and Kapur 2011; Balsalobre et al. 2015). Under an imperfect competition scenario, which is compatible with the existence of financing activities able to generate technological progress (Romer 1987; Grossman and Helpman 1991), the endogenous postulate of economic growth establishes the existence of a relationship between economic growth and the environment, where hi-tech energy advancements would reduce the necessary hazards and allow sustainable growth rates to be reached (Gradus and Smulders 1993).

Furthermore, innovation achievements aimed at environmental correction measures are premised on the idea that the expansion of clean technologies will promote a reduction in environmental pollution levels (Aghion and Howitt 1992).

On the other hand, sectoral *diminishing returns* would result from the positioning of imperfect competition, making it unfeasible to achieve economic growth long-term without avoiding a certain level of increased pollution. Sustainable growth requires, as a prerequisite, the *existence of constant returns to scale or growth in pollution removal activities*. Consequently, without environmental intervention (as happens in models of exogenous optimal growth), environmental pollution will grow unlimited or until an unnecessarily high income level is achieved (Arrow et al. 1995). This statement, which sits within economic growth models, discerns the probability of achieving positive rates of long-term growth without having to consider any variable as exogenous to the model. The theoretical approach of the EKC model consistently supports the idea that economic growth and innovation measures can exert a resilient influence on environmental correction, where the empirical analysis of the EKC demonstrates that the impact of the energy sector has a direct influence on the correction of emission levels (Balsalobre et al. 2015).

Once we have established the theoretical framework (under the EKC hypothesis) that links economic growth, environmental degradation, and the innovation process, the next step is to develop the relations between economic growth and energy innovation that influence environmental correction processes.

3.3 Energy Innovation Within the Environmental Kuznets Curve Model

As stated above, the EKC model provides a systematic explanation of the relation between the environmental correction process and economic growth. Andreoni and Levinson (1998) showed that decontamination processes mainly depend on the technical effect and that investment in energy innovation processes will contribute to reducing environmental pollution levels. However, this theory also supports the idea that, under the pressure of increasing returns to scale, the innovation process is the key to correcting environmental pollution levels (Fig. 9).

Figure 9 illustrates how energy innovation processes and economic growth achieve a reduction in emission levels after the first stage. As previously shown, the EKC model supports the idea that once economies reach a certain level of income, economic systems begin to reduce their emission levels until they reach a second

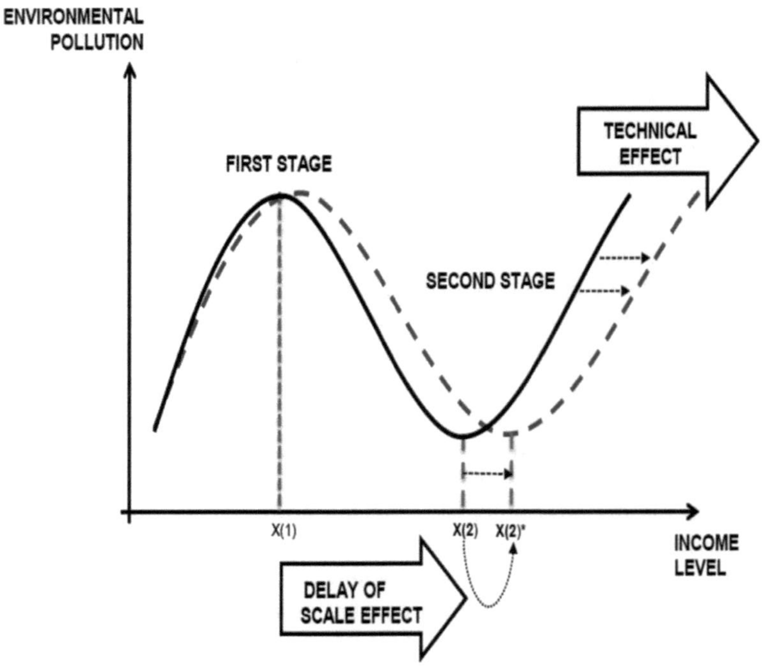

Fig. 9 Technical effect in the EKC model (*Source* Prepared by the authors)

turning point, where the scale effect overcomes the composition and technical effects. Figure 9 illustrates the positive effect that the energy innovation process exerts over the scale effect in order to delay the return to a stage of ascending pollution (Balsalobre et al. 2015). This assessment links the EKC's economic growth-environmental pollution relationship with technological development. Some authors believe that economic growth will not be able, by itself, to resolve contamination issues. In other words, in a scenario where environmental regulation is lacking or absent, given the trajectory described by the economic growth-environmental pollution relationship, the correction of contamination could still be achieved with an unnecessarily high income level (Arrow et al. 1995). Based on this fresh perspective, innovative energy regulations have become a fundamental strategy for long-term growth, and energy policies are central axes. Thus, when economies reach a developed stage, this leads to higher energy consumption and lower rates of environmental pollution because of a strategy of innovation under which increasing returns to scale are able to improve energy efficiency with lower income requirements (Balsalobre et al. 2015).

The EKC model lets us distinguish between still-developing and already-developed economic systems, with the first being an economic model that is based on the over-exploitation of fossil sources and high income rates, where the tertiary sector has low weight. Assuming that, over the course of decades, advances in the energy sector have altered economic systems, the increase in energy improvements is currently leading to the creation of a sustainable model, one that has less energy intensity and lower dependency on non-renewable sources. This new scenario reflects unanimity regarding the need to increase environmental sustainability through the use of low-carbon technologies and the implementation of new technologies. Furthermore, the effectiveness of these actions is linked with a rate of innovation that delays *technical obsolescence*.

3.4 Technical Obsolescence in the EKC Model

As explained previously, technical obsolescence can diminish the long-term efficiency of innovation efforts. Some studies, such as that by Balsalobre and Álvarez (2016), address technical obsolescence though the delay in the scale effect that occurs when improvements in energy innovation produce composition and technical effects that enable the scale effect to be overcome (Fig. 10).

Figure 10 illustrates the idea that an adequate environmental regulation policy can be effective in accelerating technological change. This process also delays the scale effect in the long-term and decreases the level of contamination (Torras and Boyce 1998; Balsalobre et al. 2015). On the other hand, a long-term return to increasing contamination (N-shaped) could be due to the scale effect overcoming both the composition and technological effects. This process is justified by diminishing returns in terms of the technological change that reduces the amount of contamination as a consequence of the collapse of technical obsolescence.

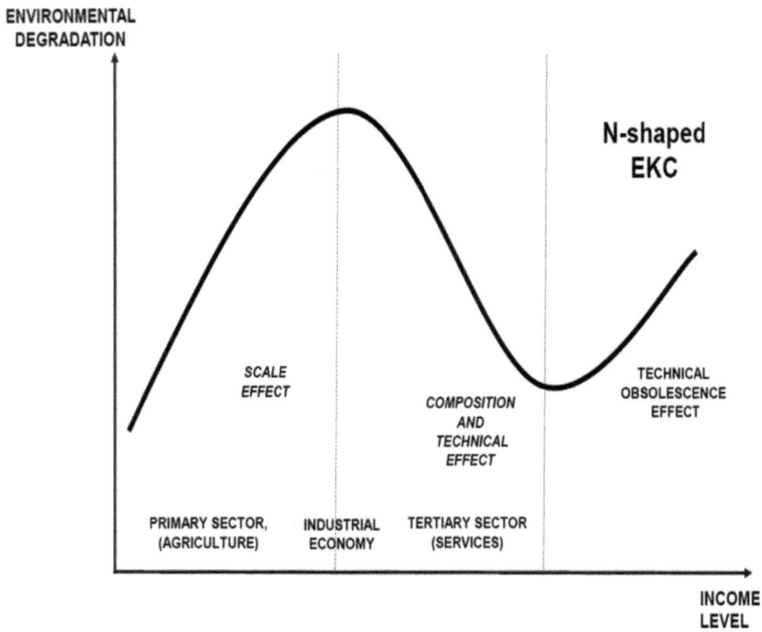

Fig. 10 The effect of technical obsolescence in the EKC model (*Source* Prepared by the authors)

Once the second turning point is reached, emissions levels increase again. This new stage, where technical obsolescence appears, indicates the effectiveness of energy-related regulatory policies and implies a return to ascending pollution levels.

Clearly, it is undeniable that energy is an influential factor in the development of economic systems. Balsalobre and Alvarez (2016) demonstrate that energy innovation measures exert a positive effect on environmental quality, thus delaying the scale effect that implies a return to ascending pollution levels (Fig. 8). Within the relationship between income level and environmental degradation, energy innovation can be considered the main variable that delays technical obsolescence (Fig. 11).

According to the EKC theoretical framework, Fig. 8 reflects the positive influence of environmental policies supporting energy innovation on the environmental correction process.[4] Furthermore, in the absence of energy regulation policies, economies reach technical obsolescence sooner.

Therefore, when Administrations implement regulatory measures that are linked with energy innovation, this delays the scale effect; additionally, a decreasing level of emissions is maintained (Balsalobre and Álvarez 2016). However, when the scale effect again overcomes the composition and technical effects, the technical

[4]In the field of technological innovation, regulation processes offer an additional explanation supported by the endogenist theory, which holds that the change in the behaviour of the 'income level/environment' relationship must be in good part due to the improvement in production processes derived from technological change (Gradus and Smulders 1993).

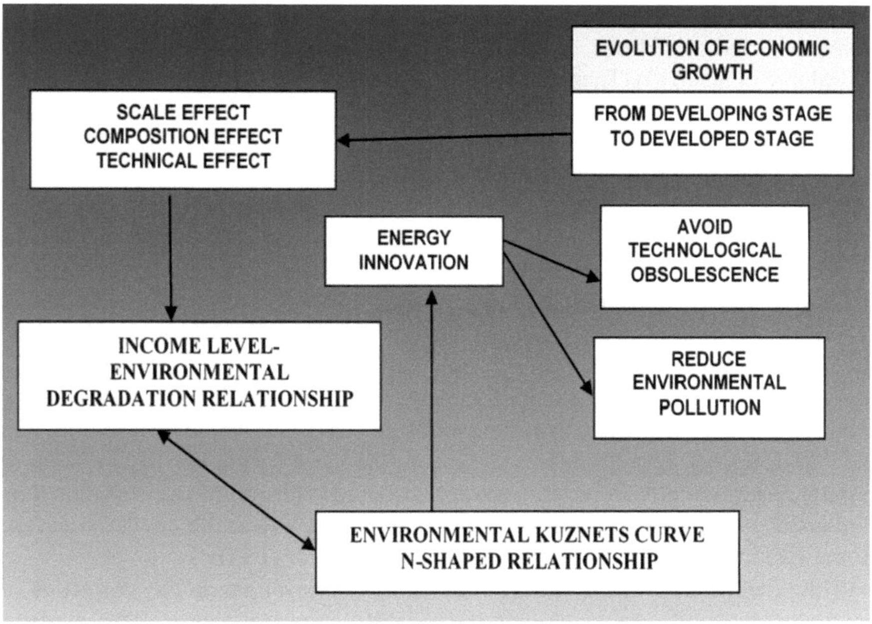

Fig. 11 Conceptual scheme (*Source* Prepared by the authors)

obsolescence effect appears. This effect was defined by Balsalobre and Alvarez (2016) and supposes that if the implemented regulatory policies are inefficient, environmental degradation will once again increase. Thus, it is necessary to maintain an ascending level of energy innovation policies to delay the increase in environmental pollution that could appear after the second turning point.

4 Summary and Final Conclusions

Throughout this chapter, we have considered the positive effect of energy improvements on the adoption of cleaner technologies that delay the technical obsolescence that has a negative effect on air pollution levels.

Furthermore, without energy innovation policies, economic growth will not, by itself, be able to solve air pollution problems. In other words, in the absence of environmental regulation, even if the "income-environmental quality" relationship were able to correct pollution, this might be achieved at an unnecessarily high-income level (Arrow et al. 1995).[5] Some studies claim that the relationship

[5]Andreoni and Levinson (1998) show that the 'income level/pollution' relation depends, fundamentally, on the technology used for the reduction of pollution and not on the number of polluters or the marginal utility of consumption and environmental quality, or even on the externalities.

between economic growth and environmental quality is the consequence of technological change (Torras and Boyce 1998; Smulders and Bretschger 2000; Balsalobre et al. 2015). These studies recognize that deficient rates of innovation in the energy sector will cause technical obsolescence. This effect implies a return to a stage of increasing environmental degradation. This empirical assessment supposes that the environmental correction process will require substantial and continuous efforts in terms of energy innovation to delay the scale effect. Heerink et al. (2001) state that the extent to which the technical effect dominates the total effect depends on incentives for policymakers. He and Zhang (2012) also found that government policies on environmental regulation are beneficial to energy savings and emission reductions. However, their research did not discuss environmental regulation's specific effect on the EKC.

In other words, energy regulation measures expand improvements in the energy sector with the aim of avoiding the trap of decreasing technological returns on a path to technical obsolescence. Moreover, the implementation of measures to promote energy innovations and renewable sources will result in a deviation from diminishing technological returns, thereby helping to reverse the upward trajectory of the EKC (Torras and Boyce 1998; Balsalobre et al. 2015).

This chapter introduces the role of energy innovation in the correction of environmental pollution levels within the EKC empirical model. This evidence distinguishes between two main elements that influence environmental correction: economic growth and energy innovation (Andreoni and Levinson 2001). This chapter presents some empirical evidence confirming that improvements in energy innovation processes exert a positive effect over the correction of air emission levels (Shafik and Bandyopadhyay 1992; Andreoni and Levinson 1998, among others). These studies also conclude that when economies are in a developing stage, it is necessary to accelerate energy innovation measures to reduce the scale effect and achieve a reduction in pollution levels, even though this entails an added cost for societies. In addition, once economies reach a developed stage, they must continue to pursue energy regulation to avoid technical obsolescence. When the total effect of the relationship between economic growth and environmental pollution is broken down, the technical effect is considered to be the main factor in the correction of environmental pollution process (Andreoni and Levinson 1998). In this regard, technical obsolescence will lead to the re-emergence of increasing pollution levels once the scale effect exceeds the composition and technical effects (Balsalobre and Álvarez 2016).

Moreover, when Administrations implement regulatory policies, those measures help to delay technical obsolescence and also control the scale effect that drives economies to return to increasing pollution levels. Although the promotion of energy innovation actions has a direct impact on the reduction of GHG emission, in the long-term it is necessary to continue implementing energy innovation measures to delay technical obsolescence and, thus, the return to a stage of increasing GHGpc emission levels.

However, some studies reveal that this effect of innovation on the environmental correction process requires a time-lag to become fully efficient. This finding reveals that all energy innovation efforts do not have short-term effects; it is therefore necessary to assume an initial energy over-cost for society, but in the long-term this will yield the highest return.

Finally, the environmental correction process is also linked to the economic cycle, such that as an economy's income level increases (developing stage), the energy demand and the energy-mix increase emissions levels. For this reason, Administrations must strengthen their efforts to promote energy innovations to reduce both energy dependency and the pollution impact of energy systems when economies' energy demands increase. This chapter illustrates the relevance of environmental regulation measures in ensuring that countries stay on the path of decreasing environmental pollution levels.

References

Aghion, P., Hepburn, C., Teytelboym, A., & Zenghelis, D. (2014). *Path-dependency, innovation and the economics of climate change. Supporting paper for New Climate Economy*. London: Grantham Research Institute on Climate Change and the Environment, London School of Economics and political Science.

Aghion, P., & Howitt, P. (1992). A model of growth trough creative destruction. *Econometrica, 60* (2), 323–351.

Andreoni, J., & Levinson, A. (2001). The simple analytics of the environmental Kuznets curve. *Journal of Public Economics, 80,* 269–286.

Andreoni, J., & Levinson, A. (1998). The simple analytics of the environmental Kuznets curve. NBER Working Papers, 6739.

Arrow, K., Bolín, B., Constanza, R., Dasgupta, P., Folke, C., Holling, C., et al. (1995). Economic growth carrying capacity, and the environmental. *Science, 268,* 520–521.

Balsalobre, D., Álvarez, A., & Cantos, J. M. (2015). Public budgets for energy RD&D and the effects on energy intensity and pollution levels. *Environmental Science and Pollution Research, 22*(7), 4881–4892.

Balsalobre, D., & Álvarez, A. P. (2016). Economic growth and energy regulation in the environmental Kuznets curve. *Environmental Science Pollution Research, 23*(16), 16478–16494.

Bergek, A., Mignon, I., & Sundberg, G. (2013). Who invests in renewable electricity production? Empirical evidence and suggestions for further research. *Energy Policy, 56,* 568–581.

Brock, W. A., & Taylor, M. S. (2005). Economic growth and the environment: A review of theory and empirics. *Handbook of Economic Growth, 1,* 1749–1821.

Chassot, S., Hampl, N., & Wüstenhagen, R. (2014). When energy policy meets free market capitalists: The moderating influence of worldviews on risk perception and renewable energy investment decisions. *Energy Research and Social Science, 3,* 143–151.

Dinda, S. (2004). Environmental Kuznets Curve Hypothesis: A survey. *Ecological Economics, 49* (4), 431–455.

Dogan, E., & Seker, F. (2016). The influence of real output, renewable and non-renewable energy, trade and financial development on carbon emissions in the top renewable. *Renewable and Sustainable Energy Reviews, 60,* 1074–1085.

Eichhammer, W., Ragwitz, M., & Schlomann, B. (2013). Introduction to the special issue: Financing instruments to promote energy efficiency and renewables in times of tight public budgets. *Energy and Environment, 24,* 1–26.

European Union. (2001). European Commission Directive (2001/77/EC) on the promotion of electricity produced from renewable energy sources in the internal electricity market. (OJ L283/33), October 27, 2001.

European Union. (2008). European Commission. Community guidelines on state aid for environmental protection (2008/C 82/01).
European Union. (2012). European commission energy efficiency directive: Directive (2012/27/EU) of the European parliament and of the council of October 25, 2012 on energy efficiency.
Foxon, T. J., & Pearson, P. J. G. (2007). Towards improved policy processes for promoting innovation in renewable electricity technologies in the UK. *Energy Policy, 35,* 1539–1550.
Friebe, C. A., Von Flotow, P., & Täube, F. A. (2014). Exploring technology diffusion in emerging markets—the role of public policy for wind energy. *Energy Policy, 70,* 217–226.
Gradus, R., & Smulders, S. (1993). The trade-off between environmental care and long-term growth pollution in three prototype growth models. *Journal of Economics, 58,* 25–51.
Greening, A. L., Greene, D. L., & Difiglio, C. (2000). Energy efficiency and consumption the rebound effect a survey. *Energy Policy, 28*(6), 389–401.
Griliches, Z. (1992). The search for rand spillovers. *Scandinavian Journal of Economics, 94,* 29–47.
Grossman, G., & Helpman, E. (1991). *Innovation and growth in the global economy.* Cambridge: MIT Press.
Grossman, G., & Krueger, E. (1991). Environmental impacts of a North American free trade agreement, NBER Working Paper, 3914.
Grossman, G., & Krueger, E. (1995). Economic growth and the environment. *Quarterly Journal of Economics, 110*(2), 353–377.
He, X., & Zhang, Y. (2012). The influence factors of carbon emissions of Chinese industry and the reorganization effects of CKC—an empirical research on the dynamic panel data of different industries based on the STIRPAT model. *China Industrial Economics, 1,* 26–35.
Heerink, N., Mulatu, A., & Bulte, E. (2001). Income inequality and the environment: Aggregation bias in environmental kuznets curves. *Ecological Economics, 38,* 359–367.
Hettige, H., Mani, M., & Wheeler, D. (1998). Industrial pollution in economic development: Kuznets revisited. World Bank working paper, 1876.
Heyes, A., & Kapur, S. (2011). Regulatory attitudes and innovation in a model combining internal and external randd. *Journal of Environmental Economics and Management, 61*(3), 327–340.
Holtz-Eakin, D., & Selden, T. M. (1995). Stoking the fires? CO_2 emissions and economic growth. *Journal of Public Economics, 57,* 85–101.
IEA. (2013a). *World Energy Outlook.* Paris: IEA.
IEA. (2013b). *CO_2 Emissions from Fuel Combustion Highlights* (2013th ed.). Paris: IEA.
IEA. (2016). *International Energy Agency.* Retrieved from www.worldenergyoutlook.org.
IPCC. (2007a). *Intergovernmental Panel on Climate Change,* Climate Change 2007—The physical science basis: Working group I. Contribution to the fourth assessment report of the IPCC, Cambridge University Press.
IPCC (2007b). Climate change 2007—impacts, adaptation and vulnerability: Working group II. Contribution to the fourth assessment report of the IPCC, Cambridge University Press.
IPCC (2013). Climate change 2013: The physical science basis. Working group I. Contribution to the fifth assessment report of the intergovernmental panel on climate change summary for policymakers.
IPCC (2014). Climate change 2014: Mitigation of climate change-IPCC. Working group III. Contribution to AR5. Int, Berlin.
IPCC. (2014). *Intergovernmental Panel on Climate Change.* (2014). Climate change 2014—impacts, adaptation and vulnerability: Regional aspects. Cambridge University Press.
Jefferson, M. (2008). Accelerating the transition to sustainable energy systems. *Energy Policy, 36,* 4116–4125.
Kuznets, S. (1955). Economic growth and income inequality. *The American Economic Review, 45* (1), 1–28.
Kuznets, S. (1998). Population, income and capital. *International Social Science Journal, 50,* 329–334.

Lin, B. Q., Jiang, Z. J., & Lin, J. (2009). The analysis and design of China's residential electricity tariff subsidies. *Financial Research Journal, 11,* 48–58.
Lüthi, S., & Prässler, T. (2011). Analyzing policy support instruments and regulatory risk factors for wind energy deployment: A developers' perspective. *Energy Policy, 39*(9), 4876–4892.
Lüthi, S., & Wüstenhagen, R. (2012). The price of policy risk: Empirical insights from choice experiments with European photovoltaic project developers. *Energy Economics, 34*(4), 1001–1011.
Marques, A. C., Fuinhas, J. A., & Pires Manso, J. R. (2010). Motivations driving renewable energy in European countries: A panel data approach. *Energy Policy, 38,* 6877–6885.
Meadows, D. L., Meadows, D. H., Randers, J., & Behrens, W. (1972). *The limits to growth: A report for the club of Rome's project on the predicament of man.* New York: Universe Books.
Moomaw, W., & Unruh, G. (1997). Are environmental Kuznets curves misleading us? The case of CO2 emissions. *Environment and Development Economics, 2,* 451–463.
Müller, S., Brown, A., & Ölz, S. (2011). Renewable energy: Policy considerations for deploying renewables. International Energy Agency.
Munasinghe, M. (1999). Is environmental degradation an inevitable consequence of economic growth? Tunnelling through the environmental Kuznets curve. *Ecological Economics, 29*(1), 89–109.
Nordhaus, W. D. (2008). *A question of balance: Weighing the options on global warming policies.* New Haven: Yale University Press.
OECD. (2013). *OECD Factbook: Economic Environmental and Social Statistics.* Paris: OECD.
OECD. (2016). IEA energy technology RD&D statics. Retrieved from http://www.oecd.org/statics
Panayotou, T. (1993). Empirical test and policy analysis of environmental degradation at different stages of economic development. Working paper, 238, technology and environment programme, international labour office, Geneva.
Romer, P. M. (1987). Growth based on increasing returns due to specialization. *The American Economic Review, 77*(2), 56–62.
Romer, P. M. (1990). Are non-convexities important for understanding growth? *AEA Papers and proceedings, 80*(2), 97–103.
Schleich, J. (2009). Barriers to energy efficiency: A comparison across the German commercial and services sector. *Ecological Economics, 68,* 2150–2159.
Selden, T., & Song, D. (1994). Environmental quality and development: Is there a kuznets curve for air pollution emissions? *Journal of Environmental Economics and Management, 27*(2), 147–162.
Selden, T. M., & Song, D. (1995). Neoclassical growth, the J curve for abatement, and the inverted U curve for pollution. *Journal of Environmental Economics and management, 29*(2), 162–168.
Shafik, N., & Bandyopadhyay, N. (1992). Economic growth and environmental quality: Time-series and Cross-Country Evidence. World bank working papers, 904, 1–6.
Shafik, N. (1994). Economic development and environmental quality: An econometric analysis. *Oxford Economic Papers: Special Issue on Environmental Economics, 46*, 757–773.
Smulders, S., & Bretschger, L. (2000). Explaining environmental Kuznets curves: How pollution induces policy and new technologies. Discussion Paper, 95 (2000). Tilburg: Macroeconomics.
Stern, D. I. (2004). The rise and fall of the environmental Kuznets curve. *World Development, 32* (8), 1419–1439.
Stern, N. (2006). *Stern review on the economics of climate change.* London: Cambridge University Press.
Szabó, S., & Jäger-Waldau, A. (2008). More competition: Threat or chance for financing renewable electricity? *Energy Policy, 36,* 1436–1447.
Torras, M., & Boyce, J. (1998). Income, inequality, and pollution: A reassessment of the environmental Kuznets curve. *Ecological Economics, 25,* 147–160.
UNEP. (1992). United Nations Environment Programme Rio declaration on environment and development.

UNFCCC. (1997). United Nations Framework Convention on Climate Change. Kyoto Protocol to the United Nations Framework Convention on Climate Change adopted at COP3 in Kyoto, Japan, on December 11, 1997.

UNEP. (2012). United Nations Environment Programme. The Emissions Gap Report 2012. UNFCCC, United Nations Framework Convention on Climate Change.

Weitzman, M. L. (2009). On modelling and interpreting the economics of catastrophic climate change. *The Review of Economics and Statistics, 91*(1), 1–19.

Wong, S. L., Chang, Y., & Chia, W. M. (2013). Energy consumption, energy R&D and real GDP in OECD countries with and without oil reserves. *Energy Economics, 40,* 51–60.

World Bank. (2010a). *World development report 2010: Development and climate change*. Washington, DC: World Bank.

World Bank. (2010b). *The synthesis report of the economics of adaption to climate change study*. Washington, DC: The World Bank Group. 2010.

WCED. (1987). *World commission on environment and development. Our common future*. Oxford: Oxford University Press. 1987.

Wüstenhagen, R., & Menichetti, E. (2012). Strategic choices for renewable energy investment: Conceptual framework and opportunities for further research. *Energy Policy, 40,* 1–10.

Chapter 12
Marginal Abatement Cost Curves (MACC): Unsolved Issues, Anomalies, and Alternative Proposals

José Luis Ponz-Tienda, Andrea Victoria Prada-Hernández, Alejandro Salcedo-Bernal and Daniel Balsalobre-Lorente

Abstract Policy makers proposed the MACC as an instrument to rank possible mitigation measures available in a market. This tool orders measures according to their cost-efficiency, taking into account only two variables: costs and emissions reductions. Although this tool has been used in relevant settings like the first treaty of the United Nations Framework Convention on Climate Change (UNFCCC), it has shown mathematical failures that might produce unreliable rankings. This chapter presents existing alternatives to the use of traditional MACC for ranking GHG abatement measures: (1) Taylor's method by the application of the dominance concept. (2) Ward's method directly related to the net benefit of each measure. (3) The GM method, which supports an environmentalist attitude and performs a direct comparison of measures with negative and positive costs. (4) An extension of traditional MACC (EMAC method), that considers the economically driven point of view of the decision maker, weighting the negative cost options according to its economic savings over its reduction potential. (5) And the BOM method, consisting of a linear-weighted combination of two discretional seed methods, allowing decision makers to take into account the goodness of multiple methods in order to create new rankings adjustable to a specific GHG policy, whether it is fully or

J.L. Ponz-Tienda (✉)
Civil and Environmental Department, Universidad de Los Andes,
Bogotá, Colombia
e-mail: jl.ponz@uniandes.edu.co

A.V. Prada-Hernández
School of Public Policy, University of Maryland, College Park, MD, USA
e-mail: aprada@umd.edu

A. Salcedo-Bernal
Department of Systems and Computing Engineering, Universidad de Los Andes,
Bogotá, Colombia
e-mail: a.salcedo49@uniandes.edu.co

D. Balsalobre-Lorente
Department of Political Economy and Public Finance Economic and Business
Statistics and Economic Policy, University of Castilla-La Mancha,
Ciudad Real, Spain
e-mail: daniel.balsalobre@uclm.es

R. Álvarez Fernández et al. (eds.), *Carbon Footprint and the Industrial Life Cycle*,
Green Energy and Technology, DOI 10.1007/978-3-319-54984-2_12

partially driven by economical or environmental positions. Finally, several case studies and discussions are presented showing the advantages of the exposed methods.

Notations

The following concepts, symbols and acronyms are used in this article

Acronyms	Meaning
ΔB_m	Economic benefit generated by the energy savings for a measure m
ΔC_m	Associated net present value associated to a measure m
ΔE_m	GHG abatement potential associated to a measure m
$BOM^{\alpha}(\mu_1, \mu_2)(m)$	Balanced ordering method for methods μ_1 and μ_2, and a balance α
$Cost_m$	Total cost of the measure m ($Cost_m = \Delta C_m$—ΔB_m)
ENV	Environmentalist benchmark
GHG	Greenhouse gases
GM(ε)	Gain maximizing method being ε a very small value
GM(1)	Gain maximizing method being ε = 1
GM_m	Gain value for a measure m
GRE	Greedy benchmark
EMAC	Extended MACC method
$EMAC_m$	Extended MACM value for a measure m
MACC	Marginal abatement costs curve
MAC_m	Marginal cost of abating a tonne of CO_2 for a measure m
m	Measure
NPV	Net present value
sign(x)	Sign of x
τ_{MAC}	Set of ordered measures applying method MACC
τ_{Ward}	Set of ordered measures applying method Ward
τ_{Taylor}	Set of ordered measures applying method Taylor
$\tau_{GM(\epsilon)}$	Set of ordered measures applying method GM(ε) being ε a very small value
$\tau_{GM(1)}$	Set of ordered measures applying method GM(ε) being ε = 1
τ_{EMAC}	Set of ordered measures applying method. EMAC
$K(\tau_{\mu_1}, \tau_{\mu_2})$	Kendall tau distance between methods μ_1 and μ_2

1 Introduction

Policy makers around the world are substantially committed to the reduction of carbon emissions, saving policies, energy efficiency and font substitution (Balsalobre Lorente et al. 2016; Cantos and Balsalobre Lorente 2011, 2013), since the global energy use and its corresponding emissions will grow. However, they face several difficulties in finding appropriate solutions to greenhouse gas (GHG) mitigation without imposing heavy economic burdens on society in the context of limited budgets and divergent interests.

Policy makers use models and tools such as Marginal Abatement Cost curves (MACC or MAC curves) to negotiate with emitting sectors and environmentalists the developing of route maps for emissions reductions with limited budgets (Moran et al. 2008; Morthorst 1994; Prada-Hernández et al. 2015; Van Odijk et al. 2012; Bockel et al. 2012; Moran et al. 2010). These help them to identify policies and appropriate instruments to justify investment decisions (Kesicki 2013) and to demonstrate how much abatement an economy can afford and the area to focus on in order to achieve target emission reductions (Vogt-Schilb and Hallegatte 2014).

In the challenge of designing decarbonizing policies in an economically efficient manner, policy makers rank and prioritise available abatement measures to minimize costs and maximize mitigation potentials through MACC. These show the economic feasibility of climate change mitigation, representing the marginal cost of emission abatement for varied technologic options (Kesicki and Ekins 2012; Kesicki 2012).

Some authors have recognized that MACC have methodological failures (Kok and Annema 2010). Which are mainly presented for negative-cost measures. Due to its relevance, we present and evaluate the methods for raking abatement measures.

In Sect. 2 we describe the fundamentals of the MACC method and its anomalies. Sections 4–8 briefly presents the alternative ranking methods for negative cost measures available in the literature. In the Sect. 9, is exposed a classification method based on the distance between environmental and economic "referential benchmarks". Several case studies and discussions are presented in Sect. 10. Finally, conclusions and limitations of the research are drawn.

2 The Traditional MACC Method and Its Anomalies

MACC were developed by Paulson (1948), transformed by Jackson (1991) to fit a climate change context, as a graphical representation of ordered energy measures that relates the potential of mitigation against its marginal costs, in such way that more cost-effective measures are on the left-hand side. The measures (m) are represented as labelled bar, in which the width represents the GHG[1] abatement potential (ΔE_m) and the height represents the marginal cost of abating a tonne of CO_2 (MAC_m). For computing the MAC_m for each measure (m), Eq. 1 is by applied using the following criteria: ΔC_m is the associated net present value (NPV) of implementing and operating it, ΔB_m is the economic benefit generated by the energy savings, ΔE_m is the abatement potential (the sum of tonnes abated during the

[1]Expressed as equivalent CO_2 in tonnes that a measure m could potentially abate.

studied time), and $Cost_m$ is the total cost of the measure, as the difference between ΔC_m and ΔB_m.

$$\mathrm{MAC}_m = \frac{\Delta \mathrm{C}_m - \Delta \mathrm{B}_m}{\Delta \mathrm{E}_m} = \frac{\mathrm{Cost}_m}{\Delta \mathrm{E}_m} \tag{1}$$

The use of MACC is generalized (Contreras 2016), being considered a standard and extremely efficient tool for analysing and communicate the results regarding the economic implications and impacts of energy policies for climate mitigation. Nevertheless, some discrepancies arose relating to the calculation, construction and interpretation of MACC in recent years.

The first warnings about MACC was stated by Kesicki and Strachan (2011) and Kesicki and Ekins (2012), stating that its simplistic approach possibly biases the decision-making, not taking into account the intersectoral, intertemporal and macroeconomic interactions, as well as the social implications related to climate change mitigation.

Nevertheless, the most controversial issue is the cost-effectiveness calculation that generates the MACC relating to win-to-win measures with negative costs (negative net present value). The problem is that MACC use the same criteria for ranking abatement measures with positive and negative costs, that favours measures with low emission reductions over options with the same negative cost but greater CO_2 reduction potential, and consequently producing unreliable rankings (Levihn 2016; Huang et al. 2016).

If a measure has a positive $Cost_m$, the marginal cost of the measure MAC_m will be positive, and smaller values of MAC_m are always desirable, but when a measure has a negative $Cost_m$, MAC_m is smaller (more desirable) with a lower $Cost_m$ or a lower ΔE_m, which means that, opposite to what is expected, a lower CO_2 reduction is preferred. In Figure 1 is shown the schematic MACC for 5 measures used by

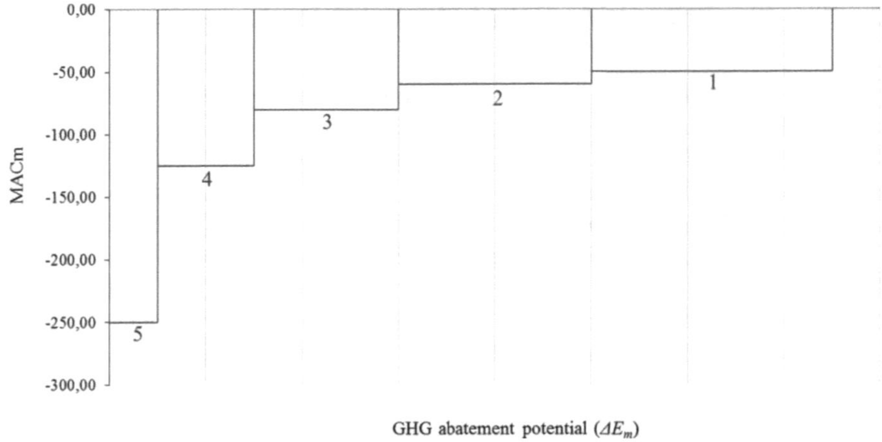

Fig. 1 Schematic MACC for five measures from Ward (2014)

Ward (2014), each with the same financial benefit ($250) but saving different amounts of CO_2 (1, 2, 3, 4 and five respectively). This example allows us to see the potential error of the conventional interpretation of MACC, because the measure with the largest negative specific marginal cost, measure 5, is preferred ($MAC_5 = -250$; Tons $CO_2 = 1$). Since they all have the same financial benefit, the measure with the largest emissions reduction, measure 1 ($MAC_1 = -50$; Tons $CO_2 = 5$), should be the preferred one.

The problem of ranking options with negative marginal cost was already identified by Wallis (1992a, b), but without a rigorous discussion by the scientific community of its methodological implications for about 20 years. Ackerman and Bueno (2011) avoided this issue by assuming a near-zero positive cost instead of negative-cost, stating that negative costs are controversial among economists and problematic for modelling purposes. Kesicki and Ekins (2012) disputed the existence of negative abatement costs, arguing these costs are not compatible with an efficient market.

Taylor (2012), Ward (2014) and Ponz-Tienda et al. (2016) do not take into account the previously exposed arguments but instead focused on the mathematical treatment and accuracy of the ranking of options that negative cost.

Recently, Levihn (2016) states that some approaches as Pareto optimization (Taylor 2012) to solve the problem of ranking GHG abatement measures would result in undesirable effects. Finally, the same author says that using economic efficiency (least cost per unit supplied) as a metric would "*fulfil the requirements of least cost integrated planning when options result in a negative marginal cost*" and that "*another metric would be needed, but would not per se result in least cost planning*".

3 The Taylor's Method for Ranking Negative Abatement Measures

Taylor (2012) was the first author to propose a mathematical solution for this issue maximizing two criteria: greater GHG reduction and lower costs by the application of the dominance concept (Goldberg and Holland 1988). The dominance concept simplifies the definition of a Pareto front (Deb 2001), in such way that a measure A dominates a measure B (quadrant 4) if the cost or abatement performance of A is better than that of B and is no worse (Eq. 2). Additionally, it neither dominates nor is dominated by the grey points in quadrants 2 and 3 (Fig. 2).

$$\begin{array}{lcl} \mathrm{Cost_A} < \mathrm{Cost_B} & \text{and} & \Delta \mathrm{E_A} \geq \Delta \mathrm{E_B} \\ \Delta \mathrm{E_A} > \Delta \mathrm{E_B} & \text{and} & \mathrm{Cost_A} \leq \mathrm{Cost_B} \end{array} \tag{2}$$

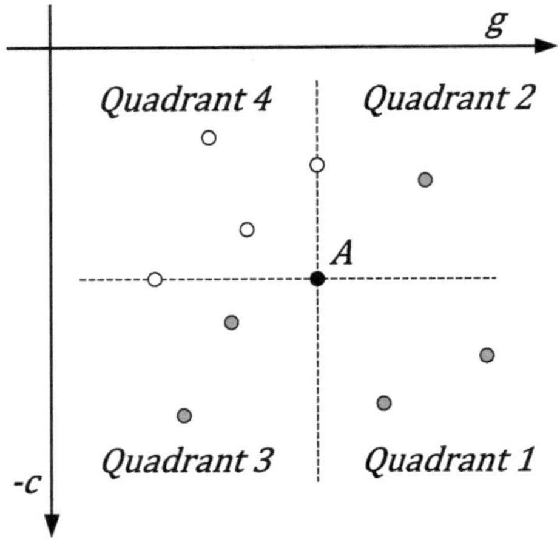

Fig. 2 Definition of a Pareto front for emissions reduction measures (Taylor 2012)

4 The Ward Method

To avoid the problem of ranking abatement measures with negative cost, Ward (2014) proposed to plot a function that is directly related to the benefit, taking a range of values for avoided CO_2 and simply plotting the net benefit of each measure. The method adopts those measures for which the area under the curve (the financial benefit) added to the width of the curve (the emissions avoided) multiplied by an assumed value of avoided emission, is maximised.

Ward (2014) recommended plotting a function more directly related to the benefit (negative costs) and correctly set priorities for negative and positive cost independently:

> In terms of policy implications it is recommended that even though MACC have a role to play in ranking efficiency measures with net positive costs, they should never include measures which have net financial benefit as the interpretation of these curves is complex and counter-intuitive, completely negating the value of this simple tool for policy decisions.

5 The Gain Maximizing (GM) Method

The Gain Maximizing (GM) method is proposed by Ponz-Tienda et al. (2016) to deal with the problem of ranking negative measures in a continuous way and favours alternatives with greater GHG reduction potential and greater Benefit/Cost (B/C) relation. This approach, unlike traditional approaches, allows comparing negative and positive cost options in a consistent way and in some cases favours greater reduction potential over low cost.

A GM_m indicator (Eq. 3), is calculated for each abatement measure. It relates the financial benefit (ΔB_m) and investment (ΔC_m) with GHG abatement potential (ΔE). In addition, to evaluate measures where $\Delta C_m = 0$, a free variable ε is added to the formula as a real number, greater than zero and lesser or equal to one. When ε is equal to one, it has no other impact than avoiding division by zero. Otherwise, the closer ε is to zero, the lower the value of GM_m and the better the measure. When the cost is equal to zero and ε is close to zero, it will favour measures with non-initial investment. This model is always negative, so it does not differ between negative and positive measures, a case where the traditional MACC model fails.

$$\mathrm{GM}_m = \frac{-\Delta \mathrm{B}_m}{\Delta \mathrm{C}_m + \varepsilon} \cdot \Delta \mathrm{E}_m \tag{3}$$

6 The Extended MACC (EMAC) Method

The Extended MACC (EMAC) method, proposed by Ponz-Tienda et al. (2016), also allows the direct and continuous ranking of both positive and negative cost measures, favouring those options with the lower cost and greater GHG reduction potential and giving priority to measures with negative total cost ($\Delta B_m - \Delta C_m < 0$) over greater reduction potential. The $EMAC_m$ indicator exposed in Eq. 4 relates the inverse of the net present value of every measure ($Cost_m = \Delta B_m - \Delta C_m$), with the GHG abatement potential (ΔE_m).

For positive cost measures, EMAC operates as the traditional MACC index, and for measures with negative costs, EMAC method prefers measures with greater cost effectiveness in the positive range (same as traditional MACC) and prefers measures with greater abatement potential and greater cost reduction on the negative side.

$$\mathrm{EMAC}_m = \mathrm{Cost}_m \cdot \Delta \mathrm{E}_m^{\mathrm{sign}(-\mathrm{Cost}_m)}, \mathrm{sign}(\mathrm{x}) = \begin{cases} \frac{\mathrm{x}}{|\mathrm{x}|}, & \mathrm{x} \neq 0 \\ 0, & \mathrm{x} = 0 \end{cases} \tag{4}$$

7 The Environmentalist (ENV) and the Greedy (GRE) Methods

The previously exposed methods rank abatement measures providing different ordered sets, some favouring measures with high environmental impacts, others favouring measures with high economical profit. These two new ranking methods (Ponz-Tienda et al. 2016) are "referential benchmarks" to establish the limit bounds under "environmentalist" or economical "greedy" attributes. In this way, the environmentalist (ENV) method ranks the abatement measures considering only the GHG abatement potential (Eq. 5), hence, it is the "environmentalist" ranking bound. On the other hand, the greedy (GRE) method (Eq. 6) orders the measures solely based on total cost of the measure ($Cost_m$), as the difference between ΔC_m and ΔB_m, hence, it is the "greedy" ranking bound.

$$\mathrm{ENV}_m = -\Delta \mathrm{E}_m \tag{5}$$

$$\mathrm{GRE}_m = \mathrm{Cost}_m = \Delta \mathrm{C}_m - \Delta \mathrm{B}_m \tag{6}$$

8 The Balanced Ordering (BOM) Method

It is possible that none of the above methods meet the particular needs of the decision maker. In order to address this issue, Ponz-Tienda et al. (2016) proposes the Balanced Ordering Method (BOM) to create alternative ranking criteria based on a combination of the previously analysed methods, gathering the goodness of present and future approaches, enabling decision makers to rank of GHG abatement measures under their specific interests, answering how "environmentalist" or "greedy" each method is.

To compare the previously exposed methods in a direct way, it is necessary to introduce some notation:

- A seed method μ is a function $\mu : \mathbb{R}^3 \to \mathbb{R}$
- A measure $m_i = (\Delta C, \Delta B, \Delta E)$ is a point on $\mathbb{R}^3$
- Given a set M with n measures, τ_μ is a function $\tau_\mu : M \to [1 \cdots n]$ such that:

$$\tau_\mu(m_1) \leq \tau_\mu(m_2) \leftrightarrow \mu(m_1) \leq \mu(m_2) \tag{7}$$

So $\tau_\mu(m_1)$ ranks the measure m_i, according to the results of the method μ.

If μ_1 and μ_2 are seed methods and M is a set of measures, the Balanced Ordering Method $(BOM^\alpha(\mu_1, \mu_2))$ is a function (Eq. 8), in such that given the ordered sets of measures $\left(\tau_{\mu_1} \text{ and } \tau_{\mu_2}\right)$ $BOM^\alpha(\mu_1, \mu_2)$ can build a new balanced method $\xi^\alpha = BOM^\alpha(\mu_1, \mu_2)$, where τ_ξ is the balanced ranking of the seed ordered sets $\tau_{\mu 1}$ and $\tau_{\mu 2}$ for a given α (Eq. 9). In other words, $BOM^\alpha(\mu_1, \mu_2)$ weights the rankings $\tau_{\mu 1}$ and $\tau_{\mu 2}$.

$$BOM^\alpha(\mu_1, \mu_2) : \mathbb{R}^3 \to \mathbb{R} \tag{8}$$

$$BOM^\alpha(\mu_1, \mu_2)(m) = \alpha \cdot \tau_{\mu 1}(m) + (1 - \alpha) \cdot \tau_{\mu 2}(m), \alpha \in [0, 1] \tag{9}$$

The α coefficient balances the obtained set, in such way that a for α = 1 the set obtained is $\tau_{\mu 1}$, for $\alpha = 0 \to \tau_{\mu 2}$, and varying α between 0 and 1 an adjusted set according to the specific interests of the decision makers.

9 The Classification of Ranking Methods (Tau Distance to ENV and GRE)

To measure how "environmentalist" or economical "greedy" is an ordered set of abatement measures, Ponz-Tienda et al. (2016) proposes the use of the Kendall tau distance $\mathrm{K}(\tau_{\mu 1}, \tau_{\mu 2})$ (Kendall 1948), in such way that the closer K is to zero, the greater the similarities between $t_{\mu 1}$ and $t_{\mu 2}$.

Given two orders for a finite set M, Kendall tau's distance measures the similarity between such orders (μ_1 and μ_2) by comparing how the methods order each pair and counting the number of inconsistences between the two orders (Eq. 10).

$$\mathrm{K}(\tau_{\mu 1}, \tau_{\mu 2}) = \frac{\{\#(m_i, m_j) \in \mathrm{M}^2 : i > j, (\tau_{\mu 1}(m_i) - \tau_{\mu 1}(m_j)) \cdot (\tau_{\mu 2}(m_i) - \tau_{\mu 2}(m_j)) < 0\}}{\binom{n}{2}}$$
$$= \frac{\{\#(m_i, m_j) \in \mathrm{M}^2 : i > j, \mu_1, \ \mu_2 \, \textit{are inconsistent with respect to} \{m_i, m_j\}\}}{\binom{n}{2}} \quad (10)$$

The Kendall tau method, indicates the percentage of unique pairs of measures [*i, j*] whose ranking differs with respect to $t_{\mu 1}$ and $t_{\mu 2}$ rankings. For μ_1 and μ_2 methods and $m_1, m_2 \in M$ measures: μ_1 and μ_2 are inconsistent with respect to $\{m_1, m_2\}$ if the order given by μ_1 over $\{m_1, m_2\}$ differ from the given by μ_2. In other words, the Kendall tau's distance method counts the number of inconsistences for two given methods with respect to all distinct pairs in M.

Comparing the results of different ordering methods for several abatement measures against the referential benchmarks (ENV and GRE), can be concluded that traditional MACC, Taylor, Ward and EMAC methods are economically driven methods (Greedy methods), and that EMAC id the most greedy. In the other hand, GM is classified as an environmental friendly method. Therefore, the selection of the ranking method depends on the specific interests of the decision maker (DM) in terms of the importance between the potential GHG reduction and the economical profit (Table 1).

Table 1 Classification of GHG abatement measures ranking methods

$\tau_{\mu 1}$	$\tau_{\mu 2}$	Kendall tau distance	Conclusion
MACC	GRE	$K(t_{\mu}, t_{GRE})$ closer to zero	Greedy
Taylor	GRE	$K(t_{\mu}, t_{GRE})$ closer to zero	Greedy
Ward	GRE	$K(t_{\mu}, t_{GRE})$ closer to zero	Greedy
EMAC	GRE	$K(t_{\mu}, t_{GRE})$ closer to zero	More greedy
GM(1)	ENV	$K(t_{\mu}, t_{ENV})$ closer to zero	Environmentally friendly
GM(ε)	ENV	$K(t_{\mu}, t_{ENV})$ closer to zero	Environmentally friendly

10 Case Studies; Comparison and Discussions of Results

To show the differences of the exposed methodologies with related applications found in the literature, we analysed the ranking proposed by Behrenz (2014) for the United Nations programme for the development of Colombia and the Final Report to the Committee on Climate Change for the Agriculture and Land Use, Land-Use Change and Forestry Sectors out to 2022, for the UK (Moran et al. 2008).

10.1 Report to the Committee on Climate Change Out to 2022 for the UK (Moran et al. 2008)

The work developed by Moran et al. (2008) describes the derivation of traditional MACC to depict abatement potential for (ALULUCF) in the UK from a bottom-up cost-effectiveness analysis of data on mitigation options within respective sectors. In the other hand, Behrenz (2014) analysed a set of mitigation measures in the context of Colombia in terms of cost-effectiveness by applying the MACC methodology for the principal emitting sectors: wastes, transportation, energy and industrial and residential sectors. For the purpose of this work, the results obtained for the industrial sector has been selected because it contains many negative cost mitigation measures, providing an illustrative scenario where MACC do not produce coherent results.

The values for the estimated CFP (central feasible potential) for 2022, Total Cost and Cost effectiveness used by (Moran et al. 2008) are exposed in Table 2 and the comparison of the ranking produced by the traditional MACC method against the exposed methodologies is presented in Table 3. Note: The BOM ordered set is obtained from the ENV and EMAC methods with $\alpha = 0.55$ (Eq. 11).

$$\mathrm{BOM}^{\alpha}(\mathrm{ENV}, \mathrm{EMac})(m) = 0.55 \cdot \tau_{\mathrm{ENV}}(m) + 0.45 \cdot \tau_{\mathrm{Emac}}(m) \tag{11}$$

The Kendall-Tau distance of the ordered sets exposed in Table 3, are compared to the ENV and GRE benchmarks (Fig. 3). The dashed red line in Fig. 3 represents the boundary established by $BOM^{\alpha}(GRE, ENV)$ by all the possible measures that can be produced varying alpha between zero (ENV) to one (GRE).

As can be seen on Fig. 3, MACC is very close to GRE, this similarity highlight MACC (and its variations) weakness to rank negative cost measures. For example, MACC method ranks better CG measure than AN, been clearly AN a better option under both environmental and economic perspectives, AN has $Cost_m = -3246$ and $\Delta E_m = 3.08$, compared to GC with $Cost_m = -1419$ and $\Delta E_m = 0.394$. In this cases are recommended to use another approach to rank the negative cost measures, for example use Taylor to rank negative cost measures and MACC to rank the positive rank ones or use another alternative like EMAC or GM.

Table 2 2022 Abatement potential CFP (Moran et al. 2008)

Code	Measure	First year gross volume abated [$ktCO_2e$]	Cumulative first year abatement [$MtCO_2e$]	Cost	Cost effectiveness [£2006/tCO_2e]
AA	Crops-Soils-BiolFix	8.49	10.83	154,696.97	14,280.16
AK	Crops-Soils-SystemsLessReliantOnInputs	10.05	10.83	48,001.73	4,434.34
CA	BeefAn-Concentrates	80.96	10.82	29,249.60	2,704.54
AB	Crops-Soils-ReduceNFert	136.20	10.73	21,952.10	2,045.10
BH	DairyAn-Transgenics	504.29	10.60	17,924.19	1,691.28
AH	Crops-Soils-ControlledRelFert	165.90	10.09	10,778.82	1,067.95
AI	Crops-Soils-Nis	603.67	9.93	2,913.87	293.50
BG	DairyAn-bST	132.31	9.32	2,089.51	224.10
AF	Crops-Soils-SpeciesIntro	365.98	9.19	1,601.43	174.22
EB	OFAD-DairyMedium	44.12	8.83	212.71	24.10
EE	OFAD-BeefMedium	50.77	8.78	148.93	16.96
AC	Crops-Soils-Drainage	1,741.02	8.73	126.08	14.44
HT	CAD-Poultry-5 MW	219.34	6.99	79.90	11.43
EC	OFAD-DairyLarge	250.81	6.77	53.89	7.96
EH	OFAD-PigsMedium	16.06	6.52	30.58	4.69
EF	OFAD-BeefLarge	97.79	6.50	16.39	2.52
EI	OFAD-PigsLarge	47.77	6.41	6.15	0.96
AM	Crops-Soils-SlurryMineralNDelayed	47.17	6.36	0.00	0.00
AO	Crops-Soils-UsingComposts	78.51	6.31	0.00	0.00
DA	Forestry-Afforestation	980.84	6.23	−44.37	−7.12
AD	Crops-Soils-AvoidNExcess	276.06	5.25	−264.07	−50.29
BB	DairyAn-MaizeSilage	95.98	4.98	−1,306.58	−262.63
AL	Crops-Soils-ImprovedN-UsePlants	331.80	4.88	−371.29	−76.10

(continued)

Table 2 (continued)

Code	Measure	First year gross volume abated [$ktCO_2e$]	Cumulative first year abatement [$MtCO_2e$]	Cost	Cost effectiveness [£2006/tCO_2e]
BI	DairyAn-ImprovedFertility	346.26	4.55	−0.18	−0.04
BE	DairyAn-Ionophores	739.66	4.20	−204.13	−48.59
BF	DairyAn-ImprovedProductivity	377.36	3.46	−0.24	−0.07
AN	Crops-Soils-ReducedTill	55.77	3.08	−3,246.31	−1,052.63
AE	Crops-Soils-FullManure	457.26	3.03	−451.05	−148.91
AJ	Crops-Soils-OrganicNTiming	1,027.16	2.57	−176.06	−68.48
AG	Crops-Soils-MineralNTiming	1,150.39	1.54	−159.62	−103.38
CG	BeefAn-ImprovedGenetics	46.32	0.39	−1,419.55	−3,602.93
CE	BeefAn-Ionophores	347.38	0.35	−606.48	−1,747.79

Table 3 Ordered measures for the 2022 abatement potential CFP of Moran et al. (2008)

Code	Measure	τ_{MAC}	τ_{Emac}	τ_{GRE}	τ_{ENV}	τ_{BOM}
AA	Crops-Soils-BiolFix	32	32	32	1	4
AK	Crops-Soils-SystemsLessReliantOnInputs	31	31	31	2	5
CA	BeefAn-Concentrates	30	30	30	3	7
AB	Crops-Soils-ReduceNFert	29	29	29	4	8
BH	DairyAn-Transgenics	28	28	28	5	10
AH	Crops-Soils-ControlledRelFert	27	27	27	6	11
AI	Crops-Soils-Nis	26	26	26	7	12
BG	DairyAn-bST	25	25	25	8	13
AF	Crops-Soils-SpeciesIntro	24	24	24	9	14
EB	OFAD-DairyMedium	23	23	23	10	15
EE	OFAD-BeefMedium	22	22	22	11	16
AC	Crops-Soils-Drainage	21	21	21	12	17
HT	CAD-Poultry-5 MW	20	20	20	13	18
EC	OFAD-DairyLarge	19	19	19	14	19
EH	OFAD-PigsMedium	18	18	18	15	20
EF	OFAD-BeefLarge	17	17	17	16	21
EI	OFAD-PigsLarge	16	16	16	17	23
AM	Crops-Soils-SlurryMineralNDelayed	15	15	15	18	24
AO	Crops-Soils-UsingComposts	14	14	14	19	25
DA	Forestry-Afforestation	11	9	11	20	6
AD	Crops-Soils-AvoidNExcess	9	4	7	21	2
BB	DairyAn-MaizeSilage	4	2	3	22	1
AL	Crops-Soils-ImprovedN-UsePlants	7	3	6	23	3
BI	DairyAn-ImprovedFertility	13	13	13	24	27
BE	DairyAn-Ionophores	10	6	8	25	22
BF	DairyAn-ImprovedProductivity	12	12	12	26	29
AN	Crops-Soils-ReducedTill	3	1	1	27	9
AE	Crops-Soils-FullManure	5	5	5	28	26
AJ	Crops-Soils-OrganicNTiming	8	8	9	29	28
AG	Crops-Soils-MineralNTiming	6	10	10	30	31
CG	BeefAn-ImprovedGenetics	1	7	2	31	30
CE	BeefAn-Ionophores	2	11	4	32	32

For this example, is computed the EMAC method for the project set. There is not an evident inconsistency in the produced ranking, but after computing Kendal-Tus distance is evident than EMAC is very close to GRE (like MACC), so, to achieve a more balanced method, ENV and EMAC are combined with $BOM^{0.55}$ resulting in a new method with an improved environmentalist perspective without neglect the economical perspective.

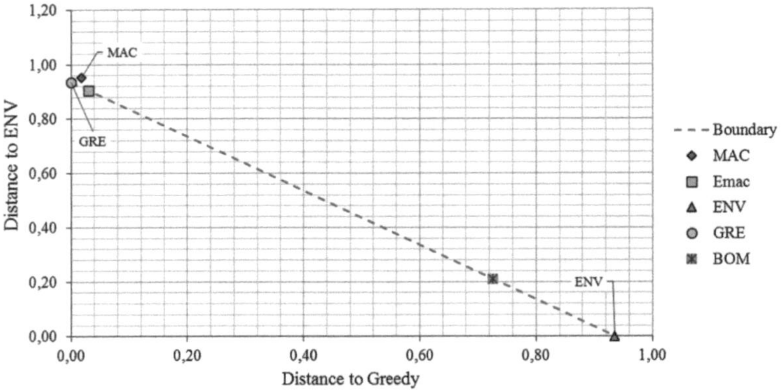

Fig. 3 Kendall tau distance between the ordered sets exposed in Table 3 compared to the ENV and GRE benchmarks

10.2 United Nations Programme for the Development of Colombia (Behrentz 2014)

The study of abatement curves and mitigation potentials in the sectors of agricultural and livestock, transport, wastes, mining and energy developed by Behrentz (2014) exposes the cost effectiveness of different mitigation options in the Colombian context by applying the traditional MACC. The measures and values for the industrial sector used by Behrentz (2014) are exposed in Table 4. Note: The BOM ordered set is obtained from the ENV and gre methods with $\alpha = 0.50$ (Eq. 12).

$$\mathrm{BOM}^{\alpha}(\mathrm{ENV},\mathrm{GRE})(m) = 0.50 \cdot \tau_{\mathrm{ENV}}(m) + 0.50 \cdot \tau_{\mathrm{GRE}}(m) \tag{12}$$

The Kendall-Tau distance of the ordered sets exposed in Table 5, are compared to the ENV and GRE benchmarks and represented in Fig. 4. The values of the traditional MACC observed above the dashed red line are far away from the ideal measures (zero distance to GRE and ENV). This implies that the ranking produced by traditional MACC could be replaced and improved by an alternative rank produced by BOM for some value of alpha.

Since in this example there are more measures with negatives cost, the erroneous ranking of MACC is more noticeable, for example the measure M16 which has $Cost_m = -9$ and $\Delta E_m = 0.08$ is ranked in first place and M9 with $Cost_m = -275$ and $\Delta E_m = 5$ is ranked sixth. In this case, oppositely to the results in the previous example, EMAC is very far away of MACC, and EMAC is no too close form GRE.

Table 4 Abatement measures of the industrial sector for Colombia (Behrentz 2014)

Code	Industry	Abatement potential	Cost	Cost effectiveness
M1	Energy use of solid-waste—transversal	69,000	160.00	2.32
M2	Recycling—transversal	55,000	230.00	4.18
M3	Direct reduction of iron ore with Hylsa technology	44,000	−965.00	−21.93
M4	Replacing coal with biomass—cement	43,000	−30.00	−0.70
M5	Reducing the proportion of clinker—cement	39,000	−15.00	−0.38
M6	CO_2 capture and geological storage—cement	27,000	1,260.00	46.67
M7	Improved efficiency of bagasse boilers—food and drinks	25,000	11.00	0.44
M8	Production change wet to dry process—cement	15,000	60.00	4.00
M9	Improving efficiency of oil-fired boilers—paper	5,000	−275.00	−55.00
M10	Improving efficiency of coal and diesel boilers—paper	3,000	−25.00	−8.33
M11	Improving efficiency of natural gas boilers—paper	2,000	−25.00	−12.50
M12	Improving efficiency of natural gas boilers—chemicals	1,000	−18.00	−18.00
M13	Improving efficiency of natural gas boilers—food and drinks	1,000	−20.00	−20.00
M14	Improving efficiency of LPG boilers—food and drinks	1,000	−20.00	−20.00
M15	Hydrogen recovery in ammonia production	0.400	13.00	32.50
M16	Improving efficiency of diesel-oil boiler—food and drinks	0.080	−9.00	−112.50
M17	Replacing coal with biomass—food and drinks	0.080	−0.60	−7.50
M18	Improving efficiency of oil-fired boilers—food and drinks	0.040	−2.00	−50.00
M19	Replacing coal with biomass—chemicals	0.030	−0.30	−10.00
M20	Improving efficiency of fuel-oil boiler—paper	0.020	−1.00	−50.00
M21	Replacing coal with biomass—paper	0.020	−0.10	−5.00
M22	Improving efficiency of LPG boilers—chemicals	0.010	−3.00	−300.00
M23	Improving efficiency of diesel-oil boiler—chemicals	0.002	−0.20	−100.00
M24	Improving efficiency of fuel-oil boiler—food and drinks	0.001	−0.10	−100.00

Table 5 Ordered mitigation options for the industrial sector in the Colombian context (Behrentz 2014)

Code	Industry	τ_{MAC}	τ_{Emac}	τ_{GRE}	τ_{ENV}	τ_{BOM}
M1	Energy use of solid-waste—transversal	20	20	22	1	10
M2	Recycling—transversal	21	22	23	2	11
M3	Direct reduction of iron ore with Hylsa technology	9	1	1	3	1
M4	Replacing coal with biomass—cement	17	3	3	4	2
M5	Reducing the proportion of clinker—cement	18	4	9	5	3
M6	CO_2 capture and geological storage—cement	24	24	24	6	17
M7	Improved efficiency of bagasse boilers—food and drinks	19	19	19	7	13
M8	Production change wet to dry process—cement	21	21	21	8	16
M9	Improving efficiency of oil-fired boilers—paper	5	2	2	9	4
M10	Improving efficiency of coal and diesel boilers—paper	14	5	4	10	5
M11	Improving efficiency of natural gas boilers—paper	10	6	4	11	6
M12	Improving efficiency of natural gas boilers—chemicals	10	9	8	12	7
M13	Improving efficiency of natural gas boilers—food and drinks	10	7	6	12	8
M14	Improving efficiency of LPG boilers—food and drinks	7	7	6	12	9
M15	Hydrogen recovery in ammonia production	23	23	20	15	22
M16	Improving efficiency of diesel-oil boiler—food and drinks	1	10	10	16	12
M17	Replacing coal with biomass—food and drinks	16	12	14	16	14
M18	Improving efficiency of oil-fired boilers—food and drinks	5	11	12	18	15
M19	Replacing coal with biomass—chemicals	13	15	15	19	19
M20	Improving efficiency of fuel-oil boiler—paper	2	14	13	20	18
M21	Replacing coal with biomass—paper	14	16	17	20	21
M22	Improving efficiency of LPG boilers—chemicals	7	13	11	22	20
M23	Improving efficiency of diesel-oil boiler—chemicals	3	17	16	23	23
M24	Improving efficiency of fuel-oil boiler—food and drinks	4	18	17	24	24

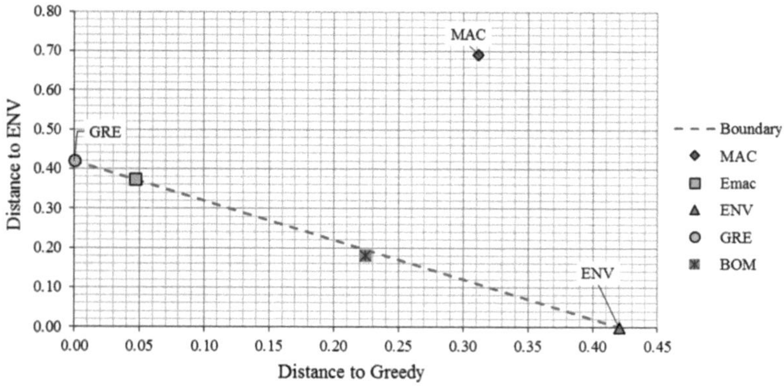

Fig. 4 Kendall tau distance between the ordered sets exposed in Table 5 compared to the ENV and GRE benchmarks

10.3 *Discussion of Results*

In the Moran et al. (2008) case study, ordered applying the traditional MACC methodology, the measure CG (BeefAn-Improved Genetics) with an abatement potential of 0.39 tons of CO_2 and a Cost of −1419.55 is preferred, since applying the BOM method, the preferred measure is BB (DairyAn-MaizeSilage) with similar negative cost (−1306.58), but nearly thirteen times of abatement potential (4.98), which is obviously better than CG measure. Similarity, applying the EMAC method, the preferred measure is the AN measure (Crops-Soils-ReducedTill) with a negative cost of −3246.31 and an abatement potential of 3.08 tons of CO_2, better values than the provided by the CG measure with a negative cost of −1419.55 and a potential of 0.39. In both cases, the BB measure applying the BOM method (more ENV) or the AN measure applying the EMAC method (more GRE), are better measures, and preferred in both cases, to the CG measure ranked by the traditional MACC.

For the Behrenz (2014) study, the EMAC method presents a more greedy and environmentalist rank than the traditional MACC, because the study includes many negative cost measures and traditional MACC prioritises measures with low abatement potential and high negative costs. The $BOM^{\alpha=0.5}(Emac, ENV)$ ranking offers a balanced ranking between the ENV and GRE benchmark methods slightly under the boundary, implying that is better than any BOM combination of ENV and GRE. In this study is noticed how the introduction of negative cost measures could impact the MACC ranking, compared to Moran et al. (2008) case study, MACC is much more distant to the reference methods and EMAC, and don't establish a clear perspective of what kind of projects is prioritizing, this implies than with more negative cost variables the error (or randomness) of MACC method increases.

11 Conclusions

As seen in this chapter there are many methodologies to address the problem of ranking GHG abatement measures, each one provides distinct results which are appropriate in different situations. So, is the decision maker liability to know and understand the methodologies for that purpose.

Traditionally, MACC are considered a standard tool for analysing the impacts of mitigation measures promoted by governments and one of the most used methodologies for evaluating which abatement options to implement by comparing their cost-effectiveness. However, in the last years some discrepancies have arisen relating to its construction and interpretation, in particular, its flaw ranking negative cost measures. So, in this chapter were exposed four alternative methodologies. (1) Taylor's method, which seeks greater GHG reduction and lower costs by the application of the dominance concept. (2) Ward proposed to plot a function that is directly related to the benefit, taking a range of values for avoided CO_2 and simply plotting the net benefit of each measure. (3) The GM method, which supports an environmentalist attitude and performs a direct comparison of measures with negative and positive costs. (4) An extension of traditional MACC (EMAC method), that considers the economically driven point of view of the decision maker, weighting the negative cost options according to its economic savings over its reduction potential.

Is possible that no one of the know methods fit the decision maker necessities. Then the combining methodology BOM, consisting of a linear-weighted combination of two discretional seed methods, allows decision makers to take into account the goodness of multiple methods in order to create new rankings adjustable to a specific GHG policy, whether it is fully or partially driven by economical or environmental positions.

Additionally to the ranking problems, decision makers need tools to compare and profile the variety of ranking methods so they can decide which methodology to apply. Kendal tau allow to measure the "distance" between two methods which indicates how similar they are, however this distance doesn't give much information if it is computed between two regular methods. Then is advisable to compute this distance respect methods with known position. ENV and GRE are reference methods, been ENV the most environment friendly and GRE the most economical driven one, this reference methods and the Kendal tau distance allow to profile other methods.

Acknowledgements The authors would like to thank the research group of Construction Engineering and Management (IN^{2}geco) of Universidad de Los Andes.

References

Ackerman, F., & Bueno, R. (2011). *Use of McKinsey abatement cost curves for climate economics modelling*. Somerville, MA: Stockholm Environment Institute.

Balsalobre Lorente, D., Álvarez-Herránz, A., & Baños Torres, J. (2016). La innovación y la sustitución energética como medidas de corrección medioambiental en países de la OCDE. *Estudios de Economía Aplicada, 34*(1), 1–26.

Behrentz, E. (2014). *Productos analíticos para apoyar la toma de decisiones sobre acciones de mitigación a nivel sectorial: Curvas de abatimiento para Colombia, Bogotá*. Colombia: Universidad de los Andes.

Bockel, L., Sutter, P., Touchemoulin, O., & Jönsson, M. (2012). Using marginal abatement cost curves to realize the economic appraisal of climate smart agriculture policy options. *Methodology, 3*(1), s.l. Food and Agriculture Organization of the United nations (FAO).

Cantos, J. M., & Balsalobre Lorente, D. (2011). Las energías renovables en la Curva de Kuznets Ambiental: Una aplicación para España. *Estudios de Economia Aplicada, 29*(2), 1–32.

Cantos, J. M., & Balsalobre Lorente, D. (2013). Incidencia del gasto público en I + D + i energético sobre la corrección medioambiental en España*. *Estucios de Economia Aplicada, 31*(1), 1–34.

Contreras, R. F. (2016). Analysis and comparison of energy saving measures through marginal abatement cost curves. En: Project Management and Engineering Research. s.l. Springer International Publishing, pp. 203–214.

Deb, K. (2001). *Multi-objective optimization using evolutionary algorithms*. Chichester, UK: Wiley.

Goldberg, D. E., & Holland, J. H. (1988). Genetic algorithms and machine learning. *Machine Learning, 3*(3), 95–99.

Huang, S. K., Kuo, L., & Chou, K. L. (2016). The applicability of marginal abatement cost approach: A comprehensive review. *Journal of Cleaner Production, Volumen, 127,* 59–71.

Jackson, T. (1991). Least-cost greenhouse planning supply curves for global warming abatement. *Energy Policy, 19*(1), 35–46.

Kendall, M. G. (1948). *Rank correlation methods*. London: Charles Griffen & Company.

Kesicki, F. (2012). Intertemporal issues and marginal abatement costs in the UK transport sector. *Transportation Research Part D: Transport and environment, 17*(5), 418–426.

Kesicki, F. (2013). What are the key drivers of MAC curves? A partial-equilibrium modelling approach for the UK. *Energy Policy, 58,* 142–151.

Kesicki, F., & Ekins, P. (2012). Marginal abatement cost curves: A call for caution. *Climate Policy, 12*(2), 219–236.

Kesicki, F., & Strachan, N. (2011). Marginal abatement cost (MAC) curves: Confronting theory and practice. *Environmental Science & Policy, 14*(8), 1195–1204.

Kok, R., & Annema, J. A. (2010). The curious case of selecting and ranking GHG mitigation measures in transport. In ETC 2010, *European Transport Conference*. Glasgow, UK, Association for European Transport (AET).

Levihn, F. (2016). On the problem of optimizing through least cost per unit, when costs are negative: Implications for cost curves and the definition of economic efficiency. *Energy, 114,* 1155–1163.

Moran, D., et al. (2008). UK marginal abatement cost curves for the agriculture and land use, land-use change and forestry sectors out to 2022, with qualitative analysis of options to 2050, s. l. s.n.

Moran, D., et al. (2010). Developing carbon budgets for UK agriculture, land-use, land-use change and forestry out to 2022. *Climatic Change, 105*(3–4), 529–553.

Morthorst, P. E. (1994). Constructing CO_2 reduction cost curves; the case of Denmark. *Energy Policy, 22*(11), 964–970.

Paulson, W. E. (1948). Characteristics of the marginal cost curve. *Journal of Farm Economics, 30* (3), 467–499.

Ponz-Tienda, J. L., Prada-Hernández, A. V., & Salcedo-Bernal, A. (2016). The problem of ranking CO_2 abatement measures: A methodological proposal. *Sustainable Cities and Society, 26,* 306–317.

Prada-Hernández, A., Vargas, H., Ozuna, A. & Ponz-Tienda, J. L. (2015). Marginal abatement costs curve (MACC) for carbon emissions reduction from Buildings: An implementation for office buildings in Colombia. *International Journal of Civil and Structural Engineering–IJCSE, 2*(1).

Taylor, S. (2012). The ranking of negative-cost emissions reduction measures. *Energy Policy, 48,* 430–438.

Van Odijk, S., et al. (2012). Utilizing marginal abatement cost curves (MAC curves) to strategically plan CO_2 reduction possibilities for the water sector: The case of watercycle organisation Waternet. s.l., s.n., pp. 13–18.

Vogt-Schilb, A., & Hallegatte, S. (2014). Marginal Abatement cost curves and the optimal timing of mitigation measures. *Energy Policy, 66,* 645–653.

Wallis, M. (1992a). Greenhouse ranking of gas-fuelling. *Energy Policy, 20*(2), 174–176.

Wallis, M. K. (1992b). Forum ranking of greenhouse gas abatement measures. *Energy Policy, 20* (12), 1130–1133.

Ward, D. J. (2014). The failure of marginal abatement cost curves in optimising a transition to a low carbon energy supply. *Energy Policy, 73,* 820–822.

Chapter 13
Global and Local: Climate Change Policies as a Paradigm of Multilevel Governance

Susana Galera

Abstract European energy policies based on energy savings and the promotion of renewables are still the main components of the most recent and ambitious climate policies. However, a distinctive key element of the current approaches is that the objectives they pursue cannot be achieved without decisive intervention by sub-state political and administrative levels, particularly the local ones. The most recent European energy and climate regulations are insistently referring specific actions to the achievement of the set goals that inescapably have to be pursued at the local level—for example, Heating and cooling energy districts or the design of infrastructures that must withstand the distributed generation or the utilization of residual heat.

1 Climate Change Goals Call for Deep Institutional Adaptation

The mitigation of human impacts on climate change that no one questions anymore is the goal of varied international strategies, which are reinforced in parallel to the progressive manifestation of their effects. The European Union has been trying to reduce its energy consumption since the 80s of the last century, initially for reasons of security and economy, given its high external dependence. Afterwards, it has been insisting on these objectives adding as well environmental considerations. Working in such way the European energy Policy has been set addressed to the reduction of energy consumption, the increase of renewable energies in the energy mix and the reduction of emissions.

Beyond the current 20-20-20 Strategy[1] the EU has already (2011) outlined a much more ambitious approach in terms of objectives and timeframes: its 2050

[1]Europe 2020 Strategy for smart, sustainable and inclusive development, COM (2010) 2020, 3 March 2010.

S. Galera (✉)
Public Law Department, Universidad Rey Juan Carlos, Madrid, Spain
e-mail: susana.galera@urjc.es

R. Álvarez Fernández et al. (eds.), *Carbon Footprint and the Industrial Life Cycle*,
Green Energy and Technology, DOI 10.1007/978-3-319-54984-2_13

Energy Roadmap envisages not only a de-carbonized economy but also a gradual energy transition having in the end the very change of the own energy model. The final goal is the replacement of the current vertical model of production, transport and distribution centrally managed by a model of distributed generation and intelligent management based on the producer/consumer (prosumer) as its main actor connected in a mega-grid. Considering the huge investment that necessarily requires the renewal of aging energy infrastructures from 2020 and on,[2] such financial efforts are now also addressed to achieve a cleaner, more efficient and more accessible energy model.

The success of both the Energy Strategy and the climate policies mainly involves local governments, since urban areas account for more than half of the world's population (60% in 2030 according to World Bank estimates), 70% of global energy consumption and account for 40–50% of greenhouse gas emissions.

Here we want to consider the coexistence of two perspectives: on the one hand, and from a factual point of view, the actions that in a global context are already shaping climate policies at the local level; On the other hand, and from a formal point of view, the legal and institutional framework that will discipline the action of local entities in this context.

This approach illustrates recent and successive conceptions of international relations—para-diplomacy, urban diplomacy, federative diplomacy (Setze 2015)—which seriously challenge the classic legal discourse, still echoing in the law classrooms, conferring the *ius legationis* and the *ius contrahendi* to the State in the international sphere in an exclusive manner. Contrary to it, the State has lost that exclusivity in favor of the emergence of other entities also playing in the international arena, which is particularly evident in environmental and, more recently, climate governance.

Setting aside the long-term climate strategies (2050), and taking now into account the current environmental and energy governance, it has to be underlined that there is a lack of correspondence between factual requirements and management rules, between what is required by an efficient environmental an energy governance and the legal and institutional framework in which it is supposed to be developed. Particularly, we will address the following issues:

[2]In the Roadmap for Energy 2050 [COM (2011) 885, p. 3], the Commission explained the ambition of its strategy, namely that "a new investment cycle will be opened in that decade (2020), while energy infrastructures built 30 or 40 years ago will need to be renewed". The costs of the energy transition do not differ substantially from these renovation costs. According to the European Investment Plan [COM (2014) 903], investment in generation, networks and energy efficiency is estimated at EUR 200 billion annually in the coming decades.

- First, and even in the absence of a specific legal basis (attribution powers), the EU has been developing Urban Policies, the *Urban Acquis*, laying on other Community provisions;
- Secondly, it has also developed local climate policies, based both on sectoral determinations and on non-legally binding models linked to financing instruments;
- Thirdly, Local entities have acquired in these strategies a growing visibility and responsibility, which goes far beyond the limited role attributed to them.

The EU has already assumed the leading role played by local authorities in some of its policies, notably those of climate and energy, reinforcing them on the basis of a twofold approach: on the one hand, establishing mechanisms for direct dialogue with cities, as will be explained later; On the other hand, reinforcing the international action of the local entities, actively supporting its international partnership and associations and the standards, instruments and models that are agreed upon, which are later implemented at the local level.

2 European Urban Policies: From the *Urban Acquis* to the European Urban Agenda

There is not a direct European configuration of an Urban Policy on urban model or standard which is consistent with the other European Policies. In fact, any Treaty provision specifically empowers the EU for doing so. However, it doesn't prevent the fact that the European law has been increasingly and fragmentarily conditioned the urban policies in the Member States on the basis of specific pieces of legislation relating water, waste, atmosphere, construction products, soils, landscape, civil protection … among others. Working in such a way, an urban standard has been set up at the EU level, the so-called Urban Acquis which is based on a triple basis:

- No-binding declarations (Charter of Leipzig 2007, declarations of Toledo 2010 and Riga 2015, urban environment strategy, European Urban Agenda), which have been outlining a model of sustainable territorial and urban development (compact city, land reuse, standard equipment and mobility …), which has been recently updated, adding resolute multilevel governance mechanisms.
- Areas of EU actions which incorporate into the European agenda local policies; from this perspective, it has to be mentioned that the current 2014–2020 funding framework links an intense urban dimension to the European regional policy,[3]

[3]The Directorate General of Regional Policy of the European Commission was renamed as "Regional and Urban Policy"; The Riga Declaration (2015) reinforces the role of small and medium-sized urban areas in the implementation of European policies, setting up the foundations for a future (June 2016) European Urban Agenda.

and for the first time creates a financial instrument which is directly managed by the cities themselves.

– Determined support from the Commission for the articulation of European thematic networks of sub-state entities, which are forums for the exchange of information and analysis of common problems and solutions.

The Commission itself underlines these results: in its Communication *The Urban Dimension of EU Policies: Key Elements of an Urban Agenda for the EU*,[4] states:

> EU cohesion policy, through the URBAN Community Initiatives 18 and the subsequent mainstreaming of integrated sustainable urban development into regional and national operational programmes 19, has fed the intergovernmental process with practical experience. Together they form what is known as the 'Urban Acquis'.
>
> At least an estimated 50% of the European Regional Development Fund (ERDF), around 80–90 billion Euro, will be invested in urban areas through the mainstream operational programmes in the 2014–2020 financial period, and a minimum 5% of national allocations of ERDF have been earmarked for integrated sustainable urban development to ensure that it is a priority in all Member States. City networking and exchange will continue to be promoted by the next generation URBACT[5] programme.

These progressive actions considered as a whole have set up an urban model having two key elements. First, it is framed into a larger model of urban environment where the urban sprawl hast to be controlled and limited.

> Urban sprawl is the most urgent of the urban design issues. Towns and cities are expanding outwards into rural areas at a faster rate than their population is growing (a 20% expansion in the last 20 years with only a 6% increase in population over the same period). Green space (valuable agricultural and natural land) is being replaced by low-density housing and commercial uses. Urban sprawl reinforces the need to travel and increases dependence upon private motorised transport to do so, leading in turn to increased traffic congestion, energy consumption and polluting emissions. These problems are most acute where residential densities are low and where daily activities (home, work, shopping) are widely separated. There is a sharp increase in car use where land use densities fall below 50–60 people per hectare[6]

Secondly, and taking into account the city's management, the EU has decisively supported the Smart City addressed to achieve higher levels of sustainability and resilience[7]:

---

[4]COM2014/490, p. 7.

[5]URBACT is an instrument of Cohesion Policy co-financed by the ERDF. In its previous two editions, it has brought together 500 cities from 29 countries integrated in 61 thematic networks of cooperation. URBACT III (2014–2020) continues and intensifies cooperation on sustainable and integrated urban development. The URBACT Secretariat is in France. http://urbact.eu.

[6]COM (2004) 60, on the Urban Environment Strategy, Sect. 2.4, p. 25.

[7]Description from the ICLEI Strategic Plan 2015–2021, Seoul Plan, p. 3. 4.

> A Smart City has embedded "smartness" into its operations, and is guided by the overarching goal of becoming more sustainable and resilient. It analyzes, monitors and optimizes its urban systems, be they physical (e.g. energy, water, waste, transportation and polluting emissions) or social (e.g. social and economic inclusion, governance, citizen participation), through transparent and inclusive information feedback mechanisms. A smart city commits to continuous learning and adaptation, and through the application of systems thinking, aspires to improve its inclusivity, cohesion, responsiveness, governance and the performance of its social, economic and physical systems.

One of the urban systems to be improved in the Smart City is energy, which leads us to the objective already outlined of the Smart Grid, which "could be described as an upgraded electricity network to which two-way digital communication between supplier and consumer, intelligent metering and monitoring systems have been added".[8] It is a non-centralized, distributed generation network, in accordance with the Energy Strategies mentioned below. These proposals, formulated by the previous Barroso Commission, have not only been maintained by the current Juncker Commission, but also take part of more ambitious and decidedly innovative projects. Smart cities and smart grids are now integrated into the European Digital Single Market,[9] a strategy whose major benefits are addressed to not only information and communication technology (ITC) providers, but also to traditional industries, and in particular energy services (e-energy).[10]

To the extent that the new programming of the European Structural and Investment Funds 2014–2020 is conceived as the financial instrument for the achievement of the Europe 2020 strategy, which is based on sustainable development and social and innovation requirements. Consequently, actions which directly or indirectly address environmental and climate goals have never had such significant funding before. A significant development of the new programming is the conception of cities as engines of the European economy, reinforcing their role in the context of cohesion policy.

Among the five European funds, the European Regional Development Fund (ERDF)[11] remains the largest allocation, highlighting, in this case, three novelties strongly supporting Climate Policies and Urban Policies.

First, *Sustainable Development in the Urban Environment* has for the first time its own and specific role[12]:

---

[8]COM (2011) 202, Smart Grids: from innovation to implementation, where is defined, following the Smart Grid Task Force as "electricity networks that can efficiently integrate the behaviour and actions of all users connected to it—generators, consumers and those that do both—in order to ensure an economically efficient, sustainable power system with low losses and high quality and security of supply and safety", p. 2.

[9]COM (2015) 192, Digital Single Market Strategy for Europe.

[10]COM (2015) 80, A Framework Strategy for a Resilient Energy Union with a Forward-Looking Climate Change Policy.

[11]The ERDF manages 183.3 million euros out of a total of 376,000 for the period 2014–2020.

[12]Regulation (EC) No 1301/2013 of 17 December 2013 on the European Regional Development Fund and on specific provisions relating to the investment objective for growth and employment, Article 7.

- Meaning that "sustainable urban development" will support "strategies that set out integrated actions to tackle the economic, environmental, climate, demographic and social challenges affecting urban areas, while taking into account the need to promote urban-rural linkages".
- These measures must be embodied in specific instruments: integrated territorial investment (direct management by the city), a specific operational program or a specific priority axis;
- Principles for the selection of urban areas where integrated measures will be implemented, and criteria for indicative allocation, lay down in the Association Agreement;
- Recognizes a new national entity, the "urban authority", meaning "cities, sub-regional or local bodies responsible for the implementation of sustainable urban strategies", to which it attributes a minimum competence ("at least the selection of the operations"), and outlines its relationship with the "managing authority".[13]

Secondly, as already mentioned, in addition to allocating a minimum percentage of funds to finance sustainable urban development actions, a new *Integrated Territorial Investment (ITI)* tool is created whose administration and execution will be delegated to the cities that, designated by States, implement integrated actions.[14]

Third, it enhances the role of the local entities both in its international action and in its direct relationship with the European institutions. In addition to the acknowledgment of "urban authority" with specific functions, this support is given by a platform in which cities interact for the first time directly with the Commission without intermediaries: it is the *Urban Development Network* (UDN)[15] which aims to improve the implementation of European Funds in cities, and to promote the exchange of information between cities involved in sustainable urban development (Article 9).

[13]Article 7, parr. 5. The managing authority shall determine, in consultation with the urban authority, the scope of tasks, to be undertaken by urban authorities, concerning the management of integrated actions for sustainable urban development. The managing authority shall formally record its decision in writing. The managing authority may retain the right to undertake a final verification of eligibility of operations before approval.

[14]Article 36 of Regulation 1303/2013 of 17 December 2013 laying down common provisions on the European Regional Development Fund, the European Social Fund, the Cohesion Fund, the European Agricultural Fund for Rural Development And the European Maritime and Fisheries Fund. These are urban development strategies or another territorial strategy involving investments by the ESF, the ERDF or the Cohesion Fund according to more than one priority axis of one or more operational programs.

[15]The UDN complements the actions of URBACT III, converging on support for cities implementing Article 7 (sustainable urban development). Recital (21) of the Regulation 1301/2013 states: "In order to reinforce capacity-building, networking and exchange of experience between programmes and bodies responsible for implementing sustainable urban development strategies and innovative actions in the area of sustainable urban development and to complement existing programmes and bodies, it is necessary to establish an urban development network at Union level".

Therefore, local entities are strongly conditioned in the setting-up of their policies by the standards and, particularly, by the financing that is established in supra-state domains. It matters little that the *European urban Acquis* lacks binding legal force, if its urban standards end up being imposed because they are linked to an important financial support. From this perspective, it has to be underlined the fact that the rules for the implementation of the European Funds has progressively narrowed the margin of national discretion in favor of a very close ex-ante monitoring of the Commission: the Program Contract and the Operational Programs verify the adequacy of the state proposals to the criteria set up on the European Strategies, and has to be accepted by the Commission before the funding is available.

The European Parliament[16] has recently emphasized the very important role of local and regional authorities in the implementation of a program of funds mobilizing huge investment, job creation and innovation at the local level. It considers that these entities should participate in the programming and execution phases of the funds, in a paradigmatic example of multilevel and bottom-up governance, underlying that just "consultations" with local entities are not sufficient for an efficient implementation and urging States to reinforce their role.

The *European Urban Agenda*, or Pact of Amsterdam,[17] reinforces these approaches, aiming the development of the concept of an intelligent, sustainable and inclusive city, providing the axes that integrate economic, social and environmental aspects from the city. This strategy is based on eight priority areas which carry on their corresponding measures: among them specifically the adaptation climate measures (10.7) and the energy transition (10.8).

The Networks or Thematic Associations of cities are still key actions for the execution of the agenda, being understood as instruments of sectorial and multilevel cooperation to ensure an integrated approach that allows articulating effective solutions for the urban areas.

3 Climate Actions as a Paradigm of Multilevel Government

It was not until the Treaty of Lisbon that the measures "combating climate change" became an objective of the European Environmental Policy (Article 191 TFEU). Until then, some measures adopted on the basis of different policies—energy, single market, territorial cohesion…—indirectly supported the climate change goals.

[16]Resolution of 4 February 2016 on the role of local and regional authorities in the European Structural and Investment Funds.

[17]Adopted on 30 May 2016 in Amsterdam at the informal meeting of Ministers responsible for urban affairs. Pending confirmation by the General Affairs Council of June 2016. http://urbanagendaforthe.eu/wp-content/uploads/2016/05/Pact-of-Amsterdam_v7_WEB.pdf.

However, from then on the energy and climate packages will be adopted in an integral way, with the previous regulations being renewed and giving rise to the current Directives with clear implications at the local level.

On the one hand, and relating to urban renewal measures, the Energy Efficiency Directive 2012/27/EU, fully applicable since January 2017,[18] imposes a minimum (3%) annual renewal of public buildings to progressively comply with minimum energy performance's standards, which requires a previous base-line inventory. The new Directive makes compulsory the energy audit for private companies and requires the public procurement being consistent with energy performance of products if stated by the EU regulations.

Furthermore, the new Directive reinforces the promotion of cogeneration lying down in a strong multilevel administrative cooperation. Among other provisions, article 14 states that:

- Member States shall adopt policies which encourage the due taking into account at *local and regional levels* of the potential of using efficient heating and cooling systems, in particular those using high-efficiency cogeneration;
- particularly, Member States shall carry out and notify (December 2015) to the Commission a comprehensive assessment of the potential for the application of high-efficiency cogeneration and efficient district heating and cooling, containing the information set out in Annex. Depending on the (positive) results of such assessment, Member States shall take adequate measures for efficient district heating and cooling infrastructure to be developed and/or to accommodate the development of high-efficiency cogeneration and the use of heating and cooling from waste heat and renewable energy sources;
- in any event, and for electricity generation and industrial installations above 20 MW there is an obligation to prepare a cost-benefit analysis on the viability of cogeneration, waste heat recovery or district heat network connection when they are built or substantially refurbished.

On the other hand, the Directive 2009/28 on the promotion of renewable energies refers to and support specific urban equipment, the district *heating and cooling systems based on renewable energy*, meaning "*the distribution of thermal energy in the form of steam, hot water or chilled liquids, from a central source of production through a network to multiple buildings or sites, for the use of space or process heating or cooling*".[19] The Directive states the obligation of States to induce local and regional entities to develop these systems (Article 13.3),[20]

[18]It replaces the previous Directive 2010/31 on energy performance of buildings and Directive 2006/32/EC on energy end-use efficiency. A description of this process of normative renewal can be found in Galera, S. "Del Ahorro de Energía a la Eficiencia Energética: objetivos e instrumentos de las Políticas Europeas", Revista de Derecho Urbanístico y Medio Ambiente no 288, Marzo 2014.

[19]Article 2, g/.

[20]3. Member States shall recommend to all actors, in particular local and regional administrative bodies to ensure equipment and systems are installed for the use of electricity, heating and cooling

obligation which specifically reaches "the planners" (Article 14.5). In a bottom-up approach, is also required that the National Renewable Plans assesses the possible need for new infrastructure for these urban systems, in particular for the production of heating and cooling from biomass, solar and geothermal (Article 16.11).

This multi-level governance approach of the Directives on which the main European climate strategies are rooted, reveals the clear intention of the European institutions not to totally ignore the participation of the sub-state entities in their application,—as theoretically would be expected according to the Principle of Institutional autonomy governing the implementation of EU law.

Reference has already been made to *urban authorities* (local and regional) and their relationship with the "management authorities" (basically state) which are now acknowledged by the new ERDF Regulation, as well as the new mechanisms for dialogue between Local Authorities. Beyond these binding provisions, the EU promotes through non-binding rules actions that specify local versions or projections of its climate policies, in order to facilitate the adaptation of these measures at local level. This promotion goes parallel with the financial support provided for by the Structural Funding.

In this context, mention should be made of local climate policy, the *Sustainable Energy and Climate Action Plans* (SECAP), which represents a minimum of local commitment in the fight against climate change (Peeters 2012, p. 285). This Plan, established within the framework of the Covenant of Mayors for Climate and Energy, must be signed by Local Authorities within the year following the accession of the Pact, which also presupposes a base inventory of emissions. The Plan has to be accepted by the Convenant, and is evaluated and monitored by the European Commission through the Joint Research Center. Failure to submit the Plan within the set timeframe or negative follow-up reports is a reason for exclusion from the Covenant. SECAPs must include:

- energy saving programs in buildings and public services, which can also be achieved through public procurement;
- mobility plans, aimed at reducing dependence on the private vehicle;
- energy performance standards and requirements for incorporating renewable energy equipment into new buildings;
- public awareness;
- promotion of local production of renewables and use of renewable sources, such as combined cycle plants.

Beyond the scope of the EU although strongly supported by its institutions, ICLEI,—*Local Government for Sustainability*—should be mentioned in this context. It is an international association of local and metropolitan governments

(Footnote 20 continued)

from renewable energy sources and for district heating and cooling when planning, designing, building and renovating industrial or residential areas. *Member States shall, in particular, encourage local and regional administrative bodies to include heating and cooling from renewable energy sources in the planning of city infrastructure, where appropriate.*

dedicated to sustainable development, which has been acquiring a growing institutional role in recent years particularly in climate conferences. At this regard, it is worth to underline:

- its observer status in the United Nations Framework Convention on Climate Change (UNFCCC) and, in this context, is the seat of representation of Local Governments and Municipal Authorities (LGMA), established in 1995 at COP1 of the Convention held In Berlin;
- holds the secretariat of the World Council of Mayors for Climate Change (WMCCC) which in 2010, in the Mexico City Pact, approved a key instrument managed by ICLEI for the standardized measurement and verification of climate commitments, *Carbon Registry of Cities* (cCCR), which is applied by about twenty programs and climate networks integrated by local entities. Furthermore, cCCR assesses and manages the climate data of non-state actors that are integrated in the NAZCA (*Non-State Actor Zone for Climate Action*) platform, a global platform where the climate commitments of non-state actors are recorded and established at the COP20 of Lime 2014 as the framework for the development of the Paris 2016 Agreement.

4 Europe 2050: Facing Energy Challenges and Shaping Cities

Climate change strategies, which the EU promotes decisively, are addressed to specific goals (reducing emissions, increasing RES, saving energy, mitigation) and will as well imply a deep transformation of the current regulatory framework for energy and energy governance. The Energy Transition Strategy, contained in the *Energy Road Map 2050*, implies, on the one hand, the appearance of new players in the scenario of the production and distribution of energy and, on the other hand, a renewed role for Local Policies. It goes ahead with the previous objectives of energy saving, increase of renewable energies and reduction of emissions, adding new and very ambitious ones. The novelty is that these objectives are now achieved in parallel with a change of model (transition) in itself, which involves a radical transformation of the sources of production, distribution and energy consumption of European citizens. The implementation of this strategy implies a substantial increase in capital investments, as expressed in the Roadmap[21]:

> The energy infrastructure which will power citizens' homes, industry and services in 2050, as well as the buildings which people will use, are being designed and built now. The pattern of energy production and use in 2050 is already being set. The task of developing

[21]COM (2011) 885, p. 2 and 6.

> post-2020 strategies is urgent. Energy investments take time to produce results. In this decade, a new investment cycle is taking place, as infrastructure built 30–40 years ago needs to be replaced. Acting now can avoid costly changes in later decades and reduces lock-in effects.
>
> … investments in power plants and grids, in industrial energy equipment, heating and cooling systems (including district heating and cooling), smart meters, insulation material, more efficient and low carbon vehicles, devices for exploiting local renewable energy sources (solar heat and photovoltaic), durable energy consuming goods etc. This has a widespread impact on the economy and jobs in manufacturing, services, construction, transport and agricultural sectors. It would create major opportunities for European industry and service providers to satisfy this increasing demand and stresses the importance of research and innovation to develop more cost-competitive technologies.

Horizon 2050 aims to reduce the number of conventional energy sources, which are temporarily considered as sources of reserves, to achieve a complete and safe supply of renewable energy sources (RES) and other unconventional sources. In this progression the target of 30% of renewables in final consumption by 2030 and 50% by 2050 has already been fulfilled, based on the purpose of correcting these percentages in the mid-term reviews.

The new model aims an structural change of the management system, which now abandons the vertical sequence of production-transport-distribution-consumption, managed by system operators, to implement a system of energy supply based on a Giant Smart and interconnected network, fueled by millions of actors—prosumer: producers/consumers—interacting with the network.

Smart management of macro-grid powered by millions of active consumer generation points, Smart Grids, will involve a deployment of novel technologies and systems, which is a strongly supported priority by the EU in ongoing research and demonstration projects,[22] for the sake of its effective implementation with the 2050 time frame.

At the time, the corresponding regulatory framework should ensure the technical functioning and non-discriminatory physical access of all parties, as well as precise rules that define the performance and responsibilities of all actors, and ensure a fair distribution of benefits and burdens among all those involved (Siano 2014, p. 476).[23] However, in the current technological stage, the first steps have already been taken at this regard: while the new smart management strategy relies on accurate information, since 2009 (the internal electricity market), smart metering by means of intelligent metering of individual energy consumption has been considered and its deployment has been required by the Energy Efficiency Directive 2012/27/EC (article 9).

---

[22]See, for example, the NOBEL GRID project (supported by Horizon 2020), which aims to develop technologies and services required by distributed generation, energy cooperatives and small-scale distribution. The results of the research will be tested in five different EU electricity cooperatives http://nobelgrid.eu/.

[23]Relating measures—technical, legal, institutional and economic—which according the Commission are required for the effective deployment of the Smart Grids, see COM (2011) 202, p. 5.

While the technical challenges for the smart management of macro-grid are a European priority, some legal issues already arose even in the current preliminary stage (Jiménez 2015, p. 73). From this approach, it has to be underlined that on 9 March 2012, the Commission adopted a Recommendation on preparations for the deployment of smart metering systems, [COM (2012) 1342], which was the subject of an Opinion of the European Data Protection Supervisor of 8 June 2012. The problem that arises is that smart counters allow the massive collection of personal data from European families and can lead to the monitoring of what members of a family do in the privacy of their homes.

Another key element of the energy transition is the urban heating and cooling district which has an insistent assertion in the European regulatory framework for renewable energies. This instrument, without being new, is now a strategic element in mitigation and adaptation actions to climate change: as nearly half of urban energy (which represents 70% of global energy consumption) is destined to supply heating and cooling, any energy transition action must explicitly include measures of sustainable urban systems of refrigeration and heating. Energy Districts add to their traditional function of energy storage and rebalancing of variable renewable energy, the matching of local production to local consumption, not only at the level of buildings but at the district level even at the city level.

Within this new framework, the local government is the relevant actor, as a planner and regulator, as a facilitator of financing, as a consumer and as a provider/facilitator of services, as a mediator of public-private partnerships.[24] But it is also the institutional level where a social and democratizing approach to energy is possible, propitiating and/or leading new community energy systems as the Danish or German experiences illustrate (Roberts 2015). And this, of course, without neglecting the state regulatory framework that is relevant to the greater or lesser degree of success of recent initiatives in countries around us and after the privatization process (Hawkey and Webb 2014).

Another new feature of the strategy, which will have a major impact on the organization and procedures, is the decision and management system for some core aspects of climate policy: the Energy Governance Strategy aims to replace the current top-down approach on Greenhouse Gases policies in an bottom-up approach for the definition of targets and measures, where local energy potential and economically efficient technologies shape GHG reduction and grid feeding targets.

In this new (bottom-up) strategy of setting new GHG emissions targets, it is essential the close cooperative involvement of all public bodies from all territorial levels, local and supra-local, regional and, finally, state. All of them have to participate in the setting up of a single strategy integrating the objectives and instruments drawn up for the each territorially levels, according to their strengths and needs.

---

[24]At the global level, see the United Nations Environment Program initiative—UNEP—Energy Districts in Cities, Sustainable Energy for All SE4ALL (http://www.se4all.org/). This initiative is coordinated by UNEP and involves municipal governments, local government networks, national representatives, international organizations, private sector entities, and clean energy associations.

Lastly, we would like to stress that the ERDF is the most important financial instrument covering these climate actions, providing "support for investments to promote energy efficiency and security of supply in the Member States through Inter alia, the development of intelligent systems for the distribution, storage and transmission of energy, including the integration of distributed generation from renewable sources". Consistently with this declaration, it links a minimum percentage of the total ERDF resources at national level to the financing of the following projects:

- energy efficiency and renewable energies, research and innovation and support to SMEs (Article 4): this chapter should be well targeted at 80% in more developed regions and regions in transition (with 20% minimum energy efficiency and renewable energy) or 50% in less developed regions (with 6% minimum energy efficiency and renewable energy);
- sustainable urban development (Article 7), with 5% of the budget being reserved within the framework of measures delegated to cities and managed through ITI (Integrated Territorial Investments).

These approaches imply a considerable change of focus both in relation to the energy market—as stated in the European Energy Transition—and in relation to the infrastructures and the design of the city. The turning point is to stop considering climate policies exclusively at a macro and micro level—which would involve state climate plans and building measures—to incorporate a meso-level of action that would pivot, mainly, in planning instruments: Urban and territorial planning, as well as energy (emission reduction and renewable) objectives at regional and local level (Hawkey et al. 2013).

This new approach will influence, in short, the instruments of territorial and urban planning, and the conception and design of the city—as previously did the urban models established by the European Urban *Acquis*, which has shaped a new urbanism more sustainable at the European level.

Measures implementing local energy and climate policies include local regulations on solar panels and efficient public lighting, progressive renovation of the public transport equipment, lanes and roads for non-motorized transport, restrictions on private traffic, and more efficient management of waste. But a more ambitious and effective approach must also include a different perspective for the qualification of the land, delimiting energy units with sufficient urban density and "mixed uses" (residential/industrial) to ensure the matching of energy demand and supply. The incorporation in planning of the renovation of buildings for reasons of energy efficiency … at least when such renewal is legally required; Local participation in energy planning—or the taking into account of local needs—in order to include such needs in the updating of energy infrastructures as foreseen by the European Renewables Directive; Participation in landscape planning and mapping, within the framework of the Florence Convention, by matching interventions with new energy infrastructures that considerably reduce visual pollution,., etc. (Rocher 2014, p. 12)

5 Conclusions

The recent European Strategies on energy and climate change have a pronounced ambition. Beyond the radical change in the traditional way of production, distribution and consumption of energy, its effects will impact on new forms of government much more collaborative between the different territorial levels, and particularly in the urban planning processes as they have to incorporate other elements from now on.

Although urban policies do not constitute an area of European competence—exclusive, shared or supportive—there are some other European Policies that indirectly impact and influence the policies of sub-state entities. Traditionally it has been the regional policy actions that have been defined a European territorial and urban model; some environmental actions have also contributed to this design. A careful reading of recent energy and climate change actions reveals the leading role played by regional and, above all, local entities, both in relation to the savings and renewables objectives and in the deployment of new energy infrastructures now much more imbricated in the urban network. This role can be seen not only in energy and climate regulation texts, but also in the financial programming of European funds for the period 2014–2020.

In this way, energy and territorial governance strategies appear as interdependent policies, where the permanent interaction of the bottom-up (local-global) and top-down (global-local) approach in decision-making and execution is required. A sophisticated and complex multi-level governance approach that requires permanent administrative collaboration between the different territorial entities.

Acknowledgements This work is an outcome of the Project, DER2012-39170, Rule of Law in Global Space, supported by the Spanish Department of Economy and Competitiveness. It has been developed during a research stay in the University of Kent. I am very grateful to both institutions.

References

European Commission COM. (2004). *Urban environment strategy*, 60.
European Commission COM. (2010). *Strategy for smart, sustainable and inclusive development*, Europe 2020.
European Commission COM. (2011). *Smart grids: From innovation to implementation*, 202.
European Commission COM. (2011). *Roadmap for energy 2050*, 885.
European Commission COM. (2012). *Recommendation on preparations for the deployment of smart metering systems*, 1342 final.
European Commission COM. (2014). *The urban dimension of EU policies. Key Features of an EU Urban Agenda*, 490.
European Commission COM. (2014). *European investment plan*, 903.
European Commission COM. (2015). *A framework strategy for a resilient energy union with a forward-looking climate change policy*, 80.
European Commission COM. (2015). *Digital single market strategy for Europe*, 192.

European Data Protection Supervisor. (2012). *Opinion of 8 June 2012 relating the deployment of smart metering systems* (O.J.C. n. 335 of 1.11.2012).

Galera, S. (2006). Urbanismo sostenible; la política europea de medio ambiente urbano. *Revista de Derecho Urbanístico y Medio Ambiente, 224.*

Galera, S. (2014). Del Ahorro de Energía a la Eficiencia Energética: Objetivos e instrumentos de las Políticas Europeas. *Revista de Derecho Urbanístico y Medio Ambiente, 288.*

Gatta Sánchez, D. F. de (2008). Incidencia del medio ambiente y el desarrollo sostenible en el urbanismo y la edificación: De la Política de la UE al Código técnico de la edificación y la nueva Ley del Suelo. en *El Derecho Urbanístico del siglo XXI: Libro homenaje al Prof. Martín Bassols Coma* (Vol. *3*). Madrid: Ed. Reus-Colegio de Registradores.

González Medina, M. (2013). La europeización urbana a través de la política de cohesión. *Revista CIDOB d'afers internacionals, 104,* 133–154 (Special Issue on Cities and Urban Areas on International Policies).

González Medina, M., & Fedeli, V. (2015). Exploring European urban policy: Towards an EU-national agenda? *Gestión y Análisis de Políticas Públicas,* 14.

Hawkey, D., & Webb, J. (2014). District energy development in liberalised markets: Situating UK heat network in comparison with Dutch and Norwegian case studies. *Technology Analysis and Strategic Management, 26*(10), 1228–1241.

Hawkey, D., Webb, J., & Winskel, M. (2013). Organisation and governance of urban energy systems: District heating and cooling in the UK. *Journal of Cleaner Production, 50,* 22–31.

Jiménez, M. L. G. (2015). Smart cities vs. smart governance: ¿dos paradigmas de interrelación administrativa no resueltos aún? *Revista de Derecho Urbanístico y Medio Ambiente,* 300.

Jiménez Jaén, A. D. (2010). El medio ambiente urbano: estrategia española para un desarrollo urbano sostenible. *Revista General de Derecho Administrativo,* 25.

Marjan Peeters, M. (2012). Local climate action in the European Union. In B. J. Richardson (Ed.), *Local climate change law. Environmental regulation in cities and other localities.* The IUCN Academy of Environmental Law Series. Edward Elgar Publishing.

Quirós, J. B. G. de (2008). La autonomía local ante el urbanismo. Incidencia del derecho de la Unión Europea. Hacia un desarrollo sostenible. In Q. Carretero & J. Pedro (Ed.), *Globalización y principio de autonomía local.* Madrid: Consejo General del Poder Judicial-Centro de Documentación.

Roberts, J. (2015). *Promoting citizen participation in the energy transition: A recommendation for an EU legal framework to support community energy.* London: ClientEarth Energy Documents (April 2015).

Rocher, L. (2014). Climate-energy policies, heat provision, and urban planning: A renewal of interest in district heating in France: Insights from national and local levels. *Journal of Urban Technology, 21*(3), 3–19.

Setze, J. (2015). Testing the boundaries of subnational diplomacy: The international climate action of local and regional governance. *Transnational Environmental Law, 4*(02), 319–337.

Siano, P. (2014). Demand response and smart grids—A survey. *Renewable and Sustainable Energy Reviews, 30,* 461–478.

Part III
Metrics for the Sustainability of Infrastructure Projects

Chapter 14
Carbon Footprint of Human Settlements in Spain

Guadalupe Arce, Jorge Enrique Zafrilla, Luis-Antonio López and María Ángeles Tobarra

Abstract The role of towns and their inhabitants in fighting climate change is becoming increasingly important (Shi et al. in Nat Clim Change 6(2):131–137, 2016). In this context, the aim of this paper is to apply a multi-regional input-output model to study the evolution of the carbon footprint for Spanish households as determined by the different type of settlement. This study analyses the household carbon footprint as a function of the municipality's population size, whether it is located in a rural or urban environment, and its relation to population density. By using a multi-regional model we are able to calculate the share of that carbon footprint that is generated within the settlement and the share that is produced around the world along global value chains. This methodology has been widely applied to study carbon footprints for households in terms of different characteristics: income levels (Duarte et al. in Energy Policy 44:441–450, 2012), age (Shigetomi et al. in Environ Sci Technol 48(11):6069–6080, 2014), consumption of agriculture products (López et al. in J Clean Prod 103:423–436, 2015), or tourism consumption (Cadarso et al. in J Clean Prod 111(Part B):529–537, 2016). The structure of household consumption as a function of the type of settlement will be used to analyse whether socio-economic features are the greatest influence in the level of carbon footprint, or by the contrary, structural, institutional or geographical factors of the settlement are more relevant. Previous literature has addressed this link in other countries, for instance Fan et al. (J Clean Prod 33:50–59, 2012), Minx et al. (Environ Res Lett 8(3):035039, 2013), Baiocchi et al. (Global Environ Change 34:13–21, 2015) or Ahmad et al. (Environ Sci Technol 49(19):

G. Arce (✉) · J.E. Zafrilla · L.-A. López · M.Á. Tobarra
Faculty of Economics and Business, University of Castilla-La Mancha,
Plaza de la Universidad, 1, Albacete 02071, Castilla-La Mancha, Spain
e-mail: guadalupe.arce@uclm.es

J.E. Zafrilla
e-mail: jorge.zafrilla@uclm.es

L.-A. López
e-mail: luis.lsantiago@uclm.es

M.Á. Tobarra
e-mail: mariaangeles.tobarra@uclm.es

R. Álvarez Fernández et al. (eds.), *Carbon Footprint and the Industrial Life Cycle*,
Green Energy and Technology, DOI 10.1007/978-3-319-54984-2_14

11312–11320, 2015), but not for the Spanish case. Regarding data sources, we propose combining the World Input-Output Database (WIOD) and the Household Budget Survey for the Spanish economy, in order to analyse the carbon footprint from household consumption for 2015.

1 Introduction

The role of towns and their inhabitants in fighting climate change is becoming increasingly important (Shi et al. 2016). Demographic trends, lifestyles and shifts in consumption patterns might have a key role in future emissions and policy decision-makers need to be aware of the environmental impact of urbanisation processes and the way settlements are built and managed.

In this context, the aim of this chapter is to apply a multi-regional input-output model to study the evolution of the carbon footprint for Spanish households as determined by the different type of settlement. This study analyses the household carbon footprint as a function of the municipality's population size, whether it is located in a rural or urban environment, and its relation to population density. By using a multi-regional model we are able to calculate the share of that carbon footprint that is generated within the settlement and the share that is produced around the world along global value chains. This methodology has been widely applied to study carbon footprints for households in terms of different characteristics: income levels (Duarte et al. 2012), age (Shigetomi et al. 2014), consumption of agriculture products (López et al. 2015), or tourism consumption (Cadarso et al. 2016).

While the topic of environmental impacts from household consumption has been widely studied from different perspectives (see Liu et al. (2011), for a review of recent papers), as they are direct or indirectly responsible for a relevant share of total production and pollution [70% of total carbon footprint in the UK, Minx et al. (2009)], the literature on the role of the spatial distribution of households is recently growing at great speed (Baiocchi et al. 2015). The number of inhabitants and the density of their distribution within a population might determine their decisions in terms of travelled distances, mean of transport used, availability of different services, or the provision of home-prepared food versus eating at restaurants. Housing, infrastructures and energy sources are generally different between large urban centres and small populations, and for the specific case of Spain, socio-economic characteristics for households in cities, towns and villages also vary. All those factors are reflected in their consumption levels and patterns and consequently in the emissions generated to produce the demanded goods and services (Duarte et al. 2012).

The structure of household consumption as a function of the type of settlement can also be used to analyse whether socio-economic features are the greatest influence in the level of carbon footprint, or by the contrary, structural, institutional or geographical factors of the settlement are more relevant. Previous literature has addressed this link in other countries, for instance Fan et al. (2012), Minx et al. (2013), Baiocchi et al. (2015), Ahmad et al. (2015), but not in so much detail for the

Spanish case (Duarte et al. 2012). While, for example, income levels are positively related to higher expenditures and total emissions, higher incomes usually present lower emission intensities. However, this last result is reversed if those higher incomes households live in outer suburbs, where distances to workplaces and services are greater and the share of detached houses is large compared to flats (Baiocchi et al. 2015; Jones and Kammen 2014). Additionally, other topic that is becoming mainstream is the carbon footprint of cities, since they have the tools required to address the climate change challenge and could simplify the implementation of low-carbon development strategies (Gouldson et al. 2016; Mi et al. 2016; Rosenzweig et al. 2010). Therefore, the quantification of the carbon footprint taking into account the type of settlement seems to be relevant.

In this study we provide a first insight into the differences in carbon footprint according to the size and density of the settlement where the households are located. In order to do so, we first present the methodology used to calculate the carbon footprint for households in different types of municipalities using a multi-regional input-output model and expenditure data from the Spanish National Institute of Statistics' Household Budget Survey for 2015. This is a basic difference with respect to (Duarte et al. 2012), as we take into account the effects on emissions from traded products and the differences in emission technologies across countries. It also differs from other studies (Ahmad et al. 2015) that calculate direct emissions from households while they do not include embedded emissions in those products consumed by them. Then we start by analysing the differences in expenditure levels and patterns for the average household living in municipalities classified by population size and density. Furthermore, we calculate the carbon footprint of the spending per consumption unit (SCU) living in different types of municipality as well as the global amount of emissions due to big, small, dense and sparse populations. We also provide an analysis of the origin of those emissions and the industries where they are generated in order to appreciate the differences due to the consumption patterns. We finally conclude by summarising our results and pointing out some consequences in terms of urban and consumption policies.

2 Methodology

Multiregional Input-Output (MRIO) models are appropriate to estimate the household's carbon footprint along the global value chains (Davis et al. 2011; Hubacek et al. 2014; Kanemoto et al. 2012; Peters et al. 2012). MRIO models will allow us to evaluate the tele-connection between local consumption and global environmental impact.

The matrix of total carbon content, direct and indirect, in an MRIO context is obtained from Eq. 1 (Miller and Blair 2009):

$$F = \hat{f}(I - A)^{-1}\hat{y} = \hat{f}L\hat{y} \tag{1}$$

where $\hat{f}$ refers to the diagonalized direct emission coefficients or factor content (CO_2 emissions, materials, added value, employment, …) per unit of sector output. Term A refers to the matrix of total (domestic and import) technical coefficients. $L = (I - A)^{-1}$ refers to the Leontief inverse and the amount of direct and indirect inputs per unit of output satisfying final demand. $\hat{y} = \hat{y}^{rr} + \hat{y}^{rs}$ is the diagonal matrix of the final demand, $\hat{y}^{rr}$ is the diagonal matrix of the domestic final goods (consumption, investment and public expenditure), and $\hat{y}^{rs}$ is the diagonal matrix of the exported final goods from country r to country s. In this chapter, we consider the factor content to be CO_2 emissions, and we discuss the carbon footprint of a region.

2.1 Household Carbon Footprint of a Region

In this paper, we calculate and assess the carbon footprints associated with household consumption for different population size and density categories in Spain. In this chapter, we estimate both direct emissions (from energy goods' combustion) and embedded carbon footprint (emissions generated along the global value chain) by SCU. The expression that assesses the total carbon footprint by household is:

$$CF = dCF + eCF \tag{2}$$

Where dCF refers to direct carbon footprint of households and eCF refers to the embedded carbon footprint. For the purpose of this chapter, information about household consumption (C_i) according to different i characteristics (i.e., size, density, income level, region, etc.) is required. This vector is decomposed into the diagonalized vector of domestic household consumption by each i characteristic (in our case, five size and three density categories) of the region r ($\widehat{C}_i^{rr}$) (in our case, r is Spain) and the diagonalized vector of imported goods from region s ($\widehat{C}_i^{sr}$). The expression that assesses the direct footprint (emissions generated by the consumption of energy goods) of the households in category i of the region r is as follow:

$$dCF_i^e = \begin{pmatrix} df^{e1} & 0 & 0 \\ 0 & df^{e2} & 0 \\ 0 & 0 & df^{e3} \end{pmatrix} \begin{pmatrix} C_i^{e1} \\ C_i^{e2} \\ C_i^{e3} \end{pmatrix} = \begin{pmatrix} df^{e1} C_i^{e1} \\ df^{e2} C_i^{e2} \\ df^{e3} C_i^{e3} \end{pmatrix} \tag{3}$$

where the term df^{e1} refers to the emissions coefficient of energy good $e1$ and C_i^{e1} denotes the household's consumption of that energy good $e1$.

Secondly, eCF_i^r shows the embedded carbon footprint of households of the households i of the region r:

$$eCF_i^r = \begin{pmatrix} f^r & 0 \\ 0 & f^s \end{pmatrix} \begin{pmatrix} L^{rr} & L^{rs} \\ L^{sr} & L^{ss} \end{pmatrix} \begin{pmatrix} \widehat{C}_i^{rr} \\ \widehat{C}_i^{sr} \end{pmatrix} = \begin{pmatrix} \overbrace{f^r L^{rr} \widehat{C}_i^{rr} + f^r L^{rs} \widehat{C}_i^{rr}}^{4.1} \\ \underbrace{f^s L^{sr} \widehat{C}_i^{sr} + f^s L^{ss} \widehat{C}_i^{sr}}_{4.2} \end{pmatrix} \quad (4)$$

expression [4.1] shows the domestic emissions embodied in the production of goods in the region *r* intended to provide the final demand of region *r*, which is the domestic carbon footprint (Yu et al. 2010). The term [4.2] refers to the imported emissions embodied in the production of imported goods from region *s*, intended to provide the imported final demand of region *r*. The sum of [4.1] and [4.2] shows the embedded household carbon footprint for region *r*.

2.2 Databases Used

We use the input-output data and environmental information from the World Input Output Database (WIOD). We use the World Input-Output Tables (WIOT) for 2009 and the original sectoral aggregation of 35 sectors and 41 regions (Timmer et al. 2015). Environmental information to estimate embedded carbon footprint was obtained from the WIOD Environmental Accounts, also aggregated in 35 industries. Specifically, the data used were CO_2 emissions for 2009. Direct emissions were estimated using the method proposed in Dejuán et al. (2013) and in Zafrilla (2014) where authors propose the estimation of emissions coefficients for different energy goods considering prices and physical carbon factors. Energy prices and the subsequent emissions coefficients have been properly updated to 2015. The input-output data have been converted to euros using exchange rates.

The final demand vector shall be composed from information about household consumption of Spanish households according to different characteristics. Therefore, Spanish households' spending patterns are provided by the Household Budget Survey (HBS) (in Spanish, Encuesta de Presupuestos Familiares) published by the Spanish National Institute of Statistics (INE 2016a), for the year 2015. We consider the spending per consumption unit (SCU), which is an indicator offered by the Spanish Household Budget Survey and defined as the ratio between total expenditure and the total number of consumption units (the number of consumption units in the household is calculated according to the modified OECD equivalence scale).[1] These economic variables present disaggregated data according to various

[1]The size of the consumption unit represented by the household-dwelling unit is indicated as the sum of the weights of its members. In accordance with international recommendations, as a standard measure, the value of each member of a household-dwelling unit is determined as follows: First adult aged 18 and over = 1.0; Subsequent adults aged 18 and over = 0.7; Each person aged under 18 = 0.5; If all persons in the household-dwelling unit are aged under 18, the weight of the first member is 1.0 and that of subsequent members 0.5.

socio-economic factors such as the size of the town of residence, educational or income level. In this paper we consider the municipality size and population density.

Regarding the municipality size, the chosen disaggregation criterion is: Municipality provincial capital; municipality non-capital with 100,000 inhabitants or more; municipality non-capital between 50,000 and 100,000 inhabitants; municipality non-capital between 20,000 and 50,000 inhabitants; municipality non-capital between 10,000 and 20,000 inhabitants and municipality with less than 10,000 inhabitants. Regarding population density, the following classification is considered: Densely populated area, intermediate area and disseminated area.

These data exclude indirect taxes on consumption, which have been removed by using data for the weighted average of VAT rate and special taxes, published by the Spanish Tax Agency (AEAT 2016). The household budget survey data have been harmonized to 2009 current prices using the Consumer Price Index for the corresponding year published by the Spanish National Statistics Institute (INE 2016b). The Spanish Household Budget Survey (EPF) provides data on household expenditure with a sectoral aggregation of 48 industries according to the COICOP classification (Classification of individual consumption by purpose), while WIOD data are presented according to the NACE rev. 1 classification (Nomenclature generale des Activités économiques dans les Communauté Européenne). Therefore, an ad hoc conversion matrix between these classifications has been constructed in order to adapt the 48-sectors aggregation of goods and services (COICOP) to the 35-sector aggregation of WIOD (NACE classification). Finally, the EPF does not distinguish between domestic consumption and imported consumption. This information is obtained by proportionally distributing the consumption using the information provided by the WIOD input-output framework and assuming that this distribution is independent of household characteristics.

3 Results

3.1 The Consumption of Spanish Families by Settlement

The distribution of households according to population size and density is the main determinant for the volume of consumption in those settlements (Fig. 1). In terms of global consumption per size of the municipality, 43% of that consumption corresponds to urban settlements above 100,000 inhabitants, followed by consumption in settlements under 10,000 (19%). For the case of density, most densely populated areas concentrate 53% of total consumption, while intermediate and low density areas have an equal share of the remaining expenditure.

The consumption pattern per standard consumption unit in the different urban settlements varies significantly from that of the total economy (Fig. 2). In this case, the two biggest types of settlement (above 100,000 and between 50,000 and

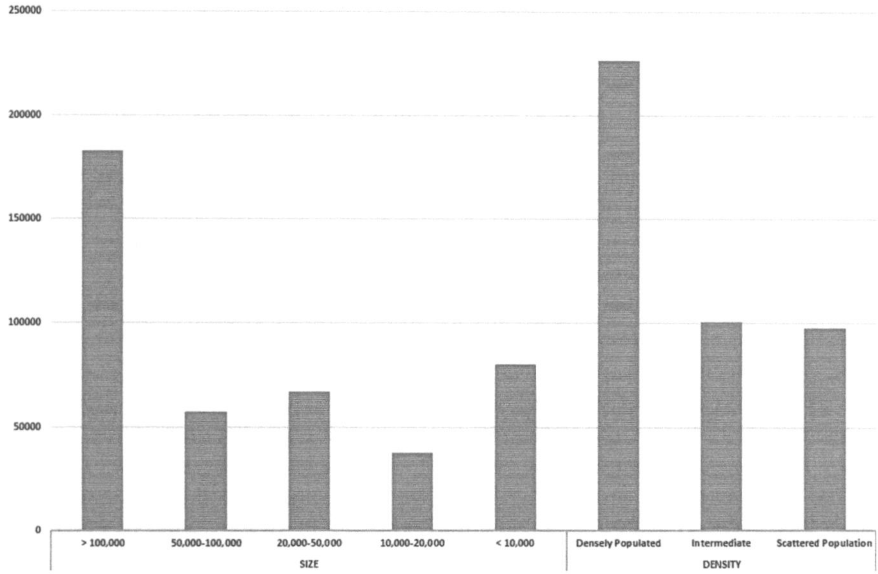

Fig. 1 Total consumption spending by size and population density in 2015 (million €). *Source* Own elaboration

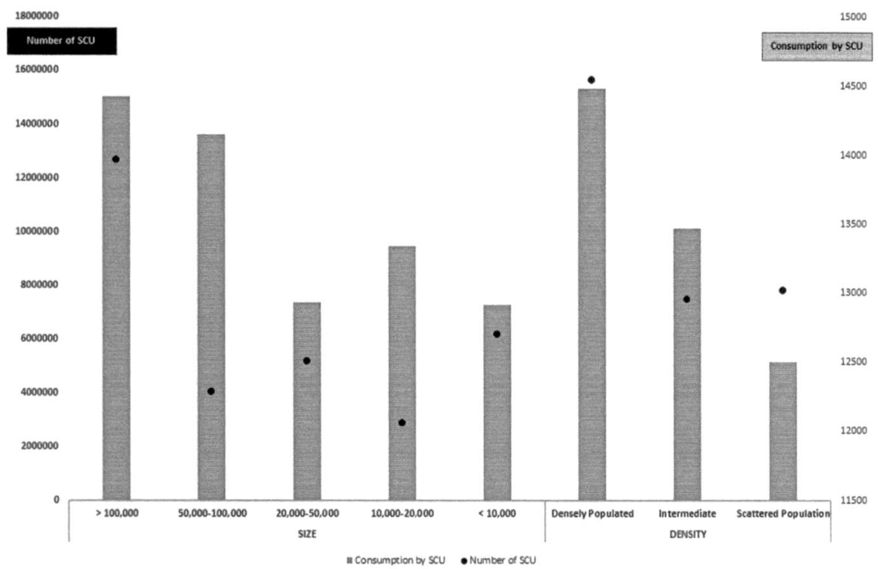

Fig. 2 SCU consumption spending (€) and number of SCU by size and population density in 2015. *Source* Own elaboration

100,000 inhabitants) and the most densely populated areas show a higher SCU consumption. This is due to the higher income per capita for households in the cities, compared to other settlements. In terms of size, there is also a significant difference between the top two biggest settlements and the rest, as there is only 267 € between number 1 and number 2 but 1332 € between number 2 and number 3. There are no relevant differences among the three smallest types of settlement, as the group of 10,000–20,000 inhabitants shows a higher consumption than for those below 10,000 and between 20,000 and 50,000. As for density, consumption per household does increase with density.

The pattern of per capita consumption by product is the second element, together with the level of consumption, which determines the households' carbon footprint (Fig. 3). Households' expenditures are in general highly concentrated on private services, real estate activities (25.9%, includes imputed rents), and hotels and restaurants (10%), post and telecommunications (5%) or other community services (4%). Purchases of manufactures account for food and beverages (10.5%), textiles (3.7%) or agriculture products (3.1%). We must highlight the expenditures in electricity, gas and water (5.4%), and transport (2.4%), as they will become key in terms of carbon footprint: despite being relatively small, expenditures in these sectors are highly carbon intensive.

While consumption patterns by product for the Spanish households are not significantly different depending on the population size or density, we can point out that families in big cities and highly dense areas spend a higher income share in everyone services (financial, hotels and restaurants, trade, personal services and education)

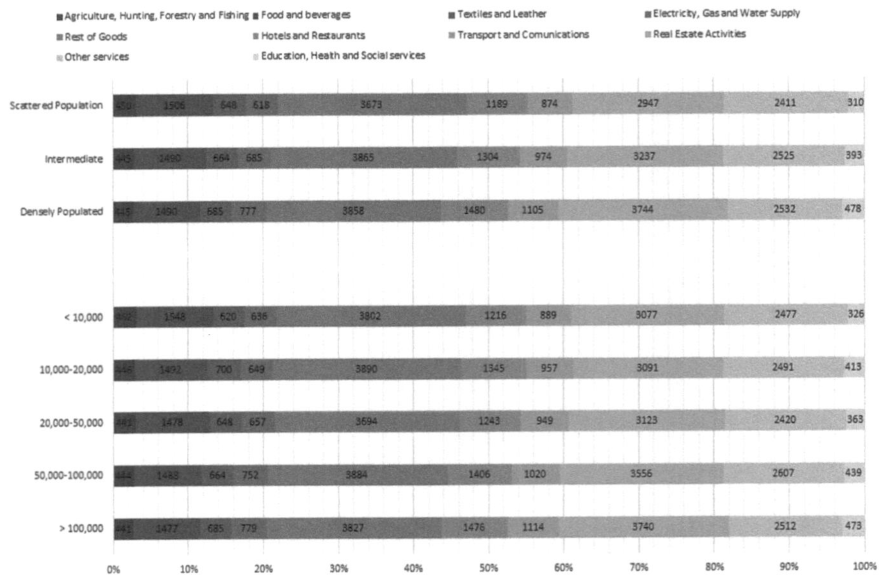

Fig. 3 SCU consumption spending (€) by products and different sizes and densities in 2015. *Source* Own elaboration

except on sale, maintenance of vehicles and electricity. They also spend more in electricity gas and water, in textiles and other manufacturing. On the other hand, households in rural and sparse areas devote a higher percentage of their consumption expenditure to manufactures, agricultural products, food and beverages, transports equipment, sale, maintenance and repair motor vehicles and fuels (gasoil and gasoline).

3.2 *Carbon Footprint of Spanish Families by Settlement*

The total carbon footprint for the Spanish households in 2015 varies between 84,684 kt CO_2 and 38,282 kt CO_2 depending on the municipality size and between 104,878 kt CO_2 and 46,927 kt CO_2 growing with density (Fig. 4). Most of those emissions are virtual carbon embedded in industrial production (75%) while direct emissions from households account for the remaining 25%. Consumption influences emissions from production process linearly, so settlements with higher total consumption also generate a bigger carbon footprint (as can be seen from comparing Figs. 2 and 4). Municipalities above 50,000 inhabitants and highly dense areas are responsible for most of the carbon footprint.

The carbon footprint by SCU is also higher in bigger and most populated settlements, as they are endowed with higher income and/or different consumption patterns (Fig. 5). The total carbon footprint by SCU varies between 6.68 tCO_2 and

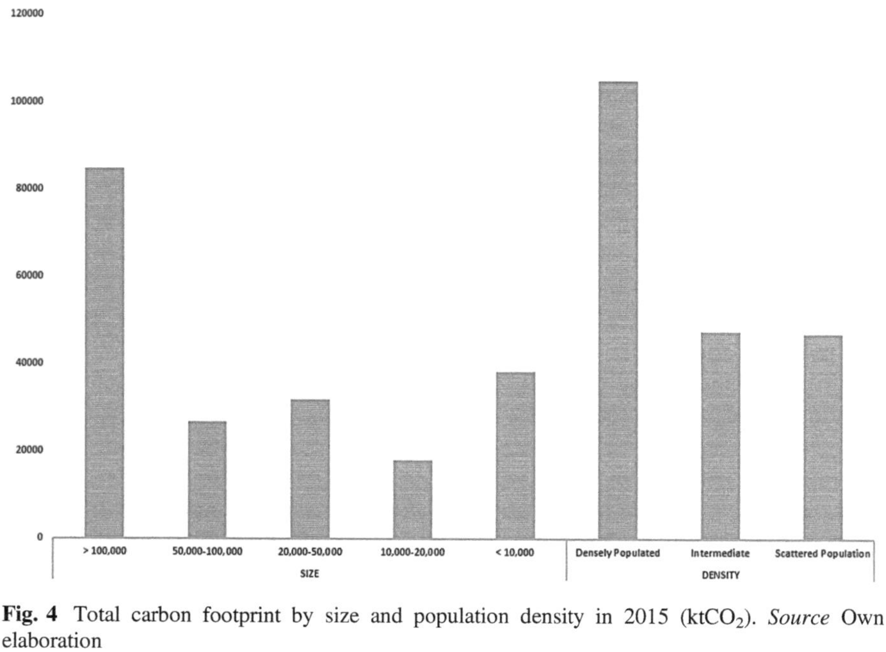

Fig. 4 Total carbon footprint by size and population density in 2015 ($ktCO_2$). *Source* Own elaboration

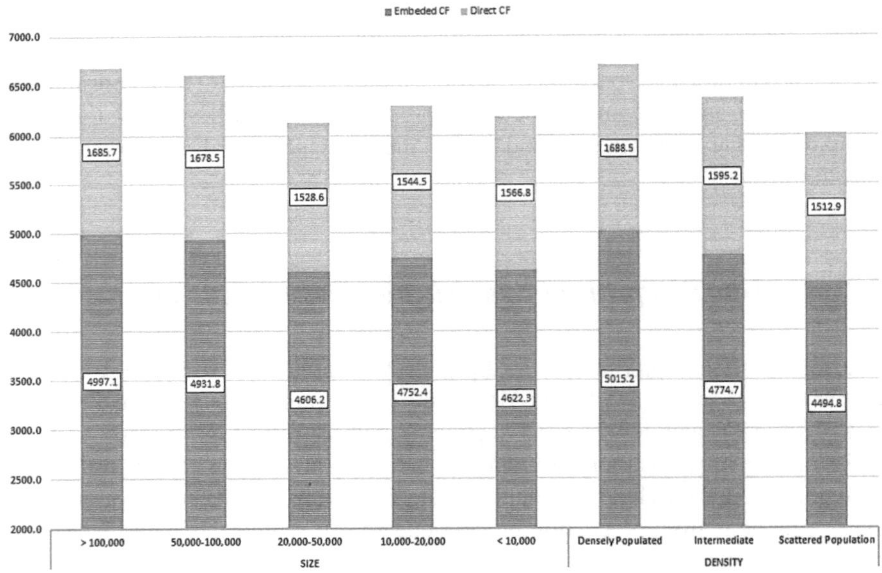

Fig. 5 SCU carbon footprint by size and population density in 2015 (tCO_2). *Source* Own elaboration

6.13 tCO_2 depending on the municipality size and between 6.00 tCO_2 and 6.70 tCO_2 with density (Fig. 4). This result agree with previous studies, like Duarte et al. (2012) for Spain, and Larsen and Hertwich (2010) for Norwegian municipalities. Nevertheless, that last paper also showed that the increase was linear up to 50,000 inhabitants, while after that the carbon footprint per capita increases (what could be used to define an objective in terms of urban policy to minimise emissions). On the other hand, for municipalities in the US, (Jones and Kammen 2014) found that, as in that country large metropolitan areas include both central cities and suburbs, density was inversely related to carbon footprint per household in the first case but suburbs increase carbon footprint compared to smaller urban settlements.

Being able to differentiate between households' direct carbon footprint and embedded emissions allows us to analyse the results in more detail. Focusing on the embedded carbon footprint of industrial processes, this ranges from 4.49 tCO_2 for sparse settlements to 5.01 tCO_2 for bigger and dense urban areas. These results are coherent with findings by Duarte et al. (2012) for Spanish households in 1999, as they found from 8.43 tCO_2 to 10.35 tCO_2, growing with density. It is important to note that the Spanish productive system in 1999 had not yet benefited from the mitigation policies for climate change undertaken under the Kyoto Protocol (Zafrilla et al. 2012).

By analysing the emissions intensity, that is emissions per euro spent by households, we can evaluate whether changes in consumption patterns can compensate for a higher budget in more populated areas (Fig. 6). For embedded carbon footprint from industrial processes, there is a different consumption pattern among the types of settlement, so carbon intensity, CO_2 per euro, decreases with the size (by 3.3%)

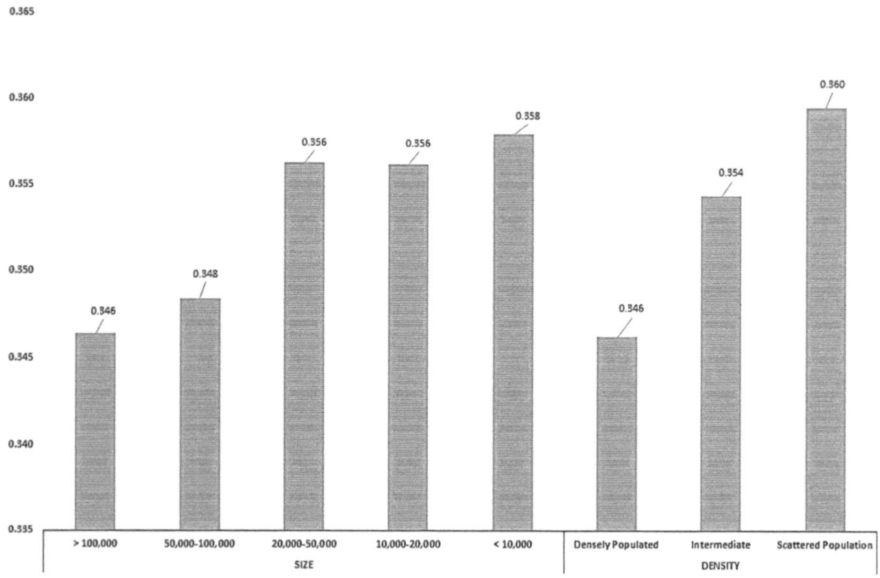

Fig. 6 SCU carbon intensity and SCU carbon footprint by size and population density in 2015 ($grCO_2$/€). *Source* Own elaboration

and density (also by 3.8%) of the municipalities. This implies that families living in big and dense cities show a consumption pattern that incorporates less CO_2 emissions than those living in rural and sparsely populated areas. For instance, one euro spent by a household in a highly dense area incorporates CO_2 emissions of 346 g versus 360 g in the case of a family living in sparsely populated areas.

However, differences found between the households' carbon footprint embedded in industrial processes and the carbon footprint per euro are less important than those found in other studies, focused on socioeconomic features of households. López et al. (2016) shows that the carbon footprint of Spanish households in 2014 ranges from 2.7 for households with monthly income below 500 € to 9.2 tCO_2 for those above 5000 €. Therefore, we can conclude that socioeconomic features are more important than the geographical distribution of population for the households, embedded carbon footprint of industrial processes.

As for the households' direct carbon footprint, there is a positive link between size/density and carbon footprint, because higher emissions from gas in big cities are only partly compensated by higher consumption of gasoil, gasoline and coal in rural areas. Total emissions increase with size and density, especially in the last case, from 1.51 tCO_2 to 1.69 tCO_2, and less significantly with size (from 1.57 to 1.69). This is again coherent with Duarte et al. (2012) for direct emissions per equivalent consumption unit in 1999, even though at that time changes were more significant, varying from 2.46 tCO_2 in dense areas to 1.90 tCO_2 in sparse areas. Our methodology, contrary to those authors, allows us to further evaluate the different energy sources responsible for those emissions.

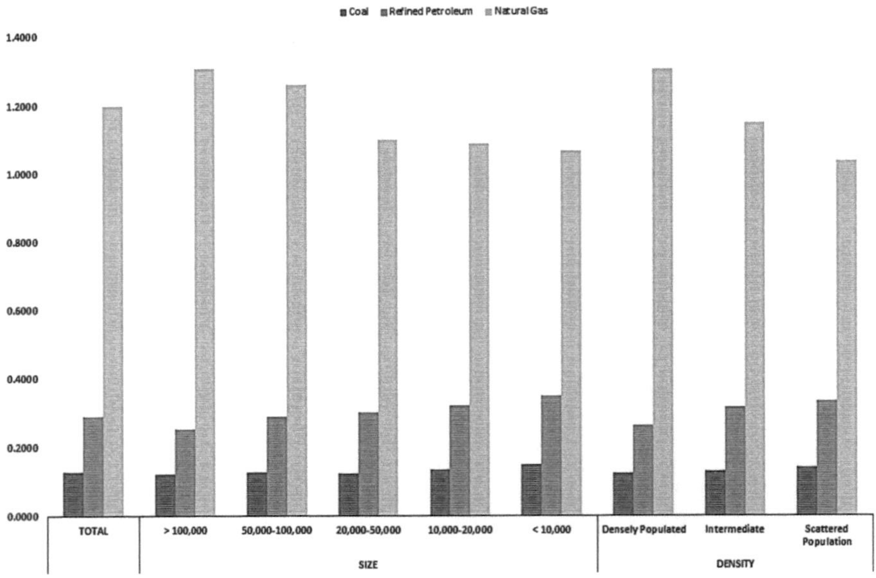

Fig. 7 Direct carbon per standard consumption unit by energy good, size and population density in 2015 (tCO_2). *Source* Own elaboration

Consumption data show that, while expenditure in transport services is higher in bigger and highly dense settlements, rural areas spend more in gasoil and gasoline for private transport, as a result of a lack of public transport infrastructures. This implies a growing household carbon footprint with the use of transport fuels from 0.254 tCO_2 in municipalities above 100,000 inhabitants to 0.349 tCO_2 in rural settlements (Fig. 7). These results are similar to those by (Jones and Kammen 2014) for US municipalities, as the extensive development of suburbs contributes to an increasing household carbon footprint due to private transport. The opposite is true for heating, as the infrastructure for natural gas, available in cities but not in rural areas, explains the higher expenditure in gas in the first type of settlement and in fuels in the second type. In terms of household carbon footprint from the use of gas for heating and cooking, it decreases from 1.308 tCO_2 in cities above 100,000 inhabitants to 1.067 tCO_2 in rural areas. These findings lead us to conclude that the structural, institutional or geographical factors or the settlements are relevant for the expenditure in heating and means of transport (public or private), highly pollutant activities that influence the households' direct carbon footprint.

A detailed analysis of consumption by type of good, on the vertical axis, and the carbon footprint embedded in the production processes associated with each settlement size, on the horizontal axis, allows us to summarise the previous results (Fig. 8). First, we must emphasise that most products exhibit a positive relationship between settlement size and household carbon footprint, with the exception of Agriculture, Food and beverages, and Coke and refined petroleum.

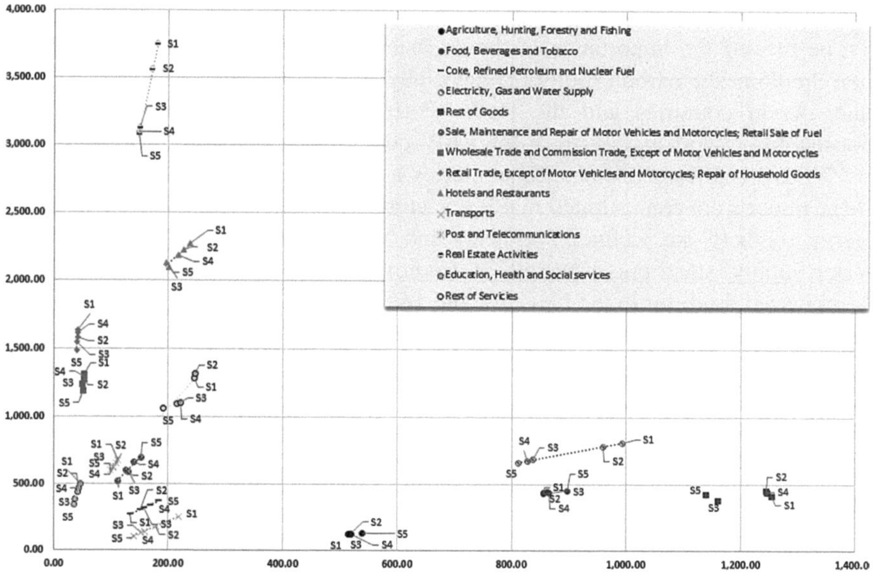

Fig. 8 Consumption patterns (€) in mayor sector by size level and embedded carbon footprint (kCO_2), 2016. *Note* Size: *S1* > 100,000; *S2* 50,000–100,000; *S3* 20,000–50,000; *S4* 10,000–20,000; *S5* < 10,000. *Source* Own elaboration

On the other hand, the products concentrating most of the households' expenditure represent, nevertheless, a small share of their carbon footprint, as they are, both directly and indirectly, low intensive in carbon emissions. Real estate stands out with an expenditure ranging between 3000 and 4000 € yearly (25.3% over total expenditure) and emissions close to 180 kCO_2 and Hotels and restaurants with an expenditure near 2500 € (16.2%) and emissions under 250 kCO_2. On the contrary, a small share of households' expenditures is linked to feeding (Agriculture products and Food and beverages) or buying other goods (textiles, computers, mobiles, etc.) representing both an expenditure under 500 € but their emissions are very different, ranging from 500 kCO_2 for the first type to over 1200 kCO_2 for the second one. Lastly, other goods and the consumption of electricity and gas represent a big share of households' carbon footprint (around 44% together), while expenditure is not that important, as they are highly carbon intensive. For these services, an average expenditure close to 900 € yearly for the most densely populated areas generate around 1 tCO_2.

3.3 *Quantifying Emissions Leakage*

Carbon footprint generated by the consumption of Spanish households in 2015 amounts to 149,201.1 kt CO_2, which 49.4% corresponds to domestic consumption and the remaining 50.6% (75,477.97 kt) are associated with imports. In this sense, it can be proved the importance of the emission leakage to other countries, higher than the domestic carbon footprint, especially to the emergent countries as China, other Asian countries and the RoW (Peters et al. 2012; Sanz et al. 2016). Households import mainly from the RoW countries (33%), China (17%), Russia (6.5%), US (4.8%) and other EU countries as Germany (4.3%) and France (4%). These imports are concentrated in a few sectors. The sectors that supply energy and energy goods (Coke, Refined Petroleum and Nuclear Fuel and Electricity, Gas and Water Supply) stand out, especially in countries such as Russia and RoW, 39% of the imported footprint in the first case and 16% in the second one. Emphasis is also placed on the imported carbon footprint in the Food, Beverages and Tobacco sector, especially in France, Germany, RoW and US. From China is important to highlight the sector of Textiles and Textile Products (42% of the total carbon leakage to China is concentrated in this industry), Leather and Footwear and Electrical and Optical Equipment. The imported carbon footprint of the sector of Agriculture, Hunting, Forestry and Fishing has an important role in countries such as US (26%), France (21%) and RoW (15%). An analysis of the sectors responsible for the imported carbon footprint allow us to state that indirect emissions are driving the leakage, more importantly than direct emissions (Arce et al. 2016).

Figure 9 shows the emissions leakage associated with the consumption of imported goods in 2015 for 3 selected municipality size (1, 3 and 5) and for the 41 regions. The figure shows how households that live in municipalities with more than 100,000 inhabitants (size 1), are the ones that generate the highest imported carbon footprint (31,628.9 kt CO_2) and these leakages go towards RoW (red color), Russia, China, EU and US. The next settlement with the largest imported carbon footprint (14,695.2 kt CO_2) is found in municipalities with less than 10,000 inhabitants (size 5). This is explained by the distribution pattern of the population by municipality, concentrated in these two types of settlements.

Nevertheless, in terms of consumption unit, the municipalities that most imported carbon footprint generate are those of size 1, 2.5 tons of CO_2, followed by the rest of settlements, size 2 (2.47 tons), 4 (2.46 tons), 5 (2.38 tons) and 3 (2.35 tons), per consumption unit. Therefore, settlements with the lowest imported carbon footprint per consumption unit are those that have between 20,000 and 50,000 inhabitants.

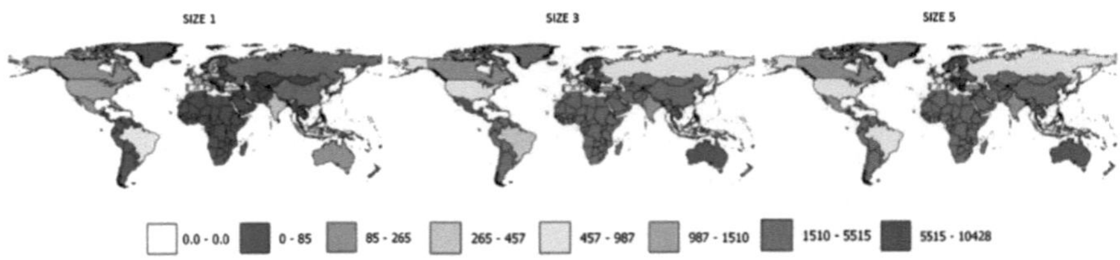

Fig. 9 Mapping imported carbon footprint of household consumption by municipality size, 2015 (kt CO_2). *Source* Own elaboration

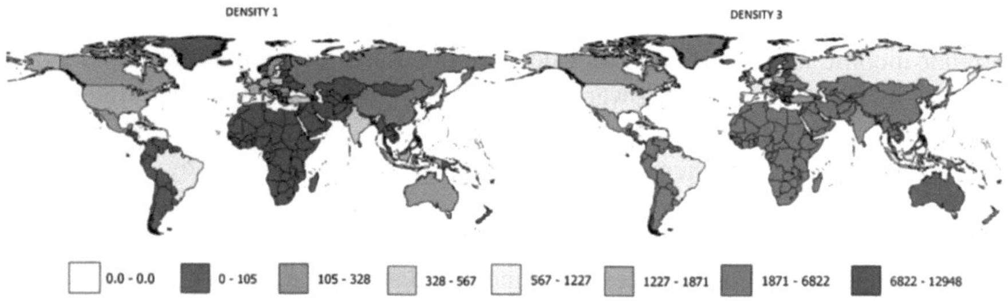

Fig. 10 Mapping imported carbon footprint of household consumption by density of population, 2015 (kt CO_2). *Source* Own elaboration

Figure 10 shows the emissions leakage associated with imports performed by household consumption depending on the density of population where they live. We have selected 2 levels because the differences between the imported consumption by households that live in a place with density level 2 and 3 are practically inappreciable. The real differences are found when passing from a densely populated municipality to an intermediate or disseminated ones. This is because densely populated areas account for 53% of consumption and those of intermediate density equitably share the rest of consumption.

The densely populated areas have higher imported carbon footprint in absolute terms (39,238.7 kt CO_2) in contrast to the intermediate zones (18,202.33 kt CO_2) and disseminated areas (18,037.8 kt CO_2). Regarding the origin of imports, the most important regions are those indicated above and that appear in Fig. 8.

Analyzing by consumption unit, although the amounts are very similar, we found that the higher population density, higher imported carbon footprint. Thus, a consumption unit of a densely populated area has a carbon footprint associated with its imports of 2.5 tons, a consumption unit of an intermediate area 2.4 tons and from a disseminated area 2.3 tons of CO_2. This can be caused by different factors. For example, in the case of Spain, the most densely populated areas are wealthier than rural areas, with higher consumption and emissions.

4 Conclusions

The objective of this study is to analyse the importance of population distribution, municipality size and density on the Spanish households' total carbon footprint. In order to do so, we have taken as a key factor the distinction within that carbon footprint between the emissions embedded in goods and services bought by the households (CO_2 emissions from industrial productive processes) and the direct emissions generated by the households (CO_2 emissions from transport, heating and cooking).

In terms of carbon footprint from goods' purchasing, our analysis leads us to conclude that the geographical, structural or institutional factors have a moderate impact, as the carbon footprint increases slightly with population size and density. Higher income levels in bigger and densely populated municipalities explains this growth, as it overcomes the lower emissions per euro as the emissions intensity is lower than in rural areas. Consumption of electricity and gas has the largest share in this carbon footprint, either directly by households or indirectly by means of the required electricity to produce goods and services. Nevertheless, when we compare our results to previous findings (Duarte et al. 2012; López et al. 2016), we can conclude that socioeconomic features (income distribution, age of the head of the household, household size) have more importance on total carbon footprint.

The mitigation alternatives focused on households would involve a reduction in consumption and/or a change towards more sustainable products, including those with eco-labelling and a higher share of personal services compared to goods. Even though households' decisions can favour this sustainability, the burden of responsibility falls mainly on energy sectors and those energy-intensive. In dynamic terms, when we compare with previous studies, we can observe that this carbon footprint due to purchased goods has reduced in time (5.01 t CO_2 in 2015 and 10.35 t CO_2 in 1999 for highly densely populated areas). Due mainly to the policies of promotion for renewable energies in electricity generation under the Kyoto protocol and the increase in the price of energy products, energy saving and efficiency have been promoted in all industries. The only flaw in this case is the growing carbon leakage towards emerging countries with high pollution intensity that has prevented a greater reduction in emissions.

The direct household carbon footprint is nevertheless affected significantly by the geographic distribution of population and by the transport and heating infrastructures available in different types of settlement, rather than by socioeconomic characteristics. This direct carbon footprint is highly concentrated on a small share of expenditure in energy products, particularly transport, heating and cooking. Reduced public services (public transport and gas pipes) in rural areas explain the higher spending in private transport and lower in gas. The trend in carbon footprint due to gasoil, gasoline and coal consumption decrease with size and density, while that related to gas increases. Furthermore, the decrease in this carbon footprint is reduced compared to that of embedded emissions from goods and products, and therefore it becomes increasingly important. Energy and related emissions efficiency policies are required in order to reduce the relative importance of direct emissions in Spain. Some examples of policies to accelerate the technological improvements could be the increasing substitution favouring more efficient heating systems (gas boilers at home versus fuel boilers), the renovation of the vehicle fleet boosting lesser emissions intensive engines or the promotion of the hybridization of the vehicle fleet, following the example of Nordic countries. Moreover, the use of public transport should be supported by the Spanish government not only via subsidies but also by increasing the quality of the services both in dense and scattered populations.

References

AEAT. (2016). Informes anuales de Recaudación Tributaria.

Ahmad, S., Baiocchi, G., & Creutzig, F. (2015). CO_2 emissions from direct energy use of urban households in India. *Environmental Science and Technology, 49*(19), 11312–11320. doi:10.1021/es505814g

Arce, G., López, L. A., & Guan, D. (2016). Carbon emissions embodied in international trade: The post-China era. *Applied Energy*. doi:10.1016/j.apenergy.2016.05.084

Baiocchi, G., Creutzig, F., Minx, J., & Pichler, P. P. (2015). A spatial typology of human settlements and their CO_2 emissions in England. *Global Environmental Change, 34,* 13–21. doi:10.1016/j.gloenvcha.2015.06.001

Cadarso, M. Á., Gómez, N., López, L. A., & Tobarra, M. Á. (2016). Calculating tourism's carbon footprint: Measuring the impact of investments. *Journal of Cleaner Production, 111*(Part B), 529–537. doi:http://dx.doi.org/10.1016/j.jclepro.2014.09.019

Davis, S. J., Peters, G. P., & Caldeira, K. (2011). The supply chain of CO_2 emissions. *PNAS, 108* (45), 18554–18559.

Dejuán, Ó., López, L. A., Tobarra, M. Á., & Zafrilla, J. (2013). A post-Keynesian age model to forecast energy demand in Spain. *Economic Systems Research, 25*(3), 321–340. doi:10.1080/09535314.2013.806294

Duarte, R., Mainar, A., & Sánchez-Chóliz, J. (2012). Social groups and CO_2 emissions in Spanish households. *Energy Policy, 44,* 441–450. doi:10.1016/j.enpol.2012.02.020

Fan, J., Guo, X., Marinova, D., Wu, Y., & Zhao, D. (2012). Embedded carbon footprint of Chinese urban households: Structure and changes. *Journal of Cleaner Production, 33,* 50–59. doi:10.1016/j.jclepro.2012.05.018

Gouldson, A., Colenbrander, S., Sudmant, A., Papargyropoulou, E., Kerr, N., McAnulla, F., et al. (2016). Cities and climate change mitigation: Economic opportunities and governance challenges in Asia. *Cities, 54,* 11–19. doi:10.1016/j.cities.2015.10.010

Hubacek, K., Feng, K., Minx, J. C., Pfister, S., & Zhou, N. (2014). Teleconnecting consumption to environmental impacts at multiple spatial scales. *Journal of Industrial Ecology, 18*(1), 7–9. doi:10.1111/jiec.12082

INE. (2016a). Encuesta de Presupuestos Familiares (Household Budget Survey). Spanish National Institute of Statistics. http://www.ine.es/dyngs/INEbase/es/operacion.htm?c=Estadistica_C&cid=1254736176806&menu=resultados&idp=1254735976608

INE. (2016b). Índice de Precios al Consumo (Consumer price index). Spanish National Institute of Statistics.

Jones, C., & Kammen, D. M. (2014). Spatial distribution of U.S. household carbon footprints reveals suburbanization undermines greenhouse gas benefits of urban population density. *Environmental Science and Technology, 48*(2), 895–902. doi:10.1021/es4034364

Kanemoto, K., Lenzen, M., Peters, G. P., Moran, D. D., & Geschke, A. (2012). Frameworks for comparing emissions associated with production, consumption and international trade. *Envionmental Science and Technology, 46,* 172–179.

Larsen, H. N., & Hertwich, E. G. (2010). Implementing carbon-footprint-based calculation tools in municipal greenhouse gas inventories. *Journal of Industrial Ecology, 14*(6), 965–977. doi:10.1111/j.1530-9290.2010.00295.x

Liu, L. C., Wu, G., Wang, J. N., & Wei, Y. M. (2011). China's carbon emissions from urban and rural households during 1992–2007. *Journal of Cleaner Production, 19*(15), 1754–1762. doi:10.1016/j.jclepro.2011.06.011

López, L. A., Arce, G., Morenate, M., & Monsalve, F. (2016). Assessing the inequality of Spanish households through the carbon footprint: The 21st century great recession effect. *Journal of Industrial Ecology, 20*(3):571–581. doi:10.1111/jiec.12466

López, L. A., Cadarso, M. A., Gómez, N., & Tobarra, M. Á. (2015). Food miles, carbon footprint and global value chains for Spanish agriculture: Assessing the impact of a carbon border tax. *Journal of Cleaner Production, 103,* 423–436. doi:10.1016/j.jclepro.2015.01.039

Mi, Z., Zhang, Y., Guan, D., Shan, Y., Liu, Z., Cong, R., . . . Wei, Y. -M. (2016). Consumption-based emission accounting for Chinese cities. *Applied Energy, 184*, 1073–1081. doi:http://dx.doi.org/10.1016/j.apenergy.2016.06.094

Miller, R. E., & Blair, P. D. (2009). *Input-output analysis: foundations and extensions* (2nd ed.). Cambridge: Cambridge University Press.

Minx, J., Giovanni, B., Thomas, W., John, B., Felix, C., Kuishuang, F., . . . Klaus, H. (2013). Carbon footprints of cities and other human settlements in the UK. *Environmental Research Letters, 8*(3), 035039.

Minx, J. C., Wiedmann, T., Wood, R., Peters, G. P., Lenzen, M., Owen, A., . . . Ackerman, F. (2009). Input-output analysis and carbon footprinting: An overview of applications. *Economic Systems Research, 21*(3), 187–216. doi:10.1080/09535310903541298

Peters, G. P., Davis, S. J., & Andrew, R. M. (2012). A synthesis of carbon in international trade. *Biogeosciences Discuss, 9*(3), 3949–4023.

Rosenzweig, C., Solecki, W., Hammer, S. A., & Mehrotra, S. (2010). Cities lead the way in climate-change action. *Nature, 467*(7318), 909–911.

Sanz, T., Yniguez, R., & Rueda-Cantuche, J. (2016). The relevance of multi-country input-output tables in measuring emissions trade balance of countries: The case of Spain. *Sort, 1*(1), 3–30.

Shi, L., Chu, E., Anguelovski, I., Aylett, A., Debats, J., Goh, K., . . . VanDeveer, S. D. (2016). Roadmap towards justice in urban climate adaptation research. *Nature Climate Change, 6*(2), 131–137. doi:10.1038/nclimate2841

Shigetomi, Y., Nansai, K., Kagawa, S., & Tohno, S. (2014). Changes in the carbon footprint of Japanese households in an aging society. *Environmental Science and Technology, 48*(11), 6069–6080. doi:10.1021/es404939d

Timmer, M. P., Dietzenbacher, E., Los, B., Stehrer, R., & de Vries, G. J. (2015). An illustrated user guide to the world input-output database: The case of global automotive production. *Review of International Economics, 23*(3), 575–605. doi:10.1111/roie.12178

Yu, Y., Hubacek, K., Feng, K., & Guan, D. (2010). Assessing regional and global water footprints for the UK. *Ecological Economics, 69*(5), 1140–1147. doi:10.1016/j.ecolecon.2009.12.008

Zafrilla, J. E. (2014). The mining industry under the thumb of politicians: The environmental consequences of the Spanish coal decree. *Journal of Cleaner Production, 84,* 715–722. doi:10.1016/j.jclepro.2014.02.031

Zafrilla, J. E., López, L. A., Cadarso, M. Á., & Dejuán, Ó. (2012). Fulfilling the Kyoto protocol in Spain: A matter of economic crisis or environmental policies? *Energy Policy, 51,* 708–719. doi:10.1016/j.enpol.2012.09.011

Chapter 15
Theoretical Analysis of the Metrics for Measuring the Sustainability of Infrastructure Projects

Olalekan Oshodi and Clinton Aigbavboa

Abstract The built environment contributes significantly to the emission of greenhouse gases. Empirical evidence suggests that these gases are principally responsible for climate change. To mitigate the impact of climate change on the environment, there is an increasing need to ensure that infrastructure projects are sustainable. Hence, this chapter presents a systematic review of extant literature on metrics for evaluating the sustainability of infrastructure projects, as there are currently limited studies that have holistically addressed the lack of theoretical information on the metrics for evaluating the sustainability of infrastructure. Indicative results from the extant review of literature revealed that the number of publications on sustainability metrics has been increasing and there are a growing number of systems and sustainable development indicators for measuring the sustainability of infrastructure projects. These findings suggest that issues relating to sustainability are becoming more important to built-environment researchers and practitioners. The results of the study provide valuable insights on the trends and gaps in theoretical knowledge on metrics for measuring the sustainability of infrastructure projects in the global South. The identified metrics in this chapter could be used in the development of an infrastructure decision support tool that can be used by for project teams to facilitate optimization of processes associated with sustainable infrastructure projects.

O. Oshodi (✉)
Department of Architecture and Civil Engineering, City University of Hong Kong, Tat Chee Avenue, Kowloon, Hong Kong
e-mail: oshodilekan2002@yahoo.com

C. Aigbavboa
Department of Construction Management and Quantity Surveying, University of Johannesburg, Johannesburg, South Africa
e-mail: caigbavboa@uj.ac.za

R. Álvarez Fernández et al. (eds.), *Carbon Footprint and the Industrial Life Cycle*, Green Energy and Technology, DOI 10.1007/978-3-319-54984-2_15

1 Introduction

The construction industry produces infrastructure that plays an important role in the process of economic development. A review of 40 years of study shows that a positive relationship exists between investment in the construction industry and economic growth (Dang and Low 2011). Recent studies have also reiterated the existence of this relationship (Chia 2012; Chiang et al. 2015). In addition, it has been shown that the construction industry provides employment (see Chiang et al. 2015). Hence, it is reasonable to suggest that sustained growth of the construction sector is a strategic component that can account for the economic success of any country.

Several problems reduce the impact of the construction industry on the economy. The problems affecting the sector have been identified to include (but are not limited to): uncontrolled construction investments (Lewis 1984); poor relationship among project parties (Meng 2012); low productivity (Abdel-Wahab and Vogl 2011) and a lack of concern for sustainable environment (Mensah 1997), among others. The need to address these problems has led to the development and use of innovative practices in the construction sector. For instance, alliancing and sustainability are good examples of such practices.

Sustainable development is an aspect of current innovative practices that is perceived as being poorly implemented by built environment professionals (Shiers et al. 2006; Prevatt et al. 2010; Ruparathna and Hewage 2015). This being despite a view heralded within the literature that advances the understanding of barriers to adopting sustainable practices, integrating sustainability into supply chains in construction projects and the need to develop green technologies for application in the built environment. This indicates that significant steps are being taken to address current challenges associated with climate change (Bourdeau 1999; Ruparathna and Hewage 2015). Climate change, which is responsible for the increase in the occurrence of extreme weather events (Van Aalst 2006), has personal, social and economic costs. For example, an increase in renal/heat/respiratory hospitalization among elderly in the US has been linked to the occurrence of extreme heat (Gronlund et al. 2016). This increases health costs and overstretches the capacity of hospitals and care homes which ultimately requires allocation of additional resources. While the importance of sustainable practices in the built environment has been acknowledge by stakeholders, there is a lack of universal metrics for measuring the sustainability of infrastructure projects.

Metrics for evaluating sustainability, whether implicit or explicit, are central to mitigating the impact of infrastructure on the environment. The metrics serve as a feedback mechanism for evaluating compliance and performance. The multiplicity of academic disciplines involved in different phases of construction makes it difficult to identify consistent metric for assessing sustainability in infrastructure projects. Against this background, this chapter presents a systematic review of extant literature on metrics for evaluating the sustainability of infrastructure projects. The need for the present is justifiable for several reasons. First, there is little published data on theoretical information on the metrics for evaluating the sustainability of infrastructure projects. Second, the trends and gaps in the metrics used in evaluating infrastructure projects. Finally, the information provided in this chapter will useful to academics and practitioners in the construction-related disciplines. It is envisaged that this chapter will inform professionals in the built environment and relevant stakeholders on the metrics for evaluating sustainability of infrastructure projects. Likewise, academics and researchers are also provided with key references on existing studies on metrics for assessing the sustainability of infrastructure projects.

2 The Concept of Sustainability

2.1 Definition of Sustainability

A considerable amount of literature has been published on sustainability. This can be found in almost every academic discipline ranging from the built environment (Ugwu et al. 2006), natural sciences (Sanders and Sheldon 2015) and social sciences (Khavul and Bruton 2013), amongst others. This has culminated in a common understanding across academic disciplines of the meaning attached to the term 'sustainability'. However, it must be noted that several different definitions of sustainability can be found in literature. A generally accepted definition for sustainability is lacking (Atkisson and Hatcher 2001; McCool and Stankey 2004) in the body of knowledge, and a primary reason for this could be linked to the existence of over 200 definitions in literature (Oltean-Dumbrava et al. 2013). From the various definitions, a common theme can be deduced which portrays sustainability (in the built environment) as consideration for the environmental, social and economic impact of a project. The origin of sustainability can be attributed to a recognition of the environmental impact of economic activities and growth that exceeds the adaptive capacity of the planet (Meadows et al. 1972). Sustainability can be defined as meeting the needs of the present generation without compromising the ability of future generations to meet their needs, as posited by the Brundtland Commission (1987). Several definitions and refinements are reported in other sources. Also, the term sustainability can refer to ensuring that energy systems are developed to synchronize most effectively with a nation's natural endowments, technological capacities, social expectations and economic development profile (Sovacool and Saunders 2014). Sustainability is a framework which integrates environmental, social and economic process/dimensions to improve quality of life without exceeding the planet earth's capacity (Adetunji et al. 2003).

2.2 Benefits of Sustainability Thinking to Infrastructure Projects

A considerable number of studies have demonstrated the importance of incorporating sustainability thinking/practices into infrastructure projects. Despite the need to comply with regulatory requirements, there is an increasing need to highlight the benefits of sustainability in order to increase its adoption by stakeholders (client, contractor, consultant, and the like) in infrastructure projects. This is because the uptake of sustainability into practice in the construction industry is relatively slow, particularly in developing countries. Based on information gleaned from literature, several benefits are associated with incorporating sustainability into the designs of in structure projects.

First, empirical evidence has shown that sustainability increases the value attached to buildings. For example, Feige et al. (2013b) show that a positive relationship exists between environmental performance and the rental price of residential properties. In a similar vein, it was found that sustainability certification (such as energy rating, environmental rating, and the like) of a property is positively related to its sales and rental value (Fuerst and McAllister 2011; Fuerst et al. 2016; Jensen et al. 2016). In particular, sustainable features integrated into buildings, especially those which enhance the water efficiency, the health and comfort level and the building's safety and security, tend to increase its rental value (Feige et al. 2013a, b). Taken together, it is evident that incorporating sustainable thinking/technologies into designs of buildings should be of interest to property investors owing to the added value.

Second, in today's dynamic construction market sector, sustainability is fast becoming an element that gives an added edge over competition to contracting and consulting firms (i.e. construction business organisation). For instance, Porter (1997) identified cost leadership, differentiation and focus as generic strategies that generate competitive advantage for business organisations. Thus, incorporating sustainability into business operations ensures that construction businesses create a product differentiation focus strategy. Although business owners may argue that adopting sustainability practices in their business operations may lead to an increase in cost, thereby possibly reducing the competitive edge, the situation is quite different from the current operations of corporations that engage construction businesses. However, empirical research has shown that sustainability is associated with reduction in waste, added value for clients, generating revenue from new streams (e.g. sales of carbon credit), improved employee working conditions, improved environmental performance of new product, and improved project performance (Baloi 2003; Jaillon and Poon 2008; Jensen et al. 2016; Kamali and Hewage 2016). Based on the foregoing, it is evident that sustainability is beneficial to construction business organisations.

Third, sustainability contributes to the economy of nations and could be a source of added revenue to governments. For instance, sustainable construction is related to building occupants' satisfaction, productivity, and health (Smith and Pitt 2009; Feige et al. 2013a). Also, the practice of sustainability in the built environment could generate revenue via tourism (Yang et al. 2010). Based on the information presented in this section, there is an obvious need to incorporate sustainability into infrastructure projects because it is beneficial to all stakeholders.

3 Research Method

To identify the sustainability metrics used in evaluating infrastructure projects, a systematic review of literature was conducted. The focus of the present study is limited to published journal papers. Several variants of review can be found in literature (e.g. narrative review). The validity of the outcomes of review studies has been the subject of academic debates. Tranfield et al. (2003), for instance, point out that systematic review ensures that the process is structured, transparent and replicable. A major advantage of systematic review is the methodological rigour of the whole process as has been affirmed in literature (see Thorpe et al. 2005; Yi and Yang 2014).

The process of conducting the review entailed three distinct but interrelated phases. At the first stage, a search was conducted using the SCOPUS database. As suggested by Falagas et al. (2008), SCOPUS provides a platform for a comprehensive search of extant literature. A comprehensive search was conducted using different keywords. The keyword combination used in the present study is (no. of articles found in brackets): "Sustainability indicator*" AND "construction" AND "Project*"; "Sustainability metric*" AND "construction" AND "Project*"; "Sustainability criteria" AND "construction" AND "Project*"; "Sustainability indicator*" AND "Infrastructure"; "Sustainability metric*" AND "Infrastructure" and "Sustainability criteria" AND "Infrastructure". A total of 253 articles was mined at the end of the initial search. Subsequently, an additional search was carried out in other databases of target journals. This is similar to the approach used in previous systematic reviews found in literature (Yi and Yang 2014). This was done to ensure that all relevant published studies were identified and considered for inclusion. The journals selected for the targeted search meet one of the following criteria: (1) the journal falls within the top four based on Chau's (1997) ranking of construction management journals; (2) at least two relevant papers were found in the journal at the initial search; and (3) key journals focused on sustainability in the built environment. Based on these criteria, another round of search was conducted in the database of the following journals: Construction Management and Economics, Journal of Construction Engineering and Management, Engineering, Construction and Architectural Management, Journal of Management in Engineering, Habitat International, and Journal of Cleaner Production. At the end of the second round of search, 265 additional articles were retrieved.

Upon completing the two stages of search, a total of 518 articles was found. The information relating to these articles was exported to a spreadsheet application. From these 518 articles, 423 were excluded. These did not meet the selection criteria for the study (e.g. aim of the study, books, editorials, conference papers and articles in press). The abstracts of these articles were carefully read and in some cases the conclusion was also examined. This process ensured that non-relevant articles were excluded. Overall, 72 additional articles were excluded: this is because the articles did not fall within the scope of this review. Lastly, 23 articles met the selection criteria and were used for final analysis. The inclusion criteria as defined in the purpose of the study serve as a basis for analysing the selected articles.

4 Metrics for Evaluating Sustainability of Infrastructure Projects

A review of peer-reviewed articles was carried out to understand the sustainability metrics for infrastructure projects that have been discussed in construction-related literature. The papers selected for this review were published between 1997 and 2015. Four major themes emerged from the studies selected for the review: (1) In order to assess the level of sustainability, a metric is usually stated; (2) there are many metrics used in literature; (3) the metrics are classified based on the elements of sustainability; and (4) different weighting systems have been developed for optimizing the sustainability performance of infrastructure projects. The results for these themes are presented and discussed in three sub-sections later in this chapter as follows: first, a frequency analysis of all metrics that appeared in literature is provided. Second, an analysis of the metrics based on elements of sustainability is presented; and lastly, an analysis of sustainability metrics by classification of infrastructure project is provided. It is relevant to note in this chapter that the terms ‘metrics’, ‘indicators’ and ‘criteria’ are used interchangeably in evaluating the sustainability performance of infrastructure projects. The results for the metrics as found from the systematic literature review are presented and discussed in the following four sub-sections: First, there is background information on the papers selected for analysis; second, a frequency analysis of all metrics that appeared in literature is provided; third, an analysis of the metrics by key feature of sustainability is presented, and fourth, an analysis of sustainability metrics by theme is provided.

4.1 Background of Information on the Sample

As stated in the method section, an exhaustive search was carried out via SCOPUS search engine which covered a wide range of peer-reviewed journals. Table 1 shows a summary of the 16 journals that have published at least one (1) article on sustainability measurement for infrastructure projects. Based on the data presented in Table 1, it is evident that the systematic literature review, reported here, is multidisciplinary in nature. The majority of the papers were published in either sustainability or construction management journals. However, it is important to note

Table 1 Distribution of the articles analysed for metrics of sustainability of infrastructure projects by journal

Journal name	Number of articles analysed
Journal of Construction Engineering and Management	3
Habitat International	2
Journal of Management in Engineering	2
Journal of Cleaner Production	2
Sustainability	2
Construction Management and Economics	2
Canadian Journal of Civil Engineering	1
Proceedings of the Institution of Civil Engineers—Engineering Sustainability	1
Architectural Science Review	1
Transport	1
International Journal of River Basin Management	1
Built Environment Project and Asset Management	1
Ecological Indicators	1
Building and Environment	1
Journal of Infrastructure Systems	1
Journal of Environmental Management	1

Source Authors' review of literature

that the selected papers appeared in a publication covering a broad field of interest which is also applicable to the current scope of this chapter.

4.2 Frequency Analysis

A total of 508 unique metrics were identified from the papers selected for the review. The frequency of use of the metrics is summarized in Table 2. A little over half of the metrics (260) appeared in the literature only once. Also, the number of metrics that appear twice and three times are 43 and 11 respectively. Overall, it was observed that approximately 75% of the metrics appeared fewer than four times in the studies reviewed.

4.3 Analysis of Metrics by Key Characteristics

Based on a framework proposed in Sahely et al. (2005) for categorising the four key characteristics of sustainability, the metrics identified in the present study were classified. A summary of the results of this analysis is provided in Table 3.

Table 2 Frequency rates of use of the identified metrics

Frequency of use	No. of unique metrics
1	260
2	43
3	11
4	11
5	5
6	4
7	1
9	1
10	2
Total	508

Table 3 Distribution of sustainability key characteristics by the identified metrics

Metrics[a]	Frequency rate				
		Economic focus	Social focus	Environmental focus	Engineering focus
Noise pollution	10			√	
Job opportunities	10		√		
Land use	9			√	
Energy consumption	7			√	
Greenhouse gas emissions	6			√	
Renewable energy	6			√	
Flexibility and adaptability	6			√	
Cultural heritage	6		√		
Cost of operation and maintenance	5	√			
Lifecycle cost	5	√			
Safety and security	5				√
Waste management	5			√	
Community engagement	5		√		

[a]Only metrics with the frequency rate of 5 or more are provided

Analysis of the results reveals that a majority of the metrics are focused on a single characteristic of sustainability. The reason for this could be linked to the fact that the environment (ecosystem) is a central issue guiding decisions relating to sustainability. For instance, "noise pollution" (10 times), "land use" (9 times), and "energy consumption" (7 times) were some of the highly frequent environmental metrics utilized in the measurement of sustainability for infrastructure projects.

High frequency social metrics are "job opportunities" (10 times) and "cultural heritage" (6 times). Job opportunity, for instance, relates to the aspect of clean job creation which is a major consideration in current developmental (infrastructure) projects.

4.4 Analysis of Metrics by Theme

A review of Table 2 shows that a number of metrics have been used for evaluating sustainable infrastructure projects. To determine the key themes that were addressed by the metrics, a keyword analysis was conducted. The identified metric were categorized into 150 themes. A summary of major themes addressed by 10 or more metrics is provided in Table 4. In addition, examples of the metrics addressed by each major theme are provided in Table 4. Table 4 shows that different themes were addressed to varying degrees. This is not surprising given the large number of identified metrics collected from different sources. It is clear that the metrics tended to measure a relatively small number of core themes from a variety of perspectives. For example, of the 29 metrics that specifically focused on energy, the majority focused on issues related to renewable energy and energy consumption. Overall, the findings as shown on Table 4 further reinforce the general lack of agreement on how the key issues relating to sustainability should be measured. Although the metrics that focused on the same theme appear repeatedly in the literature, the measure used for quantifying the theme varies.

5 Lessons Learnt

Based on the findings presented in the earlier sections, it is evident that it is important to incorporate sustainability into the designs of infrastructure projects. One of the ways of assessing levels of sustainability adopted in an infrastructure project is via 'metrics'. The ability to properly define the metrics for quantifying sustainability is the first step towards ensuring that infrastructure is developed in a viable manner. From a broad viewpoint, it is evident that sustainable infrastructure development is beneficial not only for mitigating the impact of climate change but it also benefits the project owners and the economy at large. Moreover, the identified metrics are advantageous in controlling the environmental impact of infrastructure-building construction activities in particular.

The present study focuses on a review of metrics used to measure performance in the sustainability of infrastructure projects. In summary, it is evident that several metrics used for measuring the sustainability of infrastructure projects are found in literature. There are several reasons that could be a possible explanation for this. First, the purpose and characteristics of infrastructure are heterogeneous in nature. Second, the parties involved in infrastructure projects are diverse. Differences in the

Table 4 Representative examples of major themes addressed by the identified metrics of sustainability

Major themes[a]	No. of metrics	Examples of metrics (frequency rates)
Water	36	Water (3), recycled water (1)
Cost	29	Lifecycle cost (4), compensation cost (2)
Material	29	Recycled materials (4), durable material (3)
Energy	29	Renewable energy (5), energy consumption (4)
Pollution	28	Noise pollution (4), greenhouse gas emission (3)
Waste	21	Waste management (3), waste generation (2)
Health and safety	20	Health and safety (7), health and well-being (2)
Community/public	20	Community engagement (3), public perception (1)
Labour/employment/workforce	19	Local employment and engagement with local business (2), direct job opportunities (2)
Preservation/conservation	14	Neighbourhood preservation (1), preserve prime farmland (1)
Revenue/investment/expenditure/finance/fund	14	Financial sources (4), user fees (3), revenue (1)
Land use	11	Land use (3), mixed use/high density layout (1), need for rezoning is minimal (1)
Environment	11	Indoor environmental quality (2), environment management (2), proportion of budget devoted to environmental protection (1)
Green technology/green concept	10	Green value (2), natural ventilation system (1), passive thermal design (1)
Hazard/effects of climate change	10	Sea level rise (1), emergency evacuation routes (1), disaster risks (1)

[a]Only the major themes addressed by 10 or more metrics are presented

perspectives of these professionals could be linked to the academic training of the professionals involved in projects. Finally, there is a lack of a unified body of knowledge that explicitly provides details of metrics to be used for evaluating sustainability of infrastructure projects. Although a large number of metrics were identified, there seem to be little agreement on the specific metrics (themes) that should be utilized to measure performance in terms of the sustainability of infrastructure projects. This shows that the metrics can possibly be explored and consolidated to guide infrastructure projects' delivery.

6 Conclusions

This chapter contributed to the body of knowledge by providing a comprehensive analysis of metrics for evaluating the sustainability of infrastructure projects through a systematic review of literature. A review of extant literatures was carried out to provide a reference point on the variety of metrics highlighted on the elements of sustainability. The results showed that a total of 508 unique metrics were found in literature. The metrics were analysed in depth from different perspectives. In addition, it was found that the metrics centre on major themes. This clearly shows that the metrics can be consolidated into a manageable framework that can be used for measuring the performance of sustainable infrastructure projects. A major limitation of this study is the search criteria. It is possible that the selection criteria did not include all relevant journal papers. Hence, the use of different search terms and additional databases may have resulted in the identification of more or reduced metrics. Despite this limitation, the analysis provided insights into the metrics used in the measurement of the sustainability of infrastructure projects. The information provided by this chapter is an important take-off point for stakeholders interested in measuring the sustainability of infrastructure projects.

References

Abdel-Wahab, M., & Vogl, B. (2011). Trends of productivity growth in the construction industry across Europe, US and Japan. *Construction Management and Economics, 29*(6), 635–644.

Adetunji, I., Price, A., Fleming, P., & Kemp, P. (2003). The application of systems thinking to the concept of sustainability. In D. J. Greenwood, (Ed.), *Proceedings of the 19th Annual ARCOM Conference*. University of Brighton: Association of Researchers in Construction Management (Vol. 1, pp. 161–170).

Atkisson, A., & Hatcher, R. L. (2001). The compass index of sustainability: Prototype for a comprehensive sustainability information system. *Journal of Environmental Assessment Policy and Management, 3*(4), 509–532.

Baloi, D. (2003) Sustainable construction: Challenges and opportunities. In D. J. Greenwood (Ed.), *Proceedings of the 19th Annual ARCOM Conference*. University of Brighton: Association of Researchers in Construction Management (Vol. 1, pp. 289–297).

Bourdeau, L. (1999). Sustainable development and the future of construction: A comparison of visions from various countries. *Building Research and Information, 27*(6), 354–366.

Brundtland Commission. (1987). Our common future. Oxford, UK: Oxford University Press.

Chau, K. W. (1997). The ranking of construction management journals. *Construction Management and Economics, 15*(4), 387–398.

Chia, F. C. (2012). Construction and economic development: The case of Malaysia. *International Journal of Construction Management, 12*(1), 23–35.

Chiang, Y. H., Tao, L., & Wong, F. K. (2015). Causal relationship between construction activities, employment and GDP: The case of Hong Kong. *Habitat International, 46,* 1–12.

Dang, T. H. G., & Low, S. P. (2011). Role of construction in economic development: Review of key concepts in the past 40 years. *Habitat International, 35*(1), 118–125.

Falagas, M. E., Pitsouni, E. I., Malietzis, G. A., & Pappas, G. (2008). Comparison of PubMed, Scopus, web of science, and Google Scholar: Strengths and weaknesses. *The FASEB Journal, 22*(2), 338–342.

Feige, A., Mcallister, P., & Wallbaum, H. (2013b). Rental price and sustainability ratings: Which sustainability criteria are really paying back? *Construction Management and Economics, 31*(4), 322–334.

Feige, A., Wallbaum, H., Janser, M., & Windlinger, L. (2013a). Impact of sustainable office buildings on occupants' comfort and productivity. *Journal of Corporate Real Estate, 15*(1), 7–34.

Fuerst, F., & McAllister, P. (2011). Green noise or green value? Measuring the effects of environmental certification on office values. *Real Estate Economics, 39*(1), 45–69.

Fuerst, F., McAllister, P., Nanda, A., & Wyatt, P. (2016). Energy performance ratings and house prices in Wales: An empirical study. *Energy Policy, 92,* 20–33.

Gronlund, C. J., Zanobetti, A., Wellenius, G. A., Schwartz, J. D., & O'Neill, M. S. (2016). Vulnerability to renal, heat and respiratory hospitalizations during extreme heat among US elderly. *Climatic Change, 136*(3), 631–645.

Jaillon, L., & Poon, C. S. (2008). Sustainable construction aspects of using prefabrication in dense urban environment: A Hong Kong case study. *Construction Management and Economics, 26*(9), 953–966.

Jensen, O. M., Hansen, A. R., & Kragh, J. (2016). Market response to the public display of energy performance rating at property sales. *Energy Policy, 93,* 229–235.

Kamali, M., & Hewage, K. (2016). Life cycle performance of modular buildings: A critical review. *Renewable and Sustainable Energy Reviews, 62,* 1171–1183.

Khavul, S., & Bruton, G. D. (2013). Harnessing innovation for change: Sustainability and poverty in developing countries. *Journal of Management Studies, 50*(2), 285–306.

Lewis, T. M. (1984). A review of the causes of recent problems in the construction industry of Trinidad and Tobago. *Construction Management and Economics, 2*(1), 37–48.

McCool, S. F., & Stankey, G. H. (2004). Indicators of sustainability: Challenges and opportunities at the interface of science and policy. *Environmental Management, 33*(3), 294–305.

Meadows, D. H., Meadows, D. L., Randers, J., & Behrens, W. (1972). *The limits to growth.* New York: Universe Books.

Meng, X. (2012). The effect of relationship management on project performance in construction. *International Journal of Project Management, 30*(2), 188–198.

Mensah, J. V. (1997). Causes and effects of coastal sand mining in Ghana. *Singapore Journal of Tropical Geography, 18*(1), 69–88.

Oltean-Dumbrava, C., Watts, G. R., & Miah, A. H. S. (2013). "Top-Down-Bottom-Up" methodology as a common approach to defining bespoke sets of sustainability assessment criteria for the built environment. *Journal of Management in Engineering, 30*(1), 19–31.

Porter, M. E. (1997). Competitive strategy. *Measuring Business Excellence, 1*(2), 12–17.

Prevatt, D. O., Dupigny-Giroux, L. A., & Masters, F. J. (2010). Engineering perspectives on reducing hurricane damage to housing in CARICOM Caribbean Islands. *Natural Hazards Review, 11*(4), 140–150.

Ruparathna, R., & Hewage, K. (2015). Sustainable procurement in the Canadian construction industry: Current practices, drivers and opportunities. *Journal of Cleaner Production, 109,* 305–314.

Sahely, H. R., Kennedy, C. A., & Adams, B. J. (2005). Developing sustainability criteria for urban infrastructure systems. *Canadian Journal of Civil Engineering, 32*(1), 72–85.

Sanders, J. P., & Sheldon, R. A. (2015). Comparison of the sustainability metrics of the petrochemical and biomass-based routes to methionine. *Catalysis Today, 239,* 44–49.

Shiers, D., Rapson, D., Roberts, C., & Keeping, M. (2006). Sustainable construction: The development and evaluation of an environmental profiling system for construction products. *Construction Management and Economics, 24*(11), 1177–1184.

Smith, A., & Pitt, M. (2009). Sustainable workplaces: Improving staff health and well-being using plants. *Journal of Corporate Real Estate, 11*(1), 52–63.

Sovacool, B. K., & Saunders, H. (2014). Competing policy packages and the complexity of energy security. *Energy, 67,* 641–651.

Thorpe, R., Holt, R., Macpherson, A., & Pittaway, L. (2005). Using knowledge within small and medium-sized firms: A systematic review of the evidence. *International Journal of Management Reviews, 7*(4), 257–281.

Tranfield, D., Denyer, D., & Smart, P. (2003). Towards a methodology for developing evidence-informed management knowledge by means of systematic review. *British Journal of Management, 14*(3), 207–222.

Ugwu, O. O., Kumaraswamy, M. M., Wong, A., & Ng, S. T. (2006). Sustainability appraisal in infrastructure projects (SUSAIP): Part 2: A case study in bridge design. *Automation in Construction, 15*(2), 229–238.

Van Aalst, M. K. (2006). The impacts of climate change on the risk of natural disasters. *Disasters, 30*(1), 5–18.

Yang, C. H., Lin, H. L., & Han, C. C. (2010). Analysis of international tourist arrivals in China: The role of world heritage sites. *Tourism Management, 31*(6), 827–837.

Yi, H., & Yang, J. (2014). Research trends of post disaster reconstruction: The past and the future. *Habitat International, 42,* 21–29.

Chapter 16
Impact of Urban Policy on Public Transportation in Gauteng, South Africa: Smart or Dumb City Systems Is the Question

Walter Musakwa and Trynos Gumbo

Abstract Policy on public transport often directs where infrastructure and investment is directed. Currently, the discourse is towards transport infrastructure investments that facilitate the attainment of the so-called smart city and smart mobility status. This status is often seen as the panacea towards all the public transport problems that among others include traffic congestion and unreliability. This chapter grapples with the question; to what extent have the urban planning policies in South Africa and Gauteng province been instrumental in the pursuit of efficient, effective and responsive public transport systems? Have the transport systems led to either smart or dumb city systems. The Gauteng province has put in place policies such as the Gauteng 25 year integrated master plan (ITMP 25) that has a vision to better the lives of Gauteng residents through the establishment of a smart and efficient public transport system. The ITMP 25 also seeks to attract foreign investments and boost tourism through land use densification that supports the use and efficiency of public transport systems. The policy also aims to reinforce the passenger rail-network as the backbone of the public transport system in Gauteng, and to extend the integrated rapid and road-based public transport networks that assist to strengthen freight hubs; thus ensuring effective travel demand management and mainstreaming non-motorized transport. As a result, Gauteng has invested in bus rapid infrastructure (Reya-Vaya within the City of Johannesburg, the Gautrain which is a high-speed rail network that caters for all three metro municipalities) and investments in non-motorized transport lanes in Johannesburg. The study applies smart city and smart mobility indicators to determine the level of smartness of the Reya Vaya, Gautarin and cycling infrastructure. The results indicate a steady uptake in public transport and use of cycling as a means of transport as well as a paradigm shift towards smart mobility by Johannesburg and

W. Musakwa (✉) · T. Gumbo
Town and Regional Planning Department, Faculty of Engineering and the Built Environment, University of Johannesburg, Johannesburg, South Africa
e-mail: wmusakwa@uj.ac.za

T. Gumbo
e-mail: tgumbo@uj.ac.za

R. Álvarez Fernández et al. (eds.), *Carbon Footprint and the Industrial Life Cycle*, Green Energy and Technology, DOI 10.1007/978-3-319-54984-2_16

Gauteng residents. Nevertheless, this has yielded unintended consequences such as the reinforcement of spatial segregation and inadequate use of new transport infrastructure. Parts of the challenges are a direct result of weak policy formulation and implementation strategies at both national and provincial levels as well as a deep culture that prefers private automobiles to public transport. There is therefore need to improve transportation policy and promote evidence based transportation policy.

1 Introduction

Smart cities and smart mobility have recently dominated the urban planning discourse. It appears like the majority of cities the world over are in a race to become smart yet there is little or no consensus on what constitutes a smart city (Angelidou 2015; Caragliu et al. 2009). There are four key components or concepts that are commonly used to define and identify a smart city and these include (1) "smart machines" and informated organizations, (2) engaging communities, technology providers and research institutions (3) (re)-learning and adaptation, and (4) investing for the future (Ching and Ferreira 2015). Ostensibly, for a city to be smart, there is need for the adoption and implementation of information communication technologies (ICTs), hence smart cities are often equated with highly technological cities. According to Caragliu and Del Bo (2015) ICTs enable cities to foster urban development, economize time, improve individual mobility, improve access to information and services, save energy and resources as well as improving citizenry participation in urban decision making processes. Although smart and informated machines are the most well known aspects of smart cities, it is important to note that for a city to be smart, ICTs are not the only essential cogs since ingredients such as collaboration between community, technology providers and research institutes are also key to the realisation of smart cities. Similarly, continuous (re)-learning and adaptation are crucial to sustain smart cities. (Re)-Learning can be from other cities for example; the European smart cities network (Giffinger et al. 2007; Neirotti et al. 2014). The (re)-learning and adapting is the ability to assess performance using measurable metrics or performance indicators (Giffinger and Gudrun 2010). In developing counties cities are often hesitant to assess performance as a result of political pressure and fear of public outcry. Lastly smart cities invest in the future if they are to continually survive.

An important component and topical issue in the smart cities debate is transportation, that is often referred to as smart mobility (Calabrese 2013). Chun and Lee (2015) note that smart mobility is a concept of comprehensive and smarter future traffic service in combination with smart technology. Similarly, transferring demand from private cars to public transport is an integral part of smart mobility (Siemens 2015). However, the level of smartness depends on the transport mode that is in use, for example, combustion buses or taxis and or electric trains differ in their levels of smartness. The lack of or absence of smart transport modes that lead to a deplorable commuting experience has resulted in the general public shunning public transport and this is particularly so in developing countries such as South Africa (Allen 2013). Papa and Lauwers (2015) argue that smart mobility has evolved to being more than the technology used to optimize transport planning to more about improving the consumer and or customer experiences. Such innovations and the realisation of improved commuter conveniences emanating from smart transport systems have resulted in the shift in mind-sets within local, provincial and national governments in South Africa's approaches to infrastructure provision for public transit purposes.

This chapter focuses on the concept of smart cities, applying the urban planning discourse and perspective to the current debates, consequently extending the frontiers of knowledge in this domain. The paper determines the level of smartness of the relatively new public transport infrastructure in Johannesburg against the backdrop of a myriad of policies and legislation that have been formulated and enacted to support the innovative developments. The chapter starts by reviewing and discussing the policy and legislative frameworks around smart transport planning and provision. It then focuses on the methodology that was adopted in collecting, analysing and reporting the research results. The chapter discusses the level of smartness of public transport systems within the study area and then it ends by proffering conclusions and recommendations on the best way to achieve smart mobility within South Africa and other developing countries in general.

2 Policies and Legislation Governing Public Transportation

The provision and management of affordable, well-connected and reliable modern public transportation systems, particularly in cities of the developing world are critical in ensuring both smart cities and mobility within them. Several scholars have for a long time observed that efficient and effective urban public transport is not only important for city dwellers but also assist in facilitating the functioning of cities (Cardinale et al. 2014; Potter and Skinner 2000; Perl and Goetz 2015; Schwaberger 2014). It goes without saying that an accessible, reliable, convenient and affordable urban public transport system enables the swift movement of commuters from one location to another, thus promoting the physical and socio-economic development of cities and their residents (Tan et al. 2008; Tanahashi et al. 2012; Scoppetta 2014a, b; Szczech 2014; Tillner 2014; Zhukova and Smirnova 2014; Zhou 2014; Zhukova 2014).

Notwithstanding the evident benefits of modern urban public transport systems, governments of the developing world at all levels, have been struggling to plan for, develop and manage public transport systems of acceptable standards (Musakwa 2014).

African governments in particular, have been struggling to provide well-coordinated, efficient, effective, reliable and affordable public transport systems in their ever-growing cities spatially and demographically (Cervero 2013). Consequently, there have been spirited efforts in most African countries to not only enact relevant urban transport legislative frameworks and formulate policies but also their implementation to facilitate the development and management of efficient and effective modern public transport systems.

Notably, South Africa is one of the few African countries that have made concerted efforts to improve their public transport systems. The economies and populations in South African cities are ever growing hence efficient and effective urban public transport systems have become integral for the daily lives of South Africans, hence the Bus Rapid Transit (BRT) system and the Rapid Railway Systems (Gautrain) have been implemented since the beginning of the new millennium. Several Metropolitan cities in South Africa that have been adopting and introducing modern urban public transport systems include the City of Tshwane, City of Johannesburg, City of Ekurhuleni, City of Cape Town and the City of Ethekwini metropolitan municipality. The innovations have been implemented to reduce travel time, lessen traffic congestion, create employment opportunities for citizens and minimize greenhouse emissions. Consequently, South Africa has been adopting several enabling policies and legislative instruments to promote innovative urban public transport systems since the realization of the democratic dispensation in 1994 and these commenced with the National Constitution of 1996; The Green Paper on National Transport Policy that was launched in early 1996 that culminated in the adoption of the National Transport Policy White Paper later in the same year.

Recently there has the National Rail Policy Green paper was launched in 2015 to solicit views and ways of facilitating the planning and development of improved railway transport systems within the country. In particular, the province of Gauteng, which is the economic hub of the country, is experiencing an ever-increasing demand for public transport hence the adoption of the innovations in public transport systems. Consequently, the province adopted the Gauteng's 25-year Integrated Transport Master Plan that seeks to improve urban public transport systems within the province. Other policies that sought to promote and support public transport in Johannesburg include, the National Rail Policy Green Paper of 2015, National Transport Policy Green and White Papers of 1996 and the Gauteng's 25-year Integrated Transport Master Plan of 2013 (ITMP 25) (Gauteng Government 2013). The ITMP 25 plan seeks to achieve several objectives through the adoption and implementation of several strategies and among others they include the provision of responsive and efficient urban public transport systems that is well linked and connected to promote its use and reduce reliance on private modes of transport. Consequently this has given rise to BRT systems such as the Are yeng in Pretoria, Reya Vaya in Johannesburg and the high-speed railway (Gautrain).

2.1 Rea Vaya, Johannesburg Bus Rapid System

Rea Vaya is Africa's first full Bus Rapid Transit (BRT) and is located in the Gauteng province of South Africa where it operates only in the Johannesburg metropolitan city. It was between 2006 and 2007 when the City of Johannesburg decided on the approval of the Rea Vaya BRT project, a project that was mainly aimed at improving the quality of life of the city's residence through a public transport system. In 2006, a BRT system feasibility study was conducted with most importance focusing on integrating the Rea Vaya with Gautrain and Metrorail. The City of Johannesburg send representatives to Bogota in Latin America in 2007 for further study and tour and in the same year a memorandum of understanding was signed by Top 6, Greater Johannesburg Regional Taxy Council, Putco and Metro bus which all form part of the Rea Vaya BRT system stakeholders. Through thorough research and consultation with many stakeholders in the private and public sector, the City of Johannesburg finally implemented the first phase of the project in the year 2009. In 2010 during the FIFA world cup, Rea Vaya BRT was operating to transport people from the inner city to the Soweto area were Soccer city is located.

Rea Vaya operates in Region A to F in the Johannesburg Metropolitan City. It operates in different phases and has systematic hierarchical routes that connect micro city centres in the Johannesburg Metropolitan City. It has completed the construction of Phase 1A and 1B and currently developing Phase 1C. Rea Vaya's Phase 1A has a trunk route operating between Ellis Park in Doornfontein and Thokoza Park in Soweto, linking with several feeder routes in Soweto (Reya Vaya 2016) (Fig. 1).

Feeder buses run from Protea Glen to Thokoza Park and from Eldorado Park to Lakeview (Reya Vaya 2016). The route covers 325 km of special lanes and intersections while feeder and complementary buses carry passengers to the trunk route stations.

The inner city circular route (Fig. 2) travels around the CBD from Hillbrow and Braamfontein, to Ellis Park in the east and Chancellor House on the western edge of the city (Reya Vaya 2015). The Phase 1B has routes that operate through Cresta, Windsor West, Parktown, Yeoville; in addition, routes that operate to and from the University of Johannesburg Soweto are being added. The route starts in Noordgesig in Soweto and travels through Pennyville, New Canada, Highgate, Auckland Park and Braamfontein to Parktown, Metro Centre and Rissik Street in the CBD.

The route has made it possible for commuters to easily reach key public healthcare centres such as the Rahima Moosa, Helen Joseph and Charlotte Maxeke hospitals as well as educational institutions such as the University of Johannesburg, Wits University, Milpark College, Parktown Boys' High School and Barnato Park High School. Feeders run to and from Leaglen, Stormhill, Florida, Cresta, Yeoville and Parktown. There are also additional feeders in Soweto from Pimville and Mapetla. These routes are now linked to the Metro Centre Rea Vaya loop, which travels to the inner city through Braamfontein.

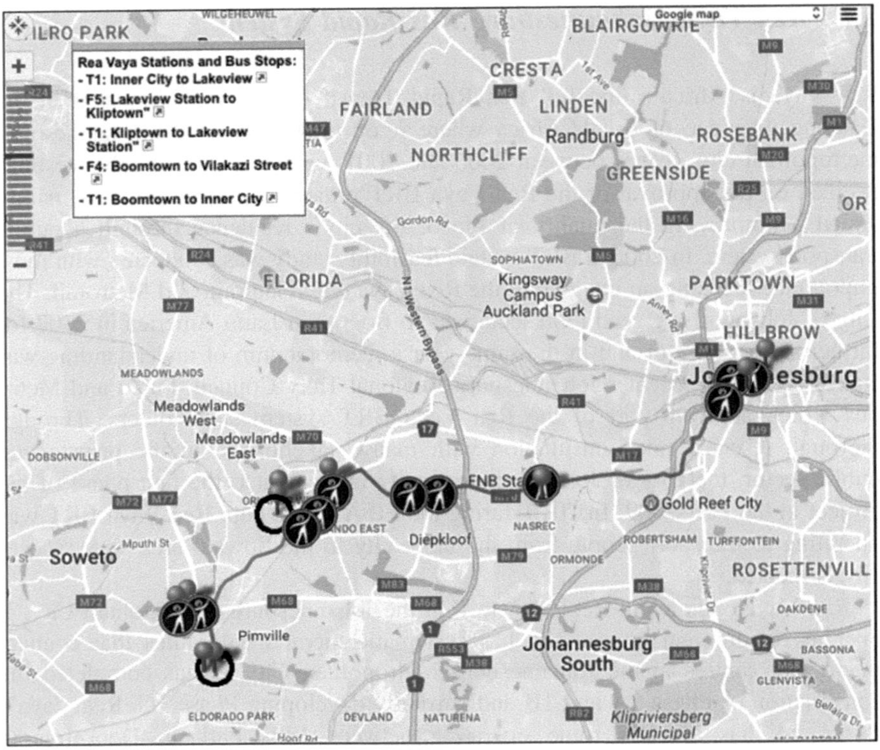

Fig. 1 Reya Vaya Soweto route. *Source* https://www.reavaya.org.za/consumer-information/the-routes

Rea Vaya's current focus is the development of Phase 1C following the completion of Phase 1B. Phase 1C will run from: Parktown to Alexandra; then Alexandra to Sandton, with complementary services between the CBD and Ivory Park; and from the CBD to Sunninghill on Oxford/Rivonia roads. Future plans also include extending the Phase 1C route from Sandton to Randburg by 2018, and possibly extending the trunk route from Soweto Highway to Dobsonville, enabling feeders to service areas such as Braamfisherville. The Rea Vaya trunk routes from the CBD to Sunninghill through Oxford Road and Ivory Park to Sunninghill will be prioritized after 2018. The three interchanges will be at Sandton, Alexandra and Westgate, where a number of station modules will be clustered and there will be integration with other modes of transport, including walking and cycling.

2.2 *The Gautrain System*

The Gautrain project Gautrain is Africa's first world-class, modern rapid rail and bus service for Gauteng, a province regarded as the economic heartland of

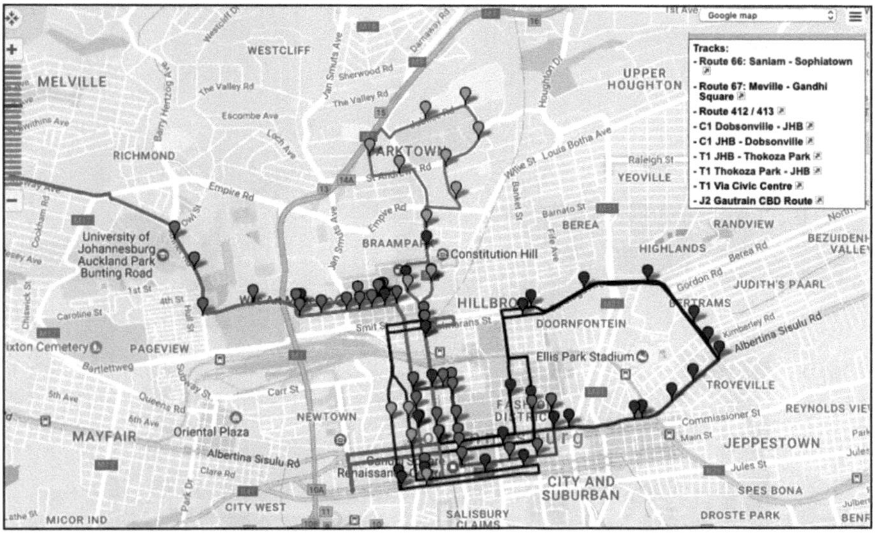

Fig. 2 Inner city route map. *Source* https://www.reavaya.org.za/consumer-information/the-routes

South Africa (Fig. 3). Gautrain is more than just a train. It is one of several strategically integrated Gauteng Provincial Government projects to meet future transport demands anticipated as a result of economic and population growth (Gautrain Management Agency (GMA) 2010a; Musakwa 2014).

It is also referred to as a mega-engineering project. It is a state-of-the-art rapid rail connection between Johannesburg (Africa's business capital) and Pretoria (Donaldson and Van De Westhuizen 2011). Gauteng, the country's economic hub currently experiences traffic congestion on its major routes, especially between Pretoria and Johannesburg. The current transport facilities and services between these two cities are mainly road based. Accordingly the Gautrain was supposed to ease this traffic congestion, in an attempt to create a smart city based on mixed land uses and development corridors (Musakwa 2014). The Gautrain project is also meant to promote rejuvenation of central Johannesburg and Pretoria (GMA 2010b). Construction of the Gautrain is informed by spatial planning is embedded in two parallel strategies that were initiated by the Gauteng Provincial Government namely the Gauteng Spatial Development Framework (GSDF) 2000 and the Gauteng Spatial Development Initiatives (SDIs). Consequently it is envisaged that the Gautrain will promote, smart mobility and accessibility, redirection of urban growth, contained urban growth, resource based economic development and rural development beyond the urban edge.

The Gautrain has two routes the South-North and West-East routes (Fig. 4). The North South route begins at Johannesburg park station in central Johannesburg to, Sandton and Pretoria and Hatfield in the north cutting across Johannesburg and Pretoria metropolitan municipalities. The West-East route takes passengers from Sandton Station, via Marlboro, to Rhodesfield Station in Kempton Park. From there

Fig. 3 The Gautrain. *Source* http://gma.gautrain.co.za (Gautrain Management Agency (GMA) 2013)

it connects to a station built within the airport terminal complex at OR Tambo International Airport (GMA 2010b).

Gautrain has proposed to expand routes as from August 2013 and this project forms part of the Gauteng Province's 25 year Integrated Transport Master Plan. There are 7 new proposed routes that will link with the existing routes from Park to Pretoria, Sandton, Marlboro and Midrand covering the South, North, East and West of Gauteng. The new proposed routes are prioritized according to phases.

3 Methodology

The chapter focuses on the new bus rapid system, the Reya Vaya in Johannesburg, the Gautrian high-speed train in Gauteng province and cycling in Johannesburg. Accordingly, smart mobility indicators where selected and divided into three categories namely (1) cycling (2) smart mobility systems which determine how smart ICT has been infused in the public transit system and (3) indicators pertaining the public transit system (bus rapid system known as Reya Vaya and the Gautrain high speed train system) (Table 1).

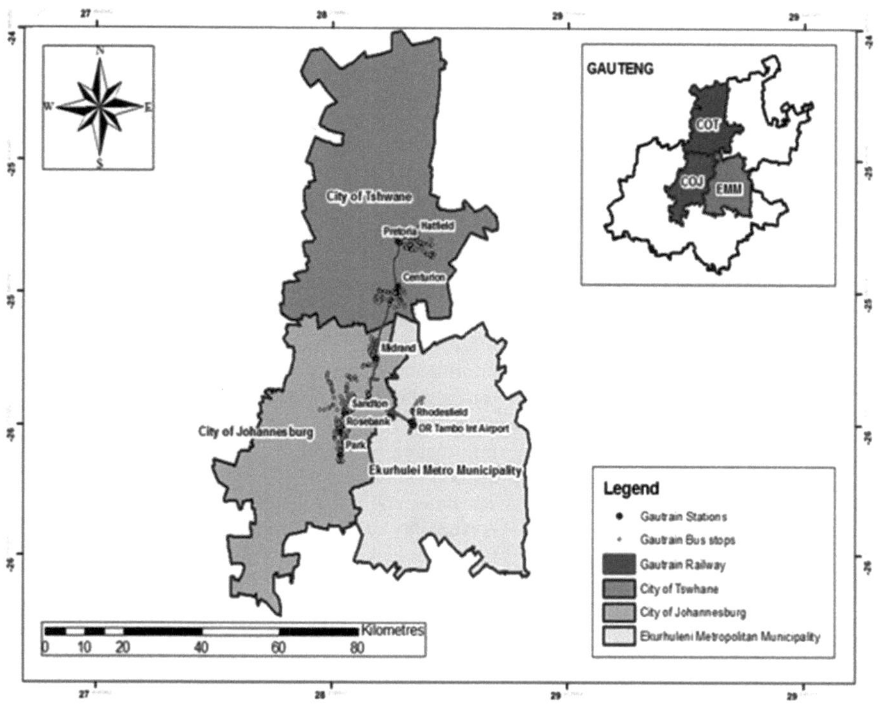

Fig. 4 Gautrain routes

Smart mobility indicators where selected because they determine how far people have shifted from using private cars to public transport (Calabrase 2013), how efficient and reliable the public transport is and how smart the public transit system is in the use of ICT to enhance the overall commuting experience. This would determine if the investments in public transport infrastructure has been smart. The indicators on smart mobility were mostly derived from observation, service provider websites, secondary data, and studies by Allen (2013), Garau et al. (2016) as well as Musakwa and Selala (2016).

The cycling indicators (total cycling trips, commuting and recreation trips) were derived for the year 2014 using a methodology described in Musakwa and Selala (2016). Data from Strava Metro was obtained from Strava for Johannesburg for the year 2014. Strava Metro utilizes data from the Strava mobile application, which is a global positioning system (GPS) enabled smartphone application that tracks bicycle rides and uploads the data to an online community of other users. Data purchased from Strava was in database (dbf), Microsoft Excel and shape file (shp) format. Cycling patterns where analysed on the basis of the type (recreational or commuting), temporal and spatial coverage. The analysis was also at city and neighbourhood level. Geospatial modelling software (GME) as well as the spatial analyst and map algebra functions of ArcGIS software were utilized to calculate the descriptive statistics (median) of cycling patterns (Musakwa and Selala 2016).

Table 1 Indicator selection

Variable	Indicators	Source
Cycling	– Total cycling activities recorded from Strava data – Total cycling trips – Total commuting and recreational cycling trips	– Musakwa and Selala (2016) – Strava data
	– Cycle lane density	– Garau et al. (2016)
Smart mobility support system	– Automated and electronic ticket system – Information on public information displays on routes schedules and waiting times – Information on mobile application on routes schedules and waiting times – Availability of tickets online – Electronic ticketing systems – Electronic bus stop signs – SMS services – Presence on social media – One ticket system	– Fieldwork observation – Garau et al. (2016) – Websites and reports from Gautrain and Reya Vaya
Public transport uptake; high speed train (The Gautrain) and bus rapid system (Reya Vaya)	– Bus network density – Demand for public transit – Passenger numbers – Fares – Level of integration	– Website and reports from service providers – Observation – Garau et al. (2016)

Sources Allen (2013), Garau et al. (2016), Musakwa and Selala (2016)

The smart mobility support system indicators and the public transport uptake where mostly analysed using semantic analysis of emoticons (Gal-Tzur et al. 2014) and quantitative analysis. Indicators that where not quantitative such as, automated and electronic automatic systems, information displays and mobile application where classified as yes if they where available and no if not. Emoticons where then used to describe the level of service (Table 2).

4 Results and Discussion

This section is structured as follows; the cycling in Johannesburg is discussed first followed by a discussion on the smart mobility support system and lastly the findings on the level of uptake of the public transport system in Johannesburg is discussed.

Table 2 Indicators on smart mobility and uptake of public transport in Johannesburg

Theme	Indicator	Reya Vaya	Gautrain
Smart mobility support system	Automated ticketing	☺	☺☺☺☺
	Electronic ticketing	☺	☺☺☺
	Information on public information displays (PID) on routes and waiting times	☹	☹
	Mobile application that shows routes and waiting times	☹	☺☺
	Tickets online	☹	☹
	Electronic bus stops	☹	☹
	SMS services	☹	☺☺
Public transport uptake; high speed train (The Gautrain) and bus rapid system (Reya Vaya)	Bus network density	0.30	N/A
	Demand	0.53	☺☺☺
	Passenger numbers	25,000/day	580,000 per month
	Fares	☺☺☺	☹
	Level of integration	☹	☹
	One ticket system	☹	☹

4.1 Cycling in Johannesburg

From the Strava analysis it is indicated that the number of cycling trips for Johannesburg in 2014 was 84,297. Only 20% of the cycling trips are for commuting whereas recreational trips accounts for 80% of the cycling trips (Fig. 5). Although millions of Rands where invested in providing cycling infrastructure in Johannesburg, it appears that the money has been wasted as people in Johannesburg are not using cycling as means to commute to places of work. Perhaps there is need to invest in cycling infrastructure for recreation as the cycling for recreational purposes is very high.

Moreover it appears that although the municipal and provincial government are promoting cycling through a myriad of initiatives such as the eco-mobility festival 2015 which closed of private automobile use in Sandton for a month (October) whilst encouraging use of public transport and non-motorized transport (NMT), the public hardly utilizes cycling as a means of commuting. Therefore there is a strong need to understand the needs, perception, behaviours and characteristics of cycling to inform smart policy and planning. A lesson to be learn is that, it is impossible to promote smart mobility without informative data that can guide decisions, otherwise it will promote planning of 'dumb' infrastructure.

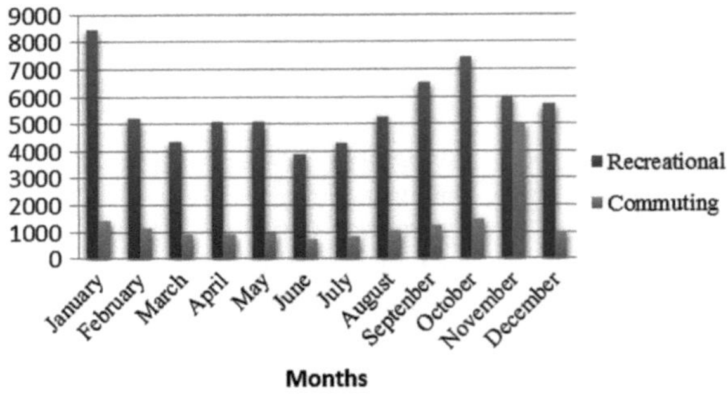

Fig. 5 Monthly commuting and recreational trips in Johannesburg for the year 2014. *Source* Musakwa and Selala (2016)

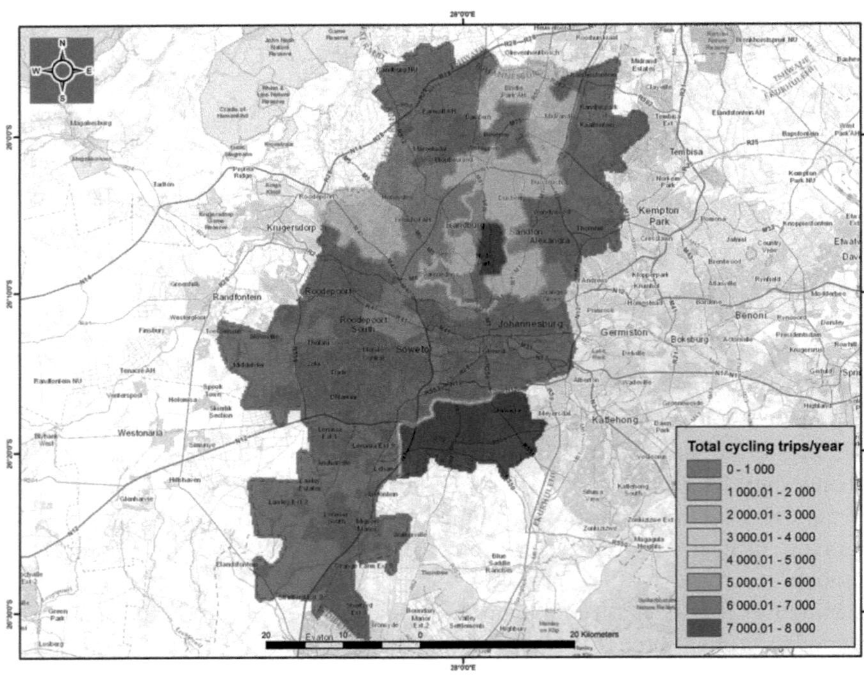

Fig. 6 Cycling trips per suburb in Johannesburg for 2014. *Source* Musakwa and Selala (2016)

Figure 6 shows where cycling is common in Johannesburg and the map can be used to argue for investing in cycling infrastructure north of Johannesburg central business district (CBD) in suburbs such as Sandton and Midrand. With ubiquitous big data such as Strava, smart mobility can be supported by making smart and evidence based decision making in cycling infrastructure will save already strained resources in local government.

4.2 *Smart Mobility Support System for the Reya Vaya and Gautrain*

Regarding the automated and electronic ticketing system, the Gautrain offers an average of four automated machines per station where commuters can purchase their train tickets or load money onto their smart card (Table 2). However, Reya Vaya does not have a fully automated system as commuters purchase at counters at the bus stations. Nevertheless commuters have an option of purchasing or uploading money onto their smart card at bank ATMs. Therefore, the infrastructure of the Reya Vaya is limited and not as smart as the Gautrain's automated ticketing machines. There is need to invest in automated ticketing infrastructure for the BRT Reya Vaya system and the infrastructure is now being put in place. What is common with the Reya Vaya and Guatrian is that they use a smart card for ticketing and this system is smart because it reduces paperwork and ensures that if lost, the smart card cannot be reused if the commuter registered it. Moreover the smart card is a source of valuable passenger data that can be used in mobility studies.

A major drawback and inconvenience with both systems is the inability of commuters to purchase their tickets or smartcards online which many commuters find frustrating in this world of e-commerce.

Another drawback is the use of two smart cards for use in Reya Vaya and Gautrain. This significantly hampers integration that is so passionately promoted in the public transit system in Johannesburg and this may inconvenience commuters particularly tourists and overseas customers who are used to the one ticketing system in their countries of origin. Although the ITMP 2030 policy documents targets having a one-ticketing system, this system is likely to prove a challenge given the various acts and legislation that separately govern the public transport operators. For example the Reya ownership is at municipal level whereas Gautrain is at provincial level. The public transport system can only be fully smart if there is real partnership and collaboration between agencies that goes beyond superficial technical optimization. Currently, it may appear there is a level of mistrust between service providers that hinders establishing smart infrastructure to enable a one smart card system. Furthermore, in a world of ubiquitous smartphones the service providers need to explore mobile ticketing options (Cheng and Huang 2013).

Provision of information is key in determining the smartness of transport infrastructure given the demanding needs of today's commuters to plan their journeys on information provided by service provides. Both the Gautrain and the Reya Vaya have public information displays (PID's) that show waiting times and the direction which the train and bus are headed. Nevertheless these PID's are static as they do not show route maps and commuters cannot use them to plan their routes. Hence there is need to relearn on the part of service providers to provide smart, collaborative and interactive PIDs which makes a public transport system smarter. The Gautrain has heeded to the commuters calls and they are in the process of installing these interactive PID's that show routes on their train stations.

This goes on to show that smart mobility is much more than infrastructure but there is a constant need to relearn and adapt to improve the public transport system.

Similarly, the public transport industry is being rapidly shaped by intelligent mobile applications. Gautrain does have a mobile application that shows timetables and the routes whereas the Reya Vaya did not have a mobile application until the launch of its application in June 2016. Commuters can use the Gautrain's mobile application to plan their journeys and it is very useful in conveying information on unexpected events, delays and service disruptions (Brugleiri et al. 2015; Alves et al. 2012) enabling commuters to re-plan their trips. Nonetheless, the Gautrain application is static and it is not real time and not on demand. Real-time information on mobile application has shaped public transit systems (Brugleiri et al. 2015) and poor information particularly in South Africa is a major deterrent on use of public transport. Reya Vaya did not have a mobile application until June 2016 and prior to that the unavailability of the mobile application inconvenienced commuters in route planning as well as information provision. The Gautrain and the Reya Vaya mobile applications are not intelligent enough because they do not allow commuters to decide on the best routes to make the trip they want, when they want and what are the expected travel times, based on the actual locations of the public transport vehicles and the travel speeds that can be estimated for the various relevant road segments for the next hour (Alves et al. 2012). Hence, there is need to make the application real-time, intelligent and more interactive.

Concerning electronic bus stops signs, both Reya Vaya and Gautrain buses posses no electronic bus stops that can enable commuters to view specific information about their journey. It has been noted that commuters, particularly the Reya Vaya wait indefinitely and frustrated because they are unable to access or locate their exact location of the incoming bus service. Hence, for the public transit system in Johannesburg to reach smart city or mobility status there is an urgent need to invest in these smart and informated electronic bus stop signs.

Regarding the SMS service, the Gautrain has fully-fledged SMS services that notify commuters of any major disruptions or any information such as delays, which assists commuters in planning their journeys. Nevertheless, the BRT Reya Vaya system do not inform its commuters on any delays or disruptions through an SMS service. Moreover, the Gautrain has a huge presence on social media platforms such as Twitter and Facebook whereas the Reya Vaya system is limited or non-existent. The Gautrain uses social media to position and portray itself as a smart transport alternative, efficient, safe, reliable, secure, predictable and comfortable transit system (Musakwa 2014). Gautrain commuters use social media in judging whether the Gautrain is a safer, reliable, dependable, and smart transit system. Likewise the GMA and commuters post useful information such as emergences, and delays that affect commuter's journeys on social media. Henceforth, the social media communication by the GMA is smart and it is a two-way system, which ensures relearning, and feedback mechanisms that can enable the system to be better in future. Overall, the Gautrain's communication strategy through social media and the sms services enables collaboration and engaged citizens who seek to improve the services.

4.3 *Public Transport Uptake Levels of the Reya Vaya and Gautrain*

Pertaining the bus network density and passenger demand the Reya Vaya scores 0.3 and 0.53 out of 10 (Gauru et al. 2016). This shows that the BRT infrastructure is very limited and that there is little uptake of the BRT services in Johannesburg. Perhaps the reason behind this is a culture where people are not used to BRT and the limited investment in BRT infrastructure. This is in stark comparison to Barcelona that scores 10 and 9 respectively in the bus network density and passenger demand category (Gauru et al. 2016). This shows that Johannesburg's public transit system is not yet smart despite an overload of numerous policy documents, which promote and advocate for the use of public transports. Perhaps the reason is also because there was lack of collaboration in building the BRT infrastructure with commuters and other service providers.

Regarding passenger numbers, the Reya Vaya transports almost 25,000 commuters per day and this number is very low given the population of over three million people in Johannesburg (Reya Vaya 2015). The figures are well below the targeted 150,000 passengers a day Allen (2013). Passenger numbers of the Gautrain on the other hand indicate a steady rise (Fig. 7). The numbers started at around 300,000 per month to almost a million users per month. Perhaps the perceived smartness and the Gautrain massive awareness campaigns are bearing fruit. Nevertheless, the commuting numbers still need to be increase to fully realize return on investment.

Pertaining to fares, the Reya Vaya is generally affordable (Allen 2013) given that an average trip cost 8 Rand. Therefore the Reya Vaya resonates with the majority, as transport is often expensive in a fragmented city like Johannesburg. Conversely, the Gautrain is commonly regarded as expensive given that a 40 km trip from Pretoria to Johannesburg costs 76 Rand (6 Euro). This is very expensive given that a similar trip will cost 2 Euro in countries in the European Union. As a result the Gautrain mostly appeals to the middle and affluent classes. Notwithstanding the high fares, the Gautrain has promoted smart mobility because people who would have used commuted to work on private automobiles now use public transit. This has resulted in less congestion and less greenhouse house gas emissions,

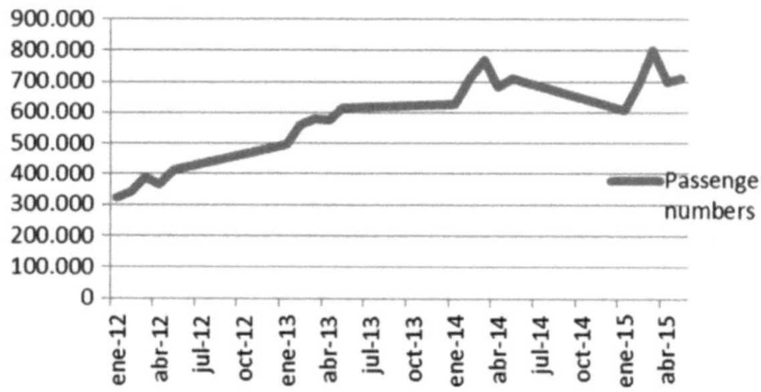

Fig. 7 Gautrain passenger numbers per month from January 2012 to May 2015

which bodes well for well for climate change. Nevertheless the Guatrain needs to be more accessible to the low income and expand its routes as it currently intends for it be smarter and possibly have greater impact in alleviating climate change.

5 Conclusion

This chapter has demonstrated that the Republic of South Africa has been investing in urban public transport systems for several years now. The work has revealed that several enabling policies and legislative instruments to promote innovative urban public transport systems were formulated and adopted at all the three levels of government from national, to provincial and local tiers. At national level, the frameworks and instruments range from National Constitution of 1996; the Green and white Papers on National Transport Policy that were adopted in 1996; the National Development Plan of 2012; the National Rail Policy Green paper of 2015 and many others. At provincial levels, the Gauteng province developed and adopted the Gauteng's 25-year Integrated Transport Master Plan in 2013. The metropolitan cities such as the City of Johannesburg, the city of Tshwane and the city of Ekurhuleni have also devised plans and formulated policies that seek to guide the planning and development of efficient and affordable urban public transport systems within the province. These instruments have been assisting in the provision and management of affordable, connected and reliable modern public transportation systems within South African cities particularly the metropolitan cities. For example the Gauteng has invested in bus rapid infrastructure in the form of Reya-Vaya within the City of Johannesburg, Are-yeng in the City of Tshwane and Harambee in the City of Ekurhuleni as well as high-speed rail network known as the Gautrain that caters for all three metro municipalities. The policies have also been assisting in reinforcing the passenger rail-network as the backbone of the public transport system in Gauteng, and to extend the integrated rapid and road-based public transport networks that assist to strengthen freight hubs; thus ensuring effective travel demand management and mainstreaming non-motorized transport. There are also investments in non-motorized transport patched throughout the province within specific metros. The quest is to improve and realise accessible, reliable, convenient and affordable urban public transport systems that enable the swift movement of commuters from one location to another, thus promoting the physical and socio-economic development of cities and their residents. However there are some unintended consequences such as the reinforcement of spatial segregation and inadequate uses of new transport infrastructure. Parts of the challenges are a direct result of policy formulation and implementation strategies at national, provincial and local levels. Pursuant to that there is need to make improve the implementation of the policies so that outcomes support connectivity and mobility of all residents and also assist in the realization of the development of egalitarian societies by linking previously disadvantaged societies.

Lastly although the city of Johannesburg has tried to improve public transport through smart mobility, it appears that the new public transit systems such as Reya Vaya and Gautrain are not very smart. There is need to realize that smart mobility is not merely a buzzword or the use of smart machines and informated organizations. It is more about collaboration, investing in the future and re-learning. Furthermore investing in smart public transportation requires evidence-based decision-making that can be obtained from ubiquitous smart and big data. Finally a culture that is geared towards public transport has to develop if the smart mobility is to be realized in South Africa.

References

Allen, H. (2013). *Africa's first full rapid bus system: The Rea Vaya bus system in Johannesburg, Republic of South Africa* (Unpublished case study prepared for the Global Report of Human Settlements).

Alves, D., Martinez, L. M., & Viegas, J. M. (2012). Retrieving real-time information to users in public transport networks: An application to the Lisbon bus system. *Procedia-Social and Behavioural Sciences, 54,* 470–482.

Angelidou, M. (2015). Smart cities: A conjuncture of four forces. *Cities, 47,* 95–106.

Bruglieri, M., Bruschi, F., Colorni, A., Luè, A., Nocerino, R., & Rana, V. (2015). A real-time information system for public transport in case of delays and service disruptions. *Transportation Research Procedia, 10,* 493–502.

Calabrese, L. M. (2013). Smart mobility: The cases of Hong Kong and the Netherlands. *GSTF Journal of Engineering Technology (JET), 2*(1), 145.

Caragliu, A., Del Bo, C., & Nijkamp, P. (2009). Smart Cities in Europe, Series Research Memoranda 0048. VU University Amsterdam, Faculty of Economics, Business Administration and Econometrics, Amsterdam.

Caragliu, A., & Del Bo, C. F. (2015). Smart specialization strategies and smart cities: An evidence-based assessment of European Union policies. *The rise of the city. Spatial dynamics in the urban century* (pp. 55–84).

Cardinale, T., Paula, L., & Zucchi, G. (2014). The city of Matera and the Sass: Smart places with a Dantean attraction. *Journal of Real Corp*, 665–674.

Cervero, R. B. (2013). Linking urban transport and land use in developing countries. *Journal of Transport and Land Use, 6*(1), 7–24.

Cheng, H. H., & Huang, S. W. (2013). Exploring antecedents and consequence of online group-buying intention: An extended perspective on theory of planned behavior. *International Journal of Information Management, 33*(1), 185–198.

Ching, T. Y., & Ferreira, J. (2015). Smart cities: Concepts, perceptions and lessons for planners. In S. Geertman, J. J. Ferreira, R. Goodspeed, & J. Stillwell (Eds.), *Planning support systems and smart cities* (pp. 145–168). Cham: Springer International Publishing.

Chun, B. T., & Lee, S. H. (2015). Review on ITS in smart city. *Advanced Science and Technology Letters, 98,* 52–54.

Donaldson, R., & van der Westhuizen, J. (2011). Built in a field of dreams? Spatial engineering and political symbolism of South Africa's rapid rail link development, Gautrain. In *Engineering earth* (pp. 683–695). Netherlands: Springer.

Gal-Tzur, A., Grant-Muller, S. M., Kuflik, T., Minkov, E., Nocera, S., & Shoor, I. (2014). The potential of social media in delivering transport policy goals. *Transport Policy, 32*, 115–123.

Garau, C., Masala, F., & Pinna, F. (2016). Cagliari and smart urban mobility: Analysis and comparison. *Cities, 56,* 35–46. doi:10.1016/j.cities.2016.02.012

Gauteng Government. (2013). *Gauteng 25-year integrated transport master plan*. Gauteng: Gauteng Government. http://www.itmp25.gpg.gov.za

Gautrain Management Agency. (2010a). *Socio-economic development progress*. http://www.gautrain.co.za/contents/brochures/sed_brochure_final_print.pdf

Gautrain Management Agency. (2010b). *Spatial development*. http://www.gautrain.co.za/about/about-gautrain/studies-documents/spatial-development/

Gautrain Management Agency. (2013). Socio-economic development progress 2006 to 2012. http://gma.gautrain.co.za/uploads/brochure/Gautrain_SED_2013.pdf

Giffinger, R., Fertner, C., Kramar, H., Kalasek, R., Pichler-Milanović, N., et al. (2007). *Smart cities: Ranking of European medium-sized cities*. Vienna, Austria: Centre of regional science (SRF), Vienna University of Technology. www.smart-cities.eu/download/smart_cities_final_report.pdf

Giffinger, R., & Gudrun, H. (2010). Smart cities ranking: An effective instrument for the positioning of the cities? *ACE: Architecture, City and Environment, 4*(12), 7–26.

Musakwa, W. (2014). The use of social media in public transit systems: The case of the Gautrain, Gauteng province, South Africa: Analysis and lessons learnt. In *Proceedings of REAL CORP*.

Musakwa, W., & Selala, K. M. (2016). Mapping cycling patterns and trends using Strava Metro data in the city of Johannesburg, South Africa. *Data in Brief, 9*, 898–905.

Neirotti, P., De Marco, A., Cagliano, A. C., Mangano, G., & Scorrano, F. (2014). Current trends in smart city initiatives: Some stylised facts. *Cities, 38*, 25–36.

Papa, E., & Lauwers, D. (2015). Smart mobility: Opportunity or threat to innovate places and cities. Web 2.0. *Australian Planner, 50*, 244–256.

Perl, A. D., & Goetz, A. R. (2015). Corridors, hybrids and networks: Three global development strategies for high-speed rail. *Journal of Transport Geography, 42*, 134–144.

Potter, S., & Skinner, M. J. (2000). On transport integration a contribution to better understanding. *Futures, 32*(3–4), 275–287.

Reya Vaya. (2015). City aims to double passenger numbers. https://www.reavaya.org.za/news-archive/may-2015/1180-city-aims-to-double-rea-vaya-passenger-numbers

Reya Vaya. (2016). *The routes*. http://www.itmp25.gpg.gov.za

Schwaberger, C. (2014). Will the guidebook green and blue spatial planning be a value help for Styrian cities to become a smart city. *Journal of Real Corp*, 1047–1049.

Scoppetta, C. (2014a). Ancient smartness of tomorrow. *Journal of Real Corp*, 65–74.

Scoppetta, C. (2014b). Checking smartness on the ground. Historically rooted dilemmas, future challenges and visions for a smarter metropolitan area of Rome. *Journal of Real Corp*, 131–140.

Siemens, (2015). Smart Mobility—A tool to achieve sustainable cities. http://www.vt.bgu.tum.de/fileadmin/w00bnf/www/VKA/2014_15/150212_Smart_Mobility_v5_TUM.pdf

Szczech, E. (2014). Concept of smart city and its practice in Poland. Case study of Lodz city. *Journal of Real Corp*, 169–180.

Tan, A., Sun, G., Krummert, K., & Ely, M. (2008). *Journeys. Land Transport Authority Singapore* (Issue 1). www.LTAacademy.gov.sg. Accessed April 09, 2015.

Tanahashi, Y., Rowland, J. R., North, S., Ma, K. L. (2012). Inferring human mobility patterns from anonymized mobile communication usage. In *Proceedings of the 10th International Conference on Advances in Mobile Computing and Multimedia* (pp. 151–160). New York, NY: ACM.

Tillner, S. (2014). More green open space in a densified city. *Journal of Real Corp*, 407–415.

White Paper on National Transport Policy. (1996). Government of South Africa.

Zhou, Y. (2014). The path towards smart cities in China: From the case of Shanghai Expo 2010. *Journal of Real Corp*, 1023–1027.

Zhukova, N. (2014). Technological solutions for knowledge management in smart cities. *Journal of Real Corp*, 653–664.

Zhukova, N., & Smirnova, O. (2014). Smart navigation for modern cities. *Journal of Real Corp*, 593–602.

Part IV
Methodology for Calculating Industrial Carbon Footprints

Chapter 17
Technologies for the Bio-conversion of GHGs into High Added Value Products: Current State and Future Prospects

Sara Cantera, Osvaldo D. Frutos, Juan Carlos López, Raquel Lebrero and Raúl Muñoz Torre

Abstract Today, methane (CH_4) and nitrous oxide (N_2O) emissions represent 20% of the total greenhouse gas (GHG) inventory worldwide. CH_4 is the second most important GHG emitted nowadays based on both its global warming potential (25 times higher than that of CO_2) and its emission rates, while N_2O is the main O_3-depleting substance emitted in this 21st century. However, despite their environmental relevance and the forthcoming stricter legislation on atmospheric GHG emissions, the development of cost-efficient and environmentally friendly GHG treatment technologies is still limited. In this context, an active bio-technological abatement of CH_4 and N_2O emissions combined with the production of high added value products can become a profitable alternative to mitigate GHGs emissions. The feasible revalorization of diluted CH_4 emissions from landfills has been recently tested in bioreactors with the production of ectoine, a microbial molecule with a high retail value in the cosmetic industry (approximately \$1300 kg^{-1}), as well as with the generation of polyhydroxyalkanoates (PHAs), a commodity with potential to replace conventional petroleum-derived polymers. This CH_4 bio-refinery approach can be also based on the biogas produced from anaerobic digestion, therefore improving the economic viability of this waste management technology. The N_2O contained in emissions from nitric acid production processes can be also considered as a potential substrate for the production of PHAs, with the subsequent increase in the cost-effectiveness of the abatement strategies of this GHG.

S. Cantera · O.D. Frutos · J.C. López · R. Lebrero · R.M. Torre (✉)
Department of Chemical Engineering and Environmental Technology, Faculty of Engineering, University of Valladolid, Valladolid, Spain
e-mail: mutora@iq.uva.es

S. Cantera
e-mail: sara.cantera@iq.uva.es

O.D. Frutos
e-mail: osvaldo.frutos@iq.uva.es

J.C. López
e-mail: jclneila@iq.uva.es

R. Lebrero
e-mail: raquel.lebrero@iq.uva.es

R. Álvarez Fernández et al. (eds.), *Carbon Footprint and the Industrial Life Cycle*, Green Energy and Technology, DOI 10.1007/978-3-319-54984-2_17

On the other hand, the off-gas N_2O abatement from diluted wastewater treatment plant emissions has been recently confirmed, although at the expense of a high input of electron donor due to the need to first deplete the O_2 transferred from the emission. This chapter constitutes a critical review of the state-of-the-art of the potential and research niches of bio-technologies applied in a CH_4 and N_2O bio-refinery approach.

1 Introduction

The steady rise in the average temperature of the Earth within the last decades and its associated detrimental impacts in the environment have resulted in an increasing concern on global warming and the development of political initiatives for greenhouse gases (GHGs) abatement (European Environment Agency 2015). Methane (CH_4) and nitrous oxide (N_2O), with global warming potentials 25 and 298 times higher than that of CO_2, respectively, represented 11 and 6% of the total EU-28 GHG emissions in 2014. Over the past years, CH_4 concentration in the atmosphere has increased at a rate of 0.2–1% year^{-1}, with 462 Mt CO_2 equivalents emitted in 2014. On the other hand, N_2O emissions accounted for 253 Mt CO_2 equivalent in 2014, 65% of which derived from agriculture and chemical industry (European Environment Agency 2015). N_2O, with an average atmospheric concentration increase of 0.3% year^{-1}, is also the most important O_3-depleting substance emitted in the XXI century (Ravishankara et al. 2009).

Methane produced by human activities can be used for energy recovery when the concentration in the gaseous emission is higher than 30% (v/v) (i.e. combustion of biogas in new landfills or wastewater treatment plants for electricity or steam production). However, more than 56% of anthropogenic CH_4 emissions contain concentrations below 3%, which are neither efficient nor economically suitable for energy recovery (Avalos Ramirez et al. 2012). For instance, dilute CH_4 emissions are typically found in old landfills (0–20%), ventilated coal mines (0.1–1%), covered liquid manure storage tanks (0–3%), or in confined cattle operations (<1%) (Estrada et al. 2014). In the particular case of N_2O, concentrations ranging from 100 to 2000 mg N_2O m^{-3} are commonly found in emissions from wastewater treatment plants, nitric and adipic acid production plants or livestock farming (López et al. 2013).

The European Union, within the frame of the 2030 Energy Strategy, is committed to reduce by 2030 its GHG emissions to levels below 40% of those recorded in 1990. In this context, the abatement of CH_4 and N_2O emissions becomes mandatory to meet this target emission cut objective. Unfortunately, current physical-chemical methods for CH_4 or N_2O treatment, such as adsorption, scrubbing or incineration, are inefficient, environmentally unfriendly (high CO_2 footprint as a result of their intensive energy usage) and/or present prohibitive operating costs, especially during the treatment of diluted gas emissions. These technologies possess a high CO_2 footprint as a result of their intensive energy usage (López et al. 2013).

The implementation of conventional gas treatment biotechnologies such as biofilters, biotrickling filters or bioscrubbers for CH_4 or N_2O abatement offers a potential alternative to overcome these limitations. These biological technologies are based on the biodegradation of these GHG pollutants by specific microbial communities, which transform them into less harmful products such as CO_2, N_2 or H_2O. Under optimal operating conditions, GHG-laden emissions can be also used by microorganisms as raw materials to synthesize high added value products such as biopolymers, exopolysaccharides or ectoine. The valorization of these waste gases through their bioconversion into commodities with a high market value will turn their abatement into a sustainable and profitable process.

This chapter will critically review the state-of-the-art of conventional and innovative bio-technologies applied to CH_4 and N_2O removal, and assess their potential as well as their current limitations. Moreover, the potential biological valorization of these GHGs as added value products will be discussed.

2 Biotechnologies for Methane Abatement: Biodegradation and Bioconversion into Value Added Bioproducts

2.1 Microbiology of CH_4 Treatment

Biotechnologies for the treatment of CH_4 are based on the biocatalytic action of microorganisms, mainly aerobic methane-oxidizing bacteria (MOB, also called methanotrophs), which transform methane into carbon dioxide and water using oxygen as electron acceptor. Methane oxidizing bacteria belong to the methylotrophic bacterial group, consisting of organisms that utilize reduced one-carbon substrates for their metabolism. Within this physiological group, MOB were classified and considered the only group able to use CH_4 as their single energy and carbon source (Hanson and Hanson 1996). However, recent findings demonstrated that some methanotrophs are also able to utilize multicarbon compounds as their carbon source in some environments (Semrau 2011).

Traditionally, MOB have been classified into three different groups according to their physiological and morphological characteristics: type I, type II and type X methanotrophs. Type I includes those MOBs that (a) present intracytoplasmatic membranes as bundles of vesicular discs, (b) use the ribulose monophosphate (RuMP) pathway for carbon assimilation and (c) contain phospholipid fatty acids of 14 and 16 carbons length. Type II MOB are characterized by (a) an intracytoplasmatic membrane aligned along the periphery of the cell, (b) the use of the serine pathway for carbon assimilation and (c) phospholipid fatty acids of 18-carbons length. On the other hand, type X strains share characteristics of both type I and II, including the RuMP pathway to assimilate formaldehyde and the synthesis of the enzyme ribulose biphosphate carboxylase from the serine pathway, to fix CO_2

(Hanson and Hanson 1996; Whittenbury 1981) (Fig. 1). The current classification of known aerobic methanotrophic genera based on 16S rRNA encloses a wide phylogenetic distribution within the three general groups: *Alphaproteobacteria*, *Gammaproteobacteria* and *Verrucomicrobia* (den Camp et al. 2009; Semrau et al. 2010). The alpha-proteobacterial methanotrophs can be further divided into the *Beijerinckiaceae* and *Methylocystaceae* families, while the gamma-proteobacterial methanotrophs belong to the *Methylococcaceae* family and the verrucomicrobial methanotrophs to the family *Methylacidiphilacecae*. Furthermore, the phylogenetic relationships between methanotrophs are also commonly examined considering the enzyme methane monooxygenase (MMO), used in the conversion of methane to methanol. In particular, the particulate methane monooxygenase (pMMO) is found in most known MOB and is located in the cytoplasmatic membrane, while the soluble methane monooxygenase (sMMO) is present in the cytoplasm and can be expressed either as the sole form of MMO or together with pMMO (Auman and Lidstrom 2002). The enzyme sMMO has been traditionally associated with type II methanotrophs, however it has been recently found that some genera of type I methanotrophs are also capable of synthesizing sMMO (Sazinsky and Lippard 2015).

Aerobic methanotrophs are widely distributed in environment such as wetlands, bogs, forests, rice paddies, groundwater, landfill cover soils, etc. (Semrau et al. 2010). The optimum temperature for methane oxidation has been established at

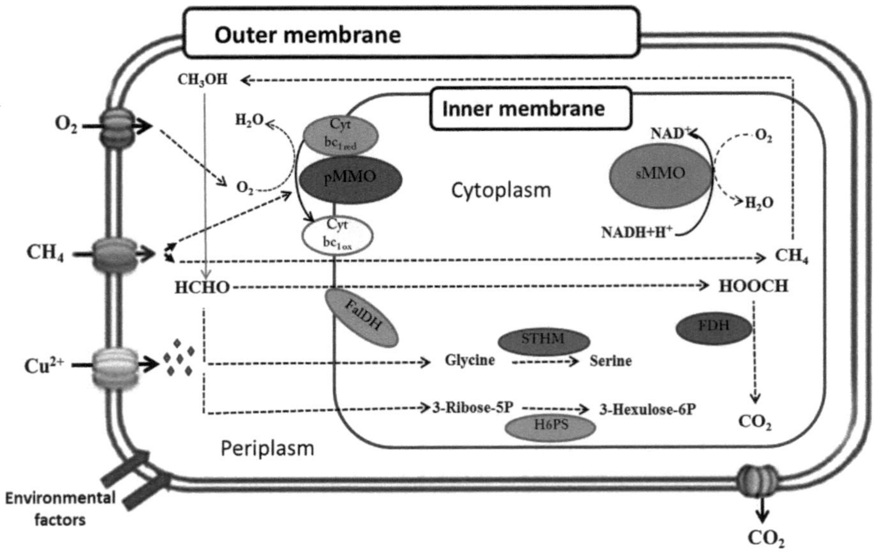

Fig. 1 Methane oxidation pathways in methanotrophic bacteria, where: *sMMO* soluble methane monooxygenase; *pMMO* particulate methane monooxygenase; *FDH* formate dehydrogenase; *FalDH* formaldehyde dehydrogenase; *Cyt red/ox* cytochrome reduced/oxidized; *STHM* serine hydroxymethyltransferase; *H6PS* hexulose-6-phosphate synthase. Adapted from Hanson and Hanson (1996) and Semrau et al. (2010)

25 °C, with an optimum pH range of 7.0–7.65, even though these values can change depending on the specie (Hanson and Hanson 1996; Knief 2015; McDonald et al. 2008) (Table 1). Methane concentrations ranging from 4 to 23% do not significantly influence the cultivation or oxidation efficiency (Bussmann et al. 2004; Cantera et al. 2016a; López et al. 2014). However, an increase in the concentration of methane over 20% leads to higher methanotrophic activity according to Hanson and Hanson (1996). The influence of the O_2 concentration on the cultivation efficiency is quite ambiguous and depends on the source of the inoculum. Bussman et al. (2004) recorded O_2 concentrations of 17% as the optimum values for methane oxidation in lakes, while 0.1–11% resulted in a decrease in methanotrophic growth. However methanotrophs from rice soil fields grew better under O_2 concentrations lower than 1% (Henckel et al. 2000; Jahnke and Nicholst 1986). Similarly, the ratio O_2/CH_4 constitutes an important parameter due to its influence on the type of MMO expressed. At low CH_4 concentrations, methanotrophic communities preferentially express pMMO (low Ks, which entails a higher affinity for the substrate), while at high concentrations of methane sMMO is preferentially expressed (high Ks, lower affinity for CH_4) since it accelerates the catalytic turnover of methane (López et al. 2013; Murrell et al. 2000; Semrau et al. 2010). High ammonium concentrations in the cultivation broth have a negative influence on the cultivation of methanotrophs. This compound is a competitive inhibitor of methane, binding on the active site of pMMO (Carlsen et al. 1991). Magnesium and sulfate inhibit microbial growth at high concentrations (>1 mM) (Bussmann et al. 2004), with the optimal reported magnesium concentration being 50 μM (Bowman et al. 1993). However, the most important nutrient affecting methane degradation is copper, due to its key role on the relative expression of sMMO and pMMO. Cu^{2+} increases the expression of pMMO at levels higher than 0.86 μmol Cu^{2+} g biomass^{-1}, thus enhancing CH_4 biodegradation(Cantera et al. 2016a; Semrau et al. 2010).

Even if aerobic oxidation is the main implemented process in biotechnologies treating methane, recent findings have demonstrated that some anaerobic archaea and bacteria are responsible for 7–25% of the total methane oxidation worldwide. The anaerobic oxidation of methane (AOM) is carried out by bacteria belonging to NC10 phylum (*Candidatus* "*Methylomirabilis oxyfera*") and three groups of archaea: ANME-1 (distantly related to the *Methanosarcinales* and *Methanomicrobiales* group), ANME-2 (within the *Methanosarcinale* groups), and ANME-3 (closely related to the *Methanococcoides* group). Some of them form consortia with sulfate-reducing bacteria (SRB) closely related to the *Desulfosarcina-Desulfococcus* branch of the *Deltaproteobacteria* (Caldwell et al. 2008; Cui et al. 2015). ANME-1 and ANME-2a, 2b and 2c oxidize methane with sulfate as electron acceptor, forming consortia with SRB, while some ANME-2 (ANME-2d) do not need a syntrophic partner and can oxidize methane alone, using nitrate (and nitrite in a minor extent) as electron acceptor. ANME-2d are not known to be able to form consortia with SRB and thus, it is not known if they can use sulfate as electron acceptor in syntrophy. Recently, it has been also proven that ANME-2a and 2d are able to oxidize methane with metals as electron acceptor, also without a syntrophic partner. In the particular case of ANME-3, little is known

Table 1 Main characteristics of methanotrophic bacteria

Phylum	*Gamma-proteobacteria*	*Alpha-proteobacteria*	*Alpha-proteobacteria*	*Verrucomicrobia*
Family	*Methylococcaceae*	*Methylocystaceae*	*Beijerinckaceae*	*Methylacidiphilaceae*
Genera	*Methylococcus*, *Methylobacter*, *Methylocaldum*, *Methylohalobius*, *Methylomicrobium*, *Methylomonas*, *Methylosarcina*, *Methylosphaera*, *Methylothermus*, *Crenothrix*, *Clonothrix*	*Methylosinus*, *Methylocystis*	*Methylocapsa*, *Methylocella*	*Methylacidiphilum*
RuMP pathway	+	–	–	–
Serine pathway	–	+	+	+
Nitrogen fixation	SD	SD	+	SD
pH growth range	4–9	6–7.5	4.2–7.5	0.8–6
Temperature growth range (°C)	0–72 (SD)	23–31	6–32 (SD)	37–65 (SD)
Salt concentration range (% NaCl)	0.1–12 (SD)	0.5	NR	0.8–6.0
sMMO	SD	SD	SD	–
pMMO	+	+	SD	+

about its metabolism (Cui et al. 2015; Timmers et al. 2016). Moreover, some fungal genera such as *Graphium* have been reported as methane oxidizers, the presence of methanol being required for its co-metabolic degradation (Lebrero et al. 2015). However, knowledge on both fields is still scarce and further research is necessary to understand these processes before their implementation as methane treatment technologies.

2.2 *Bioreactors for CH_4 Treatment*

Biological treatment technologies can become, if properly tailored, a low-cost and environmentally friendly alternative for methane abatement compared with their physical-chemical counterparts when energy recovery is not technically or economically feasible. Bioreactors for the treatment of methane have been widely studied over the last 20 years. Several bioreactor configurations have been applied at laboratory-scale, which are classified into two different groups: (i) packed bed bioreactors and (ii) suspended growth bioreactors. Among conventional biotechnologies, biofiltration and biotrickling filtration are by far the most commonly implemented packed bed technologies for methane abatement, while the most

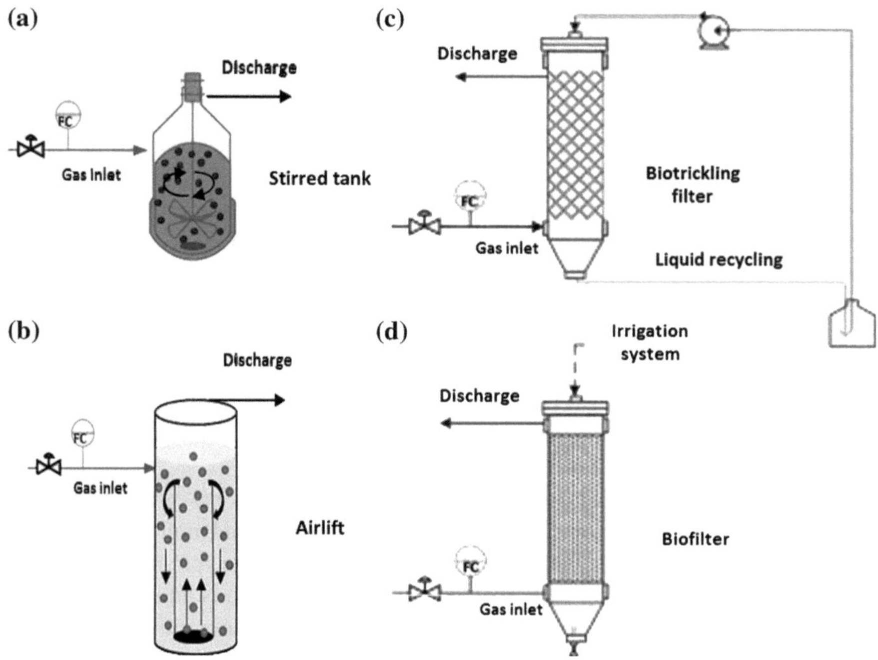

Fig. 2 Main bioreactor configurations applied for CH_4 treatment. **a** Stirred tank. **b** Airlift bioreactor. **c** Biotrickling filter. **d** Biofilter

representative suspended growth reactors are stirred tanks and airlift bioreactors (López et al. 2013). The methane-laden emission is supplied by forced ventilation in either in an upflow or downflow configuration (Fig. 2).

Methanotrophs-based biofilters are likely the most widespread bioreactor configuration, both at laboratory and field scale. Hence, outdoors open biofilters have been extensively applied to reduce CH_4 emissions from landfills, livestock farms or even coal mines (Estrada et al. 2014; Nikiema et al. 2007). In this context, passively vented biofilters are typically installed as landfill covers, where the flow of the CH_4-laden emission naturally occurs upwards, while the O_2 is supplied from the ambient air into the packed material via diffusers (when methane concentrations are low) or passively diffuses from the ambient when explosion hazards are likely to occur (concentrations of methane higher than 5%) (Ménard et al. 2012).

However, the application of GHG treatment biotechnologies is still limited by the poor mass transport of CH_4 from the gas to the microbial community. CH_4 is a hydrophobic volatile organic compound (VOC) with low solubility in the aqueous phase (dimensionless Henry's law constant $H_{CH4} = C_g/C_{aq} = 30$ at 25 °C) where the microbial community is present (Estrada et al. 2012; Kraakman et al. 2011; Muñoz et al. 2007). This low mass transport entails process operation at high empty bed gas residence times (EBRT), which significantly increases both the investment and operating costs of conventional biotechnologies (Estrada et al. 2012). Passively vented biofilters achieve CH_4 removal efficiencies (RE) of 60–90% at EBRTs of 1–14 h, while actively vented advanced biofilters require EBRTs of 8–70 min to reach similar removals (Nikiema et al. 2007; Scheutz et al. 2009). In contrast, conventional biofilters support an efficient odour and/or VOC abatement performance (RE > 90%) at considerably lower EBRTs of 30–120 s, resulting in ×16–1600 lower bioreactor volumes (Muñoz et al. 2012). In order to overcome this limitation and enhance methane transfer to the microbial community, several strategies have been tested in conventional bioreactors.

For instance, traditional packing materials such as compost, perlite or pine bark have been substituted by metallic or polymeric rings, plastic foams or custom-made nutrients containing polymeric pellets (Nikiema et al. 2007). The filter bed (the solid support where the biofilm containing the microorganisms is formed) is a key parameter in packed bed bioreactors since it must offer a high specific surface area to promote biofilm formation and gas transfer. Changes in the filter bed have resulted in reductions of EBRTs from 0.5–10 to 0.05–1.2 h, and higher elimination capacities (ECs), increasing from 25 to 65–280 g m^{-3} h^{-1} (Girard et al. 2011; Nikiema and Heitz 2009; Park et al. 2009).

More recently, two-phase partitioning bioreactors (TPPB), based on the addition to a conventional bioreactor of a non-toxic, non-volatile and non-biodegradable, non-aqueous phase (NAP) with a high affinity for the target pollutant (Daugulis 2001), have been applied for CH_4 abatement. The NAP supports an increase in both CH_4 and O_2 mass transfer from the gas phase to the microorganisms, while buffering the process against high pollutant or metabolite concentrations potentially toxic for the microbial community. Different NAPs have been used, the most common being silicone oil, with an affinity for CH_4 15 times higher than water

(Kraakman et al. 2011; Lebrero et al. 2015; Muñoz et al. 2007). TPPBs have been implemented in different bioreactor configurations for the treatment of methane with promising results. Avalos-Ramirez et al. (2012) reported ECs ranging from 3.9 to 34 g m^{-3} h^{-1} in a biofilter packed with stones and operated with a non-ionic surfactant at an EBRT of 4.25 min, while an EC of 51 g m^{-3} h^{-1} was recorded by Rocha-Ríos et al. (2009) in a two-liquid phase biotrickling filter (BTF) (10% v/v silicone oil) packed with polyurethane foam and operated at an EBRT of 4.8 min and an average inlet load (IL) of 131 g CH_4 m^{-3} h^{-1}. Lebrero et al. (2015) obtained ECs of 45 g m^{-3} h^{-1} at EBRTs of 4 min, while a BTF operated under identical conditions but without silicone oil reached a maximum EC of 35 g m^{-3} h^{-1} (Estrada et al. 2014; Lebrero et al. 2015). Thus, the addition of the NAP resulted in a remarkable improvement in the reactor performance compared to aqueous phase systems. Rocha-Ríos et al. (2009) also tested this innovative biotechnology in stirred tank reactors, achieving a maximum EC of 106 g m^{-3} h^{-1} at an EBRT of 4.8 min using an IL of 187 g m^{-3} h^{-1} in a two phase partitioning stirred tank reactor with 10% (v/v) of silicone oil and an agitation rate of 800 rpm (Rocha-Rios et al. 2009). However, Rocha-Rios et al. (2011) did not report a significant improvement in the performance of a two partitioning airlift bioreactor in the presence of 10% (v/v) silicone oil with a recirculation of 1–2 m^3_{air} $m^{-3}_{reactor}$ min^{-1} and an IL of 170 g m^{-3} h^{-1}, mainly due to the poor dispersion of silicone oil when bacteria were present in the hydrophobic phase (Rocha-Rios et al. 2011). The same result was observed by Cantera et al. (2015) operating a stirred tank reactor with 60% (v/v) silicone oil under high biomass accumulation inside the NAP (Table 2), probably due to a modification of the hydrodynamics by the biomass growth inside the NAP (Cantera et al. 2015).

Fungal-based biofilters and hydrophobic-membrane bioreactors have also emerged as potential biotechnologies for enhanced gas-liquid mass transfer in waste gas treatment, although the number of studies devoted to CH_4 abatement is still scarce. For instance, membrane bioreactors, which promote a direct gas-biofilm contact in the absence of a water layer surrounding the biofilm, have been successfully tested for the treatment of other organic compounds (Lebrero 2013), but only implemented for aerobic methane oxidation coupled to denitrification (Modin et al. 2008; Sun et al. 2013). A fungal-bacterial biofilter using the fungus *Graphium* sp. to co-metabolically biodegrade methane and methanol achieved methane ECs ~ 40 g m^{-3} h^{-1} at an EBRT of 20 min. Although the results of this pioneer study did not support higher ECs than TPPBs, this fungal-bacterial biofilter showed a more stable and enhanced performance compared with previous bacterial biofilters (Lebrero et al. 2016). Other innovative studies have focused on CH_4 mass transfer enhancement by applying complex bioreactor configurations such as horizontal biofilm or tailor flow reactors. However, both approaches have resulted in limited ECs and entailed high investment costs (Kennelly et al. 2014; Rocha-Rios et al. 2013). Therefore, the development of new technologies based on the production of high added value products such as biopolymers, exopolysaccharides and ectoine, combined with methane abatement can be, if properly tailored, the best alternative to make the process cost efficient.

Table 2 Conventional and innovative bioreactors for the treatment of methane

Ref.	Reactor Type	System Characteristics	Inoculum	EBRT (h)	CH_4 Load (g m^{-3} h^{-1})	Max. EC (g m^{-3} h^{-1})
Nikiema and Heitz (2009)	Biofilter	Biofilter packed with inorganic gravel and stones	Microorganisms from a biofilter treating CH_4 for 7 months	0.05–0.3	13–130	65
Park et al. (2009)	Biofilter	Biofilter packed with landfill cover soil mixed with earthworm (60:40, w/w)	Microorganisms from the packing material	0.07–1.2	31–560	280
Rocha-Rios et al. (2009)	Two-partitioning phase stirred tank reactor	Stirred tank reactor operated at 800 rpm with 10% SO200	Enriched methanotrophic consortium from a wastewater treatment plant	0.08	187	106
Rocha-Rios et al. (2009)	Two-partitioning phase biotrickling filter	Biotrickling filter packed with polyurethane foam, recycling rate 4–8 h^{-1}, operated with 10% silicone oil	Enriched methanotrophic consortium from a wastewater treatment plant	0.08	131	51
Rocha-Rios et al. (2011)	Multiphase airlift reactor	Concentric tube airlift reactor with recirculation of 1–2 m^3_{air} $m^{-3}_{reactor}$ min^{-1} and operated with 10% (v/v) silicone oil	Enriched methanotrophic consortium from a waste water treatment plant	0.12	170	22
Girard et al. (2011)	Biofilter	Biofilter with packing gravel (void space 40%)	Not specified	0.07	5–28	14.5
Avalos Ramirez et al. (2012)	Biotrickling filter	Biotrickling filter operated in the presence of a non-ionic surfactant and packed with stones (0.73 cm, specific surface 470 m^2 m^{-3})	Microorganisms from the lixiviate of a biofilter treating CH_4 for 1 year	0.07	62	34
Kennelly et al. (2014)	Horizontal flow biofilm reactor (HFBR)	Horizontal flow biofilm reactor constructed with terram inorganic polymer sheets	A nitrifying activated sludge from a wastewater treatment plant	0.8	8.6	8
Rocha-Rios et al. (2013)	Capillary bioreactor	Capillary bioreactor with Taylor flow formation and silicone oil 10% (v/v)	Enriched methanotrophic consortium from a waste water treatment plant	Not specified	Not specified	93

(continued)

Table 2 (continued)

Ref.	Reactor Type	System Characteristics	Inoculum	EBRT (h)	CH_4 Load (g m^{-3} h^{-1})	Max. EC (g m^{-3} h^{-1})
Estrada et al. (2014)	Biotrickling filter	Biotrickling filter packed with polyurethane foam (PUF)	*M. sporium* culture	0.065	420–450	35
Lebrero et al. (2015)	Two-partitioning phase Biotrickling filter	Biotrickling filter operated with 25% SO200 impregnated in the packing material, polyurethane foam (PUF)	*M. sporium* culture	0.065	420–450	45
Cantera et al. (2015)	Two-phase partitioning stirred tank reactors with hydrophobic biomass	Stirred tank reactor operated at 300 rpm with 60% (v/v) silicone oil	Enriched hydrophobic consortium from wastewater treatment plant and cow manure	0.16	150	54.4
	One-phase partitioning stirred tank reactor	Stirred tank reactor operated at 500 rpm	Enriched hydrophobic consortium from wastewater treatment plant and cow manure	0.16	150	90
Lebrero et al. (2016)	Fungal bacterial biofilter	Biofilter packed with compost	Pure strain of the fungus *Graphium*	0.3	40	37

2.3 Biopolymers from Methane: An Alternative to Petrochemical-Derived Plastics

Polyhydroxyalkanoates (PHAs), such as poly-3-hydroxybutyrate (PHB) and poly-3-hydroxyvalerate (PHV), are linear biopolyesters produced by a broad range of microorganisms, including *Ralstonia eutropha*, *Pseudomonas oleovorans*, *Azotobacter vinelandii*, *Bacillus megaterium*, *Methylobacterium* sp. and *Alcaligenes latus*. PHAs are the only bioplastics completely synthesized by microorganisms and are formed intracellularly under nutrient-limiting conditions (usually N, P, S, O or Mg) through three different sequential steps: (i) CH_4 is introduced into the cell, (ii) conversion of CH_4 into a hydroxyacyl (HA) coenzyme A thioester, and (iii) PHA formation through HA monomers polymerization via ester bond linking (Castilho et al. 2009; Laycock et al. 2014; Nath et al. 2008; Steinbüchel and Valentin 1995). The commercial interest on these added value bioproducts lies on their potential to substitute conventional fossil fuel-based plastics (polyethylene, polypropylene) due to their biocompatibility, biodegradability and their mechanical properties (Chanprateep 2010; Chen 2010a). With over 150 different HA monomers reported, the variable thermal and mechanical properties of PHAs greatly depend on the monomers of composition. Thus, melting temperature, glass-transition temperature and thermodegradation temperature for PHAs can range between 60 and 177 °C, −50 and 4 °C, and 227 and 256 °C, respectively. Similarly, PHAs have the most variable molecular weights (10×10^4–10×10^6 Da), with a polydispersity index ranging from 1.2 to 6.0 (Chen 2010a; Doi et al. 1995; Galego et al. 2000; Harding et al. 2007).

Since 1980s, industrial PHA production has been mainly focused on the production of PHB and the copolymers poly(3-hydroxybutyrate-*co*-3-hydroxyvalerate) (PHBV), poly(3-hydroxybutyrate-*co*-4-hydroxybutyrate) (P3HB4HB), and poly (3-hydroxybutyrate-*co*-3-hydroxyhexanoate) (PHBHx) by at least 24 companies involved in production and R&D, such as Metabolix, Biomer, ICI or Jiangmen Biotech Center. The production at large scale for commercial exploitation in these facilities has resulted in PHA accumulations ranging from 50 to >90% of the cell dry weight, productivities ranging from 400 to 2500 $g_{PHA}\ m^{-3}\ h^{-1}$ and production costs ranging from 4 to 20 € kg_{PHA}^{-1}, depending on the type of biopolymer, carbon source, microbial strain and product purity (Bugnicourt et al. 2014; Chen 2009, 2010b; Shen et al. 2009). Despite the significant decrease in PHA market prize in the past 5 years, it is generally higher than those of bio-based polyesters due to the costs associated to PHA downstream and raw material adquisition, which accounts for 30–40% of the total costs (Bugnicourt et al. 2014; Chanprateep 2010). In this regard, an intensive research evaluating the use of waste and surplus materials such as cheese whey, beet molasses, glycerol or sugarcane liquour as feedstock for PHA production has been carried out during the past 10 years to reduce PHA production costs (Koller et al. 2010; Nath et al. 2008).

The methane contained in waste gas emissions from old landfills, coal mines, etc., or in the biogas produced from the anaerobic degradation of organic waste, has recently emerged as a potential feedstock for PHA production. The use of residual methane as a C source will significantly decrease PHAs production costs while contributing to reduce the environmental impacts of GHG emissions (Rostkowski et al. 2012; Strong et al. 2015). Type II methanotrophs (*Methylocystis*, *Methylosinus* and *Methylocella* genera) are the only methanotrophs with a demonstrated metabolic capability to produce PHAs through the so-called serine pathway (Pieja et al. 2011a) (Fig. 3). When CH_4 is used as the sole C and energy source, PHB of high molecular weight is the only PHA synthesized under nutrient-limited growth (N-, P-, Mg- or S-limiting conditions) (Helm et al. 2008; Myung et al. 2015a; Pieja et al. 2012; Wendlandt et al. 2001) (Table 3). The production of PHB from biogas or natural gas has been successfully reported (Myung et al. 2016; Van Der Ha et al. 2012). However, although PHB is the most commonly produced PHA, market applications are limited by its low thermal stability, slow crystallization, narrow melt processing window and the lack of flexibility. These limitations have boosted recent research towards the use of co-substrates (citric acid, *n*-propanol, valerate, propionate) to both increase the PHA yields and produce copolymers such as PHBV with superior properties (Cal et al. 2016; Myung et al. 2015b; Zhang et al. 2008) (Table 3). Indeed, Myung et al. (2016) recently evaluated the use of several ω-hydroxyalkanoates as co-substrates for the production of PHA copolymers (PHBV, P3HB4HB, poly (3-hydroxybutyrate-*co*-6-hydroxyhexanoate-*co*-4-hydroxybutyrate) (P3HB6HHx4HB)), which not only demonstrated the possibility of using fatty acids for PHA production, but also the capability of methanotrophic bacteria to produce a wider type of biopolymers (Fig. 3).

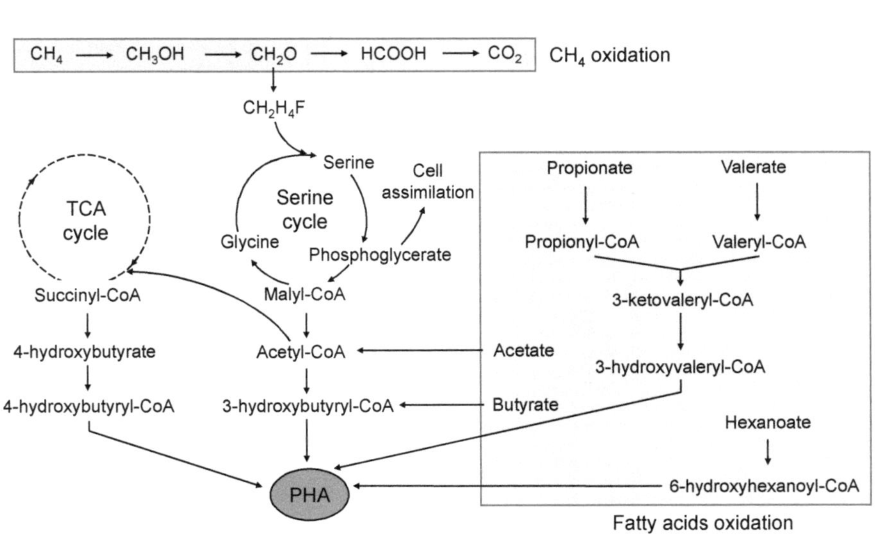

Fig. 3 PHA production pathway in obligate type II methanotrophs. *Dotted arrows* indicate the existence of intermediates not-mentioned in the figure. Adapted from Chen (2010a), Pieja et al. (2011b) Myung et al. (2015b) and Myung (2016)

Table 3 Comparison of the most significant PHA production studies by obligate methanotrophs reported to date

Ref.	System	Methanotrophic strain	C_1 source	Nutrient culture conditions[a]	Cosubstrate(s)	Maximum PHA accumulation[b] (% wt)	Other PHA characteristics[c]
Wendlandt et al. (2001)	STR[d]	*Methylocystis* sp. GB 25	CH_4	N/P/Mg limitation (growth phase on both NO_3^- and NH_4^+)	–	51/4/28	$M_w^e = 1.0–2.5 \times 10^6$ Da
Zhang et al. (2008)	Serum bottles	*Methylosinus trichosporium* IMV3011	CH_4 and CH_3OH	N limitation (Cu, Fe, Mg and P concentrations optimized; growth phase on both NO_3^- and NH_4^+)	Citric acid	No citric acid: 26; with citric acid: 38	$M_w = 1.5 \times 10^6$ Da
Helm et al. (2008)	STR	*Methylocystis* sp. GB 25 (dominant in a mixed culture)	CH_4	K/S/Fe limitation	–	33/32/10	–
Pieja et al. (2011a)	Serum bottles	*Methylocystis*/ *Methylosinus* species	CH_4	N limitation	–	9–44	–
Pieja et al. (2011b)	Serum bottles	*Methylocystis parvus* OBBP	CH_4	N limitation	–	50	–
Pieja et al. (2012)	SBR[f]	*Methylocystis* (*parvus* OBBP, 42/22, KS30) and *Methylosinus* sp. LW4 consortia	CH_4	N/N + O_2/N + CH_4 limitation	–	20/22/25	–

(continued)

Table 3 (continued)

Ref.	System	Methanotrophic strain	C_1 source	Nutrient culture conditions[a]	Cosubstrate(s)	Maximum PHA accumulation[b] (% wt)	Other PHA characteristics[c]
Rahnama et al. (2012)	BCB[g]/ VTLB[h]	*Methylocystis hirsuta*	CH_4 (natural gas)	N limitation	–	42/51	–
Van der Ha et al. (2012)	Serum bottles	*Methylocystis parvus*	CH_4 (biogas)	N limitation	–	30	–
Rostkowski et al. (2012)	Serum bottles	*Methylosinus trichosporium* OB3b/ *Methylocystis parvus* OBBP	CH_4	N limitation (growth phase on N_2 and NH_4^+, respectively)	–	45/60	–
Myung et al. (2015a)	Serum bottles	*Methylocystis* sp. (dominant in a mixed culture)	CH_4	N limitation (growth phase on NO_2^- and NH_4^+)	–	40	–
Myung et al. (2015b)	Serum bottles	*Methylocystis* sp. (dominant in a mixed culture)	CH_4	N limitation (growth phase on NH_4^+)	Valerate (low/high concentrations)	Low valerate: 43; high valerate: 30	Low valerate: 20% mol HV, $M_w = 1.2 \times 10^6$ Da; high valerate: 40% mol HV, $M_w = 9.3 \times 10^5$ Da
Sundstrom and Criddle (2015)	96-well microplates	*Methylocystis parvus* OBBP	CH_4	N limitation (growth phase on NO_3^- or NH_4^+, optimization of Cu, Ca, K and P levels)	–	49	–

(continued)

Table 3 (continued)

Ref.	System	Methanotrophic strain	C_1 source	Nutrient culture conditions[a]	Cosubstrate(s)	Maximum PHA accumulation[b] (% wt)	Other PHA characteristics[c]
Zhang et al. (2015)	Serum bottles	*Methylosinus trichosporium* OB3b	CH_4	N limitation (growth phase on NO_3^-/NH_4^+/N_2)	–	51/45/32	–
Cal et al. (2016)	Serum bottles	*Methylocystis* sp. WRRC1	CH_4	N/N + Cu limitation	*n*-pentanol/valerate	Pentanol: 41; valerate: 60 (with Cu) or 78 (no Cu)	Pentanol: 54% mol HV; valerate: 50% mol HV (with Cu) or 60% mol HV (no Cu)
Myung et al. (2016)	Serum bottles	*Methylocystis parvus* OBBP	CH_4	N limitation (growth phase on NH_4^+)	3-hydroxybutyrate/propionate/valerate	3-hydroxybutyrate: 60; propionate: 32; valerate: 60	propionate: 10% mol HV; valerate: 60% mol HV, $M_w = 1.7–3.2 \times 10^6$ Da

[a]Unless specified otherwise, the nutrient limitation was induced after a growth phase with NO_3^- as N source
[b]Unless specified otherwise, the PHA produced was P3HB (composed exclusively by 3HB monomers)
[c]Unless specified otherwise, the entirely monomers (or the remaining up to the 100% mol HA) constituting the PHA are 3HB
[d]STR: stirred tank reactor
[e]Mw: molecular weight
[f]SBR: sequencing batch reactor
[g]BCB: bubble column bioreactor
[h]VTLB: vertical tubular loop bioreactor

2.4 Production of Exopolysaccharides from CH_4

Exopolysaccharides (EPS) comprise a wide variety of polysaccharides, proteins, glycoproteins and glycolipids that provide mechanical stability to biofilms, mediate their adhesion to surfaces and form a cohesive, three-dimensional polymer network that interconnects and transiently immobilizes biofilm cells. EPS are produced by archaeal, bacterial (lactic acid bacteria such as *Leuconostoc mesenteroides*, *Lactobacillus brevis*) such as other strains *Pseudomonas aeruginosa* and *Xanthomonas campestris*) and even eukaryotic microbes (Flemming et al. 2007; Flemming and Wingender 2010; Matsumoto et al. 2014). Microorganisms generally excrete EPS under unbalanced nutrient conditions (high C/N ratios) as a nutrient or water reservoir, or as a protective barrier. These bioproducts are of interest due to their colloidal and adhesive properties, and their effects on liquid rheology, thus being applied in the food, pharmaceutical, textile and oil industries. The molecular weight and viscosity of EPS greatly vary from 1×10^4 to 1×10^9 Da and 2–1000 cP, respectively, according to the culture conditions, the strain used or the EPS type (Morris and Harding 2009; Nwodo et al. 2012; Sarwat et al. 2008).

Microbial EPS production by companies such as CPKelco, Merck, Pfizer and Prasinotech Ltd is up to date mainly focused on polysaccharides such as xanthan (4–13 € kg^{-1}), dextran (30–50 € kg^{-1}), hyaluronan, alginate or gellan. However, the limited productivities and high costs derived from EPS downstream and the cost of the C substrates still hinder their industrial production (Garcia-Ochoa et al. 2000; Morris and Harding 2009; Strong et al. 2015). Thus, the use of CH_4 as a feedstock for the production of this bioproduct has been recently evaluated as an alternative to reduce costs. In this regard, EPS production has been reported in both types of methanotrophs (I and II), though several studies suggested that the highest EPS-synthethic activity occurs in the ribulose monophosphate pathway (RuMP) of type I methanotrophs such as *Methylomonas* and *Methylobacter* (Malashenko et al. 2001; Wilshusen et al. 2004b). Among the parameters determining methanotrophic EPS production, high O_2 concentrations seem to positively influence their excretion, while EPS accumulation around the cells may inhibit further synthesis under gas diffusion limitations (Chiemchaisri et al. 2001; Hilger et al. 2000; Wei et al. 2015; Wilshusen et al. 2004a). Moreover, Hernández et al. (2015) recently confirmed that low pH and N concentrations are stress-promoting conditions that enhance the production of both protein- and polysaccharide-type EPS by type I methanotrophs in a biofilter treating CH_4 (Hernández et al. 2015).

2.5 Production of Ectoine from CH_4

Ectoine (1,4,5,6-tetrahydro-2-methyl-4-pyrimidinecarboxylic acid) is a compatible solute (water soluble with a low molecular weight and zwitterionic) that provides osmotic balance to a wide number of halotolerant bacteria, without interfering with their essential processes and cellular metabolism (Pastor et al. 2010). Due to its high effectiveness as stabilizer of enzymes, DNA-protein complexes and nucleic acids, ectoine is probably one of the most valuable products synthesized by microorganisms, retailing in the pharmaceutical industry at approximately US\$1000 kg^{-1} and with a global consumption of 15000 tones $year^{-1}$ (Strong et al. 2016).

Bacterial processes for the production of ectoine have received an increasing attention in the last two decades due to their easier and highly specific product synthesis in comparison with chemical processes (Pastor et al. 2010). However, while the increasing commercial demand for ectoine has promoted research on new bacterial methods and strains, nowadays ectoine is only industrially produced by the γ-*Proteobacteria Halomonas elongata*. This strain can be reused up to 9 times (exhibiting ectoine accumulations of $\sim$15.5% g g^{-1} biomass), has a broad salt tolerance and is able to rapidly synthesize and excrete ectoine to the medium (Galinski and Trüper 1994; Sauer and Galinski 1998). The industrial ectoine production process *-bio-milking-* consists of a long fed-batch fermentation (120 h) with two steps at different salt concentrations (12 and 0%), to obtain first a high culture density (25 g L^{-1}) and subsequently induce a hypoosmotic shock (Sauer and Galinski 1998). The sudden decrease in media salinity results in the excretion of ectoine from the cell to the culture broth, where the product is collected for its downstream purification. Nevertheless, this upstream processing is still inefficient due to the high amount of nutrients, oxygen and time required, besides entailing a complex and expensive downstream processing (Lang et al. 2011; Zhang et al. 2009). These limitations represent a challenge to its commercial large-scale production. Thus, current research efforts are focused on the use of strains able to grow in inexpensive substrates and to excrete ectoine to the medium in order to increase the cost-competitiveness of the process.

In 1997, Khemelenina and coworkers isolated and characterized the first halotolerant alkaliphilic methanotrophic strain from soda lakes, *Methylomicrobium alcaliphilum 20Z*. Further studies demonstrated that this bacterium was able to express the three specific enzymes in the ectoine pathway used by other halotolerant heterotrophic bacteria (EctA, EctB and EctC) (Khmelenina et al. 1997; Reshetnikov et al. 2005), and that the environmental conditions were crucial for ectoine production. For instance, But et al. (2012) observed a 4 times increase in the concentration of intra-cellular ectoine when salt concentration was increased from 3 to 7%, while Cantera et al. (2016b) observed that salinities higher than 8% damaged the cell culture of *Methylomicrobium alcaliphilum 20Z*, resulting in a lower ectoine production (But et al. 2012; Cantera et al. 2016b). Moreover, Khemelenina et al. (2000) demonstrated that nitrogen was a key factor to obtain high ectoine yields, while Cantera et al. (2016b) observed that high Cu^{2+} concentrations (25–50 μM) promoted the excretion of ectoine to the cultivation medium, with the subsequent enhancement in ectoin downstream (Cantera et al. 2016b; Khmelenina et al. 2000). The promising results obtained hitherto with methanotrophic bacteria encourage further research in order to implement systems capable of creating value out of GHG mitigation. Finally, the search for different methanotrophic extremophiles able to produce ectoine combined with high methane removal rates will be of major interest for the optimization of CH_4 bio-refinery concepts.

3 Biotechnologies for Nitrous Oxide Abatement and Valuable Bioproducts Synthesis

N_2O emissions are largely produced in activities such as agriculture, when nitrogen is added as a fertilizer in soils, or during organic waste and wastewater treatment processes by the transformation of nitrogen by microorganisms through nitrification and denitrification. In nitrification, ammonia (NH_3) is first oxidized to nitrite (NO_2^-) by ammonia oxidizing bacteria and then to nitrate (NO_3^-) by nitrite oxidizing bacteria. N_2O is generated from the chemical decomposition of hydroxylamine (NH_2OH), an intermediate in the first stage of nitrification, in a reaction called chemodenitrification (Wunderlin et al. 2012). On the other hand, denitrification involves the reduction of NO_3^- by heterotrophic denitrifiers to N_2 through sequential reductions to NO_2^-, nitric oxide (NO) and N_2O as intermediates. Nitric and adipic acid production plants represent another important source of N_2O, which is emitted in the oxidation (combustion) stage of ammonia (Pérez-Ramírez et al. 2003). Despite the active abatement of N_2O emissions from agriculture is hindered by the difficult collection of the emission from such a passive source, N_2O control units can be implemented in active sources like industries and wastewater treatment plants.

Biological technologies for N_2O biodegradation can be based on three potential biological mechanisms as proposed by Frutos et al. (2014): nitrification, assimilation and denitrification, although only denitrification has been hitherto applied in laboratory-scale bioreactors. In this sense, the oxidation of N_2O to NO_2^- or NO_3^- could be theoretically conducted by nitrifying bacteria under sufficient inorganic carbon, oxygen and an optimal pH based on the negative Gibbs free energy of the reactions of N_2O oxidation to NO_2^- or NO_3^- (Eqs. 1 and 2). However, there is no experimental proof confirming the feasibility of this biodegradation pathway (Frutos et al. 2014).

$$N_2O + O_2 + H_2O \rightarrow 2NO_2^- + 2H^+ \left(\Delta G^{0'} = -87.4\ \text{kJ}\right) \quad (1)$$

$$N_2O + 2O_2 + H_2O \rightarrow 2NO_3^- + 2H^+ \left(\Delta G^{0'} = -15\ \text{kJ}\right) \quad (2)$$

Likewise, an assimilation pathway where N_2O is transformed to organic nitrogen to form the building blocks of proteins, similarly to the N_2 assimilation process, has been also proposed (Burgess and Lowe 1996; Desloover et al. 2014; Vieten et al. 2008). Unfortunately, the feasibility of this mechanism has not been empirically proven yet. Denitrification is the unique biological mechanism known to be able to degrade this GHG through its reduction to N_2 using an electron donor under the appropriate conditions. This is a strongly exergonic reaction (Eq. 3), which is conducted by the nitrous oxide reductase enzyme, a dimeric multicopper protein present in most denitrifiers that catalyzes the transfer of two electrons from the electron donor to N_2O (Zumft 1997).

$$N_2O + 2H^+ + 2e^- \rightarrow N_2 + H_2O\,(\Delta G^{0'} = -341\ \text{kJ}) \tag{3}$$

Therefore, the application of biotechnologies for the abatement of N_2O is limited to the denitrification process. Thus, when the emission contains large amounts of oxygen such as those from wastewater treatment plants, a costly prior depletion of the transferred oxygen through an external organic carbon supply is required to maintain anoxic conditions in the bioreactor. N_2O denitrification has been implemented in conventional bioreactors like biofilters, biotrickling filters or bioscrubbers at laboratory-scale, although the number of studies found in the literature is still very scarce. Frutos et al. (2014, 2016b) have recently proposed a bioscrubber for the continuous abatement of N_2O ($\approx$100 ppm_v) emitted from wastewater treatment plants (Fig. 4a) (Frutos et al. 2014; Frutos et al. 2016a). The bioscrubber was a two-stage process where the N_2O was first transferred from the gas to the recycling liquid phase in a packed absorption column and then reduced in a stirred tank reactor (STR) maintained under anoxic conditions by the supply of a low-cost organic carbon such as methanol or domestic wastewater. The liquid phase of the STR was recycled to the absorption column at different liquid recycling velocities to improve the mass transfer of this GHG. N_2O REs of ~40% were recorded using methanol as the carbon source at an EBRT of 3 min and liquid recycling velocities of 8 m h^{-1}. The authors reported RE > 90% when the gas EBRT was increased up to 40 min at same liquid recycling velocity but using the organic matter present in synthetic wastewater as the electron donor. The abatement of N_2O from swine house pit exhaust (NH_3, CH_4 and 0.38–0.69 ppm_v of N_2O) was evaluated by Hood (2011) in a biofilter packed with compost and woodchips (30/70%) (Hood et al. 2011). High removals of NH_3 and CH_4 were recorded, but only 14–17% of the N_2O was removed at an EBRT of 7.6 s. Likewise, Akdeniz et al. (2011) evaluated the performance of a biofilter packed with pine nuggets and lava rock at a gas EBRT of 5 s and a relative humidity of 90% for the treatment of exhaust gases from a swine manure and wastewater storage pit (Akdeniz et al. 2011). The pollutants studied were hydrogen sulfide (H_2S), total reduced sulfur (TRS), NH_3, CH_4, N_2O and other odorous compounds. The authors observed a poor N_2O removal efficiency (~0.7%) at inlet concentrations of 428 ± 22 ppb_v. This was attributed to both the presence of oxygen in the waste gas and to the low gas EBRT, which hindered an efficient mass transfer of this poorly water soluble gas pollutant.

In the context of industrial N_2O emission abatement, Frutos et al. (2016a) evaluated a bubble column and an airlift bioreactor for the abatement of a synthetic nitric acid emission containing $\approx$3500 ppm_v of N_2O and 1.1% of O_2 using a *Paracoccus denitrificans* strain (DMS 413) (Frutos et al. 2016b). Removal efficiencies close to 90% were supported by both bioreactors. The authors proposed

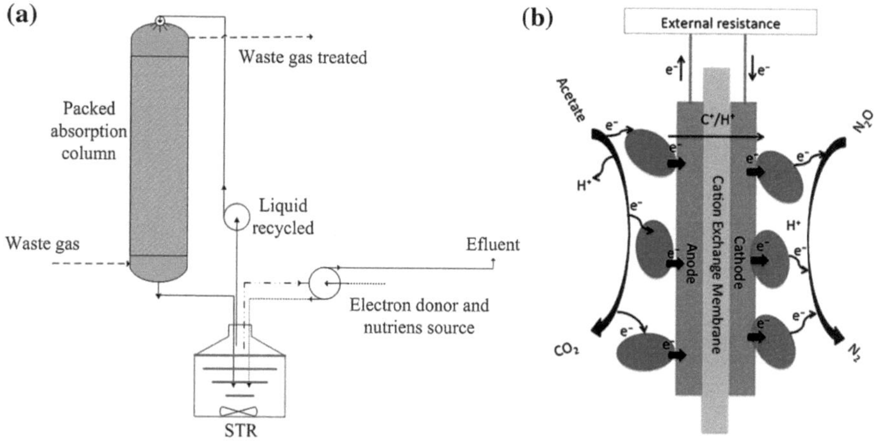

Fig. 4 Schematic representation of a bioscrubber (**a**) and a bioelectrochemical system (**b**) for N_2O removal as reported in Frutos et al. (2014, 2016b) and Desloover et al. (2011)

these biotechnologies as low-cost alternatives due to their low energy consumption (pneumatically agitated bioreactors) and their simple configuration and operation compared to conventional non-selective catalytic reduction or catalytic decomposition. However, the high gas EBRT (>15 min) required to obtain a good mass transfer of the pollutant due to the low solubility of N_2O in water (dimensionless Henry constant of 1.6 at 25 °C) might result in high reactor volumes with the corresponding increase in investment costs. In addition to these applications in conventional bioreactors, Desloover et al. (2011) designed a novel bioelectrochemical system (Fig. 4b) with an autotrophic denitrifying biocathode for the removal of N_2O. In this system, sodium acetate was used as the electron donor and NO_3^- as the cathodic electron acceptor in a first stage for the enrichment of a denitrifying culture, changing then to N_2O as the sole electron acceptor in the cathode compartment in a second stage, which resulted in an almost complete N_2O removal. However, and despite of the promising results obtained in these suspended growth bioreactors, the evaluation of their sensitiveness to O_2 concentration is crucial for the development of cost-efficient and robust processes since industrial N_2O emissions usually contain trace level of oxygen (0–4%).

While N_2O biodegradation has been scarcely studied, to the best of the authors' knowledge there are no studies devoted to the evaluation of the synthesis of added value bioproducts combined with N_2O removal. In this context, several studies have highlighted the ability of some microorganisms to produce chemicals of industrial interest, such as PHA, pigments or ubiquinone (Q10 coenzyme), using N_2O as electron acceptor (Beun et al. 2000; de Dieu Ndikubwimana and Lee 2014; Kim et al. 2008; Kirti et al. 2014; Yoshida et al. 1998). Thus, innovative operational strategies could be developed in suspended growth bioreactors to promote the co-production of added value commodities that would significantly improve the cost-competitiveness of N_2O removal biotechnologies. For instance, a recent study

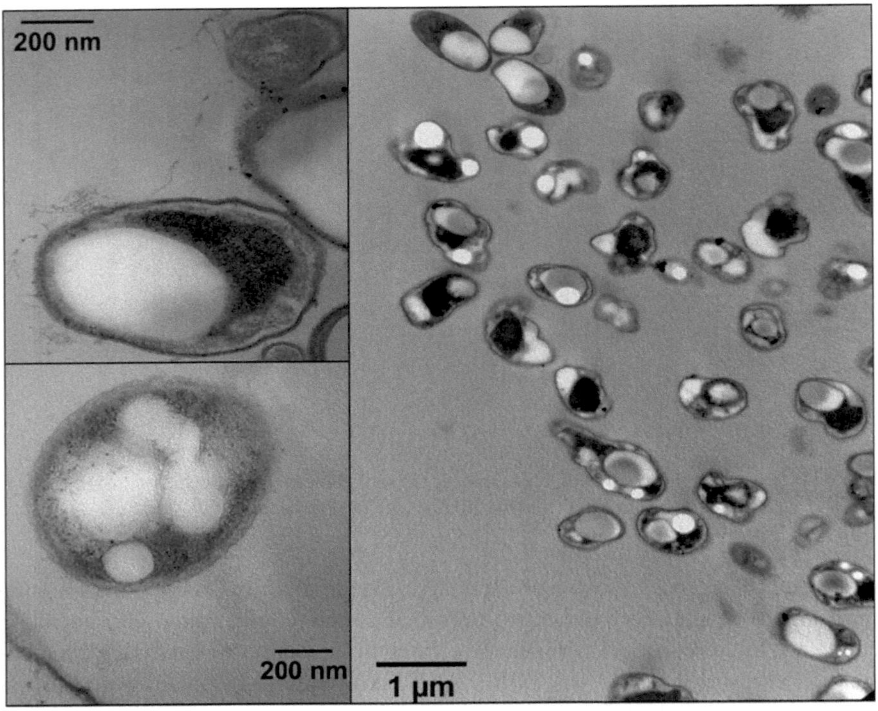

Fig. 5 Photographs from the transmission electron micrograph of *Paracoccus denitrificans* containing PHBV (Frutos et al., data not published)

conducted by Frutos et al. (data not published) demonstrated the potential of the simultaneous N_2O removal and PHBV production in airlift and bubble column bioreactors using *Paracoccus denitrificans*. The authors evaluated the effect of periodic nitrogen limitations using methanol as the carbon source and electron donor, obtaining high N_2O REs ~87% with PHBV accumulations of 38–64% (on a dry weight basis) (Fig. 5). Nevertheless, further studies are required in order to assess the effect on the type and amount of biopolymers produced of different factors such as reactor design, carbon source, oxygen concentration, microbial strain and nutrient limitations.

4 Conclusions

The mitigation of global warming via an active abatement of greenhouse gases such as CH_4 and N_2O represents nowadays a huge technological and economic challenge that must be overcome by most nations in order to comply with the recent international compromises achieved in the COP 21 United Nations Conference of

Climate Change. The high operating costs and environmental impacts of conventional physical-chemical off-gas treatment technologies requires the development of a new generation of high mass-transfer bioreactors operated with customized microbial communities capable of bioconverting these GHGs into bioproducts of commercial interest. The development of GHG biorefineries supplying both commodities and fine chemicals from CH_4 and N_2O bioconversion constitutes a promising approach to turn climate change mitigation into a profitable business, but requires an intensive research effort in fields such as microbiology, systems biology and environmental engineering.

Acknowledgements This work was supported by the Spanish Ministry of Economy and Competitiveness and the European Union through the project CTM2015-70.442-R (Retos and FEDER Programs).

References

Akdeniz, N., Janni, K. A., & Salnikov, I. A. (2011). Biofilter performance of pine nuggets and lava rock as media. *Bioresource Technology, 102*(8), 4974–4980. doi:10.1016/j.biortech.2011.01.058

Auman, A. J., & Lidstrom, M. E. (2002). Analysis of sMMO-containing type I methanotrophs in Lake Washington sediment. *Environmental Microbiology, 4*(9), 517–524. doi:10.1046/j.1462-2920.2002.00323.x

Avalos Ramirez, A., García-Aguilar, B. P., Jones, J. P., & Heitz, M. (2012). Improvement of methane biofiltration by the addition of non-ionic surfactants to biofilters packed with inert materials. *Process Biochemistry, 47*(1), 76–82. doi:10.1016/j.procbio.2011.10.007

Beun, J. J., Verhoef, E. V, Van Loosdrecht, M. C. M., & Heijnen, J. J. (2000). Stoichiometry and kinetics of poly-beta-hydroxybutyrate metabolism under denitrifying conditions in activated sludge cultures. *Biotechnology and Bioengineering, 68*(5), 496–507. doi:10.1002/(SICI)1097-0290(20000220)67:4<379::AID-BIT1>3.0.CO;2-2

Bowman, J. P., Jimenez, L., Rosario, I., Hazen, T. C., & Sayler, G. S. (1993). Characterization of the methanotrophic bacterial community present in a trichloroethylene-contaminated subsurface groundwater site. *Applied and Environmental Microbiology, 59*(8), 2380–2387.

Bugnicourt, E., Cinelli, P., Lazzeri, A., & Alvarez, V. (2014). Polyhydroxyalkanoate (PHA): Review of synthesis, characteristics, processing and potential applications in packaging. *Express Polymer Letters, 8*(11), 791–808. doi:10.3144/expresspolymlett.2014.82.

Burgess, B. K., & Lowe, D. J. (1996). Mechanism of molybdenum nitrogenase. *Chemical Reviews, 96*(7), 2983–3012. doi:10.1021/cr950055x

Bussmann, I., Pester, M., Brune, A., & Schink, B. (2004). Preferential cultivation of type II methanotrophic bacteria from littoral sediments (lake constance). *FEMS Microbiology Ecology, 47*(2), 179 LP-189. Retrieved from http://femsec.oxfordjournals.org/content/47/2/179.abstract

But, S. Y., Rozova, O. N., Khmelenina, V. N., Reshetnikov, A. S., & Trotsenko, Y. A. (2012). Properties of recombinant ATP-dependent fructokinase from the halotolerant methanotroph Methylomicrobium alcaliphilum 20Z. *Biochemistry (Moscow), 77*(4), 372–377. doi:10.1134/S0006297912040086

Cal, A. J., Sikkema, W. D., Ponce, M. I., Franqui-Villanueva, D., Riiff, T. J., Orts, W. J., & Lee, C. C. et al. (2016). Methanotrophic production of polyhydroxybutyrate-co-hydroxyvalerate with high hydroxyvalerate content. *International Journal of Biological Macromolecules, 87,* 302–307. doi:10.1016/j.ijbiomac.2016.02.056

Caldwell, S. L., Laidler, J. R., Brewer, E. A., Eberly, J. O., Sandborgh, S. C., & Colwell, F. S. (2008). Anaerobic oxidation of methane: Mechanisms, bioenergetics, and the ecology of associated microorganisms. *Environmental Science and Technology*. doi:10.1021/es800120b

Cantera, S., Estrada, J. M., Lebrero, R., García-Encina, P. A., & Muñoz, R. (2015). Comparative performance evaluation of conventional and two-phase hydrophobic stirred tank reactors for methane abatement: Mass transfer and biological considerations. *Biotechnology and Bioengineering, 113*(6), 1203–1212. doi:10.1002/bit.25897.

Cantera, S., Lebrero, R., García-Encina, P. A., & Muñoz, R. (2016a). Evaluation of the influence of methane and copper concentration and methane mass transport on the community structure and biodegradation kinetics of methanotrophic cultures. *Journal of Environmental Management, 171*, 11–20.

Cantera, S., Lebrero, R., Sadornil, L., García-Encina, P. A., & Muñoz, R. (2016b). Valorization of CH_4 emissions into high-added-value products: Assessing the production of ectoine coupled with CH_4 abatement. *Journal of Environmental Management, 182*, 160–165.

Carlsen, H. N., Joergensen, L., & Degn, H. (1991). Inhibition by ammonia of methane utilization in *Methylococcus capsulatus* (Bath). *Applied Microbiology and Biotechnology, 35*(1), 124–127. doi:10.1007/BF00180649

Castilho, L. R., Mitchell, D. A., & Freire, D. M. G. (2009). Production of polyhydroxyalkanoates (PHAs) from waste materials and by-products by submerged and solid-state fermentation. *Bioresource Technology, 100*(23), 5996–6009. doi:10.1016/j.biortech.2009.03.088

Chanprateep, S. (2010). Current trends in biodegradable polyhydroxyalkanoates. *Journal of Bioscience and Bioengineering, 110*(6), 621–632. doi:10.1016/j.jbiosc.2010.07.014

Chen, G.-Q. (2009). A microbial polyhydroxyalkanoates (PHA) based bio- and materials industry. *Chemical Society Reviews, 38*(8), 2434–2446. doi:10.1039/b812677c

Chen, G.-Q. (2010a). Plastics completely synthesized by bacteria: Polyhydroxyalkanoates. In G. Q. Chen (Ed.), *Plastics from Bacteria: Natural Functions and Applications* (1st edn, pp. 17–38). Heidelberg: Springer.

Chen, G.-Q. (2010b). Plastics completely synthesized by bacteria: Polyhydroxyalkanoates. *Plastics from Bacteria: Natural Functions and Applications, 14*, 121–132. doi:10.1007/978-3-642-03287_5

Chiemchaisri, W., Wu, J. S., & Visvanathan, C. (2001). Methanotrophic production of extracellular polysaccharide in landfill cover soils. *Water Science and Technology, 43*, 151–158.

Cui, M., Ma, A., Qi, H., Zhuang, X., & Zhuang, G. (2015). Anaerobic oxidation of methane: An "active" microbial process. *MicrobiologyOpen, 4*(1), 1–11. doi:10.1002/mbo3.232

Daugulis, A. J. (2001). Two-phase partitioning bioreactors: A new technology platform for destroying xenobiotics. *Trends in Biotechnology, 19*(11), 457–462. doi:10.1016/S0167-7799 (01)01789-9

de Dieu Ndikubwimana, J. & Lee, B. H. (2014). Enhanced production techniques, properties and uses of coenzyme Q10. *Biotechnology Letters, 36*(10), 1917–1926. doi:10.1007/S10529-014-1587-1

den Camp, H. J. M. O., Islam, T., Stott, M. B., Harhangi, H. R., Hynes, A., Schouten, S., … Dunfield, P. F. (2009). Minireview environmental, genomic and taxonomic perspectives on methanotrophic Verrucomicrobia. *Environmental Microbiology Reports, 1*(5), 293–306. doi:10.1111/j.1758-2229.2009.00022.x

Desloover, J., Puig, S., Virdis, B., Clauwaert, P., Boeckx, P., Verstraete, W., & Boon, N. (2011). Biocathodic nitrous oxide removal in bioelectrochemical systems. *Environmental Science & Technology, 45*(24), 10557–10566. doi:10.1021/es202047x

Desloover, J., Roobroeck, D., Heylen, K., Puig, S. S., Boeckx, P., Verstraete, W., & Boon, N. (2014). Pathway of nitrous oxide consumption in isolated *Pseudomonas stutzeri* strains under anoxic and oxic conditions. *Environmental Microbiology, 16*(10), 3143–3152. doi:10.1111/1462-2920.12404

Doi, Y., Kitamura, S., & Abe, H. (1995). Microbial synthesis and characterization of poly (3-hydroxybutyrate-co-3-hydroxyhexanoate). *Macromolecules, 28*(14), 4822–4828. doi:10.1021/ma00118a007

Estrada, J. M., Lebrero, R., Quijano, G., Pérez, R., Figueroa-González, I., García-Encina, P. A., & Muñoz, R. (2014). Methane abatement in a gas-recycling biotrickling filter: Evaluating innovative operational strategies to overcome mass transfer limitations. *Chemical Engineering Journal, 253*, 385–393. doi:10.1016/j.cej.2014.05.053

Estrada, J. M., Rodríguez, E., Quijano, G., & Muñoz, R. (2012). Influence of gaseous VOC concentration on the diversity and biodegradation performance of microbial communities. *Bioprocess and Biosystems Engineering, 35*(9), 1477–1488. doi:10.1007/S00449-012-0737-x

European Environment Agency. (2015). European environment—State and outlook 2015: Assessment of global megatrends. *European Environment*. doi:10.2800/45773

Flemming, H. C., Neu, T. R., & Wozniak, D. J. (2007). The EPS matrix: The "house of biofilm cells." *Journal of Bacteriology, 189*(22), 7945–7947. doi:10.1128/JB.00858-07

Flemming, H., & Wingender, J. (2010). The biofilm matrix. *Nature Reviews Microbiology, 8*(9), 623–633. doi:10.1038/nrmicro2415

Frutos, O. D., Arvelo, I. A., Pérez, R., Quijano, G., & Muñoz, R. (2014). Continuous nitrous oxide abatement in a novel denitrifying off-gas bioscrubber. *Applied Microbiology and Biotechnology, 99*(8), 3695–3706. doi:10.1007/s00253-014-6329-8

Frutos, O. D., Cortes, I., Arnaiz, E., Lebrero, R., & Muñoz, R. (2016a). Biological nitrous oxide abatement by paracoccus denitrificans in bubble column and airlift reactors, *54*, 289–294. doi:10.3303/CET1654049

Frutos, O. D., Quijano, G., Pérez, R., & Muñoz, R. (2016b). Simultaneous biological nitrous oxide abatement and wastewater treatment in a denitrifying off-gas bioscrubber. *Chemical Engineering Journal, 288*, 28–37. doi:10.1016/j.cej.2015.11.088

Frutos OD, Cortes I, Cantera S, Arnaiz E, Lebrero R, Muñoz R (Under review). Nitrous oxide abatement coupled to biopolymer production as a model GHG biorefinery for cost-effective climate change mitigation.

Galego, N., Rozsa, C., Sánchez, R., Fung, J., Vázquez, A., & Santo Tomás, J. (2000). Characterization and application of poly(3-hydroxyalkanoates) family as composite biomaterials. *Polymer Testing, 19*(5), 485–492. doi:10.1016/0142-9418(99)00011-2

Galinski, E. A., & Trüper, H. G. (1994). Microbial behaviour in salt-stressed ecosystems. *FEMS Microbiology Reviews, 15*(2–3), 95–108. doi:10.1111/j.1574-6976.1994.tb00128.x

Garcia-Ochoa, F., Santos, V. E., Casas, J. A., & Gomez, E. (2000). Xanthan gum: Production, recovery, and properties. *Biotechnology Advances, 18*(7), 549–579. doi:10.1016/S0734-9750(00)00050-1

Girard, M., Ramirez, A. A., Buelna, G., & Heitz, M. (2011). Biofiltration of methane at low concentrations representative of the piggery industry-Influence of the methane and nitrogen concentrations. *Chemical Engineering Journal, 168*(1), 151–158. doi:10.1016/j.cej.2010.12.054

Hanson, R. S., & Hanson, T. E. (1996). Methanotrophic bacteria. *Microbiological Reviews, 60*(2), 439–471.

Harding, K. G., Dennis, J. S., von Blottnitz, H., & Harrison, S. T. L. (2007). Environmental analysis of plastic production processes: Comparing petroleum-based polypropylene and polyethylene with biologically-based poly-3-hydroxybutyric acid using life cycle analysis. *Journal of Biotechnology, 130*(1), 57–66. doi:10.1016/j.jbiotec.2007.02.012

Helm, J., Wendlandt, K.-D., Jechorek, M., & Stottmeister, U. (2008). Potassium deficiency results in accumulation of ultra-high molecular weight poly-beta-hydroxybutyrate in a methane-utilizing mixed culture. *Journal of Applied Microbiology, 105*(4), 1054–1061. doi:10.1111/j.1365-2672.2008.03831.x

Henckel, T., Roslev, P., Conrad, R., & Mikrobiologie, È. (2000). Effects of O_2 and CH_4 on presence and activity of the indigenous methanotrophic community in rice field soil. *Environmental Microbiology, 2*, 666–679. doi:10.1046/j.1462-2920.2000.00149.x.

Hernández, J., Gómez-Cuervo, S., & Omil, F. (2015). EPS and SMP as stability indicators during the biofiltration of diffuse methane emissions. *Water, Air, & Soil Pollution, 226*(10), 343. doi:10.1007/s11270-015-2576-2

Hilger, H. A., Cranford, D. F., & Barlaz, M. A. (2000). Methane oxidation and microbial exopolymer production in landfill cover soil. *Soil Biology and Biochemistry, 32*(4), 457–467. doi:10.1016/S0038-0717(99)00101-7

Hood, M. C. (2011). *Design and operation of a biofilter for treatment of swine house pit ventilation exhaust*. Master science thesis, North Carolina State University

Hood, M. C., Shah, S. B., Kolar, P., & Stikeleather, L. F. (2011). Design and operation of a biofilter for treatment of swine house pit ventilation exhaust. *American Society of Agricultural and Biological Engineers Annual International Meeting 2011, ASABE 2011, 1,* 448–468. http://www.scopus.com/inward/record.url?eid=2-s2.0-81255188695&partnerID=tZOtx3y1

Jahnke, L. L., & Nicholst, P. D. (1986). Methyl sterol and cyclopropane fatty acid composition of *Methylococcus capsulatus* grown at low oxygen tensions. *Journal of Bacteriology, 167*(1), 238–242.

Kennelly, C., Gerrity, S., Collins, G., & Clifford, E. (2014). Liquid phase optimisation in a horizontal flow biofilm reactor (HFBR) technology for the removal of methane at low temperatures. *Chemical Engineering Journal, 242*, 144–154. doi:10.1016/j.cej.2013.12.071

Khmelenina, V. N., Kalyuzhnaya, M. G., Starostina, N. G., Suzina, N. E., & Trotsenko, Y. A. (1997). Isolation and characterization of halotolerant alkaliphilic methanotrophic bacteria from Tuva soda lakes. *Current Microbiology, 35*(5), 257–261. doi:10.1007/s002849900249

Khmelenina, Y. N., Eshinimaev, B. T., Kalyuzhnaya, M. G., & Trotsenko, Y. A. (2000). Potential activity of methane and ammonium oxidation by methanotrophic communities from the soda lakes of southern Transbaikal. *Microbiology, 69*(4), 460–465. doi:10.1007/BF02756771

Kim, J.-B., Cho, K.-S., Jeong, S.-K., Nam, S.-W., Do Jeong, H., & Kim, J. K. (2008). Identification and characterization of a pigment-producing denitrifying bacterium. *Biotechnology and Bioprocess Engineering, 13*(2), 217–223. doi:10.1007/s12257-007-0201-y

Kirti, K., Amita, S., Priti, S., Kumar, A. M., & Jyoti, S. (2014). Colorful world of microbes: Carotenoids and their applications. *Advances in Biology, 2014*(1), 1–13. doi:10.1155/2014/837891

Knief, C. (2015). Diversity and habitat preferences of cultivated and uncultivated aerobic methanotrophic bacteria evaluated based on pmoA as molecular marker. *Frontiers in Microbiology, 6,* 1346. doi:10.3389/fmicb.2015.01346

Koller, M., Atlić, A., Dias, M., Reiterer, A., Braunegg, G., & Atli, A. (2010). Microbial PHA production from waste raw materials. In *Plastics from bacteria: Natural functions and applications* (pp. 85–119). Heidelberg: Springer. doi:10.1007/978-3-642-03287-5_5

Kraakman, N. J. R., Rocha-Rios, J., & Van Loosdrecht, M. C. M. (2011). Review of mass transfer aspects for biological gas treatment. *Applied Microbiology and Biotechnology, 91*(4), 873. doi:10.1007/s00253-011-3365-5

Lang, Y., Bai, L., Ren, Y., Zhang, L., & Nagata, S. (2011). Production of ectoine through a combined process that uses both growing and resting cells of *Halomonas salina* DSM 5928T. *Extremophiles, 15*(2), 303–310. doi:10.1007/s00792-011-0360-9

Laycock, B., Halley, P., Pratt, S., Werker, A., & Lant, P. (2014). The chemomechanical properties of microbial polyhydroxyalkanoates. *Progress in Polymer Science, 39*(2), 397–442. doi:10.1016/j.progpolymsci.2013.06.008

Lebrero, R. (2013). *Comparative evaluation of conventional and innovative biotechnologies for odour abatement in wastewater treatment plants* (Doctoral thesis). Universidad de Valladolid, Valladolid, Spain. http://uvadoc.uva.es/handle/10324/2880

Lebrero, R., Hernández, L., Pérez, R., Estrada, J. M., & Muñoz, R. (2015). Two-liquid phase partitioning biotrickling filters for methane abatement: Exploring the potential of hydrophobic methanotrophs. *Journal of Environmental Management, 151,* 124–131. doi:10.1016/j.jenvman.2014.12.016

Lebrero, R., Lopez, J. C., Lehtinen, I., Perez, R., Quijano, G., & Muñoz, R. (2016). Exploring the potential of fungi for methane abatement: Performance evaluation of a fungal-bacterial biofilter. *Chemosphere, 144,* 97–106. doi:10.1016/j.chemosphere.2015.08.017

López, J. C., Quijano, G., Pérez, R., & Muñoz, R. (2014). Assessing the influence of CH_4 concentration during culture enrichment on the biodegradation kinetics and population structure. *Journal of Environmental Management, 146,* 116–123. doi:10.1016/j.jenvman.2014.06.026

López, J. C., Quijano, G., Souza, T. S. O., Estrada, J. M., Lebrero, R., & Muñoz, R. (2013). Biotechnologies for greenhouse gases (CH_4, N_2O, and CO_2) abatement: State of the art and challenges. *Applied Microbiology and Biotechnology, 97*(6), 2277–2303. doi:10.1007/s00253-013-4734-z

Malashenko, I. P., Pirog, T. P., Romanovskaia, V. A, Sokolov, I. G., & Gringerg, T. A. (2001). Search for methanotrophic producers of exopolysaccharides. *Prikladnaia Biokhimiia I Mikrobiologiia, 37*(6), 702–705. doi:10.1023/A:1012307202011

Matsumoto, T., Yamamura, H., Hayakawa, J., Watanabe, Y., & Harayama, S. (2014). Influence of extracellular polysaccharides (EPS) produced by two different green unicellular algae on membrane filtration in an algae-based biofuel production process. *Water Science and Technology, 69*(9), 1919–1925. http://wst.iwaponline.com/content/69/9/1919.abstract

McDonald, I. R., Bodrossy, L., Chen, Y., & Murrell, J. C. (2008). Molecular ecology techniques for the study of aerobic methanotrophs. *Applied and Environmental Microbiology, 74*(5), 1305–1315. doi:10.1128/AEM.02233-07

Ménard, C., Ramirez, A. A., Nikiema, J., & Heitz, M. (2012). Effect of trace gases, toluene and chlorobenzene, on methane biofiltration: An experimental study. *Chemical Engineering Journal, 204–205,* 8–15. doi:10.1016/j.cej.2012.07.070

Modin, O., Fukushi, K., Nakajima, F., & Yamamoto, K. (2008). A membrane biofilm reactor achieves aerobic methane oxidation coupled to denitrification (AME-D) with high efficiency. *Water Science and Technology, 58*(1), 83–87. doi:10.2166/wst.2008.648

Morris, G., & Harding, S. (2009). Polysaccharides, microbial. In M. Schaechter (Ed.), *Encyclopedia of microbiology* (3rd ed., pp. 482–494). New York: Elsevier.

Muñoz, R., Daugulis, A. J., Hernández, M., & Quijano, G. (2012). Recent advances in two-phase partitioning bioreactors for the treatment of volatile organic compounds. *Biotechnology Advances, 30*(6), 1707–1720. doi:10.1016/j.biotechadv.2012.08.009

Muñoz, R., Villaverde, S., Guieysse, B., & Revah, S. (2007). Two-phase partitioning bioreactors for treatment of volatile organic compounds. *Biotechnology Advances, 25*(4), 410–422. doi:10.1016/j.biotechadv.2007.03.005

Murrell, J. C., McDonald, I. R., & Gilbert, B. (2000). Regulation of expression of methane monooxygenases by copper ions. *Trends in Microbiology, 8*(5), 221–225. doi:10.1016/S0966-842X(00)01739-X

Myung, J., Flanagan, J. C. A., Waymouth, R. M., & Criddle, C. S. (2015a). Methane or methanol-oxidation dependent synthesis of poly(3-hydroxybutyrate-co-3-hydroxyvalerate) by obligate type II methanotrophs. *Process Biochemistry, 51*(5), 561–567. doi:10.1016/j.procbio.2016.02.005

Myung, J., Galega, W. M., Van Nostrand, J. D., Yuan, T., Zhou, J., & Criddle, C. S. (2015b). Long-term cultivation of a stable *Methylocystis*-dominated methanotrophic enrichment enabling tailored production of poly(3-hydroxybutyrate-co-3-hydroxyvalerate). *Bioresource Technology, 198,* 811–818. doi:10.1016/j.biortech.2015.09.094.

Myung, J. (2016). Expanding the range of polyhydroxyalkanoates synthesized by methanotrophic bacteria through the utilization of omega-hydroxyalkanoate co-substrates. In *Recovery of resources and energy using methane-utilizing bacteria: Synthesis and regeneration of biodegradable, tailorable bioplastics and production of nitrous oxide*. Dissertation, Stanford University (California, US).

Myung, J., Kim, M., Pan, M., Criddle, C. S., & Tang, S. (2016). Low energy emulsion-based fermentation enabling accelerated methane mass transfer and growth of poly

(3-hydroxybutyrate)-accumulating methanotrophs. *Bioresource Technology, 207*, 302–307. doi:10.1016/j.biortech.2016.02.029

Nath, A., Dixit, M., Bandiya, A., Chavda, S., & Desai, A. J. (2008). Enhanced PHB production and scale up studies using cheese whey in fed batch culture of *Methylobacterium* sp. ZP24. *Bioresource Technology, 99*(13), 5749–5755. doi:10.1016/j.biortech.2007.10.017

Nikiema, J., Brzezinski, R., & Heitz, M. (2007). Elimination of methane generated from landfills by biofiltration: a review. *Reviews in Environmental Science and Biotechnology, 6*(4), 261–284. doi:10.1007/s11157-006-9114-z

Nikiema, J., & Heitz, M. (2009). The influence of the gas flow rate during methane biofiltration on an inorganic packing material. *Canadian Journal of Chemical Engineering, 87*(1), 136–142. doi:10.1002/cjce.20131

Nwodo, U. U., Green, E., & Okoh, A. I. (2012). Bacterial exopolysaccharides: Functionality and prospects. *International Journal of Molecular Sciences, 13*(11), 14002–14015. doi:10.3390/ijms131114002

Park, S., Lee, C. H., Ryu, C. R., & Sung, K. (2009). Biofiltration for reducing methane emissions from modern sanitary landfills at the low methane generation stage. *Water, Air, and Soil Pollution, 196*(1–4), 19–27. doi:10.1007/s11270-008-9754-4

Pastor, J. M., Salvador, M., Argandoña, M., Bernal, V., Reina-Bueno, M., Csonka, L. N., et al. (2010). Ectoines in cell stress protection: Uses and biotechnological production. *Biotechnology Advances, 28*(6), 782–801. doi:10.1016/j.biotechadv.2010.06.005

Pérez-Ramírez, J., Kapteijn, F., Schöffel, K., & Moulijn, J. A. (2003). Formation and control of N_2O in nitric acid production: Where do we stand today? *Applied Catalysis B: Environmental, 44*(2), 117–151. doi:10.1016/S0926-3373(03)00026-2

Pieja, A. J., Rostkowski, K. H., & Criddle, C. S. (2011a). Distribution and selection of poly-3-hydroxybutyrate production capacity in methanotrophic proteobacteria. *Microbial Ecology, 62*(3), 564–573. doi:10.1007/s00248-011-9873-0

Pieja, A. J., Sundstrom, E. R., & Criddle, C. S. (2011b). Poly-3-hydroxybutyrate metabolism in the type II methanotroph methylocystis parvus OBBP. *Applied and Environmental Microbiology, 77*(17), 6012–6019. doi:10.1128/AEM.00509-11

Pieja, A. J., Sundstrom, E. R., & Criddle, C. S. (2012). Cyclic, alternating methane and nitrogen limitation increases PHB production in a methanotrophic community. *Bioresource Technology, 107*, 385–392. doi:10.1016/j.biortech.2011.12.044

Rahnama, F., Vasheghani-Farahania, E., Yazdian F, & Shojaosadati, S. A. (2012). PHB production by *Methylocystis hirsuta* from natural gas in a bubble column and a vertical loop bioreactor. *Biochemical Engineering Journal, 65*, 51–56. doi:10.1016/j.bej.2012.03.014

Ravishankara, A. R., Daniel, J. S., & Portmann, R. W. (2009). Nitrous oxide (N_2O): The dominant ozone-depleting substance emitted in the 21st century. *Science, 326*, 123–125. doi:10.1126/science.1176985

Reshetnikov, A. S., Mustakhimov, I. I., Khmelenina, V. N., & Trotsenko, Y. A. (2005). Cloning, purification, and characterization of diaminobutyrate acetyltransferase from the halotolerant methanotroph *Methylomicrobium alcaliphilum* 20Z. *Biochemistry (Moscow), 70*(8), 878–883. doi:10.1007/s10541-005-0197-x

Rocha-Rios, J., Bordel, S., Hernández, S., & Revah, S. (2009). Methane degradation in two-phase partition bioreactors. *Chemical Engineering Journal, 152*(1), 289–292. doi:10.1016/j.cej.2009.04.028

Rocha-Rios, J., Kraakman, N. J. R., Kleerebezem, R., Revah, S., Kreutzer, M. T., & van Loosdrecht, M. C. M. (2013). A capillary bioreactor to increase methane transfer and oxidation through Taylor flow formation and transfer vector addition. *Chemical Engineering Journal, 217*, 91–98. doi:10.1016/j.cej.2012.11.065

Rocha-Rios, J., Quijano, G., Thalasso, F., Revah, S., & Muñoz, R. (2011). Methane biodegradation in a two-phase partition internal loop airlift reactor with gas recirculation. *Journal of Chemical Technology and Biotechnology, 86*(3), 353–360. doi:10.1002/jctb.2523

Rostkowski, K. H., Criddle, C. S., & Lepech, M. D. (2012). Cradle-to-gate life cycle assessment for a cradle-to-cradle cycle: Biogas-to-bioplastic (and back). *Environmental Science and Technology, 46*(18), 9822–9829. doi:10.1021/es204541w

Sarwat, F., Qader, S. A. U., Aman, A., & Ahmed, N. (2008). Production & characterization of a unique dextran from an indigenous *Leuconostoc mesenteroides* CMG713. *International Journal of Biological Sciences, 4*(6), 379–386.

Sauer, T., & Galinski, E. A. (1998). Bacterial milking: A novel bioprocess for production of compatible solutes. *Biotechnolnogy and Bioengineering, 57*(3), 306–313.

Sazinsky, M. H., & Lippard, S. J. (2015). Methane monooxygenase: Functionalizing methane at iron and copper. *Metal Ions in Life Sciences*. doi:10.1007/978-3-319-12415-5_6

Scheutz, C., Kjeldsen, P., Bogner, J. E., De Visscher, A., Gebert, J., Hilger, H. A., et al. (2009). Microbial methane oxidation processes and technologies for mitigation of landfill gas emissions. *Waste Management & Research, 27*(5), 409–455. doi:10.1177/0734242X09339325

Semrau, J. D. (2011). Bioremediation via methanotrophy: Overview of recent findings and suggestions for future research. *Frontiers in Microbiology, 2*(10), 1–7. doi:10.3389/fmicb.2011.00209

Semrau, J. D., Dispirito, A. A., & Yoon, S. (2010). Methanotrophs and copper. *FEMS Microbiology Reviews, 34*(4), 496–531. doi:10.1111/j.1574-6976.2010.00212.x

Shen, L., Haufe, J., & Patel, M. K. (2009). Product overview and market projection of emerging bio-based plastics. *Group Science, Technology and Society, 6*, 41. doi:10.1002/9780470697474.ch1

Steinbüchel, A., & Valentin, H. E. (1995). Diversity of bacterial polyhydroxyalkanoic acids. *FEMS Microbiology Letters, 128*(3), 219–228. doi:10.1016/0378-1097(95)00125-O

Strong, P. J., Kalyuzhnaya, M., Silverman, J., & Clarke, W. P. (2016). A methanotroph-based biorefinery: Potential scenarios for generating multiple products from a single fermentation. *Bioresource Technology, 215*, 314–323. doi:10.1016/j.biortech.2016.04.099

Strong, P. J., Xie, S., & Clarke, W. P. (2015). Methane as a resource: Can the methanotrophs add value? *Environmental Science and Technology, 49*(7), 4001–4018.

Sun, F., Dong, W., Shao, M., Lv, X., Li, J., Peng, L., et al. (2013). Aerobic methane oxidation coupled to denitrification in a membrane biofilm reactor: Treatment performance and the effect of oxygen ventilation. *Bioresource Technology, 145*, 2–9. doi:10.1016/j.biortech.2013.03.115

Sundstrom, E. R., & Criddle, C. S. (2015). Optimization of methanotrophic growth and production of poly(3-hydroxybutyrate) in a high-throughput microbioreactor system. *Applied and Environmental Microbiology, 81*(14), 4767–4773. doi:10.1128/aem.00025-15

Timmers, P. H., Suarez-Zuluaga, D. A., Van Rossem, M., Diender, M., Stams, A. J., & Plugge, C. M. (2016). Suppl.: Anaerobic oxidation of methane associated with sulfate reduction in a natural freshwater source. *ISME Journal, 10*, 1400–1412. doi:10.1017/CBO9781107415324.004

Van Der Ha, D., Nachtergaele, L., Kerckhof, F. M., Rameiyanti, D., Bossier, P., Verstraete, W., & Boon, N. (2012). Conversion of biogas to bioproducts by algae and methane oxidizing bacteria. *Environmental Science and Technology, 46*(24), 13425–13431. doi:10.1021/es303929s

Vieten, B., Conen, F., Seth, B., & Alewell, C. (2008). The fate of N_2O consumed in soils. *Biogeosciences, 5*(1), 129–132. doi:10.5194/bg-5-129-2008

Wei, X. M., Su, Y., Zhang, H. T., Chen, M., & He, R. (2015). Responses of methanotrophic activity, community and EPS production to CH_4 and O_2 concentrations in waste biocover soils. *Waste Management, 42*, 118–127. doi:10.1016/j.wasman.2015.04.005

Wendlandt, K. D., Jechorek, M., Helm, J., & Stottmeister, U. (2001). Producing poly-3-hydroxybutyrate with a high molecular mass from methane. *Journal of Biotechnology, 86*(2), 127–133. doi:10.1016/S0168-1656(00)00408-9

Whittenbury, R. D. H. (1981). The methylotrophic bacteria. In M. P. Tarr, H. Stolp, H. G. Trüper, A. Balows, & H. G. Schlegel (Eds.), *The prokaryotes: A handbook on habitats, isolation, and identification of bacteria* (pp. 894–902). Berlin: Springer.

Wilshusen, J. H., Hettiaratchi, J. P. A., De Visscher, A., & Saint-Fort, R. (2004a). Methane oxidation and formation of EPS in compost: Effect of oxygen concentration. *Environmental Pollution, 129*(2), 305–314. doi:10.1016/j.envpol.2003.10.015

Wilshusen, J. H., Hettiaratchi, J. P. A., & Stein, V. B. (2004b). Long-term behavior of passively aerated compost methanotrophic biofilter columns. *Waste Management, 24*(7), 643–653. doi:10.1016/j.wasman.2003.12.006

Wunderlin, P., Mohn, J., Joss, A., Emmenegger, L., & Siegrist, H. (2012). Mechanisms of N_2O production in biological wastewater treatment under nitrifying and denitrifying conditions. *Water Research, 46*(4), 1027–1037. doi:10.1016/j.watres.2011.11.080

Yoshida, H., Kotani, Y., Ochiai, K., & Araki, K. (1998). Production of ubiquinone-10 using bacteria. *The Journal of General and Applied Microbiology, 44*(1), 19–26. doi:10.2323/jgam.44.19

Zhang, L. H., Lang, Y. J., & Nagata, S. (2009). Efficient production of ectoine using ectoine-excreting strain. *Extremophiles, 13*(4), 717–724. doi:10.1007/s00792-009-0262-2

Zhang, T., Zhou, J., Wang, X., & Zhang, Y. (2015). Coupled effects of methane monooxygenase and nitrogen source on growth and poly-β-hydroxybutyrate (PHB) production of *Methylosinus trichosporium* OB3b. *Journal of Environmental Sciences (China), 52*(2017), 49–57. doi:10.1016/j.jes.2016.03.001

Zhang, Y., Xin, J., Chen, L., Song, H., & Xia, C. (2008). Biosynthesis of poly-3-hydroxybutyrate with a high molecular weight by methanotroph from methane and methanol. *Journal of Natural Gas Chemistry, 17*(1), 103–109. doi:10.1016/S1003-9953(08)60034-1

Zumft, W. G. (1997). Cell biology and molecular basis of denitrification†. *Microbiology and Molecular Biology Reviews: MMBR, 61*(4), 533–616.

Part V
Modelling a Low-Carbon City: Eco-city and Eco-planning

Chapter 18
The Architecture and the Value of the Waste

Sara Marini

Abstract The text addresses the architecture of the WTE plant: the first part is dedicated to the relationship between architecture and waste, the second to the reading of some projects, in the last part conclusions are drawn and we outline the possible perspectives of change of this architectural machine. The relationship between architecture and waste draws various imaginary paths. The relationship between architecture and the trash is the reflection of the plot between the city and the minimum fragment of something useless and of the cultural role attributed to waste: to the city in use corresponds its double made of what is discarded, that in "less civilized" realties remains a resource and not a problem. In the vocabulary "WTE plant," one of the many structures that transform and give new meaning to waste, is inherent reason of the uncanny relationship: the industrial machine, often called to coincide with the architectural building, change the unnecessary material to look into it new scraps of necessity. The nature of the container and its contents are opposed but also in the word 'value' lies an embarrassing contradiction: the word may tell a "soul virtues" and even "the merit or the price of everything." Environmental issues remain in the background of this story: what burns inside of architecture produces powders, which change the air quality and the ecological parameters. They are the black shadow of this close relationship between waste and energy. The processing machinery of waste are part of a production process which implies the end of things: their umpteenth "enhancement" finally brings the total disappearance of the object, to its accommodative disappearance.

1 The Architecture of Waste

The relationship between architecture and waste draws various images: is the reflection of the inherent overlap between the city and the minimum fragment of something useless and of the cultural role attributed to waste. The inhabited city

S. Marini (✉)
IUAV University of Venice, Venice, Italy
e-mail: marini@iuav.it

R. Álvarez Fernández et al. (eds.), *Carbon Footprint and the Industrial Life Cycle*, Green Energy and Technology, DOI 10.1007/978-3-319-54984-2_18

corresponds with its double made of what is discarded, that in "less civilised" realties remains a resource and not a problem.[1]

The cumbersome presence of waste (Rathje and Murphy 1992) returns today (Bauman 2004) to define the urban scenario, after having defined dark visions and premonitions during the 1970s, dictated by the energy crisis and a consequent attention to ecology. In both cases, now as then, the idea of the city is subject to a radical revision imposed by the widespread weight of what remains (Oswalt 2005, 2006; Berger 2006, 2008).

In *Metropolis* (1927), Fritz Lang hypothesised that the city of 2026, with its heroic figures, monumental works of architecture and multi-level infrastructures, would have designed the future of the city. With the animated film *WALL•E* (2008), Pixar presented adults and children with a vision of planet earth in seven centuries as an uninhabitable waste dump, an endless and consistent landscape of use—less rusty metal: the technologies/archaeologies spoken of by Bruno Munari in *Fossili del 2000*. The end of things, from a simple detail that marks one of the moments in the cycle of production, becomes a characterising element of what the future has in store (Fleming 1991; McDonough and Braungart 2002), capable of prevaricating the orders and structures of the urban system (Tachieva 2010; Bertagna and Marini 2011). *WALL•E* is a *remake* of Akira Kurosawa's last film, *Dodes'ka den*. The work by the Japanese director speaks about a city-dump where, in addition to the consumption of objects and spaces, beyond scenarios of decommissioning, waste exceeds its condition of uselessness to become a pervasive presence. Even *Wasting Away*, the incomplete text to which Kevin Lynch dedicated twenty-five years of work (Lynch 1990), explores the risks of the termination of resources (and space), less for their exhaustion as much as for their incessant consumption and the consequent production of entropy (Corbellini 2016). Lynch examined the role of architecture and the possibility that it could be presented not as a *throwaway* object, but as something persistent in virtue of its farsighted ability to accept changes in its use. Paul Auster dedicated various works to waste: in *In the Country of Last Things* cities and architecture are only void and abandoned backdrops to be crossed in the search for waste (Auster 1987). Finally, it is possible, or perhaps necessary, for a problem of space and occupancy, to employ waste to construct part of the city, as hypothesised by Archizoom in their *Colline di spazzatura per la città di pianura* (1969), an image borrowed and dramatized in 1999 by MVRDV in "Landscape Waste" (MVRDV 1999). While in these latter two visions the coexistence between waste, or its development as landscape, is recounted as

[1] "Despite the presumed certainties connected with the positivist myth of unlimited progress, even from this point of view there appears, during the second half of the 20th century, the disturbing truth according to which things and substances, in particular sources of energy, are pregnant, from a qualitative point of view, with their opposite" (Emery 2011, our translation).

problematic, in the *Re(f)use City* hypothesised by Joachim Mitchell,[2] the destiny and substance of the urban system are constructed of waste, an abundant material identifiable as the sole and possible resource for the future. In this vein of investigation, the *Re(f)use City* represents a moment of surpassing the inherent tensions of the relationship between architecture/waste, in which design must confront its alter ego, achieving moments when these two conditions coincide.

2 The Problem of Value

"What of objects, of their consistency, of their duration, of their stability? Objects have always been consumed and grown useless, however, in the cycle of production-consumption, which cannot be interrupted, they are considered in relation to their rapid uselessness. In fact, not only is their transitoriness foreseen, but even their 'expiry date', which must be as short-term as possible. Thus instead of being limited to the end of their existence, the end of an object's existence is conceived from the outset as its aim. Within this process, where the principle of destruction is immanent to production, the use of objects must coincide as much as possible with their usage" (Galimberti 2003, our translation).

The substance of the relationship between architecture/waste resides in the encounter between the desire to conserve and possess objects, a form of control, and the destructive tendencies of matter itself (Viale 1994; Royte 2006; Scanlan 2005). The possibility that form and formless coexist or find modalities or necessities of encounter is the subject of this conciliation.[3]

The incinerators, one of the many facilities now existing for transforming and giving new meaning to waste (Thompson 1979; Pawley 1982), inherently contains the reasons of this disturbing relationship: the industrial machine, often required to coincide with the architectural factory, modifies useless material in order to seek remnants of necessity within it. The nature of the container and its content are

[2]"New York City is disposing of 38,000 tons of waste per day. Most of this discarded material ended up in Fresh Kills landfill before it closed. The Rapid Re(f)use project supposes an extended New York reconstituted from its own landfill material. Our concept remakes the city by utilizing the trash at Fresh Kills. With our method, we can remake seven entirely new Manhattan islands at full scale. Automated robot 3d printers are modified to process trash and complete this task within decades. These robots are based on existing techniques commonly found in industrial waste compaction devices. Instead of machines that crush objects into cubes, these devices have jaws that make simple shape grammars for assembly. Different materials serve specified purposes; plastic for fenestration, organic compounds for temporary scaffolds, metals for primary structures, and etc. Eventually, the future city makes no distinction between waste and supply." Mitchell J et al. http://www.terreform.org/projects_urbanity_rapid_refuse.html. Accessed 10 nov 2016.

[3]"The darkest desperation can exist side by side with the most striking invention; entropy and efflorescence are fused together. Because so little has remained, it is not possible to dispose of anything and new possibilities have been discovered for utilizing materials that were once scorned and considered junk" (Auster 1987).

contrasting and, at the same time, the use of the term "value" contains an inherent contradiction: it may speak of a "virtue of the soul" as well as the "merit or price of any good." The machines that transform waste are part of a cycle of production that subtends the end of all things: their umpteenth "valorisation" finally leads to a total disappearance of the object, to its accommodating disappearance. Architecture is called upon to contain the mystery of pulverisation, but "one has the impression that thermodynamics allows for the surfacing in a certain sense of the negative of energy processes and that it corresponds with a sort of initial return of what has been removed in the era of machines, where in the end the qualitative returns to disturb the quantitative" (Emery 2011, our translation).

What burns inside of incinerators produce powders that change the air quality and the ecological parameter. They are the black shadow of this close relationship between waste and energy.

Leonardo da Vinci, in his project *Il fiume Arno e la sua regolazione in canale* (1502–1503), confronted the problem of controlling matter in continuous movement and change.[4] His drawing restores the rigour of an idea that wishes to organise and bring about a new configuration, based on a consideration of the convective movements of water. In *Il diluvio* he narrates the spreading of the *ruina*, the deformation of the appearance of things, arriving at the supposition that water as destroyer was subject to the destiny of waste in contemporary society, in other words to be dissolved.[5]

Form and formless labour to encounter one another for great periods of time; the distances in meaning between the two often waver, above all in architecture. The *Merzbau* of Kurt Schwitters, a true monument to the useless given new meaning and a new conception of process, has become the system of the architecturally formless, or better yet of "fluid," "continuous" spatiality, intolerant of the rules of tectonics, which however translate conditions *in itinere* into crystallisations, designed by chance or the absence or partial control of their author (Schwitters invited his friends to participate in the definition of the piece).

The hospitality offered by architecture to waste causes its very meaning to undulate. The container, conscious of the dichotomy of containing its very opposite, is not content to offer itself as a mere box or, better yet, attempts to overcome the separation already narrated by the condition of *I'm a monument* learned from Las Vegas (Venturi et al. 1972). Architecture, when it lends itself to hosting waste, confronts its own role as a machine of transformation, a possible monument to the mystery of the separation of matter, an object that despite its own intentions is separated from urban and territorial dynamics, self-isolated as a result of normative and cultural issues. If architecture is not a part of the city, if it does not institute

[4]In Sigfried Giedion's book *Space, Time and Architecture: The Growth of a New Tradition* (Giedion 1941), the project for the Arno River is cited as a vision able to reconcile technique and evocation of the environmental system motions.

[5]So writes Leonardo: "Here, then, natural reasons fails us; and therefore to resolve such a doubt we must needs either call in a miracle to aid us, or else say that all this water was evaporated by the heat of the sun" (Richter 2008).

relationships with the territory; if it has assumed the role of a cathedral, though orphaned of its stately and symbolic content, if the process of transformation that elevates it to the role of a machine of progress testifies in reality to a dysfunction in the productive process that is unaware, unable to foresee, incapable of fully controlling its own *the end*, then the architecture of waste must respond, similar to a sort of psychoanalytic session, to its own nude condition and thus support its own reasons of necessity.

Against the background of these cathedrals—where production is washing its own conscience—is the landscape of the earth, water, air and people. There is no redemption for waste: their transformation into energy produces fine particles, killer to the environment and to humans. Today by the term Anthropocene we try to emphasize the responsibility of men in changing the planet, while the waste cathedrals (incinerators and landfills) continue to show the dark side of production. Climate change, led mainly by the manner in which energy is produced, is the most explicit link between the danger of human action and the fragility of the environment, connection underlined by the name given to the contemporary era, actually called the Anthropocene. Not by chance in the waste cathedrals is not expected the presence of man, they are empty and inaccessible places because they contain the secret of a production that continues to have a dross: the subtle and dangerous dust generated together with the energy.

3 Architectural Stratagems and Deceits

3.1 Machines

Recycling Madrid brings together projects and considerations for recycling current conventions (Ábalos and Herreros 2000); both Madrid and *recycling* represent metaphors of a transformation of the idea of architecture and also a pretext for uniting and establishing a precise moment in the work of this Spanish office. More than half of the book is dedicated to the project *New Recycling Plant for Urban Waste* (1996–1999) a waste transformation centre that is part of the more complex system of the Parque Regional del Sureste. The facility is comprised of two volumes, one dedicated to composting and the other to recycling, anticipated by an entry pavilion. The vast park, historically used to house the waste produced by the city of Madrid, is articulated in two platforms: the first housed trash for thirty years, the other was destined to do the same for another twenty-five years. The project for this vast area, with dimensions similar to those of Central Park though clearly separate from the metropolitan system, examines the arid nature of the landscape and the role of the works of architecture that will dot its surface. Other metaphors are echoed in the functional programme of the park: as it is made of waste materials expelled from the metropolis, to the same degree this space should contain uses generally in conflict with the city, but which often mark its very urban nature (in synthesis, ludic and noisy activities and spaces for the undesired); it must also be the space, given the difficult environmental conditions that characterise it, in which to test plant species that can be used to reforest parts of Madrid. The waste transformation centre is a machine the renders explicit and shares the destiny of its content: realised in recycled polycarbonate, it is designed to exist for twenty-five years, after which its buildings will be utilised to host services for the park, or dismantled, as each of its parts are in turn recyclable. Beyond its intentionally industrial appearance, as a mechanism inserted in a landscape with a given life span, further underlining the temporal nature of the product, the centre also hosts an exhibition and educational space, with the mission of raising awareness about environmental issues. The book by Abalos and Herreros, and in particular the two projects for and inside the Parque Regional del Sureste in Madrid, undoubtedly have the merit of exposing the necessity of involving architectural structures in the process of recycling, of investing it with a logic of "participation" in a context founded on a continuum, without however slipping into strategies of mimesis, but working precisely on the meaning and role of architecture in contexts without form (Fig. 1).

The year 2004 marked the completion by Yoshio Taniguchi of the *Naka Incineration Plant* in Hiroshima. This project is part of *Hiroshima 2045: The City of Peace and Creativity*. In this case, didactic intentions, architecture, museum and machine for transformation coincide 100%. Taniguchi, the author of the MoMA in New York, was invited by the government of this Japanese city precisely for his

Fig. 1 Federico Segat, *Ash Lines*, 2016. Abalos and Herreros, New Recycling Plant for Urban Waste, Madrid

renown as a designer skilled in the art of exhibiting and substantiating the nature of spaces, even internal, with urban issues. In fact, he was requested to design a building capable of communicating the enormous problem of waste production and that, at the same time, enrich the city with an idea of quality, or better yet, with a new presence able to fill the historical void written across the buildings of Hiroshima and reconcilable with its more recent vocation as a tourist destination. Taniguchi looked to the work of his mentor, Kenzo Tange, and his Hiroshima Peace Centre, in order to design a *museum of trash*: a sobriety of the volumes and materials does not conceal the industrial nature of the building but rather exalts it, declaring that the logics of architecture can interest any type of content. The structure is subdivided into two parts, separated by a tall glazed partition that can freely be crossed: the *ecorium* (a fusion of *ecology* and *atrium*). The building terminates Yoshijima street, overlooked also by the Hiroshima Peace Centre, while on the opposite side its looks toward the sea. The risk of constructing a wall between the city and its bay was avoided by the placement of the *ecorium* as a space in which to read the spectacle of the transformation of waste, but also the means of passing through the building. This crossing is repeated above grade with a walkway that extends beyond the void by some 150 m, running above the *ecoasis* (*ecology* and *oasis*), a garden space looking along the water's edge, evidence that architecture contains and modifies the street. Inside the *ecorium* what is normally concealed is stripped bare: the incineration of waste. It is also possible to observe the other phases in its treatment as part of guided tours and views through large glazed surfaces that separate machine and visitor. The architecture contemplates and substantiates two passages of state, that of waste and that of the street: both overlap in the glass box, in the void enunciated on the façade as a transparent space that confirms the nature of narration and exposure assumed by the womb of the building. This work is an intertwining of two positions on the architecture of waste, that of the machine and that of the monument, with the substantial difference that with respect to those projects that generally adopt the second strategy, here it is not the container that is monumentalised, but its content: matter is in transformation and as such it often finds its way into the museum (Fig. 2).

3.2 Monuments

In 2008 Erick van Egeraat won a competition for the design of the *Incinerator Line* in Roskilde, Denmark; the building was completed in 2014. The Dutch architect proposed a *cathedral of waste*: the structure is externally clad in micro perforated aluminium, its shell reproducing the figure of a church, deformed and extended upward into a 100 m tall spire. The object not only refers to ecclesiastical architecture, depicted in a synthetic form that is unabashedly produced by the use of virtual modelling tools, but also the roofs of the factories present in the area where the building is to be realised. The desire to produce an icon in the landscape, which negates even while plastically recalling its own concealed industrial character,

Fig. 2 Federico Segat, *Ash Lines*, 2016. Yoshio Taniguchi, Naka Incineration Plant, Hiroshima

is further exalted by the nocturnal image of this building, which enunciates the holes in its armour and thus becomes a beacon. The project on the one hand imprisons its own cumbersome content behind a façade composed of two layers, which also serves to bring light and air and, on the other hand, renders explicit the process that takes place inside it thanks to a mechanism of illumination that in part mimics the burning flame and its movement inside the building. In this case architecture is entrusted with the role of containing and enchanting, of hosting the process of producing energy but also of responding to its mandate as a landmark. Here architectural language is the principle material of the project, the true protagonist, despite its content (Fig. 3).

The year 2004 marked the completion of the *Incinerator* on the Isle of Man in the Irish Sea, home to a semi-rural community of 80,000 people. The project was defined by the Man Government's Waste Plan in 2000 with the intent of r educing the quantity of trash to be shipped to landfills. The machine handles 60,000 tonnes of domestic and commercial waste, producing 10% of the electrical energy

Fig. 3 Federico Segat, *Ash Lines*, 2016. Erick van Egeraat, Incinerator Line, Roskilde

consumed by the island. Designed by Savage and Chadwick, the facility welcomes 1000 visitors each year. The building is a sort of prehistoric animal with a steel carapace resting atop a typical Irish lawn. The iconic quality of this work of architecture in this case was not constructed as a reference to other objects or other forms present in the territory, nor as a metaphor of the process it contains; it is substantiated by exalting the anomaly of the building with respect to its context: an anthropomorphic machine in the landscape. Once again architecture is a question of language. However, while the project by Erick van Egeraat concretises a design object founded on an a-scalar logic (no detail allows for a reading of the size of the building), the structure on the Isle of Man makes use of deconstructivist afflatuses. In substance, the monumental character returns as an insistence on the formal problem of architecture, as the necessity of design to make an autonomous communicative decision (Fig. 4).

Has been under discussion for years the realization of the *Case Passerini Incinerator,* near Sesto Fiorentino and designed by Gae Aulenti. The facility is a piece of a city, or better yet a true assembly of figures that recall an urban past. The incinerators is to be located in an open industrial area, at the point of convergence of the Firenze-Mare and Bisenzio autostrade, a node of overlapping urban and infrastructural presences, a site in which it is necessary to take up a position and clearly declare it. The project is presented as an abstraction of the urban; more evidence of its absence in the area, indecisive among the fragments of landscape and structures of the exploded and disarticulated city than a true pioneer of a ne w colonization of the territory. Architecture here is not the simple memory of itself, but of its f ailed participation in the idea of the city. Its geometric simplification resonates with that of the Centro Torri in Parma by Aldo Rossi where, however, the communication written on or attached to the walls renders explicit the awareness of the problem inherent to architecture without city. In both the incinerators and the

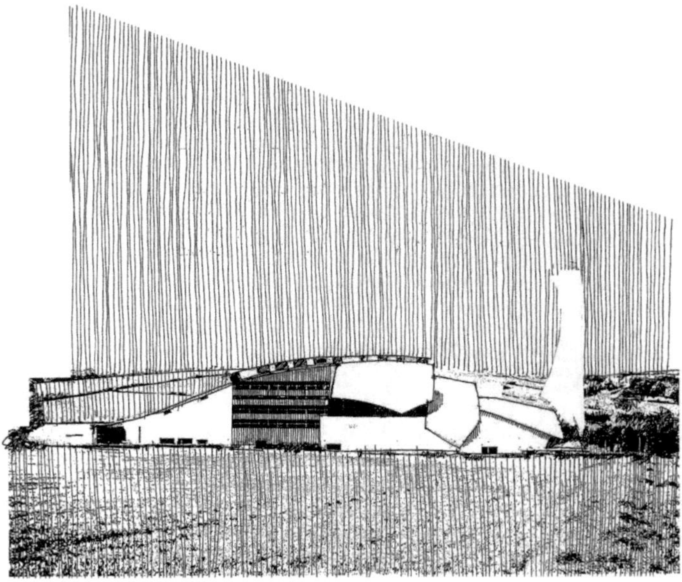

Fig. 4 Federico Segat, *Ash Lines*, 2016. Man Government's Waste Plan, Incinerator, Isle of Man

Centro Torri, design must unwillingly face up to structures that are cause for crisis in the direct correspondence between content and container because there is no possibility to make reference to other sites, to connections with fabrics and open spaces: they are urban events awaiting a sense of urbanity, from which they can draw and speak only of nostalgia (Fig. 5).

3.3 Integrated Architecture

The long-term vision for "achieving a true synthesis of the future community" requested by Walter Gropius of architects and urbanists such that they act in favour of an *integrated architecture* (Gropius 1955) echoes in design considerations related to the "problem of waste" (Hough 1995; Inoguchi et al. 1999; Heynen et al. 2005). This third section of design strategies brings together those experiences that, more or less happily, centre their work on the problem of the future community or that consider the reaction of context to the arrival of a waste transformation centre. The initial tendency is to "mitigate," or better yet "soften" the presence of waste industries through operations of superficial overwriting. When this overwriting is limited to the use of a colour, for example sky blue and the green of a freshly cut lawn compete for first place. As blue as the sky: this is how to describe the *Waste Incinerator and Power Plant* in Brescia, an architectural machine that transforms waste into value, contextualized by its ever-changing glass cladding,

Fig. 5 Federico Segat, *Ash Lines*, 2016. Gae Aulenti, Case Passerini Incinerator, Sesto Fiorentino

an intervention of color design, designed and built in 1996 by ASM in collaboration with the architect, biologist and painter Jorrit Tornquist (Fig. 6).

The *Teesside WTE Power Station* in Billingham, south of Newcastle, one of the Energy-From-Waste Plants in the United Kingdom, completed in 1998, was not content with being partially sky blue, but adopted a further strategy of integration: the building, intentionally with the form of a house with a double pitched roof, hosts other sky blue volumes that pass through it, piled one atop the other and articulating its functional programme. The architectural solution adopted, above all in the relations between the sky blue volumes, recalls the piles of housing of the *Tokyo Apartment* project by Sou Fujimoto or the more famous *Vitra House* by Herzog and de Meuron, with the substantial difference that in the plant in Billingham the articulation of the building is designed to "bring it closer" to the ordinary landscape, to render it more acceptable, while in the other two examples it is a matter of spatial considerations (Fig. 7).

The most well known case of domestication, or better yet of the addition of architectural value through an interpretation of the skin of the building remains the *Müllverbrennungsanlage Spittelau* in Wien. The building was transformed following a fire in 1987 by Friedensreich Hundertwasser into a colourful work of art (Fig. 8).

An opposing philosophy is demonstrated by the virgin white skin of the *Gaoantun Waste-to-Energy Power Plant* near Beijing, which narrates the purity

Fig. 6 Federico Segat, *Ash Lines*, 2016. ASM and Jorrit Tornquist, Waste Incinerator and Power Plant, Brescia

Fig. 7 Federico Segat, *Ash Lines*, 2016. Teesside WTE Power Station, Billingham

Fig. 8 Federico Segat, *Ash Lines*, 2016. Friedensreich Hundertwasser, Müllverbrennungsanlage Spittelau, Wien

of energy that can be drawn from waste. In synthesis, through a language written on its skin architecture seeks to re-stitch the gap that waste to-energy facilities represent in social terms: structures useful to disposing of at least part of the cumbersome leftovers produced by and in the city, which spew mysterious fumes into the environment, the object of different statistics and evaluations, but which substantially are interpreted by local communities as the black soul of waste that returns as a threat.

An extreme proposal, because in this case architecture is not content to communicate its own innocence, but instead is articulated to offer true spaces for the community, is the project for a *Waste-to-Energy Plant* in Amagerforbrænding, near Copenhagen, proposed by BIG architects, whose roof hosts three outdoor ski runs. Counting on the application of recent sustainable technologies in waste transformation and environmental respect, the new structure will be completed in 2017 and will substitute the old existing plant, which has functioned for 40 years. The proposal seeks to insert the structure as an integral part of the functional programme of the city, hosting ludic activities and assuming the form of a hill, a fragment of landscape or a ne w idea of an urban park. Diverse stratagems encountered previously return in this work of architecture: skiers reach the summit of the promontory-machine using an elevator that offers a view of the spectacle of the transformation and valorisation of waste, while the facility expels rings of CO_2 that are illuminated at night to "artisticize" these mysterious fumes (Fig. 9).

Fig. 9 Federico Segat, *Ash Lines*, 2016 BIG Architects, Waste-to-Energy Plant, Amagerforbrænding

4 Transforming and Containing the Formless

"Will one day the shells of derelict factories (…) be able to capture the imagination of tourists as the Roman ruins?" (Lynch 1990) The relationship between architecture and waste appears to gather and raise diverse questions inherent to design, intended in its vastest meaning: on the one hand it is precisely the design of the manufacturing process that declares its own congenital dysfunction, evidence of which is provided by its leftovers; on the other hand architecture wavers between its own communicative mission and the temptation to return to being part of an urban or territorial system, heavily fragmented and also rich with leftovers. The *ruins of everything* weigh on and surround architecture, by its nature a bringer of order, but perhaps this assumption is called into question by a sort of urban metabolism that today reads the urbanised system on par with its natural counterpart (Doherty and Mostafavi 2010; Hawkins 2005). Better yet, this crisis may place architecture in the condition less of being able to clarify than to comment (Hegger et al. 2008), reason and explicate even its own doubts about containing the uncontainable.

The first line of crisis leads to a substantial revision of processes, even those that precede design itself, such that a diverse relationship with things, with objects, a diverse way of producing and consuming, can now be hypothesised to face up to the crisis. Waste may assume value without undergoing transformation, once again becoming part of the cycle of life without traumas. Or, with a view to active participation by the building-machine in the transformation of waste, we may be witness to a total enslavement to the logic of stripping bare, in order to render the process of hosting completely explicit, as part of a change in direction according to which the final act resembles the spreading of ashes, only if they will not be polluted, useful to cultivating the soil. Or, once again as part of the first approach, the destiny of waste may be that of seeking to once again become part of the natural cycle, accompanied by zero volume agricultural machines (buildings-farms) capable of interacting with the soil and its processes.

The second line insinuates itself more precisely within the vision of an architecture that participates and does not accompany the logic of manufacturing, that confirms it but also denounces at the same time its "inattentions", that critically reads its role and knows how to explicate its discomfort.

Art has often used, and continues to use waste to construct considerations regarding processes of transformation to expose the dark sides or imperfections of process. In 1969, Peter Smithson put in place in a gravel pit in Rome his work called *Asphalt Rund*own: a truck poured asphalt on the ground. Both materials involved in the work needed to build roads, to waterproof the ground, to give up a primary resource, to continue to change the planet. *Asphalt Rundown* is an action of land art built to reflect on actions considered discounted. In 1975, Gordon Matta-Clark realizes *Conical Intersect*. The work consists in drilling of two adjoining buildings of the eighteenth, twins and demolition pending in the heart of Paris to leave space to the Centre Pompidou. Drilling is a void that runs through the different times of the city and its transformations that produce gentrification. The work asks you to reflect on the value of the things that disappear: disappears itself with the waste on which was impressed. In 1993 Rachel Whiteread realizes *House*, her first public commission for the London association Artangel: a monumental concrete cast of a Victorian house (demolished to make way for a new housing development in Greater London), made in situ and in turn devoted to destruction, to stigmatize the speculation affecting the neighborhood.

But in the case of incinerators, architectures of rejection and of the value, there's more: there is the assertion of a further transition of the matter that brings into being a union of construction, production process and environment. Critical architecture must respond to this logic chain without mitigate, it must be a critical act to tell the secret of entropy and avoid leaving it in the blue sky.

References

Ábalos, I., & Herreros, J. (2000). *Recycling madrid*. Barcelona: Actar.

Auster, P. (1987). *In the country of last things*. London: Faber & Faber.

Bauman, Z. (2004). *Wasted lives. Modernity and its outcasts*. Cambridge: Polity Press.

Berger, A. (2006). *Drosscape: Wasting land in urban America*. New York: Princeton Architectural Press.

Berger, A. (2008). *Designing the reclaimed landscape*. London: Taylor & Francis.

Bertagna, A., & Marini, S. (2011). *The landscape of waste*. Milan: Skira.

Corbellini, G. (2016). Entropy. In S. Marini & G. Corbellini (Eds.), *Recycled theory. Illustrated dictionary*. Macerata: Quodlibet.

Doherty, G., & Mostafavi, M. (Eds.). (2010). *Ecological urbanism*. Zurich: Lars Müller Publisher.

Emery, N. (2011). *Distruzione e progetto. L'architettura promessa*. Milan: Christian Marinotti.
Fleming, G. (Ed.). (1991). *Recycling derelict land*. London: Telford.
Galimberti, U. (2003). *I vizi capitali e i nuovi vizi*. Milan: Feltrinelli.
Giedion, S. (1941). *Space, time and architecture: The growth of a new tradition*. Cambridge: Harvard University Press.
Gropius, W. (1955). *Scope of a total architecture*. New York: Harper & Brothers Publisher.
Hawkins, G. (2005). *The ethics of waste: How we relate to rubbish*. Lanham: Rowman & Littlefield.
Hegger, M., Fuchs, M., Stark, T., & Zeumer, M. (2008). *Energy manual: Sustainable architecture*. Basel: Birkhäuser.
Heynen, N., Kaika, M., & Swyngedouw, E. (Eds.). (2005). *In the nature of cities: Urban political ecology and the politics of urban metabolism*. London: Routledge.
Hough, M. (1995). *Cities and natural process: A basis for sustainability*. London: Routledge.
Inoguchi, T., Newman, E., & Paoletto, G. (Eds.). (1999). *Cities and the environment: New approaches for eco-societies*. Tokyo: United Nations University Press.
Lynch, K. (1990). *Wasting away. An exploration of waste: What it is, how it happens, why we fear it, how to do it well*. San Francisco: Sierra Club Books.
McDonough, W., & Braungart, M. (2002). *Cradle to cradle. Remaking the way we make things*. New York: North Point Press.
MVRDV (1999) MetaCity Datatown. 010 Publishers. Rotterdam.
Oswalt, P. (Ed.). (2005). *Shrinking cities* (Vol. 1). Ostfildern-Ruit: Hatje Cantz.
Oswalt, P. (Ed.). (2006). *Shrinking cities: Interventions* (Vol. 2). Ostfildern-Ruit: Hatje Cantz.
Pawley, M. (1982). *Building for tomorrow: Putting waste to work*. San Francisco: Sierra Club Books.
Rathje, W. L., & Murphy, C. (1992). *Rubbish! The archaeology of garbage*. New York: Harper Collins.
Richter, I. A. (Ed.). (2008). *Leonardo da Vinci: Notebooks*. New York: Oxford University Press.
Royte, E. (2006). *Garbage land: On the secret trail of trash*. Boston: Back Bay Books.
Scanlan, J. (2005). *On garbage*. London: Reaktion.
Tachieva, G. (2010). *Sprawl repair manual*. Washington: Island Press.
Thompson, M. (1979). *Rubbish theory: The creation and destruction of value*. Oxford: Oxford University Press.
Venturi, R., Scott Brown, D., & Izenour, S. (1972). *Learning from Las Vegas*. Cambridge: MIT Press.
Viale, G. (1994). *Un mondo usa e getta. La civiltà dei rifiuti e i rifiuti della civiltà*. Milan: Feltrinelli.

Chapter 19
Modeling a Low-Carbon City: Eco-city and Eco-planning

Wynand Lambrechts and Saurabh Sinha

Abstract Mathematically modelling a low-carbon city in the traditional sense is a complex task and have been studied from a variety of perspectives, potential challenges and ultimately towards providing accurate models for low-carbon emissions for cities. Unknown and statistically fragmented data, future uncertainty and limited or inaccurate historical datasets complicate this task. The effects of climate change, based on models or on perceived impacts, also vary among cities. For example, cities on coastal regions experience a rise in sea levels and an increase in the frequency and severity of cyclones; whereas inland, resulting temperature rises pose significant health impacts for humans and animals. There needs to exist a mutual understanding between climate change, urban development and eco-city planning as well as the causes and effects of carbon pollution. Low-carbon cities are long-term investments in city infrastructure to create sustainable and environmentally friendly cities. Low-carbon cities can be realized through an amalgamation of smart city technologies, efficient and sustainable buildings and sustainable transport. Urbanization occurs rapidly and it is common to find infrastructure to be relatively old-fashioned; relying on increased supply rather than decreasing demand. Refurbishment of infrastructure is typically the most economically feasible and environmentally friendly solution. Accurate mathematical modelling and research into cost-effective technologies for improvements are necessary to support the business case for infrastructure overhauls. The contributed chapter provides cost-effective and technologically sustainable means to achieve efficient and low-carbon cities. Emission modeling is a dynamic research discipline; this chapter aims to highlight the considerations and concerns of generating a complete eco-city and sustainable model by identifying and understanding the characteristics of individual sectors. The chapter supplements the related body of knowledge by

W. Lambrechts (✉) · S. Sinha
Department of Electrical and Electronic Engineering Science, Faculty of Engineering and the Built Environment, Auckland Park Kingsway Campus, Auckland Park, Johannesburg, South Africa
e-mail: wynand.lambrechts@ieee.org

S. Sinha
e-mail: ssinha@ieee.org

R. Álvarez Fernández et al. (eds.), *Carbon Footprint and the Industrial Life Cycle*, Green Energy and Technology, DOI 10.1007/978-3-319-54984-2_19

thematically providing guidelines for low-carbon city modelling. The chapter investigates potential scholarly contributions by assisting researchers to theoretically identify and classify overlooked and underestimated sources of GHG emissions in urban settings. The notional overview on low-carbon cities through economic planning provides a means to identify known issues and sub-optimal eco-city infrastructure. The chapter aims to serve as a starting point for specialized research to improve upon such scenarios.

1 Introduction

As the global population is rapidly urbanizing, modern economic investments and commitments in eco-city infrastructures are defining the long-term carbon emission [the collective term for all GHG emissions, including carbon dioxide (CO_2)] profiles of current and future populations. To strive towards a lower carbon emission path with ideally zero fossil-fuel energy use and full dependence on renewable energy, zero waste (fully recyclable non-consumables) and zero (or at least very low) overall carbon emissions, these investments require prioritized, considerable and immediate consideration and devoted financial backing. Gouldson et al. (2015) point out that adopting more compact, connected and efficient strategies towards urban development predictably stimulates the economy, attracts local and foreign investments, improves air quality, is advantageous to the health of humans and animals, promotes safety, drives the reduction of global poverty and significantly reduces the rate at which climate change is occurring. Eco-urban developments should be efficient in tracking and modeling environmental impacts, economic, social and environmental benefits and impending damaging effects to the planet from processes that contribute to total energy consumption of development planning, implementation and sustainability. This chapter focuses on the modeling aspect of eco-cities and eco-planning of sustainable urbanized regions. Modeling of variables and parameters that influence pollution and lead to degradation of land is a powerful technique to understand and mitigate these adverse effects. Typically, five primary categories that allow modeling, quantifying, characterizing and monitoring of climate change can be identified, namely

- the total energy consumption within a pre-described sector, which includes all forms of non-renewable energy generation,
- electricity consumption at peak and average usage,
- contributing GHG emissions from energy generation and waste products,
- raw material consumption, and
- water consumption.

These five categories allow for the modeling of the carbon footprint of populations in a broad range of scenarios and gives insight into understanding and monitoring energy demands and supply. The critical requirements in ensuring accurate approximations are efficient, accurate and reliable activity data collection and reporting. Generating carbon emission models to contribute to eco-city planning is only as effective as the gathered data; an activity which is not properly implemented in both developing and developed counties. Data collection is an integral discipline to develop and continuously update the GHG inventory, since GHG have long atmospheric lifetimes (Dhar et al. 2013) in addition to a continuous accumulation of emissions.

The intergovernmental panel on climate change (IPCC) recommends two primary categories of data sources used for data collection: literature and surveys/census information on national and international fronts. The IPCC employs three concepts of good practice to promote the development of high-quality national GHG inventories and collection principles:

- methodological tiers to represent levels of complexity,
- readily available statistics of default data and
- key categories, which identify areas exerting a significant influence on emissions in a country (IPCC 2006).

In addition, the IPCC (2006) provides five indicators of inventory quality: transparency, completeness, consistency, comparability and accuracy. The guidelines provided to achieve qualitative inventory, adapted from the IPCC (2006), are summarized in Fig. 1.

As shown in Fig. 1, the IPCC guidelines to achieve qualitative inventory are, most importantly, data collection, uncertainty assessment (and forms of risk management), key category analysis, time series consistency of units, quality assurance and control, as well as chemical precursors and consideration of indirect emissions.

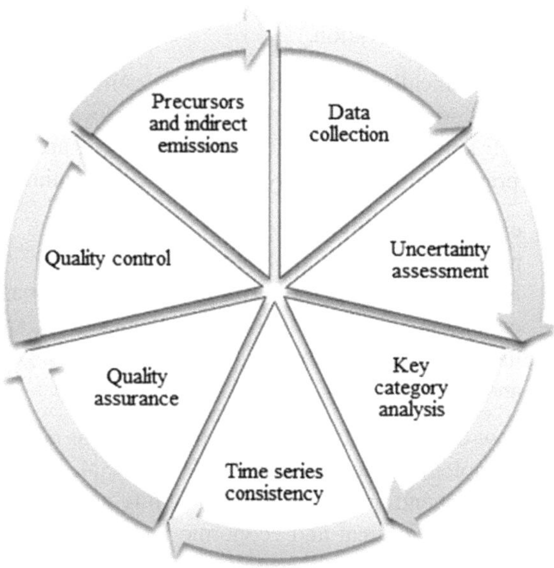

Fig. 1 The IPCC guidelines to achieve qualitative inventory reporting, adapted from the IPCC (2006)

The principles of inventory reporting presented in Fig. 1 are described in detail by the IPCC (2006). These principles are crucial to use for eco-city modeling and eco-planning as a means to systematically and logically describe the feasibility and integrity of statistical and measured data. Implementing these principles also requires planning and a holistic overview of the stages of eco-city development. Di Castri (2000) describes four stages of eco-city development, summarized by Acma (2013) and listed below. The four stages of eco-city development are

- initiating a realistic and feasible concept and doing comprehensive pre-planning,
- combining architecture and garden landscaping (eco-scape) with adequate planning and legislation considerations,
- incorporating eco-engineering energy-efficient design and long-term sustainable development, and
- continuous and meticulous ecosystem monitoring and management.

Furthermore, comprehensive human ecosystem analysis to assist in eco-city modeling and to understand relevant environmental impact factors within sustainable development is described by Grimm et al. (2000). Essentially, ecosystem analysis concerns identifying demographic patterns, understanding the economic system, implementing a power hierarchy, land use management and designing an efficient and sustainable environment. These tasks are achieved by incorporating disciplines within the engineering, built environment and architectural establishments. Acma (2013) also fittingly summarizes five motivations for eco-city developments, adapted from Grimm et al. (2000), presented as

- administrative authorizing,
- scientific supervision,
- industrial sponsoring,
- participation by citizens, and
- medium motivation.

Economically, development and investment in eco-cities and sustainable practice have different motivators, depending on local conditions and priorities within its infrastructure and policy frameworks. There are, however, common grounds that are present in most eco-city modeling and sustainable urban developments, such as

- energy-efficient buildings,
- retrofitting existing buildings,
- increasing and upgrading infrastructures for mass transit,
- endorsing short- to medium-distance cycling, and
- increasing the volume of distributed energy generation (Gouldson et al. 2015 and Erickson and Tempest 2014).

Low-carbon measures supporting different approaches, state-of-the-art technology implementations, collaboration, social interactions and renewable energy sources depend on the economic stability and energy security in a city or a country as a whole. Political and economic instability in a country typically lead to green economies and eco-city developments receiving lower investment drive from governments and private institutions; health, education, employment and equality are understandably among the higher-prioritized sectors in an unstable political or economic milieu. To ensure that sustainable development policies and measures are planned for, documented and properly promoted, Winkler et al. (2002) suggest steps to approach such a commitment in a generalized environment, based on the sustainable development policies and measures model they present. These steps include

- determining and drawing a detailed outline of future development objectives, quantifying the potential benefits and proposing comprehensive risk management policies,
- identifying factors other than only climate change that would lead to a sustainable development path (demographic equality and social interactions for example), and
- quantifying and documenting fluctuations in GHG emissions resulting from individual policies and measures to measure the probability of success or failure of these implementations (Winkler et al. 2002).

Sustainable development policies, although generalized, should be applied to specific domains or sub-sectors that have a common goal of energy efficiency and long-term sustainability. Applying models and equivalent scenario analysis to sub-sectors ensures that relevant parameters and variables, specific to each domain, are identified and leads to less-complex models with fewer input variables and disturbances. Shukla (2013) identifies and summarizes the three key modeling domains and the sub-sectors attributed to each of these key domains. The three primary modeling domains, as presented in Shukla (2013) are

- spatial domain attributes,
- temporal domains, and
- sectoral domains.

Under the spatial domain attributes, physical (such as buildings) and non-physical entities (such as borders) define urban and rural territories. Essentially, the spatial domain determines the extent of a model; in a smaller spatial domain it is generally easier to manage the data gathering and statistical updates for a defined characteristic. The spatial domain can be sub-divided into (from smallest physical size to largest) local, national, regional and global domains. Technology advances are in a sense shrinking the spatial domain boundaries through miniaturization and an improved capability to operate on renewable energy sources. Electronic equipment such as radio-frequency identification and wireless sensing networks are capable of operating at extremely low power levels and requires less energy-intensive supplies. Spatial distribution of technology nodes able to record and transmit environmental data is allowing modeling and prediction of environmental impacts on a broad level.

The temporal modeling domain concerns the time associated with data capturing, therefore the availability of historical statistics, as well as the accuracy of forecasting as a function of the period of predictions. The temporal domain, according to Shukla (2013), encompasses short-, long- and medium-term considerations. Temporal databases typically have two primary time aspects: valid time and transaction time. Valid time is the real-world time at which the phenomenon that generates the data occurs, whereas transaction time is the time when the data are recorded (Shukla 2013).

The third key modeling domain, which is also largely referenced in this chapter, is the sectoral domain. Categorizing emissions by their contributing domain simplifies the modeling process and allows cross-correlation and interaction-modeling among sectors. The sectoral domain can be divided into the sectors of

- energy,
- industry,
- agriculture,
- forestry,
- other land-use,
- transport,
- buildings, and
- waste.

Emission modeling within each sector is a dynamic research discipline; this chapter aims to highlight the considerations and concerns of generating a complete eco-city and sustainable model by identifying and understanding the characteristics of the individual sectors. As presented in Linderman et al. (2005), for example, modeling of human impacts and projected environmental disturbances by a combination of modeling domains is required to determine spatial distribution effects on temporal predictions. Linderman et al. (2005) follow the spatio-temporal integrated analysis to propose changes in household management and activities to maintain current levels of environmental stability.

Household energy demands are among the key modeling considerations presented in this chapter; the following section offers introductory contextual information about terms and concepts used to describe modeling considerations in each sector. The concepts of sustainable urbanization are presented in the following section.

2 Concepts of Sustainable Urbanization

As this chapter concerns low-carbon cities and modeling considerations of eco-cities integrated into eco-planning, distinguishing between the primary types of environmentally conscious cities is beneficial to recognize and acknowledge limitations, fundamental differences and strategies to achieve energy-efficient urban centers. The primary terms of energy-efficient city descriptions, adapted from Kim (2009), De Jong et al. (2015) and various other sources, listed in order of most often used in literature according to De Jong et al. (2015), are

- sustainable city,
- smart city,
- digital city,
- eco-city,
- green city, and
- low-carbon city.

Other concepts that describe and promote sustainable urbanization include knowledge city, resilient city, intelligent city, ubiquitous city, livable city and information city. De Jong et al. (2015) and Hassan and Lee (2015) both explain the differences among these concepts; albeit small differences, each concept focuses on different facets to obtain sustainable urban environment. City concepts are also not mutually exclusive and it is common for cities to implement a combination of these concepts, distinguishing between the concepts purely for convenience and to establish policies and frameworks for each. Since the concept of smart cities is a relatively modern invention, in certain cases the concepts are misunderstood and could be interpreted incorrectly. Typically, the effects of such malpractice are not inherently damaging and do not detract from the credibility of the goals.

Kim (2009) presents eco-city, sustainable city and low-carbon green city concepts with conceptual images for each concept. Essentially, an *eco-city* is an idea or perception of recognizing the complex relationships between humans and nature to cultivate biodiversity by integrating natural and urban environments. The aim is to decrease pollution and effects such as heat islands through a similar practice of recognizing the potential synergies between nature and anthropogenic urbanization.

A *sustainable city* concerns co-association of economic, social and cultural aspects for efficient mechanisms of production and consumption of resources, with environmental sustainability being a primary driver. The key concept of developing a sustainable city is creating an environment that has a long-term, maintainable (from a technology and human perspective) and as small as possible environmental impact in terms of GHG production and any other form of waste generation.

Because of the inherent similarities, Kim (2009) combines the concepts of a low-carbon and a green city and presents a conceptual model of a *low-carbon green city*. Through this concept, essentially a city intends to methodically integrate mitigation and adaptation strategies, which enables it to respond to climate change, primarily through eco-planned urban infrastructure. Low-carbon and green cities therefore use planning, modeling and technology to improve energy efficiency and promote cleaner energy sources, as well as lower the demand for energy. The primary driver for this is reducing or eliminating dependence on non-renewable energy.

A *digital city* refers mainly to strategies and mechanisms implemented to monitor, analyze and distribute information about the immediate environment, primarily aspects such as pollution and energy consumption. Digital cities use techniques such as internet services and networks, cloud-based services, wireless sensor networks, the internet-of-things and other forms of ubiquitous computing, essentially creating a virtual city able to gather information and provide the means, through data, to achieve a sustainable urban or rural environment.

Sustainable urbanization and the terms to describe the typical forms of environmentally conscious cities also require indicators that cannot be technically categorized and concern socio-economic and environmental protection as key descriptors. The economic policies and indicators are described in the following section.

3 Economic Policies and Indicators

In the Department of Minerals and Energy (1998) White Paper on energy, a comprehensive differentiation between two defining sectors to promote development policies are provided. Essentially, the energy sector is divided into a supply and a demand sector. This is also applicable to other sectors. To model a low-carbon city effectively, energy generation and energy expenditure must be quantifiable. Quantifying it makes it possible to develop strategies, policies and objectives to distribute resources and to promote growth and secure sustainable development. Common policies and objectives in the energy sector, for example, include increasing access to affordable energy services (through market-based models, government-led models, private company participation or subsidies), improving energy governance, stimulating economic development, managing energy-related environmental and health impacts and securing energy supply through diversity (Department of Minerals and Energy 1998).

Wang and Ye (2004) in addition summarize a similar set of three indicators for the development of an economic urban or rural area. These indicators include economic development, environmental protection and social progress. According to Wang and Ye (2004), the first indicator, economic development, concerns productivity in terms of per capita gross domestic product, the efficiency of the use of resources such as water and energy, as well as the potential for future developments and improvements on available infrastructure.

The second indicator, environmental protection, concerns areas kept and maintained as forests, per capita available public green spaces in urban areas, area conservation statistics, restoration of land degraded or destroyed by industrialization and the quality of the environment in terms of water, air and noise pollution. The environmental quality and environmental amenities such as aesthetic views or proximity to recreational sites play a significant role in eco-city modeling, as the commonly-used hedonic pricing method indicates (Rosen 1974). Freeman (2003) and Freeman et al. (2014) provide an in-depth discussion on measuring environmental and resource values and Zabel and Kiel (2000) use these methods to calculate welfare gains from non-marginal improvements in air quality.

The third indicator, social progress, is quantified by local infrastructure levels, therefore the percentage of the population that has access to drinking water, heat and gas. Social progress is also measured by factors such as the urbanization ratio, the quality of life of local residents, social equity, equality, education and awareness of the environment.

These three indicators can be addressed in both the demand and supply domains. Identifying each indicator within both domains provides a structural and logical means to perform eco-city modeling and address problem factors, risk management and enhancement potential. Modeling of carbon emissions requires determining of the amount of pollutants and GHG discharged into the atmosphere from all source categories, a process referred to as emission inventories. Geographical information and time-based statistical representation of pollutants through emission inventories are not only used scientifically to quantify air quality, but also for developing strategies and policies to manage pollutants and GHG, as described in the following section.

4 Emission Inventories

The IPCC provides a simplistic and commonly adapted methodological approach to present information on the extent of anthropogenic activity in the natural environment. This approach combines time and geographic specificity with inventory boundary activity data with coefficients that quantify emissions or removals per unit of activity, referred to as emission factors. For certain activities it is possible for cities to implement direct measurement of GHG emissions through, for example, continuous emissions monitoring at power stations, point source GHG emissions from industrial waste facilities or other technology-specific measuring techniques (WRI/WBCSD GHG Protocol 2014). For most emissions, however, cities must estimate the GHG emissions by calculating the product of activity data with the mass of the GHG emissions relative to a unit of activity. This relationship is represented simply by

$$emissions = activity\,data \times emission\,factors \tag{1}$$

where the activity data are categorized within a sector (energy generation, industry, commerce, mining, agriculture, households or transport) and the emission factors are the mass of CO_2 or other GHG emitted per unit of fossil energy consumed within the sector (g CO_2/kWh). Importantly, effective and holistic emission estimations and predictions should be implemented across all supply sectors, including

- traditional power-plant electricity and heat generation,
- nuclear energy if implemented in a country,
- oil and gas production,
- exploration energy consumption,
- all types of liquid fuels commonly used in households,

- all types of gas energy sources,
- coal production and associated energy consumption,
- renewable energy sources, and
- low-smoke transition fuels.

Emission estimations, as opposed to direct measurements of GHG emissions, require reliable and up-to-date sourced data from activities of energy consumption and other activities that contribute to carbon emissions. Adapted from IPCC (2006) and iterated in the WRI/WBCSD GHG Protocol (2014) are four data collection principles to ensure good practice of data gathering, namely

- establishing sustainable and effective processes of data collection, which also covers the prioritization of resources, proper planning of collection activities and comprehensive documentation for post-analysis,
- identifying critical datasets with the largest potential to contribute or hold the greatest uncertainty, requiring thorough statistical proof,
- not assuming that data collection principles do not change over time and regularly reviewing, adapting or changing collection principles, and
- incorporating reliable and robust data suppliers such as government and statistics agencies, research institutes or the national GHG inventory report of a country to support consistent and continuous flow of information.

Activity data, ideally from local and national sources, should be reputable and publicly available. This ensures that the data, if not from the body of knowledge in peer-reviewed journals, are extensively scrutinized. Sourced data should contain the contact information of the sourcing institution, clear and concise tracking of the dates on which the data were obtained and published, as well as unambiguous clarity about units, assumptions, uncertainties and known shortcomings (WRI/WBCSD GHG Protocol 2014). In addition, if specific data are unavailable, the IPCC provides an emission factor database with default values of carbon emissions from all sectors. Even if reliable and accurate data are available, it is a complex task to regulate a complete and clear synopsis of an entire energy sector to identify and quantify interactions and interdependencies among sectors. The Department of Minerals and Energy (1998) outlines objectives of energy sector policies, which contribute to sustainable national growth and a cohesive development strategy. These objectives are briefly summarized as

- affordability: Improving and increasing national access to sustainable, affordable and reliable energy services and aiming to provide energy to most of the population;
- governance: National and local (and in certain instances, international) management of public resources and improving on processes of energy governance;
- stimulation: Encouraging economic development by promoting competition within sectors through transparent and regulatory mechanisms that aim to provide better, more sustainable and dependable services in urbanized and rural developments;

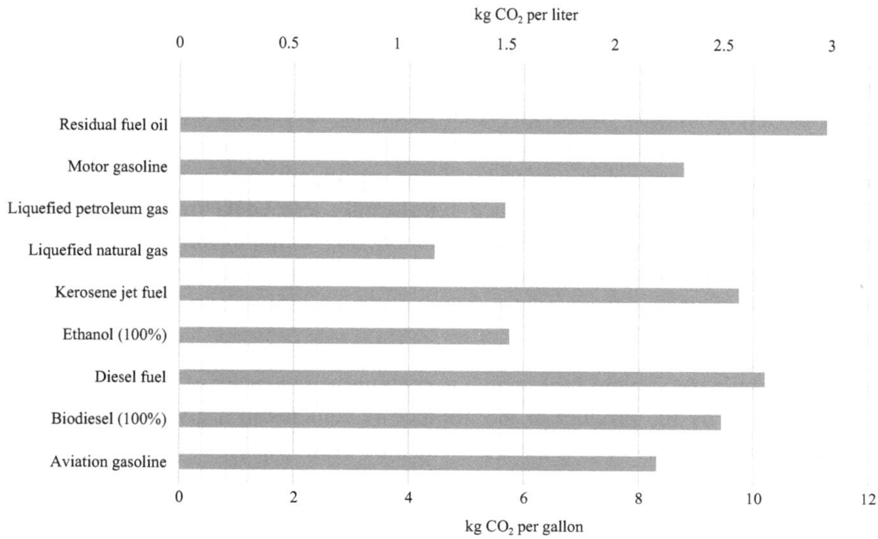

Fig. 2 The average EPA emission factors of mobile CO_2 combustion presented in kg CO_2 per liter and kg CO_2 per gallon

- assessments: Identifying, understanding and managing potential environmental and health impacts and recognizing investment potential to mitigate or reduce adverse effects; and
- diversifying: Securing energy supply through a diverse set of alternative and traditional energy sources.

If data capturing is not practiced and historical data are not available, emission factor databases provide universal data on fuel emissions. Adapted from the United States Environmental Protection Agency (EPA) emission factor database, Fig. 2 summarizes the carbon emissions of commonly used fuels. The units of measurement in Fig. 2 are both in kg of CO_2 per liter and kg of CO_2 per United States gallon. These mobile carbon-combustion emissions factors are commonly used to model the effects of carbon emissions in the environment, in cities and in rural areas, such that the measured energy consumption can be converted to the weight of CO_2 emitted into the atmosphere (EPA 2015).

Refraining from delving into the full details of the mechanisms producing carbon emissions in each fuel type, Fig. 2 aims to give a brief indication of the largest fuel-type contributors of CO_2 to the atmosphere. As seen from Fig. 2, however, most fuel types contribute significantly to carbon emissions and there are no distinct outliers in this list. Fuels omitted from Fig. 2 include vegetable oil (9.79 kg CO_2 per gallon or 2.59 kg CO_2 per liter), anthracite coal (2.602 kg CO_2 per short ton[1]), agricultural byproducts (975 kg CO_2 per short ton) and wood/wood residuals

[1]A short ton is equal to 2000 lb or 907.18474 kg.

(1.640 kg CO_2 per short ton). The EPA emission factors database provides extensive default values for most fuel sources in terms of CO_2, CH_4 and nitrous oxide (N_2O) factors.

Equally relevant and important when modeling eco-cities, whether new developments or implementing changes or retrofitting existing infrastructure, is determining or measuring carbon emissions from specifically the transportation sector. The transportation sector can produce significantly less GHG by planning routes, modeling urban and intercity travel and modeling tailpipe emissions. Transport emissions are measured and presented similarly to traditional energy consumption fuel usage, by its CO_2 factor in kg per transportation unit, where the transportation units range among

- vehicle-mile (or vehicle-km), which is essentially a measure of traffic flow and is determined by the product of the number of vehicles in a given transportation network and the average distance of each of these trips,
- passenger-mile (or passenger-km), representing the distance traveled by passengers on public transportation, calculated by the product of the number of unlinked passenger trips and the average length of these trips,
- passengers per bus hour, which quantifies the congestion of traffic based on the passengers accommodated on public transport: a high passenger count can still result in low passenger bus hours in congested areas,
- passengers per bus-mile (or bus-km), which is often used for longer-haul transport of passengers based on the number of passengers and the distance traveled; or
- ton-mile (or ton-km or kg-km) which relates the transport energy requirement to move one unit of freight across one unit of distance.

Methane (CH_4) and N_2O are non-CO_2 emissions and are measured against CO_2 based on their global warming potential (GWP), often referred to by their CO_2-equivalent (CO_2eq) emissions. The GWP of a gas is the warming caused for 1 ton of gas over a century, relative to CO_2 emissions of the same quantity and over the same period of time. CH_4 has a GWP of 25, whereas the toxic pollutant and GHG, N_2O, has a GWP of 298. The GWP of common GHG is provided in Table 1.

Table 1 presents the GWP and the average radiative forcing capacity of GHG. The radiative forcing (ΔF) capacity of GHG and aerosols is defined as the quantity of energy per unit area in a unit of time that is absorbed by a GHG; energy that would be dissipated in space instead. Radiative forcing therefore denotes an externally imposed perturbation in the radiative energy budget of the climate system of the earth (Ramaswamy et al. 2000). Mathematically, ΔF is described for CO_2 as the trace gas as

$$\Delta F = \alpha(g(C) - g(C_0)) \tag{2}$$

where

Table 1 List of the GWP and average radiative forcing of common GHG measured against CO_2 (adapted from Ramaswamy et al. 2000)

Gas	GWP	Average radiative forcing (W/m^2)
Gases relevant to radiative forcing only		
CO_2 (baseline of measurements)	1	1.46
CH_4 (methane)	21–25	0.48
N_2O (nitrous oxide)	298–310	0.15
PFC (per fluorocarbons)	6500–9200	0.34
HFC (hydro fluorocarbons)	140–11,700	0.34
SF_6 (sulfur hexafluoride)	7400–23,900	0.34
NF_3 (nitrogen trifluoride)	17,200	–
CF_4 (tetrafluoromethane, carbon tetrafluoride)	7390	0.003
C_2F_6 (hexafluoroethane)	12,200	0.001

$$g(C) = \ln\left(1 + 1.2C + 0.005C^2 + 1.4e^{-6}C^3\right) \tag{3}$$

where C is CO_2 in parts per million (ppm) and the radiative efficiency constant α is 3.35 [summarized by Ramaswamy et al. (2000) and originally determined by Hansen et al. (1988)]. For CH_4 as a trace gas, the radiative forcing capacity is given by

$$\Delta F = \alpha\left(\sqrt{M} - \sqrt{M_0}\right) - \left(f(M, N_0) - f(M_0, N_0)\right) \tag{4}$$

where α is 0.036 and

$$f(M, N) = 0.47 \ln\left[1 + 2.01e^{-5}(MN)^{0.75} + 5.31e^{-15}M(MN)^{1.52}\right] \tag{5}$$

and M is CH_4 concentration in parts per billion (ppb) and N is N_2O concentration in ppb. The subscript *0* denotes the unperturbed concentration of the gases. If N_2O is the trace gas, the radiative forcing component is given by

$$\Delta F = \alpha\left(\sqrt{N} - \sqrt{N_0}\right) - \left(f(M_0, N) - f(M_0, N_0)\right) \tag{6}$$

where the constant α is equal to 0.12 as presented by Ramaswamy et al. (2000). A detailed review and description of radiative forcing of climate change is presented by Ramaswamy et al. (2000).

Similar to Fig. 2, which presents a guideline to the most prominent fuel sources with their carbon contributions, Fig. 3 presents common modes of transport and the weight of CO_2 released into the atmosphere per specified unit. Essentially, the trend of carbon emissions for vehicles is that an increase in physical size of the propelled vehicle and therefore its engine also increases the amount of GHG produced during

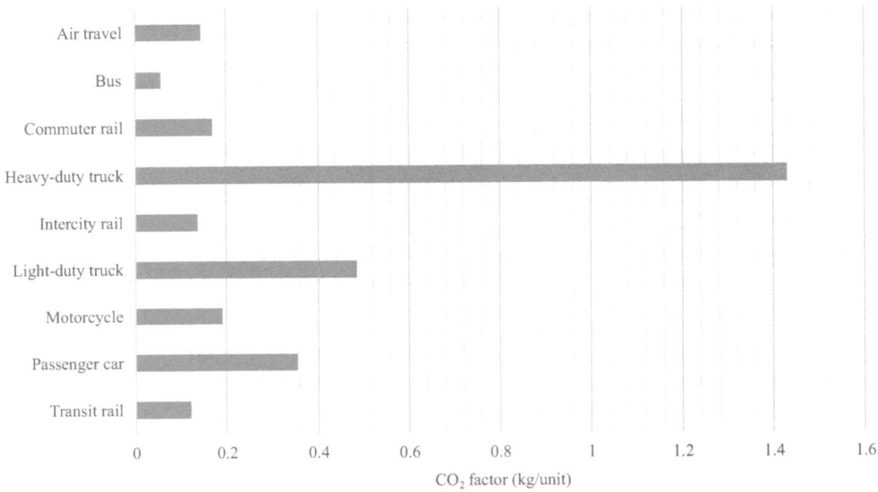

Fig. 3 The average EPA travel emission factors of CO_2 combustion presented in kg CO_2 per specific unit

travel. Fig. 3 represents the CO_2 emissions of air travel (average among short-, medium- and long-haul), bus travel, commuter rail, heavy-duty trucks, intercity rail (long-distance rail among cities), light-duty trucks (full-sized pickup trucks and vans, as well as extended length sports utility vehicles), motorcycles, passenger cars (cars, minivans, sports utility vehicles and small pickup trucks) and transit rail (rails in urban areas such as subways, elevated railways, metropolitan railways and trams). All units are in passenger-mile except for passenger cars, light-duty trucks and motorcycles, which are given in vehicle-mile. The vehicle types in Fig. 3 are listed alphabetically and in no particular order of emission averages.

As seen in Fig. 3, as expected, heavy-duty trucks, light-duty trucks and passenger cars, which are all propelled by fossil fuels, contribute the largest portion of CO_2 emissions. Railways (commuter, intercity and transit) contribute relatively low emission factors primarily owing to the passenger-mile (or passenger-km) factors in favor of transporting high volumes of passengers in single trips. Bus travel offers a similar advantage, assuming the bus system in a city is efficiently utilized and congestion levels are not above normal. Motorcycles contribute relatively low emission factors, primarily owing to the smaller fossil-fuel-burning engines.

Figure 4 shows the non-CO_2 gas emissions (specifically CH_4 and N_2O; not accounting for particulate matter, hydrofluorocarbons, aldehydes and ketones, which are formed by incomplete combustion and have carcinogenic and ozone formulation potential) of the same vehicle types, also adapted from the EPA emission factors database and used for eco-city modeling. These parameters are also used to determine effective and energy-efficient means of transport in cities, in rural areas and for intercity commuting. The emissions in Fig. 4 are given in units of equivalent CO_2 emissions weight per unit.

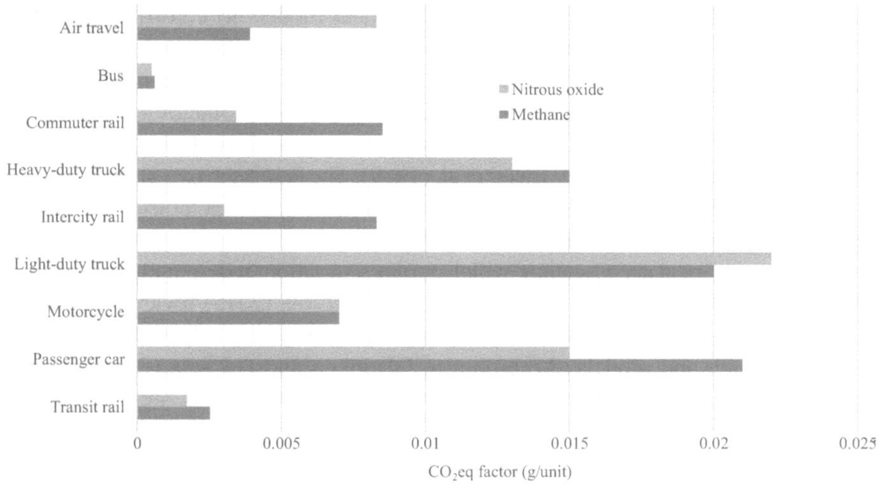

Fig. 4 The average EPA travel emission factors of non-CO_2 gases (CH_4 and N_2O) presented in kg CO_2eq per specific unit

In Fig. 4, the exhaust CH_4 and N_2O emissions of the types of vehicles show a relatively similar trend when compared to the CO_2 emissions given in Fig. 3. Road vehicles such as trucks, passenger cars and motorcycles contribute high CO_2eq emissions, whereas air travel and railways contribute significantly less. These effects are partly due to the passenger-mile factors being higher for these modes of transport, as well as the gas-burning mechanisms, where older vehicles with older technology, engine design and engine parameters typically produce higher quantities of CH_4 during internal combustion. CH_4 is emitted from vehicles resulting from incomplete combustion of fuel and the incomplete oxidation of engine-out CH_4 in current catalytic after-treatment systems (Kumar et al. 2011; Huai et al. 2004). N_2O emissions are also dependent on the combustion process of fossil fuels, but more so on the type of emission control system in engines, which is determined by the temperature, pressure, sulfur content and age/condition of the catalyst.

The emission inventories presented in this section provide a baseline for determining GHG emission from energy sources and from transportation. These parameters are useful in determining and modeling primary sectors where high carbon emissions are typically found, such as in consumer households, industry, commerce and mining, transport and agriculture. The following section reviews the energy demands of consumer households.

5 Consumer Household Energy Demands

Demands of consumer households for electricity, gas and other fuels such as forms of renewable energy are used to model the energy expenditure of individual households and identify how individual consumption changes relative to energy price and availability. Baker et al. (1989) provide theoretical models to construct a constant utility-based measure of household welfare to identify the welfare costs and potential benefits of energy price changes and subsidies paid to particular energy items. The models provided by Baker et al. (1989) in addition provide econometric specification of energy-demand equations. Two models to estimate and quantify consumer demand and energy expenditure are given by Baker et al. (1989). The first is the two-stage budgeting model, which implements a recursive structure that allocates household income between fuels and non-fuels and secondly disaggregate fuel expenditure. The second model is an empirical specification, which exploits variation in demographic and other characteristics across households. Oladokun and Odesola (2015) review how household energy consumption and carbon emission modeling have evolved compared to historic research and models. According to Oladokun and Odesola (2015), primary approaches to model energy consumption and carbon emissions are econometric, building physics and statistical models. Importantly, the limitations of these models are also highlighted and these limitations are evident in many eco-city modeling techniques, typically because of the complexity and multivariate nature of energy demand. Limitations such as lack of transparency in model algorithms, complex and interdependent environments, limited indications of occupant-dwelling interactions and lack of qualitative input data hinder many energy consumption and supply-demand interaction models. Interdisciplinary approaches are required to holistically quantify household energy demand and consumption leading to carbon emissions. Interaction among the disciplines of complex technology, society, economics, climatology and culture (Oladokun and Odesola 2015) attempts to overcome the restraints of simplified approaches. Motawa and Oladokun (2015) expand on the research methodology through problem articulation, model boundaries, dynamic hypothesis, causal loop diagrams and model formulation of complex household energy consumption. Motawa and Oladokun (2015) conclude that the judgement and reasoning of researchers and industry practitioners have proven more valuable in optimizing household energy consumption and carbon emissions compared to data collected on household demands on energy. Multivariate and interdisciplinary data on energy consumption lead to complex and scenario-specific models, which must be altered and optimized for variability of the immediate environment, social aspects and economic prosperity. Essentially, a complex model of energy dependency of a household requires model boundaries distributed along three core types of variables, namely

- variables exhibiting endogenous behavior, whose value is therefore dependent and determined by the current or previous states of additional variables within the enclosed system,
- exogenous variables, which are factors in causal models whose value is not dependent from the current or previous states of additional variables within the enclosed system, and
- excluded variables.

In household energy consumption modeling, for example, instances of endogenous variables included in model boundary definitions include the effects of insulation on energy efficiency, the effects on energy efficiency of standard appliances, total household carbon emissions and also the number of births, mortality and household size. These all determine the energy demand of the household. Motawa and Oladokun (2015) list approximately 50 additional endogenous dynamic variables taken into account during modeling of household energy demand. Examples of exogenous variables include, among the detailed list of parameters listed by Motawa and Oladokun (2015), total floor area, solar flux, external air temperature and relative humidity. Each variable is independent of the states of other variables in the system. Excluded variables may, for example, include parameters related to the behavior, culture and social preferences of occupants, certain physical parameters of dwellings, such as their orientation, as well as parameters correlated to the external environment, such as new technologies, political qualms and energy security. Household classification and related end uses in addition structure and categorize demand for energy in households based on socio-economic parameters. Household groupings into, for example, low-, middle- and high-income electrified and non-electrified datasets determine the primary consumption areas or end uses within a household. These end uses typically include, for electrified households, food refrigeration, preparing and cooking of food, inside and outside lighting, space heating of rooms and living areas, water heating for consumption or washing and other end uses such as technology, entertainment and irrigation. For non-electrified households, the primary uses of energy are lighting, cooking, space heating and water heating only.

Marszal-Pomianowska et al. (2016) present a high-resolution model of household electricity from a combination of measured and statistical data in a bottom-up approach. The effects of parameters such as the number of occupants in a household, as well as their attitude to energy use, are included in the model. The structure of the implemented model is presented by Marszal-Pomianowska et al. (2016); it highlights input data for each appliance in the household, which includes activity probability based on the time of day and the day of the week and also the frequency of use of each appliance, the number of occupants and their respective approaches to energy savings, as well as certain seasonal variations and social factors that influence the use of energy. Essentially, the model creates a list of all appliances installed in the household, determines usage for each appliance and usage of energy of the household and generates resultant load profiles of specified resolution. The model is validated by obtaining reliable data for demand on overall energy use, a more practical approach than accurate data on the demand of single appliances. Results are categorized into the annual electricity demands of the individual households, with variations in the number of occupants, seasonal demand, peak and average demand, as well as daily load profiles. Potential enhancements of this model include improvements on algorithms used for lighting, including socio-economic factors and extending the model for various building topologies.

The following section reviews modeling considerations of the industry, commerce and mining sector, a heavy carbon-emitting industry compared to household energy demands and the largest contributor to carbon pollution through the burning of fossil fuels.

6 Industry, Commerce and Mining Energy Demands

It is a well-known and undoubted fact that heavy industries such as the steel and cement sectors, the buying and selling of goods and services and the mining industry are all large contributors to anthropogenic carbon pollution (Van Ruijven et al. 2016).

Several modeling studies to estimate energy consumption, modern use and the effects of emissions on the environment have been conducted, each model typically being sector-specific and tailored to variations and improvements on production practices and technology advances in the sector. One such modeling tool is the Clean Energy Manufacturing Analysis Center (CEMAC) materials flow through industry (MFI) supply chain modeling tool. The CEMAC MFI tool tracks materials, resources, energy, water use and GHG emissions throughout commodity supply chains (CEMAC 2016). The CEMAC MFI tool is based on a database of over 650 industrial commodities and close to 1400 input-output models of material and fuel consumption by manufacturing processes. The vast number of combinations and the database of gathered information stress the complexity of modeling an economically viable and environmentally friendly industry and implementing eco-planning for new and current developments. No single model exists that accounts for all variables in all industries, but computer-aided models provide a framework to develop application-specific models.

It is necessary to analyze and record trends occurring over the long term, typically several years, to account for seasonal changes in energy use, generated carbon emissions and potential modification strategies and trade-offs/limitations for each strategy. In construction projects, for example, the workflow of the construction life cycle is identified by first defining the goal and scope of the project, then by implementing a life cycle inventory and finally by providing a detailed impact assessment. Each major stage or subsystem of the overall system can be individually characterized and quantized and the environmental impact can be determined (Li and Chen 2016). Construction projects are not only scrutinized during development, but also at their completion because they play an important role in eco-city and sustainable development. Environmental modeling and carbon emission contribution should be considered during the construction phase and based on long-term operation through sustainable practice.

Developed models are typically bottom-up approaches, which are based on material flows, macro-economic models or econometric models, which account for socioeconomic indicators and material demand energy use and emissions (Van Ruijven et al. 2016). According to Van Ruijven et al. (2016), the best models to represent patterns in historical data are linearized regression models that relate economic activity as gross domestic product per capita to material consumption per capita. Non-linear models are used by Wang and Ye (2016), who introduce a power exponent into the traditional multivariate grey model (Julong 1989) to describe the non-linear relationship between carbon emissions from fossil fuel energy consumption and economic growth. grey systems are typically used to describe systems that lack structure, operation mechanism and behavior documenting (Julong 1989). These systems have been applied effectively to various disciplines such as agriculture, ecology, economy, meteorology, medicine, history, geography, industry, geology, hydrology, irrigation, the military, traffic management, material science, the environment, biological protection and the judicial system, with many more applications applying the theory of grey systems. A complete description of the grey system is provided by Julong (1989).

Linear regression models can be used to predict variable behavior, forecast changes in processes and technology improvements or focus on error reduction by fitting a predictive model to an observed and recorded dataset. Assumptions of linear regression models, which may affect the overall accuracy and feasibility of the results, are their weak exogenous behavior, restriction of forced linearity, constant variance across response variables, the assumption of uncorrelated errors and the lack of multi-collinearity in the predictor variables. A typical linear regression model, given a dataset

$$\{y_i, x_{i1}, \ldots, x_{ip}\}_{i=1}^{n} \quad (7)$$

where y_i is the dependent variable on the linear p-vector of exogenous variable x_i. An unobserved random error variable ε_i is used to characterize the discrepancy from the linear relationships between the dependent variable and the exogenous variable. The linear regression model used to forecast economic behavior of the material consumption process in industry in its simplest form is given by

$$y_i = \beta_1 x_{i1} + \cdots + \beta_p x_{ip} + \varepsilon_i, \quad i = 1, \ldots, n \quad (8)$$

where β is a p-dimensional parameter vector representing regression coefficients for statistical estimation and interference.

In general, models are based on case studies in specific industries (again, the most typical studies are conducted in the cement and steel manufacturing industries) with geographic specificity. Many studies have been conducted on industries in rapidly growing and expanding counties such as China, Thailand, Japan and India. Rootzén and Johnson (2016) examine the impacts of carbon pricing and investments in carbon abatement in the steel industry. A case study of passenger vehicle manufacturing and the effects of steel price increases on cost structures, as well as price variations at each step in the supply chain, was assessed. Rootzén and Johnson (2016) conclude that in order to achieve a return on investment in low-carbon steelmaking processes, specific to this case study, the price of steel and inevitably the manufactured end-user products will increase dramatically, primarily owing to the large capital investment and long-term maintenance. This trend is unfortunately witnessed throughout industry, especially in established and long-running factories, mines, power plants and commerce. The trend is also seen on a lower scale, such as in households where initial capital investment in low-carbon practices still outweighs the viability of return on investment. Eco-city modeling is therefore only as effective as the potential for its uptake in industry and the commitments from several industries or individuals to absorb initial capital investments.

Gouldson et al. 2015 analyze the urban action scenario developed by Erickson and Tempest (2014) and provide a specific methodology to approximate superfluous costs as well as the benefits of exploring the potential contributions of cities in the global scenario. This methodology analysis offers useful parameters to estimate the economic impact of efficient urbanization. These parameters concern commercial and residential buildings, passenger and freight transport and waste management. To model a low-carbon city, it is important to include quantifiable parameters, which provide a positive or negative weighting factor to improvements and energy-efficient planning. In the commercial and residential building environment, concerning either new buildings or energy-efficient retrofitting of older, existing buildings, four crucial datasets should be accounted for. The economic impact of efficient buildings is determined by

- determining the cost of new buildings or retrofitting older buildings with low-carbon appliances, strategic lighting, temperature control and low-carbon materials; compared to an estimated (average) baseline for building. These baselines are commonly available based on the area where the development is planned and the type of building materials, specified in the local currency per m^2;
- secondly, the time associated with new energy-efficient developments or retrofitting of older buildings as a parameter of development area per year. This parameter can also be specified in development area per month, depending on the predicted time of completion;
- quantifying the economic impact, annual savings in energy consumption (specified in kWh per year), compared to a baseline of similar construction without additional energy-savings strategies, which provides a useful and quantifiable parameter to measure the long-term sustainability and value-added benefits of these developments; and
- finally, as specified by Gouldson et al. (2015), specifying the quantity of non-renewable fuel usage avoided through efficient developments as a percentage of the total energy consumption in a building.

These parameters are valuable contributions to determine the viability of additional expenditure on energy saving compared to the potential benefits and cost savings over the lifetime of the development or to determine the future estimated break-even point where the initial investment is superseded by annual savings in energy. Also covered in Gouldson et al. (2015) and listed under commercial and residential buildings are economic variables of appliances and lighting concerning the cost of saving a unit of electricity, specified in local currency per kWh, annual electricity savings in kWh per year and the energy savings per efficient appliance. The global drive towards photovoltaic cells to replace or contribute to energy generation also requires quantifiable parameters such as energy capacity factors, installation and maintenance cost and annual electricity savings compared to a traditional energy-generation baseline.

The following section reviews the modeling requirements and constraints used to estimate transport-based carbon emissions from various scenarios, fuel types and vehicle types.

7 Transport Energy Demands

As in the case of household and industry fuel consumption and carbon emission modeling, transport modeling of light and heavy vehicles has a large degree of uncertainty from real-world operating conditions. Travel demand models quantify traffic volumes through the interaction of transportation supply and demand. To distinguish between supply and demand, generally, travel demand is represented by *traffic zones,* which produce and attract trips, while the physical road networks represent the *travel supply*. Supply zones are defined by the population, socioeconomic status, number of households, average household size and employment ratio, among other factors. Transportation models, once they are set up, defined and producing relevant traffic information, must also be calibrated and properly analyzed to identify patterns and to improve the accuracy of the model. Transportation models are constantly changing with the introduction of new variables such as technology improvements and urban developments, and long-term adjustment of these models inevitably improves their accuracy; otherwise they may become redundant if not adapted to the changing environment.

Typically, for passenger vehicles, freight vehicles and aviation, factors such as:

- fleet configuration,
- operating conditions,
- vehicle type,
- road conditions or air traffic control, and
- driver profiles

are considered during modeling of transport emissions. Transport models provide quantitative information about travel demand and the performance of current and planned developments of transportation systems, providing valuable data and statistics to evaluate alternatives and employ informed decisions on energy-efficient eco-planning and eco-city transportation (Castiglione et al. 2015).

In the passenger and freight transport sector, Gouldson et al. (2015) also contribute significantly to awareness of reducing travel demands through urban planning and identify three main categories of energy savings in the transport sector. These are the shifting of transport modes for passenger transport, improving upon energy efficiency in the public transport sector and electrification of the public transport fleet. The associated considerations and variables of these three categories are presented in Fig. 5 (adapted from Gouldson et al. 2015).

In Fig. 5, the considerations of mode shifting, vehicle efficiency and electrification of fleets are listed. Common in the three categories is determining accurate

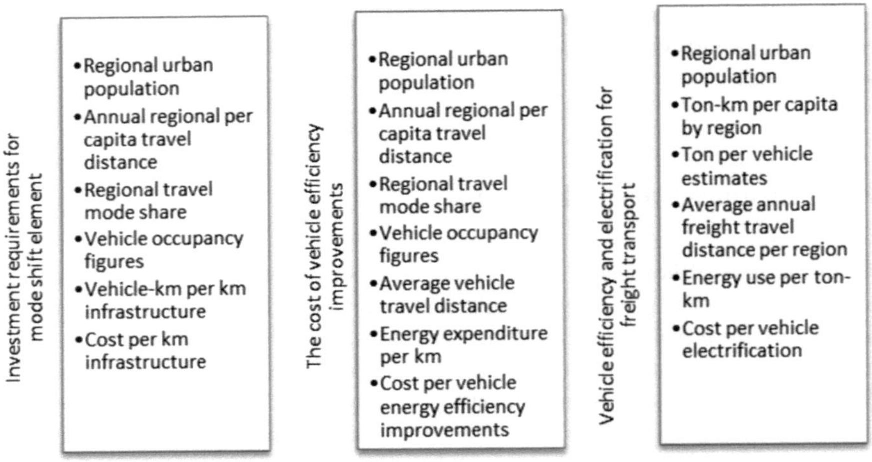

Fig. 5 Passenger and freight transport sector efficiency improvement calculation datasets as proposed and used by Gouldson et al. (2015)

statistics for urban populations by region, which also contributes to intercity travel demands and goods or service delivery among regions. Essentially, these considerations are used to determine the overall viability of upgrading or replacing existing transportation modes or technologies. The parameters can be used to plan and model feasibility studies in the passenger and freight transport sector.

Less obvious but equally relevant factors also play a role in efficiency modeling. Waygood and Avineri (2016) argue that the gender of the driver plays a significant role in the carbon emissions of vehicles. In this work several hypotheses are investigated regarding the differences between genders and their general concern about climate change. These concerns are based on education level, whether having children affects the attitude of people to climate change as well as how residents of developed and developing countries react to climate change. The results are detailed by Waygood and Avineri (2016); essentially, the results show that there is in fact correlation between socioeconomic characteristics and the concerns of individuals about climate change. These viewpoints can be linked to driver profiles, requiring complex and multivariate modeling techniques to model transport carbon emissions efficiently; again, these models are normally based on specific regions with certain de facto standards applied. In traffic-dense locations, low driving speed, driver frustration and requirements for air-conditioning typically also raise average fuel consumption significantly and lead to adaptations of transport emission modeling of carbon pollution.

Simplified representations of complex mechanisms to quantify the cause and effect of air pollution in the transportation sector require elaborate consideration of the problem. Air pollution from multiple sources, as is the case of pollution models from road vehicles and airplanes, typically entail combining

- microscopic (individual behavior needed to provide intelligence to the vehicle),
- mesoscopic (collective decision-making in groups or clusters of vehicles; physical, socio-cultural and communities), and
- macroscopic (the infrastructure of system-wide goals such as reducing congestion and efficiently using available land and roads, policies and governments)

sources along with data-backed estimations from trustworthy sources in combination with theoretical atmospheric dispersion models (Sanderson et al. 2012). The sources of emissions for on-road motor vehicles can be classified into two categories. The two categories are *exhaust emissions,* generated as by-products of the fuel combustion process, and *evaporative emissions,* which result from hydrocarbons emitted from the fuel. According to Delcan (2007), evaporative emissions are temperature-dependent, increase proportional to temperature and occur by four mechanisms, namely

- diurnal emissions, which evaporate from the fuel tank as the temperature rises,
- running loss emissions from heating of the engine,
- hot soak emissions from the heat of the engine when the vehicle is switched off, and
- refueling emissions from vapors escaping from the fuel tank during refueling.

Delcan (2007) provides a simplistic vehicle emission estimate through the product of vehicle activity data and the aggregate rate of emissions per unit of the activity; however, the simplistic estimation becomes complex depending on the required accuracy of the emissions evaluation. Activity data are categorized into macroscopic (kilometers traveled and average speed) and microscopic (real-time vehicle operation) activity. Emission factors are categorized by fleet, fuel, trip, environmental, regulatory and driver characteristics.

Road traffic air pollution can also be assessed by means of average emission estimations, as described by Marino et al. (2016). Such models necessitate simple representations to evaluate gaseous pollutants emitted by vehicles in urban environments. Vehicle emission models used to estimate emissions from sources contributing to air pollution, GHG and air toxins, especially in congested urban environments, available from the United States EPA online resource databases, include (these vehicle emission model descriptions are obtained primarily from the official online websites of each model)

- the motor vehicle emission simulator (MOVES) to estimate emissions from mobile sources at national, country and project level,
- the MOBILE vehicle emissions factor model to predict gram per mile emissions for hydrocarbons (HC), carbon monoxide (CO), oxides of nitrogen (NO_x), CO_2, PM and air toxins from cars, trucks and motorcycles under various conditions,
- assessment and reliability of transport emission models and inventory systems (ARTEMIS) for road, rail, air and ship transport at national and regional levels,
- the VERSIT+ emissions model to predict emission factors and energy use factors representative of vehicle fleets, differentiated for vehicle types and traffic situations, taking into account driving conditions,

- the computer program to calculate emissions from road transport (COPERT),
- the highway vehicle particulate emission modeling software to estimate PM emissions specifically from vehicles driving on highways, therefore at high speeds,
- the optimization model for reducing emissions of GHG from automobiles (OMEGA) to estimate technology costs for automobile manufacturers to achieve variable fleet-wide levels of vehicle GHG emissions,
- the mobile emissions assessment system for urban and regional evaluation (MEASURE) to improve predictions of CO, HC and NO_x,
- the traffic emissions and energy (TEE) model, and
- the GHG emissions model (GEM), which estimates the GHG emissions and fuel-efficiency performance of heavy-duty vehicles.

These models, as provided by the United States EPA, are used in many cases for studies on global transport emissions, typically adapted for specific situations and presented as case studies by multiple sources. The models can also be divided into categories concerning average speed models on macroscopic level (COPERT, MOBILE, EMFAC), traffic situation models such as ARTEMIS, traffic variable models such as the TEE model, cycle-variable models concerning individual driving patterns such as the VERSIT+ and MEASURE modes and finally, modal models, a microscale approach to engine operation variables. Transport vehicle pollutant emissions are typically based on exhaust gas measurements of vehicles on test benches with varying driving cycles (Marino et al. 2016). Further travel demand traffic models with an emission component, which are given by Delcan (2007), include

- the systematic traffic evaluation and planning model (STEP),
- the transportation analysis simulation systems (TRANSIMS) model,
- SYNCHRO 7,
- CORSIM,
- PARAMICS, and
- VISSIM.

Furthermore, activity-based transportation models (Castiglione et al. 2015) are derived from the daily activities and routines of people in urban, suburban or agricultural environments. Activity-based transportation models are emerging as increasingly more important and applicable for modern eco-city planning and energy-efficient and low-carbon developments. These models predict which types of activities are most likely to be conducted at specific times, with regard to the demographics of the drivers for each activity (for example distinguishing between work and leisure), typical travel times for these activities and potential alternative travelling routes based on the demands of the activity—for example, shopping and leisure activities are more likely to have several alternative route options, whereas in comparison, work activities typically follow specific and unchanged routes for drivers. Activity-based transport models are somewhat similar to more traditional four-step models in the way that for each model,

- activity types are created,
- for these activities the destinations are defined,
- the types/modes of travel are identified, and
- routes are estimated.

The four-step transportation model is derived from the transportation system analysis (Manheim 1979), expanded by Florian et al. (1988) and revised by McNally (2007). Essentially, the four steps are:

- trip generation,
- trip distribution,
- mode choice, and
- trip assignment.

A detailed description of these four steps is provided by Teodorovic (2015) and provides insightful considerations concerning the four-step model. Activity-based transport models are primarily used and specifically appropriate for permitting a continuous distribution of value-of-travel-time-savings for populations and crucial for freeway pricing assessments. They give stronger accounting for all costs and utilities involved, not only for travelling as its own entity but also for common driver activities and transit routines. The activity-based transport model is essentially based on three theoretical methodologies of activity models, namely

- constraint-based models to assess the feasibility of an individual within particular space-time constraints,
- a stream of utility-maximizing models based on the idea that individual drivers aim to capitalize on their effectiveness when planning their daily schedule, and
- models that imitate judgement heuristics to avoid uncertainties of individuals, effectively maximizing the effectiveness of their daily schedule.

A report presented by Castiglione et al. (2015) provides a detailed discussion of activity-based models and serves as an excellent primer to activity-based models and the primary characteristics that make these models commonplace and essential in eco-city planning.

Regulatory simulation test procedures of real-world environments are required to authenticate and warrant compliance across various standards, specifically focusing on emissions of PM, NO_x, HC and CO. Emissions are usually expressed in units of grams of pollutant per unit distance traveled, for example g/km or g/mile. In addition, if measurements are made based on engine dynamometer test cycles, emission units are expressed in grams of pollutant per unit of mechanical energy delivered by the engine with unit g/kWh or g/bhp-hr. High temporal and spatial resolution emissions from cycle-variable models using ubiquitous networks of sensors and nodes are required to simulate air quality in urban spaces on interpretable scales. To allow vehicle manufacturers to comply with emission standards, also known as tailpipe emission standards, government ministries such as the United States EPA certify equipment (such as internal combustion engines) if requirements are met. These requirements and standards are put in place to ensure regulations of country-specific occupational health and safety standards of ambient air quality subjected to tailpipe emissions. Regulation factors are typically sub-divided into categories for

- the vehicle type and/or size,
- its age in years and accumulated traveling distance,
- type of fuel used,
- average ambient weather conditions of its primary region, and
- the maintenance records and mechanical condition of the vehicle.

Zhang et al. (2016) employ a fuel-consumption and carbon-emission modeling methodology of distance-specific fuel consumption based on two classifications: localized type-approval fuel consumption estimated by model-year or for fleet average under a baseline cycle, which takes into account vehicle specifications, and correcting the average fuel consumption to real-world operating conditions. The modeling approach of the two classifications given by Zhang et al. (2016) are mathematically represented by

$$FC_{TA(r,y)} = FC_{TA(originalr,y)} \times C_{size(r,y)} \tag{9}$$

where $FC_{TA(r,y)}$ is the localized type-approval fuel consumption for vehicles manufactured in region r and year y, $FC_{TA(original\ r,y)}$ is the average type-approval fuel consumption of the total number of vehicles in region r and year y. $C_{size(r,y)}$ is the correction factor of vehicle sizes (Zhang et al. 2016). The mathematical representation for correcting for real-world operating conditions in the work of Zhang et al. (2016) is given as

$$FC_{real(y)} = FC_{TA(y)} \times SCF(v) \times C_{lm} + FC_{AC}(T_a) \tag{10}$$

where $FC_{real(y)}$ is the corrected real-world fuel consumption in year y, SCF is the speed (v) correction factor with the average speed of the vehicle fleet as the baseline, C_{lm} is the correction factor of the loading mass from passenger occupancy and FC_{AC} is additional fuel consumption due to air conditioning use as a function of ambient temperature T_a. Zhang et al. (2016) therefore provide an insightful investigation of environmentally specific transport modeling based on external factors such as driving speed, temperature control and average passenger occupancy in highly populated urban driving conditions.

Ehsani et al. (2016) construct a vehicle fuel consumption model for various types of vehicles and vehicle-specific features such as engine type (gasoline or diesel), road types and asphalt efficiency, a range of renewable and non-renewable fuel types used by transportation vehicles, driving style and overall fuel efficiency, as well as wind effects. The model of Ehsani et al. 2016 proposes a top-down approach of a mechanical model, which concerns losses of energy from the effects of gravity, acceleration (driving style), rolling resistance between the tires of the vehicle and the asphalt and aerodynamic resistance of the vehicle. The simplified relationship among these parameters is represented by Ehsani et al. (2016) as

$$U_{ijk} = U_g + U_i + U_r + U_d + U_c \quad (11)$$

where i represents the type of vehicle, j represents the engine type and k represents the fuel type, U_g is the energy consumed by losses due to gravity, U_i are losses from acceleration, U_r are the losses from rolling resistance, U_d are the aerodynamic losses and U_c are the cornering losses. The research conducted by Ehsani et al. (2016) is therefore specifically focused on mechanical vehicle attributes and all forces acting on the vehicle during movement; this is a predominantly physics-based approach combined with driver interaction and style.

The complexity and specificity of transport emissions modeling are somewhat relaxed in the agricultural sector, where a stricter and definite boundary can be drawn for activity data. Agricultural emissions modeling is, however, not a simple task, but allows more control over the input and output variables to a certain extent. Considerations of agricultural emissions modeling are presented in the following section.

8 Agricultural Energy Demands

Green urbanism, as described by Lehmann (2010), is holistically defined as a balanced relationship between urban and rural areas. Eco-city modeling depends on urban developments, energy-efficient practices and general support for sustainability, but it also depends on interactions with the agricultural sector on numerous facets. Farming and the livestock sector are significant contributors to global anthropogenic-induced GHG emissions. According to the IPCC, 14.5% of all human-induced GHG emissions are emitted as predominantly CH_4, N_2O and CO_2 by livestock supply chains. Based on statistics provided by the Food and Agriculture Organization (FAO) of the United Nations, CH_4 and N_2O emissions from agriculture are dependent on a combination of the development status of countries and the physical size of a country (or continent). According to the FAO statistics division (FAOSTAT) model, the average emissions by continent between 1990 and 2014 were:

- Asia (the largest continent) generated 42.3% of the total CH_4 and N_2O produced from agricultural activities; China (mainland) was the largest contributor in Asia;
- the Americas made the second-largest contribution at approximately 25.3% of the total GHG contribution from the agricultural sector, Brazil contributing more CO_2-equivalent emissions than the United States of America,
- Africa was in third place with its contribution between 1990 and 2014 estimated at 14.4%, just more than

- Europe with its 13.8% contribution to the GHG emissions average and
- Oceania having the lowest average contribution at approximately 4.1%.

The sources of CH_4 and N_2O (and CO_2) in the agricultural sector in each country or continent show a relatively comparable trend. There are typically four primary sources of emissions from livestock supply chains that contribute to the buildup of CH_4 and other gases in the atmosphere. The four primary sources as described by the FAO (2016) are

- the digestive process of ruminants and/or non-ruminants in a process called enteric fermentation,
- anaerobic decomposition of organic manure,
- aspects concerning livestock feed that include expansion of feeding crops and pastures through removal of natural land, the manufacturing of fertilizers and feed transport and processing, and
- energy consumption in all sectors of the supply chain.

Modeling GHG emissions in the agricultural sector requires gathering, analyzing and updating data in its sub-domains. The agricultural sub-domains supported by dataset models such as the FAOSTAT model are, for example

- enteric fermentation—(40.2% contribution to the total emissions from the agricultural sector, the largest contributor by a fair margin)—primary emissions: CH_4,
- manure management—(6.9% contribution to the total emissions from the agricultural sector)—primary emissions: CH_4 and N_2O,
- rice cultivation—(10.2% contribution to the total emissions from the agricultural sector)—primary emissions: CH_4,
- synthetic fertilizers—(11.5% contribution to the total emissions from the agricultural sector)—primary emissions: N_2O,
- manure applied to soils—(3.7% contribution to the total emissions from the agricultural sector)—primary emissions: N_2O,
- manure applied to pastures—(15.5% contribution to the total emissions from the agricultural sector, the second-largest contributor)—primary emissions: N_2O,
- crop residues—(3.6% contribution to the total emissions from the agricultural sector)—primary emissions: N_2O,
- cultivation of organic soils—(2.8% contribution to the total emissions from the agricultural sector)—primary emissions: N_2O,
- burning crop residues—(0.5% contribution to the total emissions from the agricultural sector, the lowest estimated contributor)—primary emissions: CH_4 and N_2O, and
- burning savannas—(5.2% contribution to the total emissions from the agricultural sector)—primary emissions: CH_4 and N_2O (FAO 2016).

In preparing national GHG estimates for the agricultural sector, national inventory compilers experience distinctive challenges in gathering and regularly updating national statistics for agriculture, forestry and other land use (cropland, grassland, wetlands and settlements) (Tubiello et al. 2015). This difficulty is especially prominent in developing countries (but also evident in developed nations) with already limited capacity to identify and collect reliable activity data and quantify emissions by sources and removals by sinks (Tubiello et al. 2015). In the agricultural sector, anthropogenic GHG emissions and removals of managed land are defined where human interventions and practices have been applied to perform production, ecological or social functions (IPCC 2006). Emission and removal processes are organized by ecosystem components of biomass, dead organic matter, soils and livestock. Agricultural emissions account for emissions produced in all the agricultural sub-domains, whereas emissions of non-CO_2 gases such as CH_4 and N_2O generated through crop and livestock production and management activities are among the predominant emissions modeled in agricultural practices.

Agricultural modeling and estimates of food production and consumption can be achieved with relatively good accuracy based on several assumptions, but determining the world average per capita availability of food for direct human consumption allows for good estimates of supply and demand, translating to a fair assumption with relatively low uncertainty on the GHG emissions associated with food production. The primary drivers of increased food production leading to higher GHG emissions from the agricultural sector are growing populations and varying income. The demand for agricultural products such as livestock (meat and milk) also increases with increases in population growth, rises in population wealth, increases in per capita consumption and changes in diets. The environmental implications associated with the expansion of livestock are deforestation, overgrazing and ground erosion and increases in CH_4 and N_2O emissions. In countries where intensive livestock operations on industrial scale are practiced, point-source pollution, such as effluents, is also evident in the agricultural and livestock sector (Alexandratos and Bruinsma 2012).

Ground water is the primary source of water for agriculture and industry; water scarcity or the depletion of underground reservoirs and environmental degradation effects in the agricultural sector are also felt in the urban areas for which it provides, which is all the more reason to include the agricultural sector in eco-city planning. In Beijing in China, for example, sustainable development among cross-disciplines is achieved by the development of urban agriculture (Wong and Yuen 2011).[2] Sustainability projects also include practices of eco-sanitation, where organic wastewater and compost are used for urban agriculture, urban farming or urban gardening. Urban agriculture also includes management and care of farm animals through the practice of animal husbandry, cultivating freshwater and saltwater aquatic populations under controlled conditions known as aquaculture, growing trees or shrubs among crops or pastureland in a practice known as agroforestry, urban beekeeping and the cultivation of crops with additional purposes related to

[2]Urban agriculture concerns the cultivation, processing and distribution of edible vegetation or livestock in an urbanized area.

art, science, technology and business, as well as in space-limited urban areas using vertical gardens for example.

Eco-city planning and sustainable city development should therefore also account for practices that remove (sink) carbon emissions from the atmosphere through urban agriculture, additionally contributing to the sustainability of these eco-cities. Watershed modeling of sustainable developments, which is also applied to forestry, is presented by Das (2008) and provides valuable insight into enhancing watershed retention to be used in agriculture and within eco-city urbanized environments. Das (2008) proposes a model for improving the hydrologic status or presence of water in the watershed through a package of interventions. The research presented by Das (2008) is another indication of the interdependence of agriculture, forestry and urbanization, which inevitably translates to eco-city modeling. The work shows how to prioritize integrated watershed development in order to attain overall sustainability for current and future populations.

The Global Livestock Environmental Assessment Model (GLEAM) developed by the FAO of the United Nations is a modeling framework that simulates the environmental impacts of the livestock sector. It represents the bio-physical processes and activities along livestock production chains under a life cycle assessment approach (FAO 2016). GLEAM is developed and designed to identify and quantify several environmental impact classifications in the agricultural sector and includes analysis of GHG emissions, land use and land degradation, nutrient and water use, as well as interaction with biodiversity. Among the described outputs of GLEAM, which include animal spatial distribution, production of manure, feed rations and livestock commodities production, the output of emissions from different stages of production (farming) provides valuable insight into the agricultural carbon footprint on micro- and macro-scale. The sources of GHG emissions in the GLEAM model are divided into three primary categories:

- upstream emissions—related to feed production, processing and transportation,
- animal production emissions from enteric fermentation, manure management and energy use on farms, and
- downstream emissions by processing and transport of livestock commodities (FAO 2016).

The GLEAM model therefore provides an evidence-based diagnostic of anthropogenic-induced effects on the environment and can be applied to create a sustainable livestock and agricultural sector. Although GHG and abundant concentrations of CO_2 present some advantages in agriculture, since a higher concentration of CO_2 would increase the yield for many crops, apart from maize, millet and sorghum, the adverse effects of loss of soil organic matter, leaching of soil nutrients, salinization and erosion, increased infestation of weeds, insects and diseases and a higher probability of extreme environmental conditions, such as droughts and floods, overshadow the miniscule benefits it may hold. Effective and holistic eco-city planning must take agricultural carbon-emission effects into account to produce sustainable growth. The following section summarizes the parameters required for carbon-emission modeling in the sectors discussed: residential households, industry, transport and agriculture.

9 Parameters Required for Modeling

This section summarizes the required parameters for modeling GHG emissions in the sectors discussed in this chapter.

Residential parameters required for modeling:

- Household sizes and types.
- Household demographics (includes multiple spatial and temporal characteristics —complexity of models considering demographics are high).
- Household wellbeing (average from residential areas/municipalities).
- The demand status and levels for energy services.
- Household ownership status.
- Energy prices and tariff structure.
- The overall efficiency levels of implemented energy-technology.
- Efficient technology shares levels.
- Renewable energy potential and predominant fuels used in area.
- Climatic considerations (indoor/outdoor temperature difference).

Industrial parameters required for modeling:

- Distributed shares of energy output within industrial subsectors.
- The demand status and levels for energy services.
- The overall efficiency levels of implemented energy-technology.
- Efficient technology shares levels.
- Carbon capture and storage levels.
- Technical and cost profiles of generating plants.
- Material demand energy use and emissions.

Transportation parameters required for modeling:

- Urban population by region.
- Demand levels of trip generation.
- Trip distances (annual km-passenger or ton-km) based on each separate mode of transport as well as the overall distance.
- Vehicle type emission factor.
- Industrial and commercial activity growth levels (for freight transport).
- The overall efficiency levels of implemented energy-technology.
- Efficient technology shares levels.
- Road conditions and traffic congestion impact.
- Vehicle occupancy.

Agricultural parameters required for modeling:

- Agricultural land use (area).
- Agricultural processes.
- Variable weather conditions (including irrigation requirements).
- Variable pest populations (fertilizer use).
- Dependence of outputs on inputs (efficiency).
- Carbon dioxide uptake by growing biomass.
- Livestock quantities and GHG emission profiles.
- Water treatment.
- Storage of manure.

10 Future Directions and Planning for Urban Growth

Eco-city planning and preparation for rapid urban growth are ongoing strategies that implement urban expansion management for the future, typically planned ahead for between 5 and 30 years. Urban expansion management accounts for issues such as:

- scenario planning for land use and housing requirements [not to be confused with policy options—a set of alternative responses by policymakers to a given outcome, forecast or scenario or by variants—the outputs from a macroeconomic model when assumptions about exogenous variables are changed (GLA Economics 2010)],
- performance indicators functioning as assessment of future urban expansion,
- land supply monitoring for urban expansion and decreasing current reliability and requirements of landfill,
- population projection to determine the additional resources required for future populations and implementing strategies to achieve them,
- economic projection and priority assessment,
- transport planning for efficient routes and amenities,
- urban planning and reducing risk,
- stakeholder engagement of local government and infrastructure agencies,
- potential prospects and limitations analysis,
- innovative financing through combinations of government, institutional investors, bonds, multilaterals and export credit agencies, outbound investments, initial public offerings, and public/private corporations,
- energy strategy and decreased dependence on non-renewable energy, and
- potential of renewable energy based on geographic location and weather conditions.

Each of these strategies contributes to a country or region's attractiveness index, as published by the renewable energy country attractiveness index (RECAI). The attractiveness of a country is primarily determined by the renewable energy investments and deployment opportunities based on five predominant factors:

- macro-fundamentals,
- energy imperatives,
- policy enablement,
- project delivery, and
- technological potential.

Countries ranked with high attractiveness indices show a strong focus of local governments on renewable energy as well as timely implementation of these projects. The aim is to find low-cost renewable energy to replace fossil fuel generation. Countries hampered by political infighting and mixed policy measures, as well as abrupt cancellation of renewable projects, typically have a lower attractiveness index. Based on RECAI (2016), the countries ranked from 1 to 40 by their attractiveness index are listed in Table 2.

Compared to historical data of the global attractiveness index ratings, recent trends show that Latin America and Africa are rising in global ranking and enjoying large investments in renewable energy projects. New energy reform policies are put in place to boost the renewable energy potential of these countries.

Table 2 Renewable energy country attractiveness index of the 40 most attractive countries based on renewable energy investment and deployment opportunities as published by RECAI (2016)

Rank	Country	Rank	Country
Rankings 1–20			
1	United States of America	11	South Africa
2	China	12	Japan
3	India	13	United Kingdom
4	Chile	14	Morocco
5	Germany	15	Denmark
6	Brazil	16	Egypt
7	Mexico	17	Netherlands
8	France	18	Argentina
9	Canada	19	Turkey
10	Australia	20	Belgium
Rankings 21–40			
21	Sweden	31	Ireland
22	Philippines	32	Jordan
23	South Korea	33	Uruguay
24	Peru	34	Norway
25	Italy	35	Poland
26	Israel	36	Finland
27	Portugal	37	Thailand
28	Spain	38	Pakistan
29	Taiwan	39	Indonesia
30	Kenya	40	Greece

Conversely, European countries have fallen in the rankings, with only Germany remaining in the top 10 countries (Denmark, Finland and the United Kingdom have all declined dramatically). It is, however, evident from the list of most attractive countries that renewable energy plays a crucial role in the future of cities. The aim, adapted from the FutureCity initiative in Japan, is to create an urban city and community with a sustainable economic and social system that can respond to the issues of aging and the environment. "*The future form, functionality, appearance and ambience of cities will have a direct impact on most people's lives, whether they live in a city or not. The future city will not only impact on society, but will also influence wider global environments and economies*" (Moir et al. 2014).

11 Conclusions

This chapter identifies the requirements for eco-city modeling and eco-planning of future low-carbon sustainable cities and ecological considerations in various sectors influencing the maintainability of a city. The concepts of sustainable urbanization are identified and listed in this chapter to differentiate between the differences in approach and implementation of models that aim to produce smarter and more ecologically friendly cities. The benefits of such a city are not only noticeable through direct measurements, but also have equally high importance for the mid- to long-term future of a city, as well as for the rural areas around a city, which are also affected by environmental degradation in the city, and vice versa. Terms such as sustainable, smart, digital, eco-, green- and low-carbon cities all have a common goal, reducing anthropogenic carbon emissions through implementation of policies, technology and concern for the environment. Choosing a strategy to achieve one or all of these concepts requires understanding of each concept, the requirements thereof and practical implementations and theoretical modeling or planning for achieving such a goal. This chapter also provides additional guidelines on economic policies and indicators that are used in current developments towards eco-cities; this material can be applied to a variety of sustainable practices.

Data are arguably the most important aspect of modeling and any predictive model is only as good as the information it is fed and its ability to learn and adapt to changing variables and scenarios. Therefore, data capturing requires reliable and accurate techniques, calibrated technology and rational interpretation of activities. The practice of data capturing is therefore not trivial and can become complex if multivariate quantities are considered and interaction between these variables have an effect on the global model. To reduce the complexity and improve on the turnaround time of predictions, data availability for generalized situations and scenarios within energy production, transport emissions, buildings and agriculture can be retrieved from numerous sources. This chapter lists some of the most commonly used carbon-emission profiles from various sources. These emission inventories are applied to models in sectors with predominant energy consumption and carbon emissions.

The sectors discussed in this chapter are consumer household energy demands, industry, commerce and mining, transport and agriculture. The consumption of energy in each of these sectors is typically not properly monitored or individuals are not aware of the energy requirements and emission profiles of the equipment they are using. This is especially dominant in industry and transport where excessive carbon emissions are released into the atmosphere and the policies that must regulate these emissions are disconnected from the experience of the individual. This chapter focuses on identifying and highlighting important considerations in each of these sectors, which benefits eco-city planning and modeling of low-carbon environments. The primary parameters required for modeling carbon emissions are listed in this chapter, recognizing the contributions of the various works listed in this chapter.

References

Acma, B. (2013). Ecosystem and ecocity planning in the Southeastern Anatolia region in Turkey. 5th *Symposium for Research in Protected Areas*. 1–5, Mittersill, 10–12 June 2013.

Alexandratos, N., & Bruinsma, J. (2012). World agriculture towards 2030/2050. The 2012 revision. *Food and Agriculture Organization of the United Nations*. ESA (Working Paper No. 12-03. June 2012).

Baker, P., Blundell, R., & Micklewright, J. (1989). Modelling household energy expenditures using micro-data. *The Economic Journal, 99*(397), 720–738.

Castiglione, J., Bradley, M., Gliebe, J. (2015). *Activity-based travel demand models: A primer. The Second Strategic Highway Research Program*. (SHRP 2 Report S2-C46-RR-1).

CEMAC. (2016). Modeling industrial materials flows to reduce energy demand and carbon emissions. Retrieved September 27, 2016 from http://www.manufacturingcleanenergy.org/

Das, D. C. (2008). System approach to watershed modelling for sustainable development, including forestry. *International Forestry Review, 10*(2), 281–291.

De Jong, M., Joss, S., Schraven, D., Zhan, C., & Weijnen, M. (2015). Sustainable-smart-resilient-low carbon-eco-knowledge cities; making sense of a multitude of concepts promoting sustainable urbanization. *Journal of Cleaner Production, 12*(109), 25–38.

Delcan. (2007). Guidelines for quantifying vehicle emissions within the ministry's multiple account evaluation framework. *A Report to British Columbia Ministry of Transportation*. November 2007.

Department of Minerals and Energy. (1998). White Paper on the energy policy of the Republic of South Africa. ISBN 0-9584235-8-X, December 1998.

Dhar, S., Pathak, M., Shukla, P. R. (2013). Low carbon city: A guidebook for city planners and practitioners. Retrieved September 5, 2016 from http://www.unep.org

Di Castri, F. (2000). Ecology in a context of economic globalization. *BioScience, 50*(4), 321–332.

Ehsani, M., Ahmadi, A., & Fadai, D. (2016). Modeling of vehicle fuel consumption and carbon dioxide emission in road transport. *Renewable and Sustainable Energy Reviews, 1*(53), 1638–1648.

EPA. (2015). Emission factors for greenhouse gas inventories. Retrieved September 19, 2016 from http://www.epa.gov

Erickson, P., & Tempest, K. (2014). Advancing climate ambition: How city-scale actions can contribute to global climate goals. (SEI Working Paper No. 2014-06). Stockholm Environment Institute, Seattle, WA, US. Retrieved September 5, 2016 from http://sei-international.org/publications?pid=2582

FAO. (2016). Global livestock environmental assessment model (GLEAM). Retrieved October 5, 2016 from http://www.fao.org

Florian, M., Gaudry, M., & Lardinois, C. (1988). A two-dimensional framework for the understanding of transportation planning models. *Transportation Research B, 22B,* 411–419.

Freeman, A. M. (2003). *The measurement of environmental and resource values: Theory and methods. Resources for the future* (2nd ed.). Washington DC: RFF Press. ISBN 1-89185-3627.

Freeman, A. M., Herriges, J. A., & Kling, C. L. (2014). *The measurement of environmental resource values. Resources for the future* (3rd ed.). Washington DC: RFF Press. ISBN 1-31770-3936.

GLA Economics. (2010). Scenarios, planning and economic outlooks. Retrieved October 11, 2016 from http://www.london.gov.uk

Gouldson, A., Colenbrander, S., Sudmant, A., Godfrey, N., Millward-Hopkins, J., Fang, W., et al. (2015). Accelerating low-carbon development in the world's cities. Contributing paper for seizing the global opportunity: Partnerships for better growth and a better climate. New climate economy, London and Washington, DC. Retrieved September 5, 2016 from http://newclimateeconomy.report/misc/working-papers

Grimm, N. B., Grove, M., Pickett, S. T. A., & Redman, C. L. (2000). Integrated approaches to long-term studies of urban ecological systems. *Bioscience, 50*(7), 571–584.

Hansen, J., Fung, I., Lacis, A., Rind, D., Lebedeff, S., Ruedy, R., et al. (1988). Global climate changes as forecast by Goddard institute for space studies three-dimensional model. *Journal of Geophysical Research, 93*(D8), 9341–9364.

Hassan, A. M., & Lee, H. (2015). The paradox of the sustainable city: Definitions and examples. *Environment Development and Sustainability, 17,* 1267–1285.

Huai, T., Durbin, D., Miller, J. W., & Norbeck, J. M. (2004). Estimates of the emission rates of nitrous oxide from light-duty vehicles using difference chassis dynamometer test cycles. *Atmospheric Environment, 38,* 6621–6629.

IPCC. (2006). 2006 IPCC Guidelines for national greenhouse gas inventories. In S. Eggleston, L. Buendia, K. Miwa, T. Ngara, & K. Tanabe (Eds.). *Intergovernmental Panel on Climate Change (IPCC), IPCC/OECD/IEA/IGES,* Hayama, Japan, ISBN 4-88788-032-4.

Julong, D. (1989). Introduction to grey system theory. *The Journal of Grey System, 1*(1), 1–24.

Kim, K. (2009). Urban development model for the low-carbon green city: The case of Gangneung. Retrieved October 8, 2016 from http://www.weitz-center.org

Kumar, S., Nayek, M., Kumar, A., Tandon, A., Mondal, P., Vijay, P., et al. (2011). Aldehyde, ketone and methane emissions from motor vehicle exhaust. A critical review. *American Chemical Science Journal, 1*(1), 1–27.

Lehmann, S. (2010). Green urbanism: Formulating a series of holistic principles. S.A.P.I.EN.S. Retrieved September 6, 2016 from http://sapiens.revues.org/1057

Li, L., & Chen, K. (2016). Quantitative assessment of carbon dioxide emissions in construction projects: A case study in Shenzhen. *Journal of Cleaner Production, 141,* 394–408.

Linderman, M. A., An, L., Bearer, S., He, G., Ouyang, Z., & Liu, J. (2005). Modeling the spatio-temporal dynamics and interactions of household, landscapes and giant panda habitat. *Ecological Modelling, 183,* 47–65.

Manheim, M. L. (1979). *Fundamentals of transportation systems analysis*. Cambridge, MA: MIT Press.

Marino, C., Nucara, A., Pietrafesa, M., & Pudano, A. (2016). The assessment of road traffic air pollution by means of an average emission parameter. *Environmental Modeling & Assessment, 21*(1), 53–69.

Marszal-Pomianowska, A., Heiselberg, P., & Larsen, O. K. (2016). Household electricity demand profiles—A high resolution load model to facilitate modelling of energy flexible buildings. *Energy, 5*(103), 478–501.

McNally, M. G. (2007). The four step model. *Institute of Transportation Studies*. UCI-ITS-WP-07-2.

Moir, E., Moonen, T., & Clark, G. (2014, June). What are future cities? Origins, meaning and uses. Compiled by the business of cities for the foresight future of cities project and the future cities catapult.

Motawa, I., & Oladokun, M. G. (2015). A model for the complexity of household energy consumption. *Energy and Buildings, 1*(87), 313–323.

Oladokun, M. G., & Odesola, I. A. (2015). Household energy consumption and carbon emissions for sustainable cities—A critical review of modelling approaches. *International Journal of Sustainable Built Environment, 4*(2), 231–247.

Ramaswamy, V., Boucher, O., Haigh, J., Hauglustaine, D., Haywood, J., Myhre, G., et al. (2000). Radiative forcing of climate change. *Evolution*. Chapter 6.

Rootzén, J., & Johnsson, F. (2016). Paying the full price of steel—Perspectives on the cost of reducing carbon dioxide emissions from the steel industry. *Energy Policy, 11*(98), 459–469.

Rosen, S. (1974). Hedonic prices and implicit markets: Product differentiation in pure competition. *Journal of Political Economy, 82*(1), 34–55.

Sanderson, D., Busquets, D., & Pitt, J. (2012). A micro-meso-macro approach to intelligent transportation systems. *IEEE 6th International Conference on Self-Adaptive and Self-Organizing Systems Workshops (SASOW)*. Lyon, France, 77–82, 10–14 September 2012.

Shukla, P. R. (2013). Review of linked modelling of low-carbon development, mitigation and its full costs and benefits. *Indian Institute of Management, Mitigation Action Plans & Scenarios (MAPS) Research Paper*.

Teodorovic, D. (2015). Routledge handbook of transportation. Taylor and Francis. ISBN 9-7813-176309-13.

Tubiello, F. N., Cóndor-Golec, R. D., Salvatore, M., Piersante, A., Federici, S., Ferrara, A., et al. (2015). Estimating greenhouse gas emissions in agriculture. A manual to address data requirements for developing countries. *Food and Agriculture Organization of the United Nations*. Rome. ISBN 978-92-5-108674-2.

Van Ruijven, B. J., Van Vuuren, D. P., Boskaljon, W., Neelis, M. L., Saygin, D., & Patel, M. K. (2016). Long-term model-based projections of energy use and CO_2 emissions from the global steel and cement industries. *Resources, Conservation and Recycling, 5*(112), 15–36.

Wang, R., & Ye, Y. (2004). Eco-city development in China. *Royal Swedish Academy of Sciences, 33*(6), 341–342.

Wang, Z., & Ye, D. (2016). Forecasting Chinese carbon emissions from fossil fuel energy consumption using non-linear Grey multivariable models. *Journal of Cleaner Production*. Article in Press, 1–13, August 2016.

Waygood, E. O. D., & Avineri, E. (2016). Communicating transportation carbon dioxide emissions information: Does gender impact behavioral response? *Transportation Research Part D, 9*(48), 187–202.

Winkler, H., Spalding-Fecher, R., Mwakasonda, S., & Davidson, O. (2002). *Sustainable development policies and measures. Options for protecting the climate*, Washington, DC: World Resource Institute.

Wong, T., & Yuen, B. (2011). *Eco-city planning: Policies, practice and design*. Earth Sciences: Springer Science & Business Media. ISBN 940-070383-X.

WRI/WBCSD GHG Protocol. (2014). Global protocol for community-scale greenhouse gas emission inventories. An accounting and reporting standard for cities. Retrieved September 14, 2016 from http://www.ghgprotocol.org/

Zabel, J. E., & Kiel, K. A. (2000). Estimating the demand for air quality in four US cities. *Land Economics, 76*(2), 174–194.

Zhang, S., Wu, Y., Un, L., & Hao, J. (2016). Modeling real-world fuel consumption and carbon dioxide emissions with high resolution for light-duty passenger vehicles in a traffic populated city. *Energy, 7*(113), 461–471.

Part VI
Characterization of Industrial GHG Emission Sources in Urban Planning

Chapter 20
Characterization of Industrial GHG Emission Sources in Urban Planning

Wynand Lambrechts and Saurabh Sinha

Abstract Urbanization produces large amounts of non-natural greenhouse gases (GHG), leading to air pollution, health hazards and unsightly fog lingering above cities. Rapid growth of cities is creating opportunities for development, new jobs and improving the quality of life of inhabitants; unfortunately, generally at the expense of carbon pollution. A severe effect is global warming, also called the greenhouse effect: the heating of the earth's atmosphere by re-radiated heat. The burning of fossil fuels to generate electricity and heat, as well as pollution in the transport sector, are among the largest contributors to the greenhouse effect. Research on minimizing this pollution is two-fold: reducing emissions through advances in technology and more efficient urban planning. An efficient built environment through appropriate urban planning, supported by energy-efficient vehicles, buildings, appliances and power generation by alternative and renewable sources, can reduce GHG emissions substantially. Urban planning can be used to create more resourceful micro-climates within cities, on roads and extending to rural areas. The increasing rate of materials and goods production worldwide means that this number will almost certainly remain constant or worse, increase, but doubtfully decrease if no intervention is posed towards decreasing its emissions. A single solution to minimize GHG emissions in micro-climates does not exist, but applying energy-saving measures, incorporating low-energy technologies, intellectually empowering workers and learning from experience can collectively lead to optimized solutions for various situations. Cities can reduce wasted energy through two main categories of planning initiatives: energy-efficient building standards for new urban constructions or energy retrofits for existing buildings and efficient urban

W. Lambrechts (✉)
Department of Electrical and Electronic Engineering Science, Faculty of Engineering and the Built Environment, Auckland Park Kingsway Campus, Auckland Park, Johannesburg, South Africa
e-mail: wynand.lambrechts@ieee.org

S. Sinha
Faculty of Engineering and the Built Environment, Auckland Park Kingsway Campus, Auckland Park, Johannesburg, South Africa
e-mail: ssinha@ieee.org

R. Álvarez Fernández et al. (eds.), *Carbon Footprint and the Industrial Life Cycle*, Green Energy and Technology, DOI 10.1007/978-3-319-54984-2_20

infrastructure planning of transportation (public and cargo), communications and distribution networks. Identification and characterization of all industrial GHG emission sources is critical to achieve these two goals. The contributed chapter, supplements the related body of knowledge by thematically combining efficient urban planning and reducing GHG emissions in the electricity and heat generation sector and in the transport sector. The chapter investigates potential scholarly contributions by assisting researchers to theoretically identify and classify overlooked and underestimated sources of GHG emissions in urban settings. The notional overview on low-carbon cities through economic planning ties in with urban planning and provides a means to identify known issues and sub-optimal infrastructure. The chapter aims to serve as a starting point for specialized research to improve upon such scenarios. Readers are encouraged to apply the academic principles recognized in this chapter to devote intellectual resources to innovating efficient urban planning and eco-planning for future sustainable cities.

1 Introduction

Approximately one fifth of the total contribution of non-natural greenhouse gases (GHGs) is produced through industrialization, characterized as a purely anthropogenic source. The primary purpose of industrialization is changing an economy to increase its capability to manufacture goods; this is advantageous for economic growth and industrial investments, but has detrimental consequences for the environment. An important negative effect is the warming of the earth's atmosphere in a process often referred to as the greenhouse effect. Human activities contribute to climate change by changing the amounts of GHGs, aerosols and ozone opacity. A GHG is capable of absorbing thermal radiation from the surface of the earth, preventing the heat generated by the sun from leaving the earth's atmosphere to allow the earth to cool down. The absorbed heat is isotopically re-radiated, parts of it downwards back to the surface of the earth to heat the atmosphere further. This effect leads to global warming, which in its natural capacity is a necessary effect to sustain life on earth, but dangerous if amplified by anthropogenic sources, for example by the burning of carbon-emitting fossil fuels and deforestation. Effects such as rising sea levels, flooding, drought, frequent wildfires and stronger ocean currents leading to more destructive storms have increasingly been witnessed in the past three decades; these are commonly accredited to global warming. In its natural form, the earth experiences changes in climate due to changes in solar energy, volcanic eruption and natural changes in GHG concentrations. Carbon dioxide (CO_2) levels typically vary with glacial cycles over several hundred thousand years, but in recent years, human activities have aggravated and amplified the effects. Different economic sectors, such as electricity and heat generation, agriculture, industry, transport and building, contribute to GHG emissions. This chapter recognizes the fact that these sectors are not mutually exclusive and aims to distinguish problem areas and potential solutions relating to industrialization and its holistic effect on GHG emissions. Industrialization leads to urbanization which can be managed and controlled in terms of the environmental impact. A common problem is, however, the unprecedented growth rate of urbanization from industrialization and planning for sustainable long-term growth is often overlooked. This chapter aims to identify contributing factors towards planning for sustainable urbanization by identifying and characterizing of industrial GHG emission sources. Reducing GHG emissions, from any sector and in any country, should be treated with highest priority, as recent studies (Friedrich et al. 2016) shows that the earth could experience a hike in temperature of between 4.8 and 7.4 °C (40 and 45 F).

The following section summarizes the largest contributing sectors towards climate change. Identifying the primary contributors is an important step in characterizing and essentially lowering carbon emissions within the earth's atmosphere.

1.1 Main Contributors to Climate Change

The Intergovernmental Panel on Climate Change (IPCC) estimates the total contribution of economic sectors to GHG emissions as follows:

1. Electricity and heat production (25%),
2. Agriculture, forestry and other land use (24%),
3. Industry (21%),
4. Transportation (14%),
5. Other energy (10%), and
6. Buildings (6%).

Electricity and heat production, agriculture, forestry and other types of land use, transportation, as well as industrial, commercial and residential use of energy undeniably contribute to global warming and to climate change. Electricity and heat production account for a quarter of total GHG emissions, according to the IPCC, and agriculture for approximately another quarter. A large contributing factor to this is the transformation of forests to urban or rural settlements, effectively reducing the environment's natural potential to absorb CO_2 from the atmosphere. Most of the efforts to reduce GHG emissions from agriculture and other land use involve improving on management practices to reduce the areas of natural land lost to developments.

Industry is responsible for approximately one-fifth of the total GHG emissions, again according to the IPCC. Industrialization is typically defined as a process that converts an economy functioning primarily through agriculture to an economy based on the manufacturing of goods. Industrialization inevitably also leads to urbanization, populations clustering to cities and effectively occupying a relatively small land area in close proximity to one another. Industrialization presents a danger to the natural environment and the earth's ability to regenerate its resources naturally. Its effects, although promoting economic growth, should ideally not dominate the destruction of forests and other natural land. The collective effect of

GHG emissions from manufacturing goods and the reduction of plants and trees, which absorb carbon from the environment, is substantial. Land-use changes also affect the amount of sunlight that is reflected from the ground back into space (the surface albedo effect), further affecting climate change.

Transportation is an expected result from industrialization and urbanization. The transport of goods and people has become a dominant industry since the industrial revolution. In modern times, to minimize the impact of transportation on the environment (due to internal combustion), fuel-efficient vehicles and the move to electric vehicles are not the only solutions; proper urban planning is equally important. Accessibility of public services, entertainment, public and private sector offices and manufacturing plants is crucial to reduce the need for fossil fuel-based vehicles and travelling demands. To decrease the carbon footprint of transportation, increased accessibility to urban infrastructure can dramatically reduce climate change, paired with fuel-efficient vehicles and the drive towards electric and solar power.

Commercial and residential use of energy is also dependent on planning, but in many circumstances even more on the efficient use of energy. The information provided in this chapter therefore also aims to differentiate between characterizing and planning for energy use, as well as for efficient use of current resources. Energy in the sector not directly associated with electricity and heat generation, such as fuel extraction, refining of raw materials, processing and transportation between power plants, accounts for a further 10% of GHG emissions. The emissions from buildings, contributing 6% of GHG according to the IPCC, originate from external burning of fossil fuels for domestic use, excluding the use of electricity and heat supplied through the grid by power utilities.

This chapter focuses on cities and how urban planning can contribute to the reduction of energy consumption in the sectors of electricity and heat production, land use to a certain extent, industry, transportation, buildings and other energy sources. Considerable focus can however be placed on electricity and heat production, transportation and buildings when considering the more practical approach or refurbishment of urban infrastructure as opposed to envisioning and realizing new developments from the ground up.

Energy generation from alternative sources has become a prominent factor in reducing GHG emissions in all sectors; where a combination of energy sources is becoming a more probable solution to achieving reduced global carbon emissions.

1.2 Alternative Urban Energies

It has been acknowledged worldwide that there is no single solution to reducing the negative effects of anthropogenic sources of global warming (UNON 2004). Decreasing global dependence on fossil fuels to generate electricity and produce carbon-based fuels can be realized through a multitude of techniques, including biofuels, solar, wind, hydro, nuclear, geothermal, and hydrogen power.

At the epicenter of realizing these alternative energies is planning to implement, expand, overcome limitations and adapt to using these substitutive sources in urban and rural settings.

Energy-conscious departments and institutions often provide resources to individuals to reduce their daily energy consumption, techniques to lessen the burden on power utilities and ways to invest in energy-efficient solutions at home, at the workplace and on the road. This information is typically invaluable for individuals not only to decrease their own carbon footprint but also to save on energy expenditure. This chapter acknowledges that these techniques are crucial for educating urban and rural inhabitants on reducing their energy dependence; however, it focuses on how effective urban planning and infrastructure development contribute to a large-scale reduction in emission of GHG in urban settlements. Countries with extremely high air pollution from a combination of transport and industrialization include Pakistan, Qatar, Afghanistan, Bangladesh, Iran, Egypt, Mongolia, the United Arab Emirates, India and Bahrain (Ryan 2014). According to the World Health Organization (WHO), 98% of cities in low- and middle-income countries with more than 100,000 inhabitants do not meet the WHO air quality guidelines; in high-income countries, this percentage decreases to 56%. In China, 4000 people die each day from cardio-pulmonary and neurovascular diseases due to air pollution, particularly small particles of haze (Associated Press 2015). Indirect industrial carbon footprint, the major source of which is consumption of electricity, contributes approximately 87% of the total industrial carbon footprint, while direct industrial carbon footprint contributes the remaining 13% (Yan et al. 2016) according to a study performed on the industrial carbon footprint of Chinese (the largest textile and garment producer and consumer globally) textile fabrics. Much can be learned from the cities having the highest, and lowest, amounts of air pollution, what they are doing to minimize their energy consumption in future and how new developments are planned to be more energy-efficient.

Careful planning and management of future land developments and upgrading existing infrastructure can contribute a significant amount to lowering carbon emissions. An efficient built environment can achieve the goals of low carbon emissions, as expressed in the following section.

2 Efficient Built Environment

Benfield (2014) highlights the fact that a large percentage (over 70%) of carbon emissions in cities in the United States of America, as in other countries, are generated by a combination primarily in commercial and residential buildings (including electricity and heat generation) as well as through transportation. The building sector contributes the largest portion of carbon pollution in most cities, according to the United States Natural Resources Defense Council (NRDC). The energy requirements are primarily for heating, cooling and lighting the buildings, with a substantial amount of energy being lost (wasted) because of ineffective techniques to retain heat, cold air or to light open areas—ideally achieved through natural light.

The built environment is a powerful tool to create more efficient micro-climates, over and above adopting energy-efficient habits and investing in fuel-efficient vehicles and appliances. Planning at a higher level, during urban development, can dramatically reduce energy consumption in the long term. The NRDC is continually researching and testing innovative strategies that aim to strengthen communities by addressing energy-related issues, encouraging the growth and future development of smart and sustainable cities and providing more efficient access to public transportation.

2.1 Importance of Efficient Urban Planning

Efficient urban planning is not a step-by-step formula and requires thorough research, modeling, analysis and understanding of the issues experienced in a city, potential solutions as well as financial and geographical limitations. Each project of efficient urban planning to reduce GHG is unique, notwithstanding having several common characteristics. Because of the dominance of carbon pollution, the focus is placed on resourceful planning with regard to buildings and transportation in cities. Essentially, as outlined in Benfield (2014), this includes

- considering the location of new developments or improving existing ones to optimize driving distances,
- efficient connectivity between developments,
- convenience forecasting to encourage walking between amenities,
- energy-efficient sourcing of building materials,
- energy-efficient architectures,
- and incorporating vegetation in urban areas to absorb carbon-based emissions.

It is more difficult to reduce carbon intensity (decarbonizing) of buildings and in industry and the transport sector compared to modifying or improving the means used in the generation of electricity (IPCC Synthesis Report 2014). The effects of reducing carbon emission of buildings and transport (direct and indirect emissions) are mostly only observed to have a significant impact over longer periods of time and this technique requires substantially more planning and inputs from several disciplines. Ultimately, the goal of efficient urban planning is achieving a completely green infrastructure.

2.2 Green Infrastructure

The term green infrastructure is commonly used to describe a low-carbon-focused urban environment and is considered for urban renovations and at the start of new urban developments. Modern developments rely on technology for sustainable urban planning, whereas smart cities incorporate technology to enhance the performance and efficiency of their inhabitants, reduce costs and resource consumption and engage actively with their citizens. Singapore, for example, has introduced multiple initiatives towards environmentally friendly buildings and sustainable infrastructure. The country provides standards and benchmarks to follow for constructing new buildings or retrofitting existing buildings. Singapore's research and green infrastructure techniques are unique to the country, since guidelines for sustainable cities were mostly developed for Europe and the United States prior to the 2000s. Singapore's tropical climate furthermore prevented it from simply adapting these guidelines. Its Building and Construction Authority (BCA) introduced the Green Mark certification program in 2005 to certify building as being green, therefore being efficient in the means that it produces and uses energy and distributes water, high-quality air and efficient extraction; thus a healthy indoor environment, which is combined with physical green spaces and built form environmentally-friendly and responsibly-sourced materials. In the city of Singapore, buildings alone account for:

- one third of GHG emissions,
- 40% of resource and energy consumption,
- as well as 25% of water consumption.

The BCA claims that energy consumption in buildings can be reduced by 30–80% using proven and commercially available technologies. This initiative has been successful in 2700 Green Mark building projects in Singapore as of May 2016. The BCA conducts regular follow-up sessions and post-occupation verifications to ensure that green buildings are not only properly implemented, but also sufficiently maintained. The BCA Green Mark rating system differs from similar international certification in that it

- places stronger emphasis on overall energy efficiency as opposed to focusing on specific resources,
- is tailored to a tropical climate where heat gain and cooling of the inside of buildings by using air-conditioning are among the key design considerations, and
- employs high standards of both measurement and verification instrumentation for continuous indoor air-conditioning performance monitoring.

Through its initial phases, the BCA Green Mark initiative adopted essential strategic emphases towards existing buildings achieving certification, including:

- insistence on the public sector to take the lead in development programs,
- furthering the development of green building technology and raising awareness, including international profiling,

- encouraging the private sector to follow suit,
- training workers and building an entire industry with capabilities to develop greener infrastructure, and
- legislating minimum standards for buildings within the city.

Furthermore, the BCA Green Mark highlights the importance of research and development to encourage industries to adopt green building technology. The focus on research and development strengthens the future of sustainable buildings and the methods of quantifying the performance of these buildings and infrastructure. The BCA annual awards recognize excellence in the built environment for green buildings and these awards highlight the accomplishments of green buildings within the city, providing valuable information on the initiatives of each building project towards sustainability and green design for other international projects. The final section in this chapter, *Reducing GHG emissions through efficient urban planning*, addresses key performance factors considered to achieve green building status and assured sustainability.

Industrialization combined with urbanization leads to higher demand for transport of people and goods, inevitably leading to the burning of gasoline and oil and releasing GHG emissions in the environment. The transportation sector as a source of pollution in industrial and urban environments is discussed in the following section.

3 Transportation Sector

The United States Environmental Protection Agency (EPA) lists the movement of people and goods by cars, trucks, ships, airplanes and other vehicles under the transportation sector. Internal combustion engines of transportation vehicles emit most GHG in the form of CO_2, with comparatively minor quantities of methane (CH_4), nitrous oxide (N_2O) and hydrofluorocarbon (HFC) also emitted by internal combustion engines. Most internal combustion engines use octane (C_8H_{18}) as a fuel source. Octane is a hydrocarbon and is a component of gasoline, with the octane rating (95 for example) being the measure of the amount of compression that the fuel can tolerate before self-igniting under high-pressure and high-temperature conditions, therefore correlating to its activation energy. The compression ratio directly relates to engine power and thermodynamic efficiency. A higher octane rating typically translates to higher fuel efficiency and higher power transferred to the engine. Exhaust emissions of CO_2 gas are also generally lower in high-octane fuels, since the fuel can burn more completely. The balanced chemical equation of octane mixed with oxygen (O_2) and ignited in internal combustion engines is

$$2(C_8H_{18}) + 25O_2 \rightarrow 16CO_2 + 18H_2O \tag{1}$$

where CO_2 and water (H_2O) are produced and the equation is representative of a stoichiometric air-to-fuel ratio of 14.7:1. These performance and efficiency increases translate to a higher price for fuel. High-octane fuel is more difficult to manufacture and requires larger amounts of energy to be expended during its production. Premium fuel can also be mixed with crude blends to increase its octane value compared to the pure fuel from the refinery. In addition, premium, high-performance fuel has gained popularity in many markets. Since it is a luxury item, a price-premium is added to the cost of the fuel. Octane ratings vary significantly between countries, regions and suppliers, measured and characterized by its research octane number (RON) and motor octane number (MON). In certain countries, such as the United States, Canada and Brazil, the anti-knock index (AKI) is the preferred measurement method. RON is measured in test engines running at 600 revolutions per minute (rpm) with variable compression ratio, whereas MON is tested and measured at 900 rpm with a pre-heated fuel mixture and variable ignition timing. AKI is determined by computing the average of (RON + MON)/2 and is also referred to as the pump octane number. Essentially, octane remains the primary determiner of the price and the value of commercial fuel. In the transportation sector, adapting efficient driving behavior and optimizing vehicle operation are equally important as fuel use and fuel quality.

3.1 Energy Efficiency Through Adapting Behavior

Globally, various initiatives and online sources present methods to encourage drivers to improve the fuel efficiency of cars, trucks and airplanes. These methods fundamentally summarize four main categories:

- choosing the correct octane for the vehicle or converting to alternative energies such as hybrid vehicles, fuel-cell powered or pure-electric vehicles,
- incorporating advanced designs, materials and technologies in vehicle design,
- improving operating practices and
- adapting driving style to increase fuel economy and reducing travel demands.

Differently put, energy efficiency can be achieved through vehicle efficiency, travel efficiency of single or combined trips and system efficiency of entire infrastructures. Automotive pollutant sources are not limited to carbon emissions from internal combustion engines. Storm water pollutants originating from automobile activities and atmospheric deposition also contribute to environmental pollution (Pitt et al. 2004). Tire wear on vehicles, for example, contributes to significant concentrations of zinc in urban runoff. Urban runoff involves the distribution of rainwater, contaminated with hazardous or toxic materials, from roads and other asphalt or concrete surfaces. Ensuring that urban runoff does not enter large bodies of water used as drinking water supplies or housing aquatic life is commonly overlooked or not planned for during urbanization. In addition, exhaust particulates, mechanical wear and fluid spills and leaks from typically older vehicles add to polluted urban runoff (Pitt 1979).

The IPCC document on mitigation of climate change (2014) states: "*without aggressive and sustained mitigation policies being implemented, transport emissions could increase at a faster rate than emissions from the other energy end-use sectors.*" The effort to reduce global transport GHG emissions remains challenging, especially in view of the continual growth in passenger and freight transit demands, potentially outweighing alleviation measures if emissions are not decoupled from gross domestic product (GDP) growth (IPCC Mitigation of Climate Change 2014). In Sao Paulo, Brazil, traffic congestion can extend for 180 km (112 miles) on average and up to 295 km (183 miles) in extreme cases. Not only does such traffic congestion lead to frustration and negative psychological impacts on drivers (nervousness, tension, irritability, fatigue and poor performance at work) as well as excessive time delays, but also the emissions from thousands of internal combustion engines in a constant stop-go milieu raises GHG emissions to unsafe levels in the immediate surroundings. Evidently, applying urban planning in these extreme cases involves much more than merely adding road surface or alternative routes; a complete mind shift for developing urbanization during initial growth is critical. Efforts to provide real-time traffic congestion models and information on air pollution, as proposed by Lehmann and Gross (2016), can provide drivers with updated information on congested routes to encourage them to choose alternative routes or transportation. Additionally, a total reduction in travel demands through effective urban planning can substantially reduce carbon emissions in the transportation sector.

3.2 Reducing Travel Demands

This section focuses on a category of reducing travel demands and maintaining sustainable transport to satisfy rapidly increasing demands for transport (especially in developing countries, but equally important in developed countries showing similar increases in transportation demands), as this encompasses a significant contribution in various disciplines to achieve the aim. Efficient urban planning is at the heart of this. Urban planning and optimizing urban mobility are effective when striving to decrease the stresses incurred on the environment and its inhabitants (humans and animals). System-infrastructure multi-modal choice (private and public transport alternatives) in urban settings (Rakkesh et al. 2016) encompasses transport infrastructure of roads, railways and airports, as well as behavioral choices among modes based on travelling speed (therefore the time required to reach a destination), comfort, costs incurred and the convenience factor. Drivers of vehicles on urban, rural and cross-country roads, and even airplanes, make travel demands directly related to the infrastructure of the environment.

The correlations between income levels and ownership of light-duty vehicles (LDVs) mean that for growing economies, there are a significant increase in travel demands as LDV ownership increases. In China, for example, at the end of 2014, the number of motor-vehicle drivers exceeded 300 million, excessively contributing to the country's industrial pollution. Intelligent traffic management technologies proposed for China (Hui et al. 2014) and fleet management (Mingrone et al. 2015) can alleviate congestion to an extent; however, in a country with high growth and increasingly demanding transportation requirements, urban planning of future developments should also receive substantial consideration. Intelligent traffic management furthermore requires favorable environments (Dridi 2007) relying on multiple sensors and information transfer and analysis to operate effectively. In poor countries and in some developing countries, creating favorable environments may be difficult from funding and government backing perspectives. Energy-efficient vehicles also play a dominant role in the potential to reduce the demand for oil and fossil fuels, together with urban planning. As electric vehicles are gaining traction and popularity, strategically placed charging stations in urban environments are becoming more important to consider for growth in the transportation sector dedicated to reducing GHG emissions.

Citizens are limited in planning for their travel demands by the available resources of alternative transportation, distances between amenities and the availability of public transportation in their immediate environment. Town and regional planning encourages and manages efficiency through the preparation, design, execution and administration of public involvement in the development and use of land, including ordinances for the transportation sector. It is also crucial to combine the built environment with socioeconomic characteristics to systematically understand and reduce travel demands (Yao and Chen 2015) of everyday trips, whereas travel demands for exploration (Liang 2010) require high quality, the shortest route and lowest risks, which could lead to accidents. Tax incentives, implemented in Hong Kong for example, aim to improve air quality by encouraging the use of low-emission and high fuel-efficiency vehicles to encourage drivers to consider purchasing these vehicles, as opposed to older or larger engine-capacity vehicles.

An effect of excessive transportation not associated with a decrease in air and water quality is noise pollution in the transportation sector. Various studies have been conducted on the effects of noise pollution, outlined and summarized in the following section.

3.3 Noise Pollution in the Transport Sector

Noise pollution in the transport sector has negative psychological and physiological effects on humans. A study presented by Tabraiz et al. (2015) investigated the acute and chronic physio-psychological effects of noise pollution in urban areas in Pakistan. Tabraiz et al. (2015) concluded that at elevated noise levels of between 85 and 106 dB, traffic wardens exposed to the noise for extended periods presented psychologically elevated signs of aggravated depression, stress, public conflict, irritation, annoyance, behavioral affects and speech interference. In addition, physiological effects of hypertension, muscle tension, exhaustion, low performance levels, concentration loss, hearing impairment, headache and cardiovascular issues were evident. In an earlier article presented by Stokols et al. (1978), similar evidence was found relating traffic congestion in urban environments to heightened systolic and diastolic blood pressure. Kristiansen et al. (2011) used individual exposure to traffic noise, employing a geographical information system and available exposure data, to predict noise levels at geocoded residential addresses. Kristiansen et al. (2011) presented evidence of sleep problems in occupationally active citizens, which were attributed to traffic noise. Urbanization introduces traffic noise and pollution closer to residential areas, increasing the negative impacts on residents even when not actively travelling. Globally, numerous studies have shown that there is a correlation among traffic congestion (air and noise pollution), social behavior (Rotton and Frey 1984) and cognition (Bos et al. 2014) and the effects are increasing with rapid urbanization and further advances in industrialization.

Another dominant source of GHG emissions resulting from industrialization and urbanization is the generation of electricity and heat in commercial, residential and industrial buildings. The sources, effects and proposed techniques to limit pollution from electricity and heat production are described in the following section.

4 Electricity and Heat Production

The generation of electricity for commercial and industrial use, directly coupled with heat production, is an economic sector that releases global emissions of GHG and other volatile gases into the environment. At its source is the burning of coal, natural gas and oil to generate electricity and heat. The United States EPA estimates its contribution at 25% of all global GHG emissions, the largest single contributor. The most prominent GHG due to energy production is CO_2, combined with smaller amounts of CH_4, N_2O and a fraction of sulfur hexafluoride (SF_6) also emitted during the burning process. The Inventory of US Greenhouse Gas Emissions and Sinks (1990–2014) shows an increase in CO_2 emissions from electricity production between 1990 and 2007 from 1900 million metric tons to just below 2500 million metric tons, specifically in the United States. Since 2007 and up to the availability of data given in 2014, there has been a decline in CO_2 GHG emissions from its peak in 2007, to just above 2000 million metric tons. A similar trend is evident in Europe, where the Eurostat database in March 2016 showed that the 28 European Union (EU) member states (EU-28) exhibited a decline of 11.8% of CO_2 emitted between 2008 and 2013. GHG emissions from the supply of electricity, gas, steam and air conditioning also declined by approximately 12.6% over the same period. Although the total GHG emissions globally are still high, these statistics show a significant improvement in leading nations. Several developments, including urban planning, are responsible for the decrease in carbon emissions. The most substantial contribution is from the manufacturing sector, with a 20.1% reduction in GHG emissions. Electricity and heat generation demands determine the production and emissions profile of fossil-fuel burning.

4.1 Electricity and Heat Generation Demands

Developments in the consumption, investments and export of electricity show continual declines in overall electricity and heat generation, but global awareness and a high priority being placed on further developments are typically lacking in developing countries and rapidly growing economies such as China. Supporting growth with sustainable development is not easy to implement, especially since planning and infrastructural changes generally take time, resources and skilled practitioners to realize and it is not always practical to implement such measures to accommodate the incoming streams of inhabitants into urban environments. The carbon footprint of individuals and organizations in urbanized locations directly triggers demands from the energy producers and peak energy consumption remains high in non-optimized environments.

4.2 Scope of GHG Emissions

The complex challenge of reducing electricity and heat production is further increased by the multifaceted milieu that consumes energy. This includes the transport sector, most industrial and commercial activities, established and new buildings and city infrastructure, water distribution in commercial, residential and agricultural lands and the production and preparation of food. GHG emissions are categorized into three scopes of emission, scopes 1, 2 and 3. These categories can be summarized as:

- Scope 1 emissions (direct GHG emissions): These emissions are defined by GHG sources, which are owned or controlled by an organization or individual. These direct emissions emanate primarily from internal fuel combustion vehicles owned by individuals or companies, used to transport goods or people. Direct emissions also include stationary combustion of fossil fuels for comfort or industrial applications, process emission from manufacturing of products from raw materials and rogue/unintentional GHG emissions from sources which require a constant energy feed such as refrigeration.
- Scope 2 emissions (energy indirect GHG emissions): Indirect energy emissions emanate from the use of procured electricity, heat or steam used to power devices, equipment, air conditioning and all other electricity-requiring components in commercial, industrial and agricultural environments.

- Scope 3 emissions (other indirect GHG emissions): These other indirect emissions include activities such as the purchasing and use of goods and services that consumed electricity during manufacturing, business travelling and employee commuting for business activities, waste disposal, investment in goods and services and the leasing of assets or franchises on commercial or industrial scale.

Through the clearly defined activities and their contribution to the categorized scopes of GHG emissions, a typical model to measure direct and indirect GHG emissions by individuals or organizations can be established. The model, the value chain footprint, shows how emissions are categorized within an organization or private value chain and includes upstream and downstream activities.

4.3 Upstream and Downstream Activities

Upstream and downstream are general terms that refer to the location of a company or supplier in the supply chain. It is a general measure of how close the company is to the end-user function; the closer the company and its profile is to the end-use product, the further downstream it is. Oil and natural gas suppliers are typical upstream companies, using raw materials to supply a product (gas or petrol) to a vastly diverse market and their marketing strategies are not focused on a single type of user or demographic. Furthermore, midstream entities are associated by means of their ability to transport or distribute upstream activities, such as oil refineries, to downstream entities, such as distributors of petroleum to the end-users.

4.4 The Value Chain Footprint

The value chain footprint is an effective model to guide companies and individuals in determining their carbon footprint based on upstream, downstream and midstream activities involved in the selling of a product or service. Carbon Trust® lists several benefits for organizations when measuring and determining its scope 3 emissions, including:

- enabling the organization to evaluate emission hotspots within its own supply chain,
- identifying any resource and energy risks that can affect the flow of business within the supply chain,
- incorporating sustainability in the larger value chain by effectively identifying sub-contractors or higher level suppliers, which also maintain and control its scope 3 emissions,
- implementing cost savings and energy-efficient practices within its own supply chain by adopting this practice,

- allowing organizational leading operations on measuring scope 3 emissions to engage suppliers and sub-contractors to implement sustainable business practices as well,
- working towards manufacturing and outsourcing goods and services to produce energy-efficient and responsibly sourced products, and
- reducing unnecessary travel and transport within the organization to reduce its overall carbon footprint.

Implementing strategies and energy-efficient practices within an organization will trickle down to the habits of its employees on a personal level and into multidisciplinary sectors as more individuals become aware of effective practices in the supply chain. Ultimately, a reduction in indirect GHG emissions from electricity and heat production will follow as these practices become more commonplace. As urbanization progresses rapidly, as is currently the case, improved urban planning for cleaner energy solutions can be incorporated in developments and infrastructure enhancements.

4.5 Scope 2 Emissions

This chapter focuses on scope 2 indirect emissions, resulting from the activities of an organization and emitted from sources owned by other entities (EPA 2016). Activities emanating from purchased electricity contribute indirectly to GHG emissions and pollution where the emissions depend on the total amount of energy used to produce electricity from mixing and burning of fossil fuels. It is important for organizations and individuals to account for the carbon, methane and nitrogen oxides emitted during fuel combustion; understanding the sources and effects can dramatically improve usage of electricity. EPA (2016) provides simplified location-based and market-based calculations to determine emissions (E) from purchased electricity. The calculations are based on

$$E = PE_F \tag{2}$$

where P is the quantity of purchased electricity, E_F is the emissions factor for CO_2, CH_4 or N_2O and E is the physical mass of CO_2, CH_4 or N_2O. The efficiency factor reflects the amount of useful output energy produced from the renewable or fossil energy input. EPA (2016) provides guidelines to determine location-based emission factors (direct line, regional and national) as well as market-based emission factors (energy attribute certificates, contracts, supplier-specific, residual mix, regional and national). The emissions equation presented in (2) is furthermore used to calculate the global warming potential (GWP). The GWP allows comparisons of global warming impacts of different GHG emissions. It is a measure of the amount of energy that 1 ton of gas absorbs over a specified timeframe, relative to 1 ton of CO_2 emissions. It is therefore a relative measure of the amount of heat that a GHG traps within the atmosphere, compared to CO_2.

4.6 Characteristics of Sustainable Energy Planning

Sotos (2012) provides a corporate, project and product-level approach to quantify and control scope 2 indirect emissions, highlighting the fact that scope 2 emissions are the largest contributor to GHG emission into the atmosphere. Its aim is to inform consumers of electricity to decrease their demand and shift the focus of energy supply to alternative low-carbon resources. In UNON (2004), the stated mission is to endorse socially and environmentally sustainable urban development to assist governments to cultivate sustainable energy and climate action plans as well as endorse implementation programs. The published work provides a large range of case studies from developed and developing urban centers, which aim to cover the areas of local government responsibilities. Important considerations in UNON (2004), which tie in with this chapter, address certain key characteristics of urban developments backed by local governments to achieve sustainable energy planning. These characteristics include:

- Treating all energy systems and supplies as an entirety. This is helpful since it relieves the complexity of urban planning for energy transport and distribution to private, commercial and industrial users. The holistic view of energy requirements can aid and benefit planners to optimize the generation and distribution of energy and determine solutions for alternative forms of energy.
- Projects and development plans should be measured against carbon mitigation through efficient energy generation and distribution, compared to planning solely for providing electricity and/or heat from traditional power plants.
- Focus should be placed on reducing the demand for energy services as opposed to increasing the capacity of traditional power plants or considering additional alternative energy sources. Informing and empowering urban (and rural) inhabitants with knowledge and understanding of their average expenditure and persuading them to reduce demand can potentially signify lower average consumption and lower GHG emissions. This would shift the focus from supplier-specific sources to alternative energy sources that will allow focus on energy conservation, efficiency and overall lower demand.
- Cognizance should be taken of the environmental costs and repercussions of any new developments, leading to higher demand and increases in electricity supply.
- Planning for change and anticipating change through new technologies, technology improvements and increased efficiency of current alternative energy solutions are essential. The economic feasibility of non-traditional energy supplies is still one of the main factors hindering growth in this sector; consideration of economic viability towards technology maturity should remain a key characteristic of sustainable energy supply.

Furthermore, UNON (2004) establishes a vision listing the key fundamentals of a sustainable system, particularly catering for the energy sector—electricity and heat generation—that is responsible for the bulk of carbon emissions in urban milieus. These fundamentals are crucial during urban planning of improvements and infrastructure enhancements, described as consistent, renewable, diverse, inclusive and interdependent. These principles are represented in Fig. 1.

In Fig. 1, the five key principles to consider during urban planning in the energy sector include consistency of short-term and long-term goals, proven to be viable and attainable over both time periods. Importantly, renewable resources are a major driver towards low-carbon emissions and should be included as a viable alternative in any urban planning; even if not deemed feasible, considering it is crucial. Diversity and adaptability of energy supplies allow for interchangeability during peak and off-peak demand periods, as well as considering future upgrades. Respecting these principles can improve the attainability of inclusive sub-systems operating to complement each other and operate interdependently.

4.7 Alternative Energies for Electricity and Heat Generation

The introduction of alternative renewable energy sources for electricity and heat generation, such as hydrogen, synthetic fuels, wind turbines or photovoltaic cells, requires assessment of the environmental impact and economics of the overall production and utilization life cycle, including the construction and operational stages of renewable plants (Granovskii et al. 2006). Life cycle assessment (LCA) of energy technologies, both renewable and conventional, is a standardized method to track all material, energy and pollutant flows of an energy-generation system to facilitate comparison of energy technologies (Granovskii et al. 2006). These are important considerations to ensure that renewable and alternative energy sources

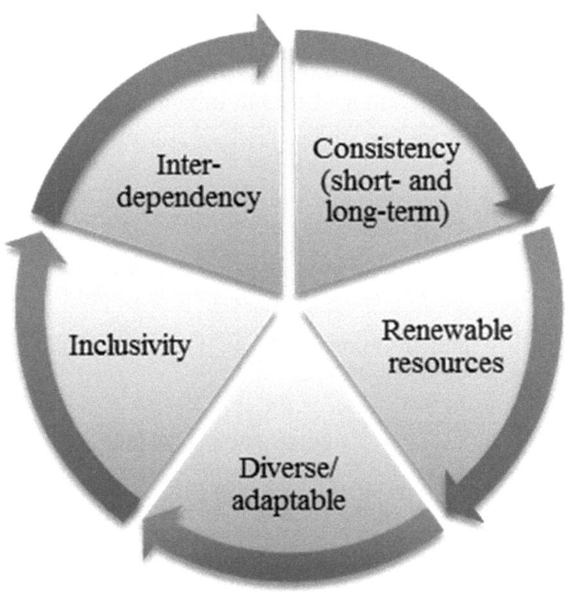

Fig. 1 The five key principles to consider during urban planning within the energy sector, adapted from UNON (2004)

produce less GHG emissions compared to traditional power plants, as the process in its entirety is often overlooked when arguing for renewable sources.

4.8 Electricity and Heat Transport Losses

Vithayasrichareon and MacGill (2016) estimate the cost of transporting photovoltaic transmission at approximately US $700/MW/km, which is indicative of the expenditure associated with alternative energies. Generated power passes through complex distribution networks of power transformers, overhead power lines and cabling, with additional equipment, such as distribution boards, all contributing to losses of generated power. Transmission and distribution losses quantify the percentage of power that reaches consumers compared to the generated power at the power stations, or in the case of renewable sources, the point of generation. The amount of power lost during transmission and distribution adds to GHG emissions and power plant overheads and is not recoverable. Losses can be categorized as technical losses and non-technical or commercial losses. A detailed discussion of technical and commercial losses is presented by Parmar (2016); the primary categories are summarized and presented in Fig. 2.

As seen in Fig. 2, adapted from Parmar (2016), technical losses are attributed to permanent/fixed losses and variable losses. Permanent losses include losses directly associated with electrical lines and distribution of power, such as current leakage, open-circuit losses and heat and noise generation by excessive current flow through cables and transformers. Variable technical losses depend on impedances of electrical cables and transmission lines, which incur losses proportional to the current flow. Commercial losses are dependent on factors such as theft, inaccurate measuring, incorrect billing by automated systems and substandard accuracy of measurement equipment; aggravated by unplanned human error in energy consumption readings and erroneous billing. Efficient urban and rural planning can potentially minimize

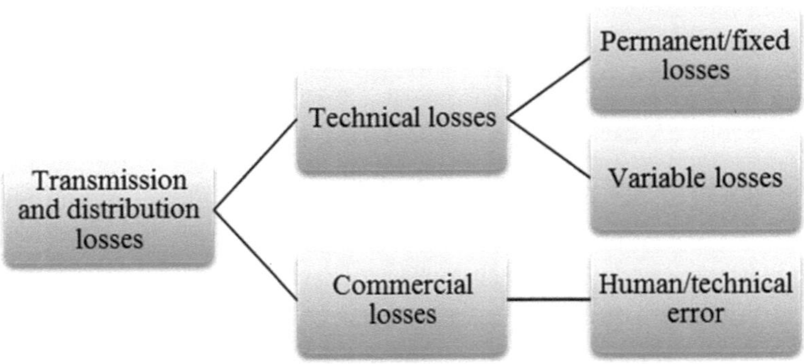

Fig. 2 The two primary categories of transmission and distribution losses; technical and non-technical (commercial) losses (Adapted from Parmar 2016)

technical and commercial transmission and distribution losses. The key reasons for both technical and commercial losses are outlined and described in Parmar (2016); the following list highlights the possible solutions through planning of urbanization and extending power transmission lines and distribution to rural areas:

4.9 Power Losses

The associated high resistance of long distribution lines that are extended to rural areas or dispersed within urbanized environments causes high I^2R losses (copper losses). I^2R losses emanate from the equation for power, given as

$$P = I^2R \tag{3}$$

where I is the electrical current flowing to the conductor and R is the electrical resistance to the current flow. The power losses in high-voltage transmission lines can be described by

$$P_{loss} = \frac{P^2R}{V^2} \tag{4}$$

where V is the voltage supplied by the power station. The power, P, is a fixed quantity for a specific municipal demand and the electrical resistance of the wire can be held constant for a specific material type and geometry. From (4) it is therefore evident that the power losses in high-voltage transmission lines decrease with increased voltage; typically power plants generate much higher voltages than the required 220 V/110 V and down-convert the voltage at supply stations. Since power losses are proportional to the square of the current flowing, losses also increase as demand increases. Typically, sparsely populated regions tend to have lower total power losses because they have fewer low-voltage, high-loss distribution lines and more high-voltage, low-loss transmission lines (Wirfs-Brock 2015). Planning for strategic placement of traditional power plants and alternative sources closer to the end-users of electricity or heat can lessen these losses dramatically. Power line distribution networks typically operate at relatively low utility frequencies (50 or 60 Hz); for this reason, (as opposed to additional parasitic losses in high-frequency applications), power loss can be minimized with conductors with large cross-sectional area that are manufactured using low-resistivity materials. Utilities often review the trade-off between increasing/decreasing the cross-sectional area versus the total cost of the conductor, based on the required length. The resistance (R) of a conductor is calculated by

$$R = \frac{\rho L}{A} \tag{5}$$

where ρ is the electrical resistivity of the material (1.68×10^{-8} Ω/m for copper), L is the length of the conductor and A is the cross-sectional area of the conductor. Urban planning can benefit from knowledge and information about power losses in transmission and distribution lines, not only between power utilities and buildings, but also within buildings and houses. In addition, weather conditions affect power losses; warmer weather is typically associated with higher losses. Power utilities charge users for the amounts of power delivered to their homes or to commercial/industrial buildings, but further losses due to suboptimal distribution boards and low-quality wiring increase the cost for the users as well as the perceived demand from the power utility. During planning of new developments, these factors must be accounted for to ensure efficiency between utilities and the end-user. Ramani and Baharin (2010) present loss allocation methods in order to assign the overall costs of power losses fairly across various market segments and demand intensity. The equations proposed by Ramani and Baharin (2010) include

$$P_G = P_D + L \tag{6}$$

$$P_G = \sum\nolimits_{i=1}^{N_G} P_{Gi} \tag{7}$$

$$P_D = \sum\nolimits_{j=1}^{N_D} P_{Dj} \tag{8}$$

where P_G is the total amount of power generated by utilities or through alternative energies, P_{Gi} represents the output power of individual generators, P_D is the current demand for electricity, P_{Dj} is the demand for electricity from each generated source, L is the losses within the high-voltage transmission lines, N_G is the total number of generator buses and N_D is the total number of power demand buses. Ramani and Baharin (2010) use these equations to determine the viability of three methods presented in literature on allocation strategies; namely

- pro rata energy allocations,
- proportional sharing, and
- novel pricing structures, which determine the gross nodal demand consisting of the losses attracted by a node as a result of demand (current flow).

Power losses are not unique to traditional power plants and also occur in alternative and renewable energy generation. Minimizing losses is especially crucial in low-power applications such as energy harvesting, requiring complex power allocation strategies. For high-voltage demands where power plants are generally capable of producing more power as demands increase, the environment suffers from additional GHG emissions. Renewable sources typically produce less GHG emission through their LCA since fossil-fueled technologies and fuel combustion emit the vast majority of GHG pollutants. Nuclear and renewable energy technologies typically emit most GHG pollutants during their upstream activities and manufacturing; nuclear activities, to construct and decommission nuclear power plants, in addition make a substantial contribution to energy expenditure.

Biomass energy emissions are mainly determined with respect to associated agricultural activities to feed and nurture livestock, according to the National Renewable Energy Laboratory. Granovskii et al. (2006) provide a method to determine the environmental impact of renewable energies compared to traditional fossil-fuel burning generation through

$$C_{GHG} = 1000\frac{(C_R - C_F)}{GHG_F - GHG_R} \tag{9}$$

where C_{GHG} is the cost of GHG emission reduction, GHG_F and GHG_R are emissions of fossil-fuel burning gas and alternative energies, respectively, both measured in grams per MJ of electricity, and C_R and C_F are the costs per MJ of electricity for renewable and fossil-fuel technologies, respectively. From (9), a negative value of C_{GHG} represents a cost-saving through implementation of alternative energies. The following section identifies several global endeavors undertaken to reduce the anthropogenic carbon footprint.

4.10 Endeavors to Reduce Carbon Footprint

The European electricity system, like many international markets in developed countries, is faced with major challenges in attempting to achieve its goal to reach 20% of electricity generation by renewable sources by 2020 and reduce GHG emissions by 80% by 2050 (Strbac et al. 2015). To realize these goals, a significant proportion of the heat and transport sectors will be electrified and local energy sources are required to reduce imports to satisfy demands. Heat generation, a major contributor of GHG emissions from energy production, can fundamentally become more efficient through co-generation and resourceful district heating and cooling (DHC) (IEA 2014). Essentially, by simultaneously generating heat and electricity, deemed co-generation, the overall energy efficiency of the conversion process can be greatly increased, as opposed to traditional thermal generation technologies (IEA 2014). DHC applies effective means of transporting heat through networks of pipes to distribute heat in the form of steam or hot water, from the initial point of generation and the primary place of energy generation towards the end-user which utilizes the energy for various applications. This removes or lessens the requirement to re-generate heat from electricity at the end-user and reduces the load on power generation facilities, effectively reducing GHG emissions from upstream and downstream users. System optimization, best practices, integration of renewables and innovative piping technologies are among the policies and deployment strategies the International Energy Agency (IEA) set out to achieve. By-products of heat during electricity generation can be distributed to end-users for heat generation applications. Its uptake, however, has been slow globally and the IEA aims to prioritize DHC in the near future. The IEA argues that whereas in 2013 global conversion efficiency in thermal power plants was at approximately 36%, from co-generation techniques this conversion efficiency was increased to 58% of energy produced and distributed as electricity and heat. As with many new innovations and high initial capital investments, the IEA highlights overall market conditions and lowering energy prices, lack of infrastructure and lack of knowledge within societies of co-generation benefits and savings. Its economic feasibility is hindered by the large capital investment and long-term uncertainties. In an IEA publication (2014), co-generation and DHC case studies are analyzed to determine the CO_2 savings compared to conventional generation technologies. The analyzed projects include the Markrinch project in the United Kingdom, the Eresma project in Spain, Nuevo Pemex in Mexico, the Mastral project in Denmark, the Bercy project in France as well as the Princess Noura University for Women project in Saudi Arabia. These projects employ biomass, gas, renewable energy, natural cooling and auxiliary boilers to realize co-generation and DHC and a savings range between 5 and 430 kt of CO_2 per year.

The EU suggests that the difficulty of displacing traditional energy generation and reducing GHG emissions from power plants can be overcome through smart grid technologies and micro-grids to enhance cost-effectiveness and resilience (Strbac 2015). Asset security can be achieved through innovative technologies and refined control strategies to implement wireless sensor networks, heat and transport demand-side response for real-time management, efficient energy-storage technologies such as pumped-storage hydroelectricity and improving the flexibility of distributed and backup energy generation (Strbac 2015). Essentially, a micro-grid is different from a traditional distribution system in its controllability and coordination of energy to particular loads, depending on demand and energy balance. Hydroelectric dams, however, have been identified as large sources (larger than lakes and wetlands) of methane and CO_2 emissions and future ways of reducing the proliferation of plant matter and bacteria in these dams should be considered (Hurtado 2016).

Emission allowances (pro rata allocations), as described by Kuri and Li (2009), also tap into opportunities to reduce GHG emission within the electricity and heat generation sector to mitigate climate change. Through this endeavor, the European Union Emissions Trading Scheme allocates national emission quotas to member countries and the allocated energy to each country is locally distributed among individual industry sectors (Kuri and Li 2009). This forces local industry sectors to invest in alternative and renewable energies to supply peak demands for electricity and heat. Carbon emissions can thus be monitored, controlled and estimated with the goal to be systematically reduced while promoting alternative energies available to specific industries. Allocation of emission quotas can be realized by considering historical emission levels and/or auctioning small percentages of allocation to industries capable and willing to invest in larger quotas.

Economic growth and career opportunities are created by industrialization, which leads to urbanization, encouraging more people to move to cities. Industry itself, typically associated with large factories and a high demand for skilled and unskilled workers, is not generally regarded as creating ideal living conditions, in view of high levels of air pollution causing a hazy and smoky atmosphere often carrying the smell of sulfur, methane and other gases. Workers tend to migrate to nearby cities as the demand for housing, retail and other services increases in modern households. Cities, as a result, are gaining popularity and the rate of urbanization over the past 50 years, especially in developing countries, shows that cities are becoming increasingly overpopulated and equally pollution-filled. Industry's demand for improved education to empower its workforce and develop new technologies, potentially pleasant living conditions and accessibility to amenities and services all contribute to urban areas forming as a result of industrialization. Economic prosperity from industrialization and inevitable urbanization should be ensured by addressing the living standards in cities, especially when considering the availability of electricity, heat, clean water and food. Energy consumption in cities is high because of the dense population. Urban planning facilitates a means to reduce energy demand from traditional power utilities and effectively reducing GHG emissions primarily resulting from industrialization; notwithstanding the fact that equal emphasis on addressing the environmental impact of industrial processes leads to cleaner and more environmentally friendly practices. The following section describes potential solutions to decrease energy demand and energy consumption through efficient and effective urban planning, specifically focused on the development of new buildings or the retrofitting of existing ones.

5 Reducing GHG Emissions Through Efficient Urban Planning

The specific areas which contribute significantly to reducing GHG emissions through urban planning described in this section include intelligent and smart lighting, materials used for walls and roofing, energy-efficient elevators, sustainable flooring, certified insulation, energy-efficient heating and cooling and a list of additional factors which also have substantial impact on lowering carbon emissions. These key factors contribute significantly to electricity and heat generation in urban environments since industrialization inevitably leads to urbanization and effective urban planning can minimize and even eliminate indirect sources of GHG emissions. The techniques and proposed solutions presented in this section are not necessarily specific to buildings and developments in urban areas, but also apply to buildings and structures (such as factories) driving industry and targeting positive economic growth in a variety of situations.

5.1 Intelligent and Smart Lighting

Shared, band-limited communication networks with efficient lighting, sensors and controllers can be used to realize spatially distributed lighting systems, called networked lighting control systems (NLCSs) (Schaeper et al. 2013). An NLCS provides convenience and at the same time, aims to reduce energy consumption of lighting in buildings and in car parks. The use of the internet-of-things (IoT) to create smart lighting is projected to stimulate growth from 46 million units in 2015 to 2.54 billion units in 2020 (Gartner 2015). In 2014, 300–500 million square feet (28–47 million m^2) of commercial space globally could be considered to have intelligent or smart lighting (Gartner 2015).

In a building, smart lighting can be used in several spaces, including restrooms, corridors, changing rooms, lobbies, staircases and in some cases small offices. Combining smart lighting such as NLCS and energy-efficient solid-state lighting[1] such as light-emitting diodes (LEDs) and compact fluorescent (CFL) bulbs can further decrease the energy consumption in buildings. A similar strategy can be implemented in car parks, with light sensors that sense the movement of people of cars and only switch on lighting during this time, ensuring that no unnecessary electricity is used when the car parks are either empty or no movement is detected. Smart lighting installations can be challenging, prone to incorrect detection of movement, or detecting animals such as cats in car parks, and require a significant capital investment for large buildings and spaces. Installations also require additional maintenance, since extra electronic equipment in the form of sensors and actuators is introduced. Automatic installation algorithms can be used (Schaeper et al. 2013), but require skilled personnel at additional cost. During urban planning of new developments or retrofitting of existing spaces, additional time and financial budgeting should be in place to account for smart and efficient lighting strategies; this includes determining the estimated savings of electricity compared to the initial capital investment.

The projected lifespan/longevity of light bulbs also influences the decision on the type used. Typically, fluorescent and LED lights have lifespans of approximately 6000–15,000 and 25,000–50,000 h, respectively. The longevity also depends on

- the quality and consistency of the operating voltage,
- the reputability of the manufacturer,
- exposure to mechanical shock or vibrations,
- the frequency at which the light is turned on and off per day, and
- the ambient temperature.

In certain cases, manufacturers supply a guarantee on the usage time of the light bulbs, generally for LEDs, so it is important to keep proof of purchase of LED bulbs in case the light breaks down prematurely. Rehman et al. (2015) list the most significant comparison parameters of light bulbs as

[1]Solid-state lighting refers to sources of solid-state lighting for electroluminescence as opposed to thermal radiation typically used in incandescent lights.

- efficiency of converting electrical energy to light as a percentage,
- luminous efficacy in lumens per watt (lm/W),
- electricity consumption over a specified period of hours in kilowatt-hours (kWh),
- projected lifespan in hours, and
- energy used per bulb to produce similar light intensity in watts.

Furthermore, Rehman et al. (2015) list additional technical electrical parameters, which are considered during a technical audit of a lighting strategy. These parameters include real power, reactive power, rated power, total harmonic distortion, apparent power and power factor. Incandescent light bulbs are capable of connecting directly to the alternating current power source, whereas CFL and LED bulbs typically require signal conditioning electronics. The listed parameters are used to determine the efficiency and quality of the supply circuitry for CFL and LED lights and quantify the losses of energy in the supporting circuitry to ensure that there are no significant losses between the power provided from the utility and the physical bulb.

In creating intelligent and smart lighting, newer technologies and the IoT also allow network- and cloud-based centralized systems to control smart lighting strategies. The five key strategic phases of smart lighting, adapted from the work of Dean Freeman, research vice-president at Gartner, include

- LED lighting—also used as a replacement for fluorescent lighting,
- smart, compact and power-efficient sensors, actuators and control circuitry,
- wireless connectivity to a central system and being controllable by mobile devices such as the Android® or Apple® frameworks,
- statistical analysis of historical event data to identify anomalies in energy patterns and create predictive models, and
- some form of intelligence, either programmed by human operators or artificially adaptable intelligence.

Wireless sensor networks (WSNs), which use spatially distributed sensors (smart sensors), actuators as well as controller circuitry offer huge prospective to improve system efficiency, allow various implementation of system automation and a variety of process control systems (Sharma et al. 2013). WSNs are not discussed in further detail in this chapter and various online references can be consulted for further reading. The following section summarizes the considerations for materials used for building exterior features and roofing.

5.2 *Materials Used for Building Exterior Features and Roofing*

The roof of a building is typically underestimated as a large contributor to energy-efficient buildings and sustainable urban planning. The roof of a building provides a large available space to utilize for energy-harvesting, water distribution or additional vegetation. An important consideration during retrofitting or reconstruction of the roof of a building is to abide by building standards. There are several mandatory provisions to be taken into account when building or changing the design of a roof. For example, the Toronto Green Roof Construction Standard in the United States lists mandatory provisions for green roof construction. These provisions ensure that no hazardous modifications are allowed, which can decrease the structural integrity of the roof and endanger the occupants of the building. Typical provisions include:

- ensuring structural integrity through gravity of loads analysis,
- stability of sloped designs,
- minimizing the potential for wind uplift,
- certifying fire safety,
- appropriate waterproofing, drainage and guaranteeing no water-retention,
- apt vegetation alternatives and the selection of plants, including certified irrigation standards, as well as
- regular and certified maintenance of the roof structure, its integrity and the integrity of any modifications or additions to the roof.

Conforming to building standards when implementing energy-saving or energy-harvesting measures is crucial. The process of changing an existing roof involves submitting applications, site plans, declaration forms and inspection checklists; once the application has been approved, modification of a roof can commence and certification can be requested. The performance and energy savings of each roof are unique, depending on the region, climate, type of building, green roof type and green roof design.

A roof that is designed to reflect more sunlight will absorb less heat compared to traditional roofs. This is especially effective in warmer climates and during summer. Various methods can be used to create the outer layer of reflected material, including highly reflective paint, covered sheets or highly reflective tiles/shingles. Because less heat is absorbed, the indoor temperature of the building will be lower, leading to a lower requirement for air-conditioning units, effectively creating a more energy efficient building. Rooftop gardens, exterior greenery and green walls reduce the ambient temperature of buildings. For colder climates, the roof and walls are typically insulated to ensure no heat generated in the building escapes through cracks, openings and ventilation shafts.

A green roof provides multiple public benefits and advantages, including, in some circumstances, aesthetic improvement, waste diversion, storm-water management, moderation of the urban heat island effect, improved air quality, new amenity spaces and job creation. In the private sector, green roofs improve energy efficiency, increase roofing membrane durability and offer advanced fire retardation, reduced electromagnetic radiation and noise reduction of insulated roofs. Environmental advantages are measured through increased biodiversity, establishment of urban agriculture and self-sustaining buildings and educational opportunities through awareness and training, according to the Green Roofs for Healthy Cities non-profit initiative in the United States.

5.3 Energy-Efficient Elevators for Multiple-Story Buildings

In multiple-story buildings, elevators (lifts) use a significant amount of power to operate on a daily basis. Where elevators are used, the variable voltage variable frequency (VVVF) lift driver offers efficient technology. The VVVF driver uses inverter technology to regulate the input voltage and frequency to reduce current drawn during acceleration and deceleration of an elevator. The primary categories associated with elevator or escalator power consumption are due to friction losses during travelling, dynamics losses while starting or stopping, potential energy transfer and regeneration into the supply system. Furthermore, integrating a sleep mode for an elevator when it has not been in use for a specified time decreases the energy consumption of the elevator (or escalator).

Lift systems in multiple-story buildings must follow guidelines on its energy efficiency. These guidelines take into account the characteristics of the equipment used, characteristics of the premises and the practical configuration of the lift or escalator system. EMSD (2007) lists the general principles to achieve energy efficiency as

- researching and stipulating the advantages or disadvantages and the energy efficiency of the desired equipment,
- keeping the design minimalistic and not over-complicating the requirements,
- determining appropriate zoning strategies of the system by subdividing the floors on the premises into clusters of stops for different lift cars,
- implementing appropriate system control and energy management measures,
- not using copious amounts of heavy materials for informative or warning signs and implementing energy-efficient lighting inside the lift car, and
- regularly maintaining and inspecting the system to ensure all moving mechanical parts have low friction losses.

EMSD (2007) provides a detailed guideline for elevator and escalator installations and should be referenced as a starting point for urban planners integrating elevator systems in new buildings or replacing older, less energy-efficient systems. Modeling of the round-trip time (RTT) during peak traffic can be done, as shown in EMSD (2007), through

$$\mathrm{RTT} = 2Ht_v + (S+1)t_s + 2Pt_p \tag{10}$$

where H is the highest call reversal floor, S is the average number of stops, t_v is the time to transit two adjacent floors at the rated speed, given in seconds, t_s is the time consumed when making a stop, also in seconds, t_p is the passenger transfer time for entering or exiting the lift, in seconds, and P is 80% of the contract capacity (of persons) of the lift car (EMSD 2007). The highest call reversal floor and average number of stops are determined by

$$H = N - \sum_{j=1}^{N-1} \left(\sum_{i=1}^{j} \frac{U_i}{U} \right)^p \tag{11}$$

and

$$S = N - \sum_{i=1}^{N} \left(1 - \frac{U_i}{U} \right)^p \tag{12}$$

where N is the number of floors above the main terminal floor, U is the total population of the zone above the terminal floor and U_i is the population at the ith floor. In addition, the total time of a single stop (t_s), given in EMSD (2007), is calculated by

$$t_s = t_{f1} - t_v + t_o + t_c \tag{13}$$

where t_{fl} is the time, in seconds, of a single floor jump, t_o and t_c are the door opening and closing times, respectively, both measured in seconds. Critical analysis and statistical modeling of the elevator or escalator system in high buildings provides data to predict and adjust the operation times effectively in order to save energy. In the list of Green Mark achievers issued by the BCA of Singapore, for example, most buildings achieving a platinum or gold rating for green operation are awarded merits for their energy-saving elevators or escalators. The following section describes the considerations for certified flooring in buildings.

5.4 Certified Flooring (Carpeting, Tiling or Wood)

Flooring is another topic receiving more attention and becoming a more important factor in sustainable building, especially in large surface areas and in areas with high quantities of occupant traffic. According to the Green Building Alliance in Western Pennsylvania, United States, the main parameters of distinction of flooring are the material extraction and production methods and indoor air quality effects.

The most common flooring materials are vinyl, wood, tile, bamboo, linoleum and carpeting. By investigating the material extraction and production methods of these alternatives, sustainability in the form of environmentally friendly and renewable sources can be achieved. For example, vinyl is produced from petroleum as opposed to recyclable materials; in addition, its transport costs are relatively high and it is known to emit chemicals after being manufactured. Vinyl is also not biodegradable. Stone flooring, although durable, requires high energy to extract and transport and its production causes high GHG emissions. Natural linoleum (marmoleum, made from 97% natural and biodegradable materials) contains natural ingredients such as linseed oil, limestone, tree rosin, wood flour, natural mineral pigments to vary color and jute (a vegetable fiber). Figure 3 summarizes the commonly used flooring materials and remarks in terms of sustainability and environmentally friendly installations.

From Fig. 3 it is therefore evident that the pricing of flooring in new or existing building developments is not the only factor that influences its sustainability and environmentally friendly status. Generally, lower-cost flooring relates to lower durability and sustainability over the longer term and significant initial capital investment in flooring can have advantageous long-term characteristics, both relating to the cost of repair or maintenance and its indirect GHG emissions. Figure 4, which has been adapted from the World Floor Covering Association, lists the average prices of the commonly used flooring alternatives.

The price comparison of Fig. 4 is an indicative and normalized average value that could fluctuate depending on demand and availability and should only be used as an approximate guideline. As seen in Fig. 4, although vinyl is typically the least expensive material used for flooring, according to Fig. 3 it has a significant impact on the environment and indirect GHG emissions, since it is partly made of petroleum. Other commonly used materials are laminate, ceramic and engineered wood; stone is typically the most expensive flooring material because of the high costs associated with its extraction and transport.

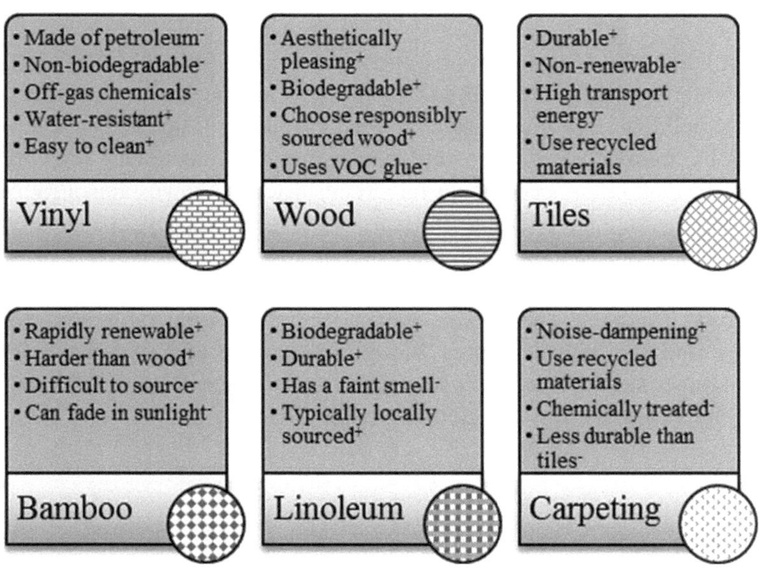

Fig. 3 Brief overview of commonly used flooring materials in buildings, adapted from the Green Building Alliance in Western Pennsylvania, United States. The superscript "+" and "–" next to each comment indicates an advantage or disadvantage, respectively

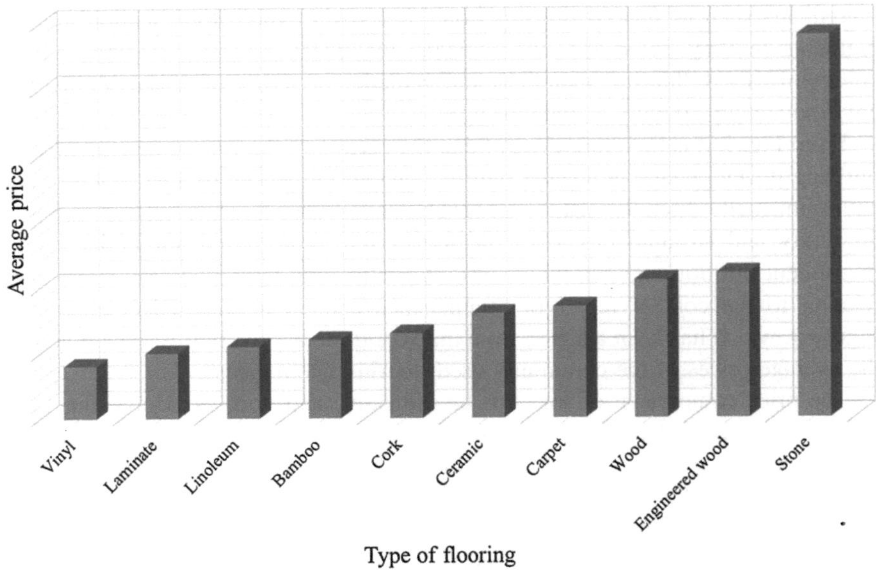

Fig. 4 Normalized price comparison of commonly used flooring materials, adapted and normalized from the World Floor Covering Association

5.5 Insulation of Walls, Floors, Foundations and Ventilation Shafts

Green building insulation of walls, floors, foundations and ventilation shafts ensures better thermal protection and energy performance of heating and cooling within the building. Historical insulators such as mud, asbestos and cork have been replaced by more cost-effective and better insulation materials such as polyurethane. Insulation has become an important factor in the sustainability of commercial and institutional buildings in terms of energy efficiency, occupant comfort, material durability and increase in the value of a property for resale.

Insulation materials are typically rated by an R-value, which gives a measure of their thermal resistance. The R-value is expressed as the temperature difference that will cause one unit of heat to pass through one unit of area of a material over a period of time. It is derived from the thermal resistance R_{TH}, which can mathematically be determined by (Chakravorty et al. 2016)

$$R_{TH} = \int_0^h \frac{dz}{\kappa(T(z))A(x,y,z)} \tag{14}$$

where h is the thickness of the material, $A(x, y, z)$ is the position-dependent heat flow and $\kappa(T(z))$ is the position-dependent thermal conductivity of the material. The R-value of an insulator is simply determined by

$$R = \frac{h}{\kappa} \tag{15}$$

and is specified in (m^2K)/W, where K is the unit of temperature, Kelvin, and W is the unit of watts. In the United States and in Canada, the U-factor is typically used to express the thermal insulation characteristics of a material; it is the inverse of the R-value and is expressed in W/(m^2K). Higher quality insulators, therefore having better (higher) thermal resistance, have a higher R-value. Indirectly, the R-value is an indication of the density of the insulating material. Thermal insulation is typically required in buildings to isolate indoor air-conditioned spaces from outdoor ambient temperature; whether warmer or colder, the indoor temperature should only be regulated by the energy expenditure within the building. In addition, thermal bridging should be controlled and insulated. It is the effect of thermally conductive materials allowing heat or cold to conduct through thermal barriers between two materials.

A higher R-value is associated with a higher initial cost of installation but will effect cost-savings in terms of energy use in a building over a longer term. According to the Green Building Alliance in Western Pennsylvania, United States, thermal insulation is achieved through many different types of materials, including loose fills, spray foam, rigid boards and fiberglass. The most commonly used insulation building materials are summarized in the following list.

- Cellulose can be used in wall cavities, attic floors or in ceilings. Cellulose typically requires very low energy to produce compared to fiberglass or mineral wool and is manufactured using mostly recycled content. It is advantageous to use in terms of indoor air quality, since it emits no hazardous gases or particles. Its major disadvantage is the fact that it can absorb moisture, although if installed in a low-moisture environment it is a cost-effective option with relatively good R-value.
- Cotton can also absorb moisture and care should be taken when installing cotton as an insulator material. In term of sustainable buildings and low-energy consumption during manufacture, cotton is a good alternative, since it is made from renewable plant-based materials, it is recyclable and non-toxic, with a relatively high R-value. Flammable materials such as cotton must be separated from living spaces by fire-resistant materials such as drywall or plaster.
- Fiberglass is a common material used for insulation, especially in older buildings, and is made from silica, an abundant material found in the earth's crust. Up to half of the material composition is recyclable although fiberglass has certain negative attributes, which deter its use for large installations. A well-known characteristic of fiberglass is its release of particles that irritate the skin, eyes and throat and make it difficult to work with. To produce fiberglass involves energy-intensive processing steps, relating to high-energy use during manufacturing.
- Polyisocyanurate foam is a popular material that can be used for interior basement walls and beneath attic ceilings, since it absorbs very little moisture. It contains no hydro-chlorofluorocarbons and typically has no detrimental effects on indoor air quality. To its disadvantage, it is made from petrochemicals and is

not recyclable. It does, however, have the largest R-value compared to the other materials in this list.

- Polystyrene expanded foam also contains no hydro-chlorofluorocarbons but it is recyclable, poses no threat to indoor air quality and has an average R-value. However, the process to manufacture it is energy-intensive, it contains petrochemicals and may contain toxic brominated flame retardants and other toxins after post-processing to reduce its moisture absorption.
- Polystyrene extruded foam is typically more moisture-resistant compared to expanded polystyrene and is also recyclable. Similar to expanded polystyrene, it is made from petrochemicals and may contain toxic materials when processed to increase fire and moisture resistance. Its R-value is relatively similar to that of expanded polystyrene, depending on the coatings applied in post-processing.
- Spray polyurethane foam has a relatively high R-value, which makes it a good contender to polyisocyanurate, also containing no hydro-chlorofluorocarbons; in addition, it will not settle into cracks and crevices, making it easy to remove or replace. It is not recyclable and most spray polyurethane is produced using petrochemicals.
- Mineral wool (rock wool) has a relatively low R-value and releases some particles that can irritate the eyes, skin and throat. To its advantage, it is made from abundant natural rock, it is moisture-resistant and requires no additional flame-retardant chemicals post-processing and most of the material is recyclable. It is a cost-effective alternative, although not as effective as most other materials on the list, but more effective compared to concrete, for example.

Figure 5 summarizes the performance of the insulator materials listed above and gives a relatively normalized performance value to each material.

As seen in Fig. 5, concrete typically has the lowest R-value, relating to the lowest performance of insulating heat or cold between adjacent sections in a building. Concrete is typically a relatively easy alternative owing to its availability during the construction of remodeling of buildings, but it is not the most effective option and also requires high energy inputs during its preparation and transport. Polyisocyanurate foam has recently been in common use in green buildings and sustainable projects, offering a manageable and effective means of insulating adjacent sections with relatively high moisture content, such as basement walls.

5.6 *Heating, Ventilation and Air-Conditioning*

For new energy-efficient green building constructions, it is important to incorporate heating, ventilation and air-conditioning (HVAC) systems early in the design phase; this makes installation less complex, decreases labor costs and installations can be planned to be most effective for each project, as opposed to retrofitting existing buildings and compromising, in view of constraints and limitations. The heating and cooling loads of the HVAC equipment are affected by solar orientation if

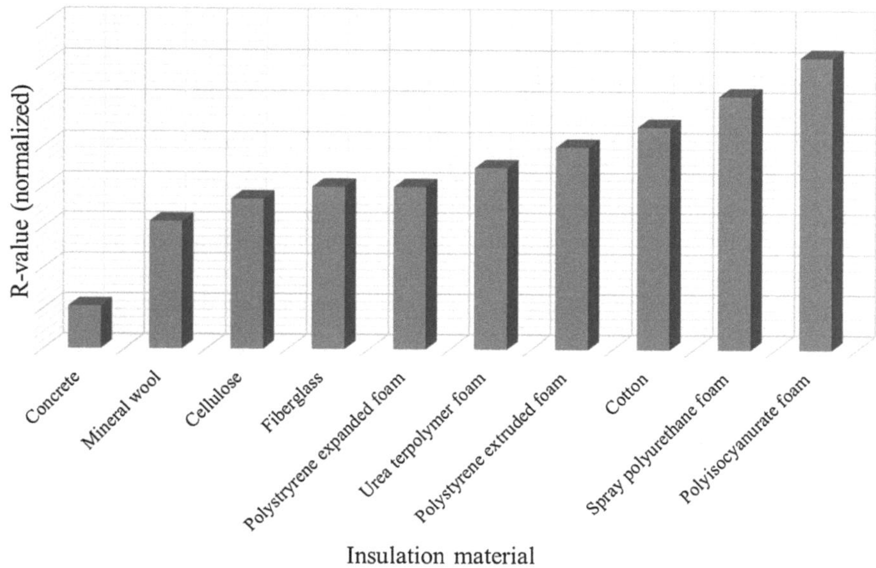

Fig. 5 Normalized performance comparison of commonly used insulator materials based on its R-value, adapted from various sources

alternative energy such as solar power is used. In some circumstances it is also affected by nearby vegetation, which can have a natural cooling effect. High-efficiency furnaces or air conditioner units should be used, which remove pollution from the air. Typically, for Energy Star® International Efficiency Marking Protocol certified units, an energy efficiency level is allocated and displayed on the informative label on many branded units, containing a Roman numeral indicating the efficiency level. After installation of HVAC equipment, all electrical outlets or wall penetrations should be sealed and insulated by using the most effective and practically appropriate insulator materials. Hui (2001) lists various design strategies and new technologies to be considered when installing HVAC systems in new sustainable developments or existing retrofitted designs.

The following section summarizes additional techniques to apply during urban planning to achieve energy-conscious and sustainable buildings for new developments and, where practical and possible, retrofitting existing buildings.

5.7 *Additional Urban Planning Strategies and Techniques to Achieve Energy-Efficient, Smart and Sustainable Buildings*

To realize sustainable and energy-efficient buildings, Fig. 6 summarizes additional considerations to be kept in mind during urban planning.

According to Fig. 6, the type, size and facing direction of windows are important considerations to reduce energy consumption in buildings. Large windows, ideally facing north with a north-south cross ventilation scheme in the southern hemisphere and facing south in the northern hemisphere, will ensure that abundant natural light enters the building to heat the environment naturally during cold months. A north-south cross-ventilated scheme will allow wind to pass through the building and cool down the environment during warmer months. A naturally light environment and clear sight to the outdoors also help improve the morale of occupants.

Water efficiency is the smart use of water resources by using water-savings technologies and ensuring reliable water supplies in commercial and residential buildings. Accompanied by rainwater harvesting and leak detection, efficient water-savings strategies include:

- installing efficient and reputable plumbing fixtures, pipes and distribution points,
- using non-potable water where possible,
- installing sub-meters to ration and distribute water effectively, based on varying demand,
- implementing rainwater harvesting for non-drinkable water uses,
- choosing drought-tolerant and slow-growing vegetation (xeriscaping) for rooftop gardens or vertical gardens to decrease ambient temperature and promote cleaner (CO_2 absorbed) air,
- ensuring proper leak detection mechanisms and implementing regular maintenance of water distribution pipes,

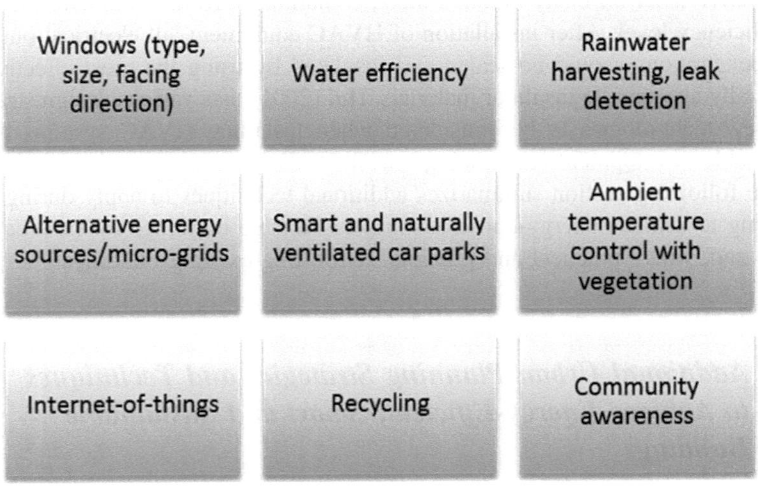

Fig. 6 Additional urban planning strategies and techniques to achieve energy-efficient, smart and sustainable buildings

- installing appropriate irrigation technologies and using non-potable water for landscape irrigation, and
- installing heat pumps as opposed to traditional warm-water geyser systems and using in-line water transfer pumps.

Also shown in Fig. 6 is consideration of feasible and environmentally practical implementations of alternative energy sources and consideration for micro-grids. Alternative energy sources include photovoltaic solar cells, wind energy or hydro energy. Capital investments in alternative energy remain large, with long-term maintenance and replacement of energy storage equipment adding to the overall cost of such an endeavor. Micro-grids are local energy grids independent from traditional power grids, consisting of energy-generation facilities, demand management and energy storage capabilities. Micro-grids aim to lower the demand for traditional transmission and distribution systems and ensure locally generated and reliable energy supply in urban (and rural) settings and therefore lowering overall GHG emissions. Since a micro-grid operates independently from utility grids and is smaller in terms of physical size, a larger number of these micro-grids can be spatially distributed in an urban setting, enabling integrated renewable alternative energy sources at each micro-grid to minimize the carbon footprint within cities. Sustainability and reducing dependence on power utility grids, as well as long-term cost savings and reduced environmental impact, are among the advantages of using alternative energy sources. Urban developments are increasingly using micro-grids because of these advantages and to ensure long-term sustainability.

Installations of smart car parks and planning of naturally ventilated ones can also lead to significant energy savings for buildings and factories, since large numbers of the workforce commute to and from these areas. High-efficiency, variable speed extractor fans with carbon monoxide sensors can reduce the toxicity typically experienced in closed spaces with high movement of cars and other vehicles. These installations allow for ductless car parks, leading to savings on materials and space, as well as savings on operational cost due to the lower power consumption of axial fans and elimination of ductwork.

Jonsson (2004) studied the influence of vegetation on the urban climate in the subtropical and rapidly expanding city of Gaborone, Botswana. The study found an apparent opposit effect of rural and urban vegetation, where the former was hindering the temperature from falling and the latter was cooling the environment through evapotranspiration. The extent to which vegetation cools the urban climate depends on species selection and strategic placement (Doick and Hutchings 2013). Urban vegetation can change the localized air temperature and air quality through several mechanisms, including evaporative cooling and evapotranspiration, reflectance of radiation, which is linked to surface albedo, as well as through shading (Doick and Hutchings 2013).

The IoT and other technology implementations such as WSNs contribute in many forms and applications to reducing pollution in urban planning strategies. Through the IoT, the physical world is becoming an information system with sensors and actuators embedded in physical objects and connected to one another and to central servers wirelessly or through wired connections. The IoT provides the potential of improving the productivity of the production processes in industry by governing and adapting to take corrective action automatically and in much shorter times (a few milliseconds potentially) compared to human operators and maintenance workers.

Further honorable mentions, as presented in Fig. 6, include effective implementation and sustainable strategies of recycling, endeavors to increase community awareness of sustainable urban planning to reduce GHG emissions resulting from increases in industry and positive economic growth.

6 Conclusion

This chapter reviews and discusses the most important anthropogenic sources of GHG emissions across the industrial economic sectors. These sectors include electricity and heat production, agriculture and other land use, industry, transportation, other sources of energy and buildings. The focus of this chapter is identifying the sources of GHG emissions and pollution leading to climate change and addressing effective strategies of urban planning to reduce and possibly eliminate these causes of pollution. In urban planning, energy distribution, industry requirements, effective transportation and logistics, as well as energy consumption in buildings, are potential areas to minimize GHG emissions. An efficient built environment can directly and indirectly be linked to conscious reduction and energy-saving routines that reduce the demand on energy. Addressing multiple industry sectors to contribute to the overall reduction in energy consumption, as well as identifying the dominant sources of pollution, empowers researchers, urban planners and technological disciplines to work together in enforcing sustainable urbanization, environmentally friendly economic growth and energy-conscious industrialization.

References

Associated Press. (2015). Air pollution in China is killing 4000 people every day, a new study finds. *The Guardian*. Retrieved August 4, 2016 from http://www.theguardian.com

Benfield, F. K. (2014). Fighting climate change with sensible city planning. Retrieved August 3, 2016 from http://www.huffingtonpost.com

Bos, I., De Boever, P., Panis, L., & Meeusen, R. (2014). Physical activity, air pollution and the brain. *Sports Medicine*, *44*(11), 1505–1518.

Chakravorty, A., D'Esposito, R., Balanethiram, S., Fregonese, S., & Zimmer, T. (2016). Analytic estimation of thermal resistance in HBTs. *IEEE Transactions on Electron Devices*, *63*(8), 2994–2998.

Doick, K., & Hutchings, T. (2013). Air temperature regulation by urban trees and green infrastructure. *Forestry Commission: Research Note*. FCRN012, February 1–10, 2013.

Dridi, M. (2007). Urban transport planning for ubiquitous environments. In *IEEE International Conference on Pervasive Services*, Istanbul, Turkey (pp. 421–426). July 15–20, 2007.

EMSD. (2007). Guidelines on energy efficiency of lift and escalator installations. Retrieved August 24, 2016 from http://www.emsd.gov.hk

EPA. (2016). Greenhouse gas inventory guidance. Indirect emissions from purchased electricity. *United States EPA Center for Corporate Climate Leadership*. Retrieved August 18, 2016 from http://www.epa.gov

Friedrich, T., Timmermann, A., Tigchelaar, M., Timm, O. E., & Ganopolski, A. (2016). Nonlinear climate sensitivity and its implications for future greenhouse warming. *Science Advances, 2* (11), November 2016.

Gartner. (2015). Gartner says smart lighting has the potential to reduce energy costs by 90 percent. *Press Release*. Retrieved August 24, 2016 from http://www.gartner.com

Granovskii, M., Dincer, I., & Rosen, M. A. (2006). Economic aspects of greenhouse gas emissions reduction by utilisation of wind and solar energies to produce electricity and hydrogen. In *IEEE EIC Climate Change Conference* (pp. 1–5). May 10–12, 2006.

Hui, S. C. M. (2001). HVAC design and operation for green buildings. In *Proceedings of the Shaanxi-Hong Kong refrigeration and HVAC seminar*, Xian, China (pp. A44–51). June 11–13, 2001.

Hui, C., Xianghui, W., Xiqiang, Z., & Shaoli, Z. (2014). The evaluation of Chinese urban traffic management system application based on intelligent traffic control technology. In *7th International Conference on Intelligent Computation Technology and Automation (ICICTA),* Hunan, China (pp. 791–795). October 25–26, 2014.

Hurtado, M. E. (2016). Dams raise global warming gas. *Science and Development Network.* Retrieved November 7, 2016 from http://www.scidev.net

IEA. (2014). Linking heat and electricity systems. Co-generation and district heating and cooling for a clean energy future. *The International Energy Agency*. Retrieved August 14, 2016 from http://www.iea.org

IPCC Mitigation of Climate Change. (2014). Climate change 2014: Mitigation of climate change. In O. Edenhofer, R. Pichs-Madruga, Y. Sokona, E. Farahani, S. Kadner, & K. Seyboth (Eds.), *Contribution of Working Group iii to the Fifth Assessment Report of the Intergovernmental Panel on Climate Change*. Cambridge: Cambridge University Press.

IPCC Synthesis Report. (2014). Climate change 2014: Synthesis report. In R. K. Pachauri & L. A. Meyer (Eds.), *Contribution of Working Groups I, II and III to the Fifth Assessment Report of the Intergovernmental Panel On Climate Change* (p. 151). IPCC, Geneva.

Jonsson, P. (2004). Vegetation as an urban climate control in the subtropical city of Gaborone, Botswana. *International Journal of Climatology, 24,* 1307–1322.

Kristiansen, J., Persson, R., Björk, J., Albin, M., Jakobsson, K., Östergren, P., & Ardö, J. (2011). Work stress, worries, and pain interact synergistically with modelled traffic noise on cross-sectional associations with self-reported sleep problems. *International Archives of Occupational and Environmental Health, 84*(2), 211–224.

Kuri, B., & Li, F. (2009). Allocation of emission allowances to effectively reduce emissions in electricity generation. *IEEE Power and Energy Society General Meeting* (pp. 1–8). July 26–30, 2009.

Lehmann, A., & Gross, A. (2016). Using crowd sensed data as input to congestion model. In *IEEE International Conference on Pervasive Computing and Communications Workshop (PerCom Workshops)* (pp. 1–6). March 14–18, 2016.

Liang, Y. (2010). The analysis of our urban transportation problem and the research of road construction and planning management. In *International conference on E-product E-service and E-entertainment*, Henan, China (pp. 1–4). November 7–9, 2010.

Mingrone, L., Pignataro, G., & Roscia, M. (2015). Smart urban electric transport system: An innovative real model. In *International Conference on Renewable Energy Research and Applications (ICRERA),* Palermo, Italy (pp. 1457–1462). November 22–25, 2015.

Parmar, J. (2016). Total losses in power distribution and transmission lines. Retrieved August 19, 2016 from http://electrical-engineering-portal.com

Pitt, R. (1979). Demonstration of nonpoint pollution abatement through improved street cleaning practices. In *EPA-600/2-79-161 US Environmental Protection Agency*, Cincinnati, August 1979.

Pitt, R., Bannerman, R., Clark, S., & Williamson, D. (2004). Sources of pollutants in urban areas (Part 1). In *Effective Modeling of Urban Water Systems* pp. 465–484.

Rakkesh, S. T., Weerasinghe, A. R., & Ranasinghe, R. A. C. (2016). Effective urban transport planning using multi-modal traffic simulations approach. In *Moratuwa Engineering Research Conference (MERCon),* (pp. 303–308) Moratuwa, Sri Lanka. April 5–6, 2016.

Ramani, A. N., & Baharin, K. A. (2010). Transmission losses allocation in deregulation electricity market. In *IEEE Conference on Power and Energy (PECon),* (pp. 841–845) Kuala Lumpur, Malaysia. 29 November–1 December 2010.

Rehman, M., Pasha, M., & Hassan, N. U. (2015). Clean consumer energy technologies in developing countries: A case study of energy efficient lights in Pakistan. In *International conference on smart grid and clean energy technologies (ICSGCE)* (pp. 85–89) Offenburg, Germany. October 20–23, 2015.

Rotton, J., & Frey, J. (1984). Psychological costs of air Pollution: Atmospheric conditions, seasonal trends, and psychiatric emergencies. *Population and Environment, 7*(1), 3–16, March 1984.

Ryan, R. (2014). Countries with the worst air pollution ranked by World Health Organization. Retrieved August 5, 2016 from http://www.news.com.au

Schaeper, A., Palazuelos, C., Denteneer, D., & Garcia-Morchon, O. (2013). Intelligent lighting control using sensor networks. In *10th IEEE International Conference on Networking, Sensing and Control (ICNSC)* (pp. 170–175) Evry, France. April 10–12, 2013.

Sharma, D., Verma, S., & Sharma, L. (2013). Network topologies in wireless sensor networks: A review. *The International Journal of Electronics and Communication Technology (IJECT), 4*(3), 93–97. ISSN: 2230–7190 (Online) 2230–9543 (Print).

Sotos, M. (2012). *GHG protocol scope 2 guidance. An Amendment to the GHG protocol corporate standard.* Retrieved August 14, 2016 from http://www.ghgprotocol.org

Stokols, D., Novaco, R. W., & Stokols, J. (1978). Traffic congestion, type A behavior, and stress. *Journal of Applied Psychology, 63*(4), 467–480.

Strbac, G., Hatziagyriou, N., Lopes, J. P., Moreira, C., Dimeas, A., & Papadaskalopoulos, D. (2015). Microgrids: enhancing the resilience of the European megagrid. *IEEE Power and Energy Magazine, 13*(3), 35–43.

Tabraiz, S., Ahmad, S., Shehzadi, I., & Asif, M. (2015). Study of physio-psychological effects on traffic wardens due to traffic noise pollution; exposure-effect relation. *Journal of Environmental Health Science & Engineering* (Open Access), 1–8. April 16, 2016.

UNON. (2004). Sustainable urban energy planning: A handbook for cities and towns in developing countries. Retrieved August 5, 2016 from http://www.ec.europa.eu

Vithayasrichareon, P., & MacGill, I. F. (2016). Valuing large-scale solar photovoltaics in future electricity generation portfolios and its implications for energy and climate policies. *IET Renewable Power Generation, 10*(1), 79–87.

Wirfs-Brock, J. (2015). *Lost in transmission: How much electricity disappears between a power plant and your plug?* Retrieved August 22, 2016 from http://insideenergy.org

Yan, Y., Wang, C., Ding, D., Zhang, Y., Wu, G., Wang, L., et al. (2016). Industrial carbon footprint of several typical Chinese textile fabric. *Acta Ecologica Sinica, 36*(3), 119–125.

Yao, H., & Chen, D. (2015). A system dynamics model for urban sustainable transportation planning. In *23rd International Conference on Geoinformatics* (pp. 1–5) Wuhan, China. June 19–21, 2015.

Chapter 21
From Grey Towards Green. About the Urban Energy Fold at Symbiont City

Nieves Mestre, Lucelia Rodrigues, Eva Hurtado and Eduardo Roig

Abstract Instead of the energy and ecological relocation, SYMBIONT City detects energy opportunities and possible urban folding to achieve thermodynamic benefits. Although some agendas have already fostered the concept of symbiotic planning, neither current infrastructural systems nor urban regulatory frameworks allow for its real implementation. SYMBIONT is a set of local laboratories designed to enable new synergies between waste, energy and information flows on existing urban waste transfer facilities. It pretends to raise the level of urban resilience in cities by acting on existing urban facilities and adjacent urban setting through the implementation of local laboratories able to monitor, process, and reconnect existing waste, energy and information flows while recovering the notion of infrastructure as public space through social engagement actions. These spatial facilities have a strategic value as nodal urban locations—with potential phase-change capacity—for neighbourhood waste and energy flows. These micro-infrastructural interventions will help in the aforementioned transition allowing for a turn from "grey" towards "green" infrastructures, with capacity to provide social, ecological and economic benefits to urban communities such as reduction of waste disposal, local energy generation and storage, improvement of air quality, reduction of energy costs and new opportunities to social cohesion and engagement.

N. Mestre (✉) · E. Hurtado
Department of Architecture, Universidad Europea, Madrid, Spain
e-mail: nieves.mestre@universidadeuropea.es

E. Hurtado
e-mail: eva.hurtado@universidadeuropea.es

L. Rodrigues
Department of Architecture and Built Environment, Faculty of Engineering, University of Nottingham, Nottingham, UK
e-mail: Lucelia.Rodrigues@nottingham.ac.uk

E. Roig
School of Architecture, Universidad de Alcalá de Henares, Madrid, Spain
e-mail: eduardo.roig@uah.es

R. Álvarez Fernández et al. (eds.), *Carbon Footprint and the Industrial Life Cycle*, Green Energy and Technology, DOI 10.1007/978-3-319-54984-2_21

Acronyms

CF	Carbon footprint
EPP	Energy plant prothesis
EU	European Union
MEU	Mixed-use ecosystem entity
MUD	Mixed-use development
MWF	Municipal waste management facilities
SLS	Symbiont local laboratory
SSP	Social symbiont platform
WMP/MWP	Waste Municipal Plan
WP	Waste protocol

1 Introduction

European cities are forerunners in the transition towards a low carbon and resource efficient economy. The current call on Sustainable cities though Nature-based solutions (SSC2–4) inside Horizon 2020 focuses on providing evidence that re-naturing of cities through the deployment of innovative, locally adapted, systemic solutions—that are inspired or supported by nature—can be a cost-effective and economically viable way to make cities more sustainable, resilient, greener, and healthier.

The project is inserted in the field of urban ecology. The city's energy and informational flows can be redesigned to motivate their interconnection into more operational networks: an infrastructure which promotes mixed clusters favoring, ultimately, greater urban resilience. The proposal currently addresses energy-information interchange protocols between different uses—residential, leisure and service buildings—able to function as mixed-use labs. This protocol will be tested on different urban locations (Madrid, Nottingham, Venice) aiming to achieve the maximum range of exportability. The results will be analyzed to define a more precise and performative approach to the concept of mixed-use aggregation —in quantitative and qualitative terms- and how it could cooperate towards urban resilience. The obtained evidences will finally serve as a basis for an urgent review on the archaic disconnection between urban policies and technical building codes.

Although some agendas have already fostered the concept of symbiotic planning, neither current infrastructural systems nor urban-building regulatory frameworks allow for its real implementation.

- Although there are existing planning figures that enable the multifunctional addition in cities, there are still no standards/evidences of what a sustainable urban mixed-use implies in energetic terms (Evans et al. 2007).
- Neither current building codes nor existing energy assessment tools contemplate energy exchange protocols between adjacent buildings/uses, and therefore we lack protocols for design, location and impact assessment of the mixed-use. The

few existing tools are mainly experimental and have low Technology Readiness Levels.

- The measuring-mapping of surplus energy flows has not been implemented so far in the strategic plans of our cities, in many cases belonging to theoretical investigations of little applicability. Some of them are Regionmaker and Function Mixer by MVRDV (2002), Spacemate at TU Delft (2004) or Cyclifier by 2012 Architecten or Rotterdam Energy Approach and Planning REAP methodology developed by Andy Van den Dobbelsteen from TU Delft.

Instead of the energy and ecological relocation, SYMBIONT City detects energy opportunities and possible urban folding to achieve thermodynamic benefits.

SYMBIONT enables a set of local laboratories designed to empower new synergies between waste, energy and information flows on existing urban waste transfer facilities. It pretends to raise the level of urban resilience in cities by acting on existing urban facilities—such as street cleaning depots, collection points, recycling/clean points—and adjacent urban setting through the implementation of local laboratories able to monitor, process, and reconnect existing waste, energy and information flows while recovering the notion of infrastructure as public space through social engagement actions. These spatial facilities have a strategic value as nodal urban locations—with potential phase-change capacity—for neighborhood waste and energy flows.

SYMBIONT sets up a protocol about intervention in the city. It increases the standard of resilience of the urban pre-existences. The impact of this action generates a symbiotic transformation within simultaneously 'energy' and 'information' urban vectors. The proposal drafts a theoretical model which will become applied in a set of case studies located in the 'leading' cities. Furthermore, the action will be monitored; a conclusive analysis will help to adjust the model in order to increase the exportability of results. The scientific evidence finally obtained will serve as a basis for a normative revision to tackle overlaps and correlations between the urban regulations and the building technical codes at European level.

2 Relation with Nature-Based Protocols

Biological analogies have been largely used by urban theory, but while the concept of urban ecosystems can be accepted as a scientific fact, urban metabolism still relies on a misused metaphorical analogy without sufficient evidences. The failing metaphor nowadays challenges the invention of new methodological tools to bridge the gap between ecological, thermodynamic and socio-economic approaches to urban systems theory.

The concept of technical metabolism was First stated by Ian McHarg in Design with Nature (1969) and later by McDonough and Braungart in Cradle to Cradle (2002), both proposing an overlapping of biological and artificial cycles defined as waste-less interdependent flows. In the early 1970s, the concept of material flows for Urban Metabolism analysis was formally incorporated as part of the UNESCO Division of Ecological Science. Since then, many agendas have noted the obsolescence of the current energy model—based on large Power plants and a unidirectional distribution network—and the need to build a new energy network based on distributed local production and capacity for bidirectional transfer. Apart from the aforementioned infrastructural obsolescence, there is another fact blocking this desirable urban resiliency in physical and procedural terms: the archaic disconnection between urban policies and technical building codes.

This hypothesis requires two substantial changes in our current models: the establishment of building entities as energy producers; and the existence of a small-scale infrastructural network[1] of energy/waste/information exchange based on symbiotic agreements between production and waste flows (Van den Dobbelsteen 2010).

As it has been scientifically proved, the high resilience of natural ecosystems is not based on energy-efficient individual protocols, but on multiple metabolic affiliations. Resiliency is a function of diversity and degree of functional connectivity between species (Odum 1992, p. 196). Spatial limits and decentralized patterns empower the increase of inner efficiency in our cities: the closer the cycle, the more proficient the exploitation of material and energy sources. The higher is the diversity of program and scale of the entities implied, the greater the potential for these local transfers to happen. This combination of diversity and hyperstatic connectivity is defined as functional redundancy.

Natural ecosystems do not follow the principle of economy: they grow till exhausting the limits of their immediate resources. To succeed and endure, ecosystems increase the wealth of their decentralized interconnections and agreements, involving in all cases in waste recycling. As the system evolves, waste becomes the main resource input: in old-growth forests, less than 10% of net production is consumed as living matter—e.g. grass—(Goldsmith 1996). In environments of scarce local sources, the relative low efficiency of certain biological processes from primary resources—such as photosynthesis or digestion- is surprising compared to the highly optimized metabolism of waste from other parts of the system (Marsh and Khan 2011).

SYMBIONT proposes a model change in terms of methodology of study approach, based on the definition of a new ecosystem urban unit beyond the classic property-based urban structure. One of the conditions of this new model is the need for programmatic hybridization (mixed-use) increasing spatial complexity and therefore the possibilities of exchange flows; as opposed to classic urban entities, Mixed-use Ecosystem Unit—MEU—has a novel metabolic capacity and can manage its energy balance internally. This protocol requires envisioning the maximum level of information—on usage, demands and consumption/waste household habits, so residents acquire skills to participate locally in space use management and waste/energy generation processes.

[1]Total municipal solid waste has decreased by 2% between 2004 and 2012, but in the same period, 18 of the 28 EU countries increased the amount of municipal waste generated per capita, rising fairly steadily in 6 of these countries.

3 Objectives

SYMBIONT is a complex action protocol that contains a set of local laboratories designed to enable new synergies between waste, energy and information flows on existing urban waste transfer facilities. The obtained data will be collected and analyzed to define a more precise approach to the concept of mixed-use ecosystem unit MEU, both in quantitative and qualitative terms, and its potential contribution to the increase of urban resilience. Evidence obtained will serve as a basis for regulatory review of the overlaps and necessary links between urban regulations and Building Codes in Europe. The project will bring together the thinking behind 'zero waste' and 'zero energy', to provide more holistic symbiotic solutions to meet European targets.

There is a clear trend towards less landfilling as countries move steadily towards alternative ways of treating waste, mainly incineration, recycling or composting. Municipal Waste Management facilities in European cities are under diverse processes of technical/spatial upgrading. Thousands of new waste management facilities will be needed over the next decade to meet the requirements of waste producers and to comply with EU Law.[2] In some cases no land in urban areas will be available and in other cases, certain waste streams need to be transported over longer distances to specialist facilities. Often, they are being relocated further away from urban centers, thus increasing city's carbon footprint and reducing public awareness about the environmental impact of waste. SYMBIONT seeks to change that trend by bringing waste management processes closer to community centers and rethinking it as an opportunity to contribute to local energy generation.

4 Overview of the Action

The proposal is based on a theoretical model that will be tested and applied to a set of diverse case studies identified by 'front-runner cities' in order to achieve the maximum range of exportability of results. The obtained data will be collected and analyzed to define a more precise approach to the concept of mixed-use ecosystem unit MEU, both in quantitative and qualitative terms, and its potential contribution to the increase of urban resilience. Evidence obtained will serve as a basis for regulatory review of the overlaps and necessary links between urban regulations and technical building codes in Europe.

The project aims to specifically analyze spatial and management infrastructures that underpin municipal waste management facilities—street cleaning depots, collection/transformation points, recycling points, and other logistic spaces—as the

[2]Only in the UK the Environment Agency estimates that 2000 new waste management facilities will be required to meet the EU Landfill Directive, and many more upgrades to separate and treat waste and to reprocess recyclates.

basis to support a new protocols of matter-energy-information exchange to promote and optimize the natural (and desirable) adjacency between public facilities and residential units.

In the present project 3 cities have been proposed as representative of both urban fabric diversity patterns and different MSW infrastructures. They will be conducive to the application of different solutions to be tested and transferred to follower cities.

In particular, Madrid has 3,273,000 inhabitants and approximately 300,000 building entrances, where different fractions of waste are generated, with selective disposal and collection of wastes since 1998. Daily home collection of packaging and mixed-waste and a unique treatment plant manages 4300 tons/day, obtaining 306,000 Mwh/year from waste treatment.[3] Nottingham's urban area has approximately 730,000 inhabitants and is representative of a two-tier system (county and district have split responsibility) for waste management. Non-recyclable domestic and commercial waste is collected and burnt at the Eastcroft 'Energy from Waste' facility, producing heat and power that are used in homes and businesses in the city. The City of Venice has approximately 263,736 inhabitants, 363,468 in the entire urban area, the differentiated collection of municipal is organized by the Municipality of Venice and takes place in different ways. Each residential unit in the city of Venice is entitled to two weekly solid waste collections of an unlimited quantity of household garbage.

The front-runner cities will define strategic locations to host local laboratories able to monitor and display existing processes related to community waste management—such as separation or waste production rates—as well as energy consumption. The lab is designed as a model of productive communication—physical displays—to transform collected data into real informational interface. These interfaces will also envision environmental externalities related to local waste management. By awareness of the 'hidden' connections between waste, energy use, and consumption activity, the monitoring will be disrupting some of existing household negative routines. These labs will in implementation stage apply digital modeling software to display current and future impairments between adjacent material and energy flows. Unlike the majority of energy modeling programs—applicable when the design process is finished- these software tools are useful in providing specific qualitative and quantitative use parameters at the initial phases of design.

The role of follower cities is defined by coupling with each of front-runner cities, based on climatic and urban similarities. The impairments proposed are Nottigham-Dublin, Venice-Treviso, Madrid-Alcalá de Henares, and Cartagena de Indias-Las Palmas de Gran Canaria.

- Rapid growth process in Dublin provoked obsolescence in many urban areas. The council is working in actions on these underused spaces (Reusing Dublin), providing spaces to test SYMBIONT results.

[3]However, figures from 2012 show that 72.9% of Madrid solid waste is not recycled.

- Geographical proximity but huge urban contrast occurs between Treviso and Venice. The latter demonstrates high efficiency in waste management systems such as selective collection and recycling without incineration, so is accountable for testing the efficiency of SYMBIONT energy plan prosthesis.
- Alcala de Henares is a historical city highly linked to knowledge creation. It would contribute with interesting feedback regarding patterns of information, and public participation.
- Las Palmas de GC and Cartagena de Indias represent urban environments of desert tropical climates. Their municipal corporations are receptive to the possibilities of urban biodiversity effects on planning systems, so they are key part to motivate planning review.

The results will be exported towards follower cities as a basis for a compared cross-analysis of building codes and urban regulations, first at National and later European level. There is a procedural and conceptual disconnection between policies managing building and planning respectively, and duplication between the environmental protection and planning regimes continues to occur. Only some EU countries have coordinated strategies/policies for building and planning codes (e.g. Bulgaria, Czech Republic, Finland, Germany and Sweden). In the last phases of the project, SYMBIONT wills to make a consistent review of existing urban regulations affecting mixed-use urban entities and their overlapping with "energy efficiency" or "building envelope" definitions in European building codes and regulations. This report will be an open-ended process of surveying professionals, administrations and universities of the cities involved. The project will entangle this review through the analysis of joint concepts such as "comfort" or "energy efficiency", to find overlapping or contradictions and new balances between prescriptions performances: more flexible comfort temperature ranges, less autonomous housing insulating requirements and dynamic air change ratios, redefining thermal envelope.

The European Council has recently adapted a set of Guiding Principles for sustainable spatial development of cities on the European continent, including control measures for urban sprawl, activation of gaps, promoting measures for urban densification and mystification (Bach and Fudge 2001).

5 Methodology

The project will be developed following a two-step structure by any of the front-runner cities in monitoring and implementation, and a two-step feedback model conceived both towards and from the follower cities in dissemination and validation of results. The aim is to integrate the monitoring/information tools throughout the entire process, avoiding fragmentary software, and using continuous feedback for its onsite evolution—as in Complex Adaptive Systems. The actions will not follow sequential order but rather will be structured in iterative loop.

This will be achieved in three phases—'diagnostic', 'implementation' and 'exploitation'—in selected neighbourhoods of the participant cities. The impact of the proposed solutions will be investigated under three perspectives: society, technology and planning.

5.1 Diagnostic Phase—Urban Flows Assessment

The objectives of this phase are:

- To create and discuss a mapping of strategic MWF and adjacent neighborhoods in front-runner cities (at least one per city). This process will be developed with the engagement of local consortia amongst councils, local business, and neighborhood associations to guarantee the maximum performance of implementation results.
- To identify relevant assessment procedures for measuring the efficiency of current waste and energy management systems and their impact on neighborhood resilience. In particular, this aims to identify existing methods that account for both together and for their transferability to the selected sites.
- To implement the aforementioned assessment method as on-site laboratories in selected Municipal Waste Facilities, identifying regulatory or technical gaps. These local platforms will be collecting and displaying relevant data related to household waste disposal and energy consumption. These laboratories will produce immediate benefits, as palliating intensive household consumption or improper disposal behaviors, while communicating community best- practices (Fig. 1).

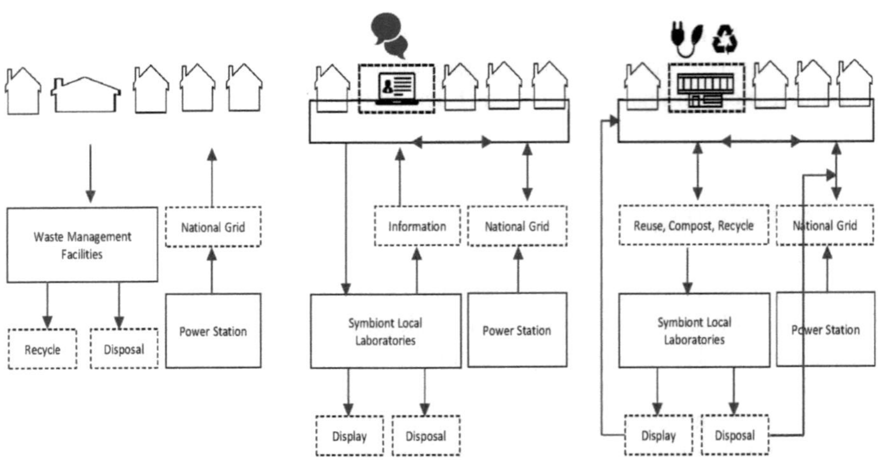

Fig. 1 SYMBIONT transition from current MWF to local laboratories and MEU

5.2 *Implementation Phase—Mixed-Use Ecosystem Environment—MEU*

The objectives of this phase are:

- To redefine impairments and propose solutions for an effective symbioses between waste and energy mixed-use systems in the selected MWF and their immediate surroundings. In doing this, SYMBIONT will apply 3D digital simulation tools to model existing performance for the MEU with especial regard to novel connections between excess and source of material and energy flows.
- To assess the efficiency of the implemented solutions in reducing waste, improving energy efficiency, increase separation rates and social empowerment in the case studies.
- To implement the proposed solutions in selected sites, according to the different needs of participant councils. Each of the cities will entail at least one of the following actions, creating a catalogue of complementary case studies (see Fig. 2):
    1. Technological (EPP + MWP): Cluster local compatible energy and waste flows through "prosthetic stations", making the existing infrastructures perform as local exchanger prosthesis to allow for new environmental supply-chains. This Hi-tech procedure directly connects to sustainable Smart City sensor-based challenges addressed by the European Commission through H2020 protocol.
    2. Urban (MUD): Develop further simulation and planned mixed-use scenarios to provide specific briefing and planning tools for the rehabilitation of MWF and make them evolve into mixed-use facilities or public areas.
    3. Social (SSP): Promote social activity inside MWF recovering their notion of public spaces through educational/didactic workshops such as innovative

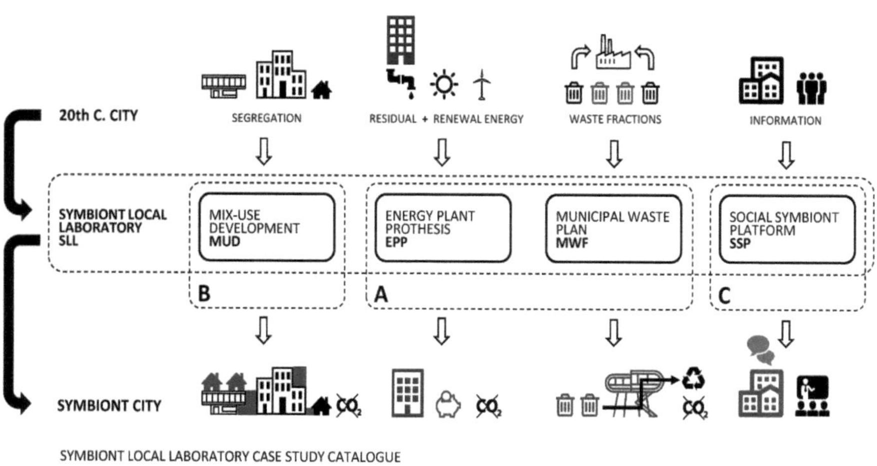

Fig. 2 SYMBIONT case study complementary catalogue

recycling, domestic compost, home gardening, conscious consumption, etc. promoting healthier and more participative neighborhood environments.

5.3 Exploitation Phase—Regulatory Revision

The objectives of this phase are:

- To define regulatory evidences of the project results engaging the collaboration of universities with external experts, professional bodies and regulatory administrations.
  - Make a Compared-analysis of Urban Regulations and Building Codes, to define cross implications of the obtained results defining ecologically sustainable mixed-use, as they are energy exchange protocols or design prescriptions.
  - Defining a specific SYMBIONT Waste Management Plan within the involvement and participation of municipal authorities, experts and stakeholders that supports the project, with the intention to create a systemic and replicable model. SIMBIONT Waste Municipal Plan criteria will be used by competent authorities as a model and guidelines for the development and/or revision of existing WMP, under the requirements of the EU Waste Framework Directive (WFD 2008/98/EC).
- To identify which of the implemented solutions are applicable to the follower cities through analysis of planning tools, with the participation of multiples stakeholder assessing the efficiency of the implemented solutions.
- To disseminate the results through the basis of existing academic networks, such as Solar Decathlon or Europan Urban Competition.

6 Expected Impacts

With this regard, SYMBIONT may achieve the following propositions:

- **Increase urban resilience in monitored case-studies**
  The project pursues the improvement of environmental quality in MWF related neighborhoods, as they turn from "grey" towards "green" infrastructures, with capacity to provide social, ecological and economic measurable benefits to urban communities. After the two-year implementation impacts, such as noise reduction (1.7 dB night), temperature regulation (−0.2 °C) and promotion of new recreational areas (+8% Floor Area Ratio) will be entangled.

- **Managing symbiosis in urban environments**
  Although clearly widespread in industry and agriculture, references to urban/architectural symbiosis are still very scarce. The project fosters evidence

on the organizational impact of urban planning models on mixed-use versus mono-functional development. In envisioning the conflicts, synergies and alliances between theoretical models and urban governance in MEU, the project provides new tools to construct sustainable relations and trigger synergies with various urban constituencies.

- **Enlighten "black-box" management protocols promote participative approach**
  Current Municipal Solid Waste management entails sometimes huge disinformation between political, administrative, professional and social agents. The selection of case-study plots at leader cities will be co-defined by local consortiums as councils, universities, neighborhood associations and private/public companies- transforming the—mainly negative—social impact of waste management infrastructures into dynamic public spaces and reducing improper waste disposal habits.
- **Reduce carbon-intensive household behavior patterns**
  Consumers energy consumption has risen by 25% since 1985, as a result of population growth, rising incomes, and households energy-intensive behavior patterns. SYMBIONT expects to reduce household solid waste production per capita in the different case studies by 3%.[4] A total increase of waste recovery rates -taken to recycle and compost- is expected in the monitored environments as an increase of 10% related to existing recycling rates per country.
- **Increase of effective exploitation of organic fraction waste**
  36% of solid municipal waste production is organic. A number of social actions related to composting of pruning waste and urban gardening will foster new market and social opportunities for this important fraction.
- **Reduction in GHG emissions from waste disposal reducing landfill disposal**
  Despite a steady increase in the overall quantity of generated waste, emissions from waste disposal will decrease in the future. This shift stems from the decrease in waste being landfilled in the EU. The waste sector accounted for about 11% of the cuts in GHG emissions in 2011, which amounts to 1.8% of total emissions. The waste management sector therefore plays a key role due to the reduction in reliance on landfill for residual disposal, a change that has been significantly influenced by EU waste legislation.

6.1 Protocol

WP 1. Urban Symbiosis: A Cross-Consolidation of State of The Art
Devoted to the consolidation of the state-of-the art of urban symbiosis best practices, looking at assets and liabilities of current regulatory systems in Europe.

---

[4]Eurostat Statistics from 2013 establishes that waste generated rates (in Kg/person year) are: Spain 449, UK 482 and Italy 491. Recycling + composting recovery rates are Spain 20 + 10%, UK 28 + 16% and Italy 26 + 15%.

- To identify and define the different parameters related to urban symbioses, energy and waste synergy management in cities, sustainability and resilience assessment methods/tools, energy and waste policies and regulations.
- To identify relevant assessment procedures, at building and city scales, for measuring the efficiency of waste and energy systems in buildings and communities and their impact on neighbourhood resilience. In particular, this aims to identify existing methods that account for both together holistically as it is *SIMUR*, designed by the *Agencia de Ecología Urbana* and already applied in many cities[5]

WP2. Case Study Collective Definition

To make a collaborative selection of strategic MWF and adjacent neighbourhoods in front-runner cities (at least one per city).

- Participant municipalities—with the help of the local academic groups- make a selection of existing sites with potential to evolve into mixed-use ecosystem units. An identification of empty plots belonging to the existing WMF should be made. The intensity of material flows (especially high water consumption and movement of goods), establishes an appropriate starting condition to entangle metabolic cycles with other uses of the neighborhood. Its location and management conditions would indicate the relevance of taking them as testing platforms to verify its prototypical capacity in terms of proving urban resilience increase.
- This process will be developed through Local Consortia amongst councils and local universities consulting with local business, neighbourhood associations and residents. Throughout the 2 years of duration of the action, Local Consortia Meetings will be held periodically at each of the leader-cities. They will facilitate the assessment of results, provide advice on developments, contribute to measure impacts against social and professional demands and delineate strategies towards the dissemination of the action's outcomes to target groups.
- The set of locations should envision the highest geographical/economic/urban/climatic diversity to maximize its exportability.

WP3. Symbiosis Assessment Lab

To implement the aforementioned assessment method, the project pursues the creation of on-site laboratories in the selected sites, identifying best practices and regulatory or technical gaps. On the technical range, this service area may face scalability difficulties as it deals with 'scaling variables'. On the societal plane, smart technology provides solutions to a multifaceted society, which may lead to ecological challenges. These local platforms will be collecting and envisioning data

[5]As they are Barcelona, Figueras, Viladecans, Manresa, Mataró, A Coruña, Mancomunidad de la Sierra de Barbanza, Ourense, Santiago de Compostela, Vitoria-Gasteiz. (San Sebastián, Usurbil, Leganés, Benicàssim, etc.). See results at http://www.bcnecologia.net/es/modelo-conceptual/simur.

related to consumption and waste in MWF and adjacent communities, such as waste disposal and energy consumption rates.

- The project runs a scalar energy structure, as much for the design of theoretical model/case studies as for the energy accounting protocol. The proposal is meant as an interface that enables a model of productive communication (feed-back based). Therefore, Symbiosis Assessment Labs have been identified as a potential alternative method for intercommunicating cities' stakeholders, whose requirements, concerns, interest and politics dictate how they interact with this technology platform and to what extends, which has a direct effect on scaling dimensions and its societal impact. Symbiont City services use sensor technology to register and monitoring data. This apparatus allows to test and monitor requirements satisfaction and to take counteractive measures when necessary.
- The three parts are designed to coexist in a performative network, participating simultaneously in the energy network and the information network. At any moment, the case studies in front-runners cities receive and report information to a common control platform. Tight control to their metabolism permits calibrating real-time efficiency or redundancy.
- The ultimate goal of these labs is thus envisioning existing processes to alleviate intensive consumption behavior, and sharing the immediate benefits of "best practices" amongst the international community.

WP4. Pilot Study Modeling of MEU

In order to define couplings and impairments for effective symbioses between waste and energy in the selected sites, at this stage the project will apply parametric design software[6] to model the selected MWF and adjacent neighborhoods as performative MEU. This 3D simulation includes all the information related to urban physical form and existing energy and matter flows taken from the on-site monitoring. It will envision possible connections between compatible waste and energy flows.

The implementation of results in selected sites, will be made according to the different needs—technological, urban or social—of participant councils. Each of the cities will entail fundamentally one of the following actions:

- Technological focus: Cluster local compatible energy and waste production through "prosthetic stations" to produce a local environmental supply-chain, making the existing MWF really perform as MEU. A connection with SENSIBLE action group could be specifically considered, through strategies of neighborhood energy collection/storage, renewable energy production, waste and secondary resource management or material reprocessing.
- Urban focus: Develop further simulation urban models to asset specific briefing for the future rehabilitation of MWF into MEU. Based on this briefing,

[6]As they are Function Mixer, Cyclifier, or more technical tools as CYPE, Design Builder or Ecotech, Vensim, Stella, Ithink, Powersim, Dynamo.

municipalities could launch an ideas-competition through the corresponding university.

- Social focus: Promote social activity inside MWF recovering the lost notion of public spaces by means of urban gardening actions, or educational/didactic workshops about innovative recycling, domestic compost, home gardening, conscious consumption, etc. promoting healthier and more participative neighborhood environments.

WP5. Environmental Monitoring

Both during the assessment process- data collection- and pilot implementation of the intervention the project promotes a real-time display of results in the selected sites, with direct consequences on consumption behavior by agents/consumers. The results will be included in a final report and possibly foster the potential to create a mobile app accessible for citizens.

- The intervention transforms thus the conventional urban rehabilitation protocols of the twentieth century in a 'living laboratory' of the XXI century.
- SYMBIONT fosters a transmission methodology using versatile real-time narratives to interest not only the partners involved, but also many other cross groups.

WP6. Participation/Validation Process by Professional Market

The report generated by universities at previous stage will undergo diagnostics with the participation of experts and agents from professional field, who will perform feasibility and potential reports. This will be an open-ended process of surveying professionals through the Council of European Architects, or National Chamber of Architects involved, aiming to gather the expertise from practitioners (a ratio of 40 per country), responsible agents from Municipal or Council Planning Areas and experts on energy assessment evaluation tools, as GBC, Passivhaus or equivalent group.

These will allow establishing significant frameworks of relationship between diverse regulatory entities, pointing in particular friction and opportunity areas. The coordination of this regulatory basis should promote definition of resource and emissions cycles in the existing city, and encourage such practices in future developments. The validation will specifically address regulatory gaps as follows:

- Compared-analysis of Urban Regulations and Building Codes. Detection of overlap/friction areas between diverse regulatory scales for the potential dissemination of results at national level.[7] This action will be carried out by analyzing specific keywords in both urban and building regulation fields, with

[7]E.g. In Spain energy efficiency requirements as described by the Basic Document HE (Energy Saving) CTE, and its overlaps with Law 8/2013 on Rehabilitation, Regeneration and Urban Renewal.

special attention to cross concepts such as envelope, surplus, storage, cooperation or energy efficiency.
- Definition of strategic action scenarios. This will encourage a first report of strategic connection nodes.

WP 7. Optimization Plan and Dissemination of Results
Reports of the experts -action 6- will be consolidated in a single final document, and circulated to the research centers mentioned and the cities defined as followers to collect their critical consideration in the following:

- To identify which of the implemented solutions is applicable to the follower cities through analysis of planning tools, with the participation of multiples stakeholder assessing the efficiency of the implemented solutions.
- To disseminate the results through the basis of existing academic networks such as Solar Decathlon or Europan Urban Competition.

Final action will undertake a critical interpretation of the obtained data in geopolitical terms, indicating the legislative framework and degree of flexibility of the diverse regulatory environments studied.

7 Conclusions

Western urban planning has historically lean on a consideration of the hygienist zoning and a strong infrastructural dependence; Although recent ordinances in North America and northern Europe promote multifunctional aggregation, there are still no specific figures to quantify the figure of "mixed-use" nor and evidence of how it enables metabolic networks in urban terms. On their side, technical building codes are highly prescriptive agendas considering the building as an autonomous isolated entity, canceling any possible—structural, environmental, energy-exchange between adjacent entities. Paradoxically, the successive revisions of respective agendas—planning and building—in terms of energy efficiency have increased further their mutual disconnection, consolidating instead a highly prescriptive culture and far from favoring a more holistic approach for urban design and rehabilitation.

Although some agendas have already fostered the concept of symbiotic planning, neither current infrastructural systems nor urban regulatory frameworks allow for its real implementation. Concepts that regulate cities and territories are currently considered independently, e.g. ICT, mobility, waste, and energy are so far not integrated parameters at urban level. Legislation is normally based on ensuring minimum standards of habitability, energy, security and accessibility, which are prescribed independently and often reduced to quantitative criteria. Reducing emissions, in an urban renovation plan, very often happens by means of adding active systems and therefore dismissing many passive strategies with lower economic and ecological cost.

The mixed-use proposed in SYMBIONT permits understanding case studies as fields of experimentation where public spaces are community managed, municipal services networks implied take on new performances and citizens acquire new agencies related to consumption and waste management protocols. This scenario turns the city into a real energy interface with optimal interconnections between urban elements capacities. Instead of a segregated approach, urban planning becomes a holistic design strategy to allow energy folding between existing mix-use structures.

These micro-infrastructural interventions will help in the urgent transition from "grey" towards "green" infrastructures, with capacity to provide social, ecological and economic benefits to urban communities such as reduction of waste disposal, local energy generation and storage, improvement of air quality, reduction of energy costs and new opportunities to social cohesion and engagement.

'If interconnectedness implies radical intimacy with other beings, then we had better start thinking about pleasure as a coordinate of the ecological thought. We must take a new path, into the vast mesh of interconnection. Who lives there?'

T. Morton, *The Ecological Thought* (Morton 2010: 37).

References

Bach, M., & Fudge, C. (2001). *Towards more sustainable land use: Advice to the European Commission for policy and action*. London: Expert Group on Urban Environment.

Evans, G., Foord, J., Porta, S., Thwaites, K., Romice, O., & Greaves, M. (2007). *The generation of diversity: Mixed-use and urban sustainability. Urban sustainability through environmental design: Approaches to time people-place responsive urban spaces* (pp. 95–101).

Goldsmith, E. (1996). *The way: An ecological world view (Rev. and enlarged ed.)*. Devon: Themis Books.

Marsh, A., & Khan, A. (2011). Simulation and the future of design tools for ecological research. *Architectural Design, 81*(6), 82–91.

McDonough, W., & Braungart, M. (2002). *Cradle to cradle*. New York: North Point Press.

McHarg, I. L. (1969). *Design with nature*. Garden City: The Natural History Press.

Morton, T. (2010). *The ecological thought*. New York: Harvard University Press.

Odum, E. P. (1992). *Ecología: Bases científicas para un nuevo paradigma*. Barcelona: Vedrá.

Van den Dobbelsteen, A. (2010). Synergy, not Autarky. In U. Hackauf & P. Haikola (Eds.), *Green dream: How cities can outsmart nature* (pp. 266–270). Rotterdam: The Why Factory & Nai Publishers.

Chapter 22
Does Urban Living Reduce Energy Use?

Arjan Harbers

Abstract The claim by Edward Glaeser and others that cities make us greener has been verified for the Netherlands, based on empirical data available on a detailed level. When considering 'greener' to mean using less domestic energy, urban areas seem to be more energy-efficient than non-urban areas, as far as mobility and energy consumption (natural gas and electricity) is concerned. Yet, for energy consumption, this depends on the unit of measurement; when calculated per dwelling and per person, urban areas consume less energy, but when calculated per square metre of residential floor space, there is no clear relationship with urban density. Type of housing and household type are better indicators of energy consumption. Renewable energy generation through the use of solar panels placed on roofs is found more often in non-urban areas where the roofs are more suited to solar panels. These are also the areas where local wind power and solar energy initiatives are found more frequently.

1 Introduction

In his book, *Triumph of the city* (Glaeser 2011), Edward Glaeser expresses his faith in the city's qualities, as he phrases in the subtitle: *How Our Greatest Invention Makes Us Richer, Smarter, Greener, Healthier, and Happier*. The objective of this chapter is to explore whether urban areas in the Netherlands are 'greener' than non-urban areas. By greener, within the context of this research, we mean using less energy and generating more renewable energy. This issue is relevant as many policies are aimed at urban investments, densification and compact cities, in order to reduce energy use and by doing so, mitigating climate change (e.g. IEA 2009; UNEP 2011; UN-HABITAT 2013; World Bank 2014; Agenda Stad 2015; European Commission 2014).

A. Harbers (✉)
PBL Netherlands Environmental Assessment Agency, The Hague, The Netherlands
e-mail: arjan.harbers@pbl.nl

R. Álvarez Fernández et al. (eds.), *Carbon Footprint and the Industrial Life Cycle*,
Green Energy and Technology, DOI 10.1007/978-3-319-54984-2_22

There seems to be overall consensus among urban planners and scientists that, in general, compact cities perform better than non-urban areas, as far as the use of energy is concerned. Empirical scientific research shows a reduced amount of energy consumption for city dwellers, compared to people that live outside the city. In general, the amount of energy used for heating/cooling (Keuning and Vroom 2012; LSE/EIFER 2014; Steemers 2003; Salat 2009; Ratti et al. 2005) and transportation is lower in more densely populated areas (Newman and Kenworthy 1999, 2015; Ewing and Cervero 2010; Bettencourt 2013). However, the possibilities for harvesting renewable energy are more present in less densely populated areas, since there is more space for photovoltaic cells and wind turbines. The latter are not even allowed within city boundaries in the Netherlands. On the other hand, the construction of a thermal energy network or district heating system and the business cases for car-sharing programmes and public transport are economically more feasible in areas of higher population densities.

Without attempting to do a comprehensive analysis on different forms of energy use, we analysed, on the basis of simple statistical rankings, whether in the Netherlands energy use in more urbanised areas differs from that in less urbanised areas.

The research was narrowed down to domestic energy use in the Netherlands. We considered four indicators for energy use: domestic energy consumption, people's mobility, solar panels, and the activity of local energy corporations. For energy consumption and solar panels, information is available about every household, aggregated to the level of neighbourhoods and local rural areas (12,000 of such neighbourhoods and areas in the Netherlands in total).

If data were available, we also considered the development over time. For every indicator, we examined whether it was affected by the degree of urbanisation.

In the Netherlands, urbanisation has been categorised into five classes (see Table 1) that relate to the average number of addresses per square kilometre within a radius of one kilometre around each address (CBS 1992). Every single address in the Netherlands can be categorised into one of the five classes. Upon starting this classification system in 1992, every class contained more or less the same number of inhabitants. Up to today, the numbers of inhabitants in first two classes have increased at the expense of the low and very low urbanisation classes. The very highly urbanised class can be considered as very densely populated inner-city neighbourhoods, and the least (very low) urbanised class can be considered to represent rural areas.

Table 1 Classification of degrees of urbanisation (CBS 1992)

Urbanisation class—degree of urbanisation	Number of surrounding addresses per km^2
1 Very high	More than 2500
2 High	1500–2500
3 Moderate	1000–1500
4 Low	500–1000
5 Very low	Fewer than 500

Twenty years ago, Vringer et al. (1997) analysed a survey on domestic energy consumption for the Netherlands, ranking the results according to degree of urbanisation (aggregated per municipality). They observed a strong relationship between income level and energy consumption. In other words, wealthier households were found to consume more energy. After adjusting for income level, the only large differences that remain between urbanisation classes relate to the consumption of natural gas, electricity and petrol. The consumption levels of these types of energy by households in very low urbanised areas, on average, are one third higher than by those in very highly urbanised areas. There seemed to be no relationship between education level and energy consumption.

2 Domestic Energy Consumption: Natural Gas and Electricity

Many studies (Ratti et al. 2005; Salat 2009; LSE/EIFER 2014) have demonstrated that energy performance, to a certain extent, is related to urban morphology. Urban morphology, including building density, shape and typology, is believed to affect energy consumption, similar to climate conditions, building physics (architecture, materials), heating and cooling systems, and occupant behaviour. Their findings are based on models derived from the existing urban fabric in various cities, worldwide.

For the Netherlands, we did an inventory based on a complete database of energy consumption per neighbourhood. On the basis of this database, we were unable to identify a causal relationship between degree of urbanisation and energy consumption (natural gas and electricity), but we did find that both dwellings and people in more densely populated areas were more economical with respect to their energy consumption.

Natural gas consumption forms a large part of Dutch domestic energy consumption. It is mainly used for room heating (78%), heating water (18%) and cooking (4%). Electricity is mainly consumed for electronics (25%), refrigeration systems (16%), cleaning appliances (15%), heating and hot water systems (13%), lighting (14%) and a number of other uses (17%) (Zitzen 2015).

Whether domestic energy consumption is related to urban density, depends on the chosen unit of measurement. There is much difference between measuring the energy consumption per dwelling, per person and per square metre (residential floor space).

Houses in very densely populated urban neighbourhoods, on average, consume 42% less natural gas than those in very sparsely populated areas (Fig. 1). This is in line with the findings from Keuning and Vroom (2012). Per person, however, this difference is only 28% (Fig. 2). This can be explained by the, on average, smaller household sizes in urban neighbourhoods, compared to those in rural areas (see Table 2, first column). In addition, household type also plays a role; in households with working persons, or children that go to school, less energy is used during working hours and/or school hours.

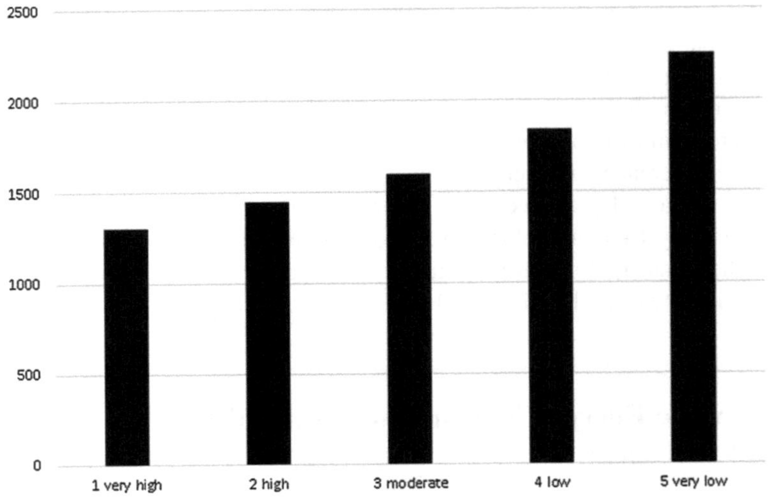

Fig. 1 Mean use of natural gas (m^3/year) **per dwelling**, sorted by degree of urbanisation. Data aggregated at neighbourhoods level, in 2013 (CBS Kerncijfers wijken en buurten 2013, edited by PBL)

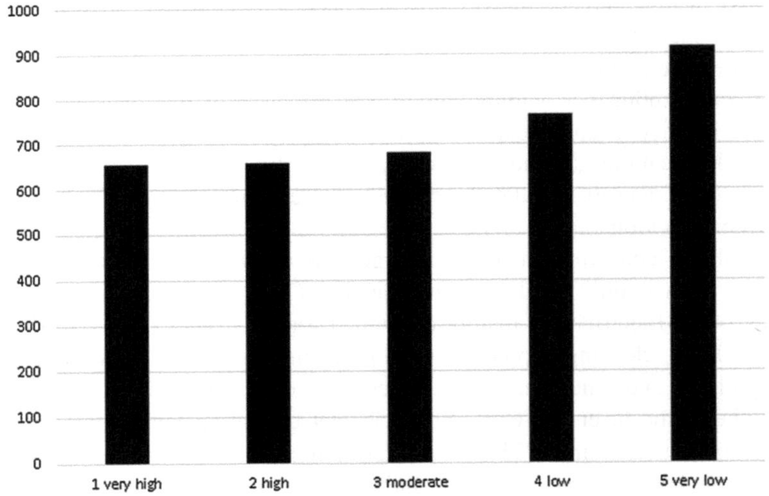

Fig. 2 Mean use of natural gas **per person** (m^3/year), sorted by degree of urbanisation. Data aggregated at neighbourhoods level, in 2013 (CBS Kerncijfers wijken en buurten 2013, edited by PBL)

For electricity consumption, we see a similar pattern. Electricity use per dwelling, on average, is 35% lower in very highly urbanised neighbourhoods compared to very low urbanised areas. When calculated per inhabitant, this difference is 20% for similar reasons as above for natural gas (Figs. 3 and 4).

Table 2 Various indicators, sorted by degree of urbanisation

Urbanisation class—degree of urbanisation	1. Mean household size (persons)	2. Share of dwellings with district heating (%)	3. Mean residential floor space, per dwelling (m^2)	4. Mean annual income, per recipient (× 1000 euros)	5. Share of dwellings built in or after 2000 (%)
1 Very high	2.0	6.4	90	30.3	8.6
2 High	2.2	7.5	107	30.5	11.3
3 Moderate	2.3	5.8	119	31.5	15.2
4 Low	2.4	2.1	132	31.7	16.4
5 Very low	2.5	0.1	142	30.3	11.0

Data aggregated at neighbourhood level, in 2013 (CBS Kerncijfers wijken en buurten 2013, Basisadministratie Adressen en Gebouwen (BAG), CBS 2014b, edited by PBL)

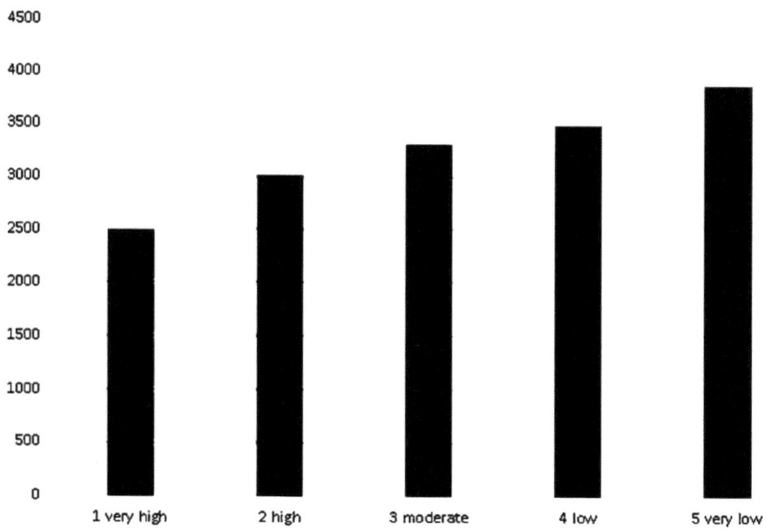

Fig. 3 Mean electricity use **per dwelling** (kWh/year), sorted by degree of urbanisation. Data aggregated at neighbourhoods level, in 2013 (CBS Kerncijfers wijken en buurten 2013, edited by PBL)

However, if we calculate household consumption of natural gas and electricity per square metre of residential floor space (Figs. 5 and 6), we no longer see a pattern indicating that urban density leads to less natural gas use; in fact, it is areas with a medium urban density where the least amount of natural gas and electricity is consumed, on average. This effect can be explained by the mean floor space of dwellings which is smaller in more urban neighbourhoods (Table 2, column 3). However, in general, in large dwellings with much floor space, not every room is heated.

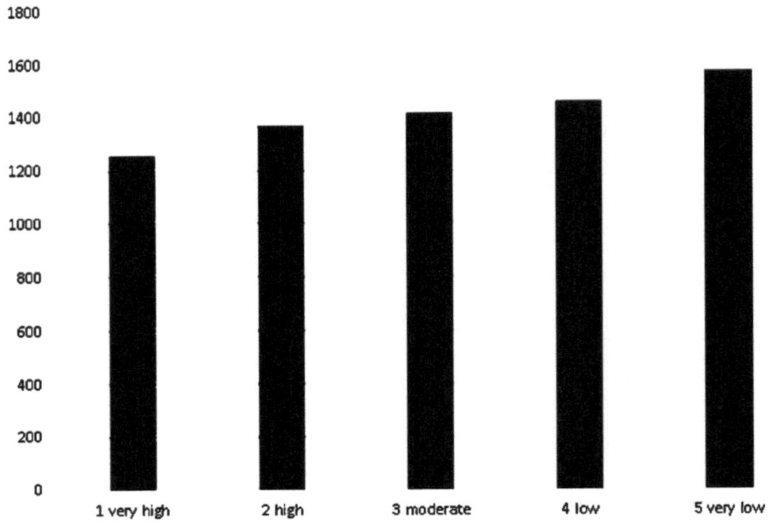

Fig. 4 Mean electricity use **per person** (kWh/year), sorted by degree of urbanisation. Data aggregated at neighbourhoods level, in 2013 (CBS Kerncijfers wijken en buurten 2013, edited by PBL)

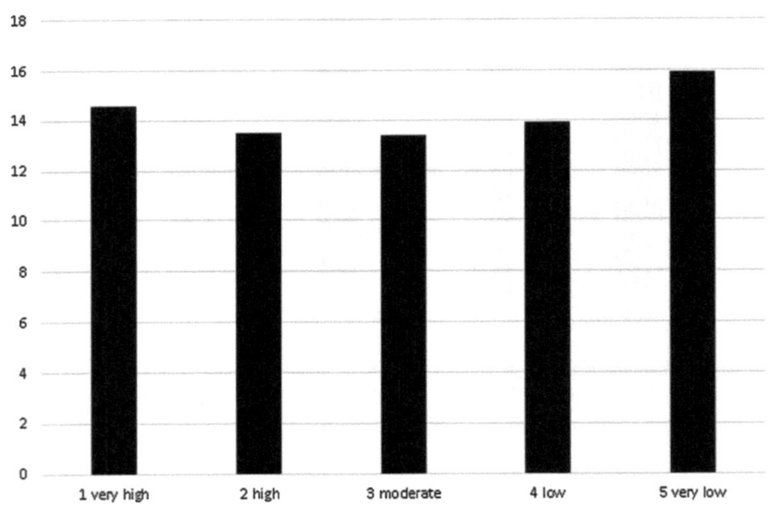

Fig. 5 Mean use of natural gas **per m^2 residential floor space** (m^3/year), sorted by degree of urbanisation. Data aggregated at neighbourhoods level, in 2013 (CBS Kerncijfers wijken en buurten 2013, CBS 2014b, Basisadministratie Adressen en Gebouwen (BAG), edited by PBL)

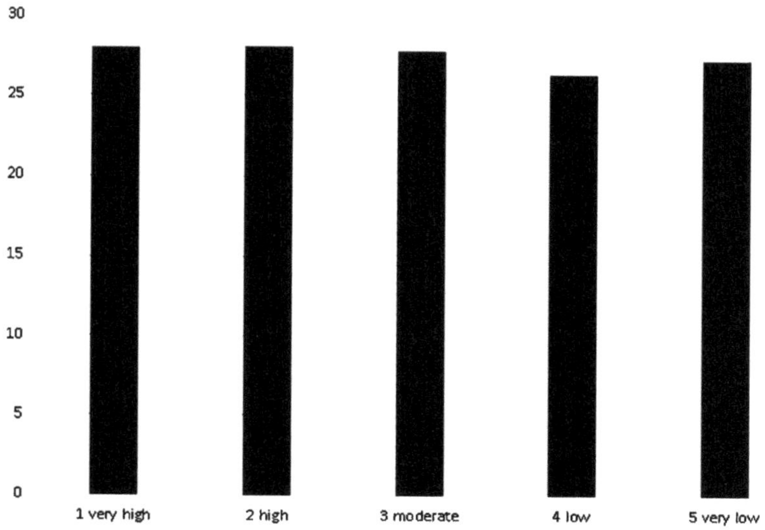

Fig. 6 Mean electricity use of natural gas **per m² residential floor space** (kWh/year), sorted by degree of urbanisation. Data aggregated at neighbourhoods level, in 2013 (CBS Kerncijfers wijken en buurten 2013, CBS 2014b, Basisadministratie Adressen en Gebouwen (BAG), edited by PBL)

Natural gas consumption depends on more indicators than on household size and floor space only; the availability of district heating, household income and thermal insulation of the building, which is often related to the year of construction, also play a role (Vringer et al. 1997; Brounen et al. 2012).

Last but not least, the housing type (e.g. detached, terraced, apartments) determines residential natural gas consumption, to a certain extent. Not only because of the size of dwelling which, on average, is larger for detached houses than for apartments, but also because of the fact that neighbouring dwellings insulate or even heat each other.

Apartments are more frequent in urban neighbourhoods, while in rural areas there are more detached dwellings, so it is not surprising that the natural gas use in urban areas is lower than in less urbanised areas. But if we sort the dwelling types by urbanisation class (Fig. 7), we do not see much difference in domestic natural gas consumption.

So, as far as the use of natural gas is concerned, it does not matter if a certain dwelling, for instance an apartment of 150 m^2, is located downtown or in a rural environment. It is the physical and sociological conditions that predominantly influence natural gas consumption. They are not fixed to a certain urban density, but the energy-saving circumstances are more prevalent in more densely populated neighbourhoods, which in the Netherlands are slightly poorer, where it is easier to install district heating systems, and they contain more apartment buildings, have smaller floor spaces and smaller household sizes. (see Table 2).

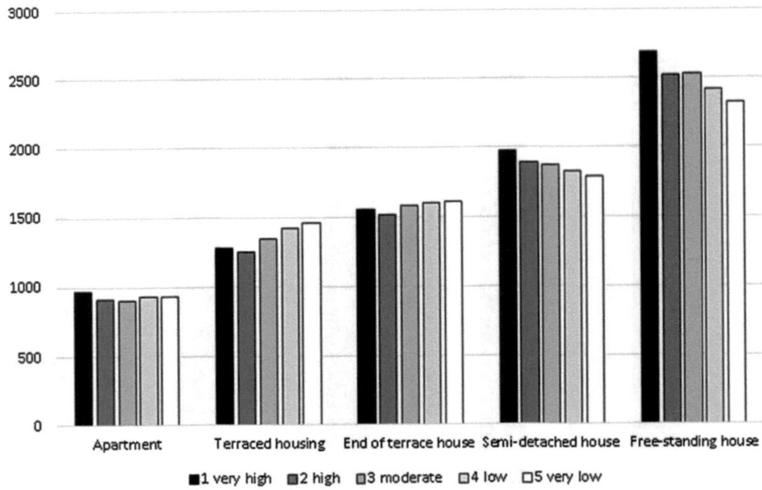

Fig. 7 Mean annual domestic natural gas use, temperature corrected (m^3), sorted by urbanisation class, for 5 dwelling types in 2014 (Klimaatmonitor)

In other words: Very urban neighbourhoods do not directly evoke saving on energy consumption, but they are the location where energy-efficient buildings and people that use less energy are to be found. The context for a more economical energy consumption is inextricably linked with highly urbanised neighbourhoods.

The newer building stock, which has better insulation conditions and better installations, most frequently found in moderate and low urbanised classes does not (yet) compensate for the higher level of energy consumption. In the near future, however, all new buildings will have to be energy neutral or generate energy, and existing buildings will have improved insulation. The different results between urbanisation classes might slowly become a thing of the past, at least where energy use is concerned.

2.1 Domestic Use of Energy Over Time

As Glaeser claims that cities make us 'greener', the comparative degree can also be explained as 'greener than before'. As we have data from 2004 onwards, we can observe domestic natural gas and electricity consumption over a 12 year period. We observe a decline in domestic natural gas use per dwelling, for all urbanisation classes (Table 3). A possible explanation for this decline is the renovation of buildings, extra thermal insulation and new central heating installations.

The decline for the more urban classes is slightly stronger, however (see Table 3).

Table 3 Mean annual natural gas use, m^3 per dwelling, in kWh (2004–2015) (CBS via Klimaatmonitor 2016)

Average natural gas use, all housing types (temperature corrected)—urbanisation class—degree of urbanisation	2004	2006	2008	2009	2010	2011	2012	2013	2014	2015	2015/2004 (%)
1 Very high	1430	1380	1320	1310	1290	1230	1170	1150	1120	1060	74
2 High	1640	1620	1540	1520	1510	1440	1390	1360	1300	1250	76
3 Moderate	1930	1900	1800	1790	1770	1690	1630	1590	1520	1480	77
4 Low	2140	2120	2010	1990	1970	1900	1820	1780	1700	1650	77
5 Very low	2270	2230	?	?	2100	2020	1950	1900	1820	1760	78

Table 4 Average annual electricity use, per dwelling, in kWh (2004–2015) (CBS via Klimaatmonitor 2016)

Average electricity use, all housing types—urbanisation class—degree of urbanisation	2004	2006	2009	2010	2011	2012	2013	2014	2015	2015/2004 (%)
1 Very high	2640	2710	2720	2720	2700	2660	2610	2530	2510	**95**
2 High	3210	3230	3260	3250	3240	3180	3130	3020	2960	**92**
3 Moderate	3500	3540	3560	3550	3540	3470	3420	3290	3220	**92**
4 Low	3630	3660	3660	3660	3650	3580	3510	3360	3290	**91**
5 Very low	3640	3660	?	3680	3660	3600	3530	3370	3290	**90**

As far as electricity is concerned (Table 4), we see an overall decrease in electricity consumption, as well, over the same period. The decrease is relatively smaller than for natural gas consumption. In contrast to natural gas consumption, the decrease in electricity consumption is lower as the level of urbanisation increases.

3 Mobility of People

Empirical research (Newman and Kenworthy 1999, 2015; Ewing and Cervero 2010; Bettencourt 2013) has shown the relationship between petrol use and population density. In areas of higher population density, the average use of petrol per person is lower, even after adjusting for fuel economy, petrol prices and income levels.

There are two explanations for the lower petrol use in more densely populated areas. First, higher densities tend to go with shorter distances, as a result of which the demand for mobility, on the whole, declines and active travel modes, such as walking and cycling, are more compatible options, compared to automobile use. Second, higher densities offer a larger basis for mass transit, leading to a more developed public transport system that can compete with automobile use.

Car traffic is reduced even further when the distances to public transport nodes are short and if land uses (e.g. housing, jobs, amenities, leisure) are laid out in a mixed configuration (Van Wee 2011; Echenique et al. 2010).

If we assume that distance travelled by car (excluding passengers) is a proxy for energy use in passenger transport, we can look at the modal split by degree of urbanisation in the Netherlands (see Table 5).

As expected, the total distance travelled decreases with increases in the degree of urbanisation. For the most urbanised neighbourhoods, this is 14% lower, compared to the areas with a very low degree of urbanisation. In neighbourhoods with a very high degree of urbanisation, the daily distance travelled by car (excluding passengers) is just 62% of the average daily distance travelled for an inhabitant of a rural area.

Reversely, the average daily distances travelled by public transport, by bicycle and on foot are considerably longer in the more urbanised neighbourhoods. People living in very highly urbanised neighbourhoods walk almost 50% more on a daily basis, compared to those in rural areas.

As presented in Table 2, highly urbanised neighbourhoods are usually less affluent and, in general, have smaller household sizes. Snellen et al. (2005) argue that the difference between urban and suburban mobility patterns are smaller when adjusted for population type (household type and labour participation), but the differences in mobility between different urbanisation classes remain significant.

If we consider the daily amount of car kilometres travelled over time (2010–2015), we do not find specific differences in development patterns in relation to the degree of urbanisation (Fig. 8).

4 Solar Panels on Roofs

As far as the presence of solar panels is concerned, we see a relationship with urban density.

First, we clearly see installed solar power increasing with decreasing urban density. A dwelling in a very low urbanised area, on average, yields over 14 times more power than those in very highly urbanised neighbourhoods (Table 6, first column).

A possible explanation for this is that the roofs in very low urbanised areas are in general more suiTable for generating solar power than those in the very highly urbanised neighbourhoods (see Table 6, columns 2–5). This suitability is determined by three physical circumstances: roof orientation, roof slope, and the shade

Table 5 Average daily travel distance in km (2010–2015), per mode of transport and degree of urbanisation (aggregated for municipalities), for persons aged 12 and over

Urbanisation class—degree of urbanisation	Automobile (driver)	Automobile (passenger)	Train	Bus/tram/metro	Moped	Bicycle	Walking	Other	Total
1 Very high	13.4	5.0	5.1	1.8	0.2	2.8	1.0	0.7	30.1
2 High	17.8	5.8	3.5	0.9	0.2	2.6	0.9	0.7	32.2
3 Moderate	18.8	5.9	2.7	0.7	0.2	2.6	0.8	1.0	32.7
4 Low	19.7	6.2	1.8	0.8	0.2	2.6	0.7	1.1	33.2
5 Very low	21.7	6.5	1.5	0.9	0.2	2.5	0.7	1.2	35.4

Air travel and travel abroad are excluded (OVIN via CBS 2016)

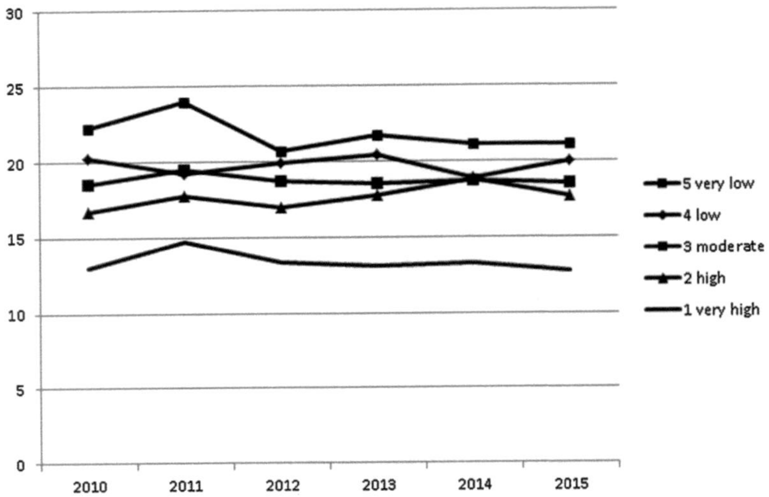

Fig. 8 Development of the average daily car (driver) kilometres (within the Netherlands) 2010–2015, by degree of urbanisation (aggregated for municipalities) (OVIN via CBS 2016)

from adjacent obstacles (buildings, trees). Roofs in very highly urbanised neighbourhoods are more affected by shade from adjacent obstacles.

In very highly urbanised neighbourhoods, only 61% of buildings is suitable for solar power. For very low urbanised areas, this is 74%.

The suitability of the roofs is not fully being exploited at any location. If we consider the installed peak power per neighbourhood and relate it to the potential peak power (Table 6, last column), we see that only a small percentage of roofs is being exploited. Solar power is exploited the most in urbanisation classes 3 and 4. The highly and very highly urbanised neighbourhoods obviously lag behind, which can be explained by the more complex roof landscapes in inner cities; roofs are smaller and more often shared by several proprietors, this last point creating legal obstacles for installing solar panels. One more possible reason is the relatively large number of rental homes in inner cities. Here, the proprietors usually do not have a direct interest in reducing the dwelling's energy bill and have no direct incentive to install solar panels.

The in Table 6 observed high suitability for solar panels in very low urbanised areas, is not being capitalised by as much as could be expected. Perhaps roof surfaces in those locations outsize the accompanying maximum quota for which energy companies offer attractive feed-in tariffs.

Table 6 Average production of solar power, per dwelling (column 1), average suitability of roofs for generating solar energy (columns 2–5), and average installed peak power for solar energy related to the potential peak power (column 6)

Urbanisation class—Degree of urbanisation	1. Average generated power/per dwelling 2015 (kilowattpeak)	2. Not suiTable (%)	3. Hardly suiTable (%)	4. SuiTable (%)	5. Highly SuiTable (%)	6. Installed to potential peak power 2015 (%)
1 Very high	0.038	34	6	39	22	0.98
2 High	0.088	34	4	33	29	1.41
3 Moderate	0.151	31	4	34	31	1.77
4 Low	0.224	26	4	34	36	1.67
5 Very low	0.548	21	5	33	41	1.60

All aggregated at neighbourhood level and sorted by urbanisation class (Zonatlas.nl 2015; CBS 2014a, CBS 2014b and Klimaatmonitor 2015, edited by PBL)

5 Energetic Society and Local Energy Initiatives

According to Agenda Stad (2015), cities, because of the concentration of breeding grounds and possibilities for scale-ups, can give an impulse to solutions to societal issues and, thus, to sustainable development in the Netherlands. On the basis of the survey *Lokale Energiemonitor* (Hier Opgewekt 2016), we have to disagree with this image, as far as local energy initiatives are concerned.

This monitor lists the development of local energy initiatives. From this monitor, we analyse solar energy projects by local energy cooperations and wind energy projects by local wind energy cooperations. The Lokale Energiemonitor registers, among other things, the domiciles of the cooperations and the capacity that is being generated.[1]

As we learned in the previous section that solar panels are more prevalent in very low urbanised areas, and as we know that wind turbines are not allowed in urban areas, we did not analyse the location of the renewable energy projects. We are only interested in the domiciles of the cooperations, in other words, where is the energetic society based?

5.1 *Solar Energy*

As indicated above, Lokale Energiemonitor (Hier Opgewekt 2016) lists solar energy projects initiated by local energy cooperations and other citizen initiatives. This concerns about 20,000 solar panels on the roofs of schools, hospitals, sports complexes and other utility buildings. Sometimes, this also includes solar parks on former agricultural land.

All together, they generate 6.7 MWp (megawatt peak), enough for 1700 households. For 2016, another 26 MWp is foreseen. Besides solar panels, local energy cooperations are also involved in wind energy, environmental awareness, building insulation and energy saving.

We did a simple analysis to obtain an indication of how the degree of urbanisation is related to the number of local energy initiatives and their capacity (Table 7).

From our analysis, it turns out that most local energy cooperations are located in very low urbanised areas, and that their generated capacity, as far as is known, is highest in the low and very low urbanised neighbourhoods (both in total and per cooperation). A possible explanation is the better conditions for solar panels in low and very low urbanised areas.

[1]The data was derived from surveys, sent out to all legal energy collectives, online research, information from regional partners and direct contact.

Table 7 Solar energy cooperations and their production (installed and planned), ranked by urbanisation class (Hier Opgewekt 2016, CBS 2014b, edited by PBL)

Urbanisation class—degree of urbanisation	Number of known cooperations	Number of known cooperations with known capacity	Total solar power (kW), if provided	Average solar power (kW), if provided
1 Very high	22	10	2576	258
2 High	22	10	3769	377
3 Moderate	22	7	1194	171
4 Low	42	13	3121	240
5 Very low	80	11	15,066	1370

5.2 *Wind Energy*

Besides solar energy, Lokale Energiemonitor also monitors wind energy projects initiated by wind cooperations. Although this part of the data set is quite small, we sorted the domiciles of the cooperations by urbanisation class in order to carefully explore trends (Table 8). In the Netherlands, 81.5 MW of wind power has been installed by wind energy cooperations, which equals 3% of total onshore wind power. Similar to the solar energy cooperations, most wind cooperations tend to have their domiciles in the low and very low urbanised areas. As far as the capacity is known, we see that the very highly urbanised class is producing relatively much renewable energy per cooperation.

Table 8 Wind energy cooperations and their installed production, ranked by urbanisation class (Hier Opgewekt 2016; CBS 2014b, Edited by PBL)

Urbanisation class—degree of urbanisation	Number of known cooperations	Number of known cooperations with known capacity	Total wind power (kW), if provided	Average wind power (kW), if provided
1 Very high	4	3	18,690	6230
2 High	3	3	6861	2287
3 Moderate	1	1	2700	2700
4 Low	8	4	48,392	12,098
5 Very low	7	3	3775	1258

6 Summary and Conclusion

As we question Glaeser's claim of cities being *greener,* we come to a differentiated answer. First, we have to mention that we only considered energy-related data, and did not take other green issues such as waste collection or water management into account.

Glaeser's statement is true for the Netherlands with respect to the *consumption* of energy. Distances travelled by car are 38% lower in very highly urbanised neighbourhoods, compared to car use in very low urbanised areas. For the use of energy (natural gas and electricity), we observe a similar tendency with 20–40% lower consumption in very highly urbanised neighbourhoods, depending on units of measurement (per dwelling or per person). However, we cannot point at a causal relationship between population density and energy consumption. A certain type of dwelling and a certain type of occupant are found more frequently in more urbanised neighbourhoods, which leads to a reduced energy consumption.

Besides, a third calculation method, that of energy consumption per square metre residential floor space, does not yield clear results towards very high or very low urbanisation classes.

Over time, we see that natural gas consumption has decreased faster in very highly urbanised neighbourhoods, compared to in low and very low urbanised areas. For electricity consumption, it is the other way round. Over time, the development of car kilometres does not reveal a clear tendency towards higher or lower population densities.

Notable in Glaeser's book is the complete absence of renewable energy generation. The production of renewable energy has its centre of gravity in very low urbanised areas. Solar panels are found more in rural areas. These areas produce 14 times more solar energy on roofs than in very highly urbanised neighbourhoods, where the roofs and property structures are less suitable for solar panels.

As far as local energy cooperations are concerned, we see a centre of gravity for cooperation domiciles, as well as for the power they generate, in the very low urbanised areas.

All in all, it is difficult to judge if living in cities is 'greener' than in rural areas. The energy saved (petrol, natural gas and electricity) in highly urbanised neighbourhoods seems to be more than the higher production level of renewable energy and activity related to renewable energy in rural areas. The solar panels installed by local cooperations, as described in the previous section, generate energy for only 1700 households (from a total of 7.7 million households in the Netherlands) and the wind energy cooperations contribute 3% to total onshore wind power (Hier Opgewekt 2016). The 2015 share in total solar energy in the Netherlands was only 4.9% of final domestic electricity consumption (ECN 2016).

That is the current situation, however. It is not hard to imagine this imbalance becoming more balanced, or even that green energy will gain the upper hand in the future, for the number of installations for renewable energy and their production capacity is rapidly increasing and new buildings have to be energy-neutral or even generate energy in the future. Besides, cars are becoming more and more economical in their use of fuel.

Acknowledgements Anne Marieke Schwencke (Asi search), Corné Wentink (TELOS), Ton Dassen, Frans Schilder, Manon van Middelkoop, Danielle Snellen, Ruud van de Wijngaart, Kees Vringer, Frank van Rijn and Hans van Amsterdam (PBL Netherlands Environmental Assessment Agency) assisted in the research, thinking and writing for this chapter.

References

Agenda stad. (2015). Samenwerken aan de toekomst van stedelijk Nederland, The Hague. http://agendastad.nl/wp-content/uploads/2015/01/agenda-stad-samenwerken-aan-de-toekomst-van-stedelijk-nederland-jan-2015.pdf. Accessed 10 January 2017.

Bettencourt, L. (2013). The origins of scaling in cities. *Science, 340*(6139), 1438–1441. doi:10.1126/science.1235823

Brounen, D., Kok, N., & Quigley, J. (2012). Residential energy use and conservation: Economics and demographics. *European Economic Review, 56*(5), 931–945. doi:10.1016/j.euroecorev.2012.02.007

CBS. (1992). *Gebiedsindelingenregister 1992*. Voorburg/Heerlen: CBS.

Echenique, M., Barton, H., Hargreaves, T., & Mitchell, G. (2010). SOLUTIONS final report: sustainability of land use and transport in outer neighbourhoods. http://www.suburbansolutions.ac.uk/findings.htm. Accessed 10 January 2017.

ECN. (2016). *Nationale Energieverkenning 2016*. Amsterdam/Petten: Energieonderzoek Centrum Nederland.

European Commission. (2014). Investment for jobs and growth, promoting development and good governance in EU regions and cities, sixth report on economic, social and territorial cohesion. Brussels. http://ec.europa.eu/regional_policy/en/information/cohesion-report/. Accessed 10 January 2017.

Ewing, R., & Cervero, R. (2010). Travel and the built environment. *Journal of the American Planning Association*. doi:10.1080/01944361003766766

Glaeser, E. (2011). *Triumph of the city: How our best invention makes us richer, smarter, greener, healthier, and happier*. New York: Penguin Press.

Hier Opgewekt. (2016). Lokale Energiemonitor 2015. https://www.hieropgewekt.nl/sites/default/files/u20232/lokale_energie_monitor_2015_-_uitgave_januari_2016.pdf. Accessed 10 January 2017.

IEA. (2009). Cities, towns & Renewable energy, Yes In My Front Yard. OECD/IEA Paris.

Keuning, H., & Vroom, J. (2012). Gas- en elektriciteitsverbruik per woning het laagst in stedelijke gebieden, Statistics Netherlands (CBS)https://www.cbs.nl/nl-nl/achtergrond/2012/03/gas-en-elektriciteitsverbruik-per-woning-het-laagst-in-stedelijke-gebieden. Accessed 10 January 2017.

LSE CITIES/EIFER (2014) Cities and energy urban morphology and heat energy demand—Final Report, LSE, London.

Newman, P., & Kenworthy, J. (1999). *Sustainability and cities, overcoming automobile dependence*. Washington: Island Press.

Newman, P., & Kenworthy, J. (2015). *The end of automobile dependence*. Washington: Island Press.

Ratti, C., Baker, N., & Steemers, K. (2005). Energy consumption and urban texture. *Energy and Buildings, 37*(7), 762–776. doi:10.1016/j.enbuild.2004.10.010

Salat, S. (2009). Energy loads, CO_2 emissions and building stocks: morphologies, typologies, energy systems and behaviour. *Building Research & Information, 37*(5–6), 598–609.

Snellen, D., Hilbers, H., Hendriks, A., Amsterdam, H., & Peeters, P. (2005). *Nieuwbouw in beweging*. The Hague/Rotterdam: RPB/NAi uitgevers.

Steemers, K. (2003). Energy and the city: Density, buildings and transport. *Energy and Buildings, 35*(1), 3–14. doi:10.1016/S0378-7788(02)00075-0

UNEP. (2011). Towards a green economy: Pathways to sustainable development and poverty eradication.http://web.unep.org/greeneconomy/resources/green-economy-report. Accessed 11 January 2017.
UN-HABITAT. (2013). *Urban planning for city leaders*. Nairobi: United Nations Human Settlements Programme (UN-Habitat).
Van Wee, B. (2011). Evaluating the impact of land use on travel behaviour: The environment versus accessibility. *Journal of Transport Geography, 19*(6), 530–1533. doi:10.1016/j.jtrangeo.2011.05.011
Vringer, K., Gerlagh, T., & Blok, K. (1997). Het directe en indirecte energiebeslag van Nederlandse huishoudens in 1995 en een vergelijking met huishoudens in 1990, Report No. 97071. Department of Science, Technology and Society (NW&S), Utrecht University (UU), Utrecht.
Worldbank. (2014). *Planning energy efficient and livable cities, Mayoral guidance note #6*. World Bank: Energy Sector Management Assistance Program (ESMAP), Washington http://www.esmap.org/node/55381
Zitzen, M. (2015). *Energiegebruik door huishoudens, brondocumentnummer 022 February 2015*. Utrecht: Milieucentraal.

Data Sources

Basisadministratie Adressen en Gebouwen (BAG).
CBS. (1992). *Gebiedsindelingenregister 1992*. Voorburg/Heerlen: CBS.
CBS. (2013). Kerncijfers wijken en buurten 2013. https://www.cbs.nl/nl-nl/maatwerk/2011/48/kerncijfers-wijken-en-buurten-2004-2015
CBS. (2014a). Kerncijfers wijken en buurten 2014. https://www.cbs.nl/nl-nl/maatwerk/2011/48/kerncijfers-wijken-en-buurten-2004-2015
CBS. (2014b). Kaart met statistieken per vierkant van 500 bij 500 meter. https://www.cbs.nl/nl-nl/dossier/nederlandregionaal/geografische%20data/kaart-met-statistieken-per-vierkant-van-500-bij-500-meter
CBS. (2016). http://statline.cbs.nl/Statweb/
Hier Opgewekt. (2016). Lokale Energiemonitor 2015.
Klimaatmonitor. (2015). Rijkswaterstaat https://klimaatmonitor.databank.nl/dashboard/
Zonatlas.nl. (2015).

Chapter 23
The Challenge of Urbanization in the Context of the New Urban Agenda: Towards a Sustainable Optimization of the Urban Standards

Alexandra Delgado-Jiménez and Jesús Arcediano

Abstract *Urbanization is one of the main challenges at present* due to the increasing urban metrics (population, economy, land/energy/resources consumption, etc.) usually associated with urban sprawl through planned developments in the metropolitan peripheries and through informal unplanned developments. In this context, the New Urban Agenda has just been adopted at the United Nations Conference on Housing and Sustainable Urban Development (Habitat III), which considers urbanization as an opportunity, readdressing the way cities and human settlements are planned, designed, financed, developed, governed and managed to achieve a sustained, inclusive and sustainable economic growth, among other issues. This approach has been questioned by some researchers and scholars proposing an Alternative Habitat III. As a contribution to this debate, this work raises the problem of urban development not only as a matter of magnitude but also of the adopted urban model, where design, that is, morphology and urban standards play a fundamental role in the achievement of objectives of sustainability. A selection of key indicators such as density, diversity of uses, open public space and the treatment of free building surfaces are analyzed, allowing *the urban model to be optimized with sustainability criteria,* being its application very simple and replicable in any territory. The way of transforming these indicators into urban standards, that are also related to the idiosyncrasy of the society is to take the needs of the citizen, not only current but future, as a unit of measure. *The citizen is at the center of the discourse of urban sustainability* in order to achieve greater and better development and wellbeing.

A. Delgado-Jiménez (✉) · J. Arcediano
Department of Architecture, Universidad Nebrija, Madrid, Spain
e-mail: adelgado@nebrija.es

J. Arcediano
e-mail: jarcedia@nebrija.es

R. Álvarez Fernández et al. (eds.), *Carbon Footprint and the Industrial Life Cycle*, Green Energy and Technology, DOI 10.1007/978-3-319-54984-2_23

1 Introduction

The discourse on urbanization has been linked historically to "How much": "How much" cities grow, "How much" urbanization increases or "How much" urban population increases worldwide, among other questions.

Every time it gets more into the speech "How" in capital letters: "How" cities should be (compact, inclusive, etc.), "How" should be urbanization (i.e. integrated in urban fabric) or "How" should be land uses (according to the characteristics of the soil).

But rarely the focus is on the detail of the urban design. The "how" lowercase: "how" to urbanize (i.e. distribution and structure of the public space), "how" is the density (with minimum and maximum thresholds), "how" the different uses are distributed in space (providing diversity and avoiding zoning), "how" is open public space or "how" is the treatment of free building surfaces (fostering the permeability of the soil).

One could ask a question: if all the land that has been developed in the last decades would have been done with a sustainable "how", that is to say, with an urban design and an optimal standards and criteria in search of sustainability, would we be talking about the same environmental, social and economic impact? Clearly not.

In spite of the fact that "How much" and "How" are always important because it tells us about the proportion and trend of the processes and about strategies, the urban planners and technicians related to the city, we also must provide simple, replicable and adequate solutions in costs, taking into account the urban model, especially in the new developments, in aspects as determinant as those previously mentioned.

There is an open debate about the right to the city (Lefebvre 1972; Purcell 2002; Harvey 2003, 2008; Marcuse 2009; Brenner et al. 2012; Delgado-Jiménez 2015), about whose the city is (Pahl 1970; Sassen 1996, 1999), about the requirements to meet the needs generated in urban life. These issues are seen in the official positioning of the United Nations at the recent United Nations Conference on Housing and Sustainable Urban Development called Habitat III (United Nations 2016), and in the responses of some sectors claiming an Alternative Habitat III (Carrión et al. 2016; Mattioli and Elorza 2016). In this last forum there are critics to the role of the technicians indicating that they are at the service of the governments and not of the citizens.

It is considered in the present research that the technicians, besides participating in the political and social debate that is open at the moment, can also provide practical solutions of urban design, that if they are not mainly focused on curbing urbanization, are oriented to improve the new developments to minimize the impact in all its dimensions.

It is therefore appropriate to open and complement the debate on urban design issues. The intervention in the existing city is so complex because we have inherited the consequences of urban policies, ways of operating and an urban design unrelated to the current needs in a context of urban transformation and demands to face challenges such as global environmental change, with the main vector of climate change.

2 The Urbanization Process: Recent Evolution and State of Art

Urbanization is one of the main challenges today. It is of interest to understand how urbanization has evolved to reach the present prominence and importance, and how it affects its increasing magnitude to the main environmental problem at present: global environmental change, and its best known vector, climate change.

Since the Industrial Revolution, it has been shown that all industrialization lead to urbanization by creating economic growth and job opportunities that draw people to cities (Investopedia 2016). Urban areas doubled, tripled, or quadrupled their size, which led to overcrowding in cities since 19th century (McCauley et al. 2016). Once an area is industrialized, the process of urbanization continues growing for a much longer period of time as the area goes through several phases of economic and social reform (Investopedia 2016). After Second World Ward, urbanization increased at an accelerated way, associated to mass consumption and suburbia development, especially in the last decades.

Recent studies show that furthermore the relationship between urbanization and industrialization, there are also a relationship between urbanization without industrialization (Gollin et al. 2015). Urbanization is usually associated with structural changes as a green revolution or industrial revolution (industrialization). But in some parts of the world (mainly in Africa), urbanization is developing without industrialization, because the structural transformation model occurred where resource exports drive urbanization (Gollin et al. 2016).

In countries that are heavily dependent on resource exports, urbanization appears to be concentrated in “consumption cities” where the economies consist primarily of non-tradable services. These contrast with “production cities” that are more dependent on manufacturing in countries that have industrialized (Gollin et al. 2016).

Urbanization promotes self-sustained growth because of returns-to-scale and agglomeration economies in developed countries and developing countries (Gollin et al. 2016), creating a virtuous circle between development and urbanization especially in Europe, United States and Asia.

In August 2016 the world population reached 7.4 billion (Population Reference Bureau 2016) implying that the world has added approximately one billion people in the span of the last twelve years. In 2015, cities harbored 54% of the world population (United Nations 2015).

The world has experienced an ongoing urbanizing process, and the urbanization level has increased from 39 to 52% in the last three decades (1980–2011) as a vast number of people migrate to urban regions (Chen et al. 2014).

Cities, especially those in developing countries and emerging economies, will face unprecedented levels of urbanization in the coming decades. This trend has not slowed in recent decades and looks like it will do in the next. In 1950, 30% of the world population was urban and around 2050 is expected to be 66% urban population, two-thirds of urban population from a rural third (Chen et al. 2014). As seen, by 2050 the world urban population is expected to nearly double, making urbanization one of the main challenges of our century (Delgado-Jiménez 2016a).

Figure 1 shows the global pattern of urbanization level in 1980, and (b) that observed in 2011. The urbanization level (0–100%) has been divided into ten categories, in blocks of 10%. Each category is denoted by a different color. World

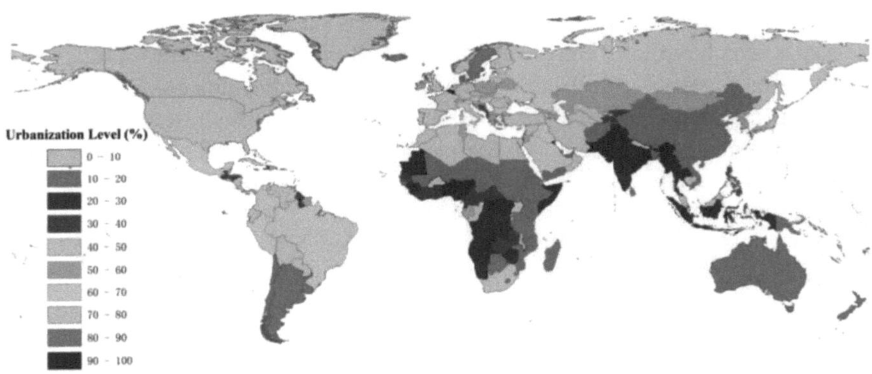

(a) 1980

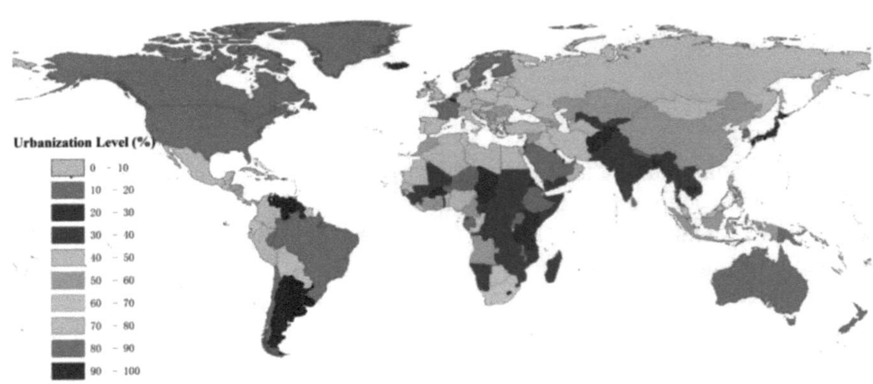

(b) 2011

Fig. 1 Global patterns of changes in urbanization 1980–2011 *Source* Chen et al. 2014, http://dx.doi.org/10.1371/journal.pone.0103799.g001 (accessed 2/09/2016)

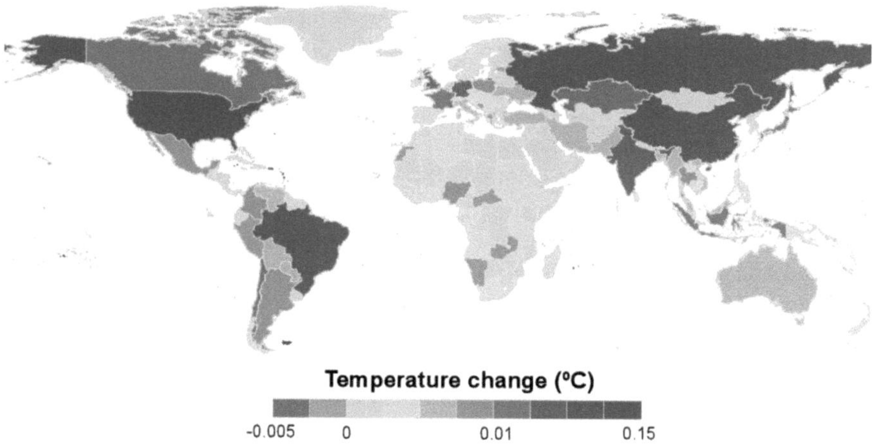

Fig. 2 National contributions to historical climate warming, including CO_2 emissions from fossil fuels and land-use change, non CO_2 greenhouse gases and sulfate aerosols. *Source* Matthew et al. 2014

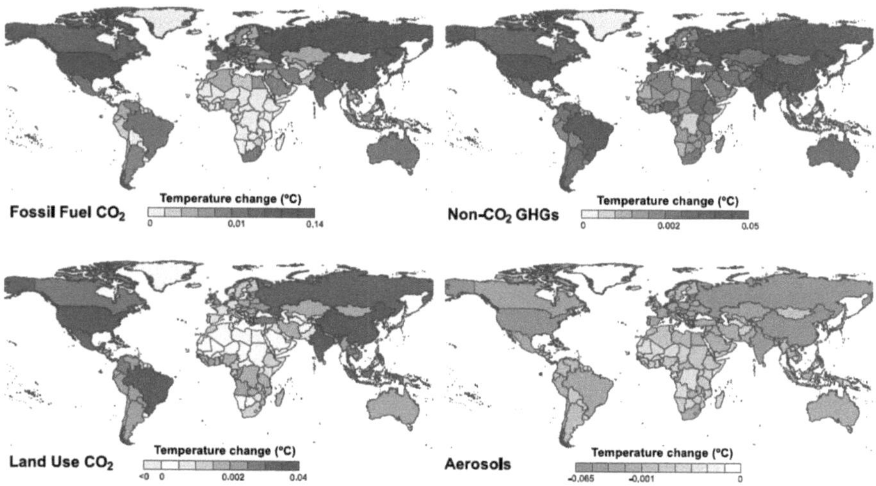

Fig. 3 National climate contributions due to each of the four components included in Fig. 1, **a** fossil fuel CO_2 emissions, **b** land-use change CO_2 emissions, **c** non CO_2 greenhouse gas emissions (methane and nitrous oxide), and **d** sulfate aerosol emissions. *Source* Matthew et al. 2014

urbanization demonstrated remarkable growth in both developed countries and developing countries during 1980–2011, especially in China, Southeast Asia, and Africa (Chen et al. 2014).

Cities are the major contributors to climate change. Although they represent less than 2% of the surface of the earth, they consume 78% of global energy and produce more than 60% of total carbon dioxide and much of the emissions of greenhouse gases (UN-HABITAT 2016). Most urbanized countries contribute more to climate change as seen in Fig. 2 (Matthew et al. 2014).

There are four main contributions to climate change: fossil fuel CO_2, Non CO_2 greenhouse gasses (GHGs), land-use changes CO_2 and aerosols (causing only in this case, cooling).

Many countries have dominant contributions from land-use CO_2 and non CO_2 greenhouse gas emissions, emphasizing the importance of both deforestation and agriculture as components of a country's contribution to climate warming as seen in Fig. 3 (Matthew et al. 2014).

Brazil, for example, has a climate contribution that is dominated by land-use CO_2 and non-CO_2 gasses, pointing to the critical importance of deforestation and agriculture in explaining their fourthplace ranking. The same is true of Indonesia, Columbia and Nigeria, whose fossil fuel CO_2 emissions are not nearly large enough to explain their position on this list of Table 1 (Matthew et al. 2014).

In the case of cities and the process of urbanization, the main issue in fossil fuel CO_2 because of the high consumption of energy as already indicated, 78% (UN-HABITAT 2016).

Table 1 Top 20 contributors to global temperature change, ranked in order of their total climate contribution, and including a breakdown of the contribution of different types of emissions

Rank	Country	Total	Fossil fuel CO_2	Land-use CO_2	All CO_2	Non-CO_2 GHG	All GHG	Aerosols
1	United States	0.151	0.143	0.026	0.170	0.044	0.213	−0.063
2	China	0.063	0.042	0.036	0.078	0.049	0.127	−0.065
3	Russia	0.059	0.059	0.014	0.072	0.020	0.092	−0.034
4	Brazil	0.049	0.004	0.032	0.036	0.018	0.054	−0.005
5	India	0.047	0.013	0.025	0.037	0.025	0.062	−0.015
6	Germany	0.033	0.035	−0.000	0.035	0.008	0.042	−0.009
7	United Kingdom	0.032	0.031	0.001	0.033	0.007	0.040	−0.007
8	France	0.016	0.014	−0.000	0.014	0.007	0.021	−0.005
9	Indonesia	0.015	0.003	0.013	0.015	0.006	0.021	−0.006
10	Canada	0.013	0.011	0.007	0.017	0.005	0.023	−0.009
11	Japan	0.013	0.021	0.001	0.022	0.002	0.024	−0.011
12	Mexico	0.010	0.006	0.008	0.014	0.003	0.017	−0.007
13	Thailand	0.009	0.002	0.006	0.008	0.004	0.012	−0.002
14	Columbia	0.009	0.001	0.006	0.007	0.003	0.010	−0.001
15	Argentina	0.009	0.002	0.003	0.005	0.005	0.010	−0.001
16	Poland	0.007	0.010	0.001	0.011	0.003	0.014	−0.007
17	Nigeria	0.007	0.001	0.001	0.002	0.005	0.007	0.000
18	Venezuela	0.007	0.002	0.002	0.004	0.003	0.008	−0.001
19	Australia	0.006	0.005	0.002	0.007	0.006	0.014	−0.007
20	Netherlands	0.006	0.004	0.000	0.004	0.002	0.006	−0.001

All values here are given in °C of global temperature change. *Source* Matthew et al. 2014

Land-use changes CO_2 is a contribution to climate change more related with deforestation using land for agriculture. But at the same time, new agriculture soils are need because urbanization covers older agriculture lands.

Urbanization is a complex phenomenon that is relation with ways of life, consumption of resources, among them energy, and as a consequence, it is related to climate change, nowadays one of the main problems worldwide.

3 The New Urban Agenda from Conference Habitat III

Urbanization has been one of the main focus of the third United Nations Conference on Housing and Sustainable Urban Development (Habitat III) in Quito in October 2016 (United Nations 2016), where the various approaches that exist on the current situation and future challenges of housing and sustainable urban development will be discussed. Habitat III inherited as field work twenty years after the second conference, a built environment in expansion and the two big issues, housing and city, turned into the subject of debate on inclusion as rights, as well as related environmental problems with a growing metric (Delgado-Jiménez 2016b).

In this scenario, the Habitat III aims as a challenge to include sustainability criteria in housing policies and city (already included in the name of sustainable urban development itself). It also has a strategy to get the mandate of the Conference aligned with the Sustainable Development Goals.

The adoption of the final outcome of the Conference, as the New Urban Agenda (United Nations 2016), supposes the establishment a common strategy in matters of housing and sustainable urban development.

The New Urban Agenda has its kick-off point in the persistence of multiple forms of poverty, growing inequalities and environmental degradation remain among the major obstacles to sustainable development worldwide, with social and economic exclusion and spatial segregation often an irrefutable reality in cities and human settlements (United Nations 2016).

But, contrary to other positions that takes urbanization as an ignored challenge (Uhel 2006) in the New Urban Agenda urbanization is considered as an opportunity, as an engine of sustained and inclusive economic growth, social and cultural development, and environmental protection, and of its potential contributions to the achievement of transformative and sustainable development (United Nations 2016).

The main objective of the New Urban Agenda is ambitious: readdressing the way cities and human settlements are *planned, designed, financed, developed, governed and managed*, and how and why is to:

- to end poverty and hunger in all its forms and dimensions;
- reduce inequalities;
- promote sustained, inclusive and sustainable economic growth;
- achieve gender equality and the empowerment of all women and girls in order to fully harness their vital contribution to sustainable development;

Table 2 Main topics introduced in the New Urban Agenda

Concepts	Classification
Quality of life (2)	Social
Inequalities (3,5)	Social
Environmental degradation (3)	Environmental
Economic exclusion (3)	Economic
Spatial segregation (3)	Social/Economic
Gender equality (5, 13)	Social
Resilience (5)	Environmental
Participation (9)	Social
Consumption and production patterns (10, 13)	Social/Economic/Environmental
Use of resources (10)	Environmental
Impact of climate change (10, 13)	Environmental
Inclusivity (11)	Social
Social and ecological function of land (12)	Social/Environmental
Right to adequate housing (13, 15)	Social
Public goods (13)	Social
Quality services (13)	Social
Safe, inclusive, accessible, green, and quality public spaces with political participation (13)	Social
Urbanization for structural transformation, high productivity, value-added activities, and resource efficiency, harnessing local economies (13)	Economic
Age- and gender-responsive planning (13)	Social
Safe, and accessible urban mobility (13)	Social
Disaster risk reduction and management (13)	Environmental
Vulnerability (13)	Social
Biodiversity (13)	Environmental
Eradication of extreme poverty (14)	Social
High productivity, competitiveness and innovation (14)	Economic
Sustained, inclusive and sustainable economic growth (43)	Economic

Source Own elaboration from United Nations 2016. *Note* Classification: Social, Environmental, Economic

- improve human health and well-being;
- foster resilience
- and protect the environment (United Nations 2016).

There are a lot of topics introduced in the New Urban Agenda. The following Table 2 is a selection of the main issues addressed in.

This Habitat III Conference that has set the New Urban Agenda has for the first time had a counter-proposal called "Manifiesto de Quito: Towards an Alternative Habitat 3".

It arises from the consideration that the Habitat Conferences are repetitive in their considerations without having visible effects or practical results. And it is considered that key issues such as the value of the land, the models of extensive urbanization without a city, the commodification of housing, the growing increase of inequalities in the city, the land ownership regime, The emerging problems, the financing of the territory, and so on (Carrión et al. 2016).

And from this Manifesto in Quito, this may have some relation to the organization of this type of Conferences by the governments of the States, when this alternative platform thinks that they should have a minority role. In order to be able to develop new contents and review their compliance, the creation of an independent agency is required. Most of the members would be partitioned between city governments and other local settlements, representatives of organizations and social movements, and collective professional or academic members.

The issues that should be addressed are housing as a public service; access to water or energy, education and health, transport; the public control of the financial system and subordinated to the state, local or cooperative sector (Carrión et al. 2016).

In addition, this alternative Manifesto points out the role of urban planners and other professionals who, while not being the main responsible for the city, legitimize the plans and projects they execute with their interventions and works addresses, city models that are not the most desirable for The signatories (Carrión et al. 2016).

It is also noted that this conference does not go into the detail of urban design, although there are two crucial issues, avoiding the models of extensive urbanization without a city and access to facilities and transport, as a guarantee of social sustainability.

4 Towards a Sustainable Optimization of Urban Standards. Key Indicators for New Developments: Density, Diversity of Uses, Open Public Space and Treatment of Free of Building Surface

Among its objectives, urban design has the configuration of a sustainable and quality urban space characterized by the urban fabrics, the network of infrastructures, equipment and services and the spatial distribution of uses; an urban space with adequate densities to allow the use of the city, with pedestrian routes for the various activities related to urban life, with a multitude of uses, fabrics and building typologies to enhance diversity, functional complexity and social interaction; an urban space with a network of communications infrastructures, urban facilities, green areas and sufficient and functional free spaces. Most of these features of urban design are included in themes of the New Urban Agenda (United Nations 2016) as those collected in Table 2.

The definition and design of this urban space are established in urban planning through the adoption of a model, and formally determined by planning, norms and urban ordinances.

This model of urban growth is determined by planners with no more limits than those established by legislation to the power of planning. These limits are, therefore, made up of a set of legal and regulatory determinations, obligatory in the elaboration of planning, and that limit the discretion that the planning legislation grants to the planner.

From the technical point of view, one of the limits with which the planner plays, and that more condition the design of the cities, are the urban standards. When these urban standards are established in legislation, they become concrete legal determinations of urban planning and act as a benchmark or binding reference and limiting the planner, guaranteeing a priori, a proper ordering in terms of quality of life for citizens. When regulation does not operate so directly in the definition of urban design through urban standards, the point is the planner who should establish them. It is necessary some knowledge of the environment in which it is operating and appropriate analysis tools is not always accessible.

It is therefore a matter of discussing and putting on the Table an accessible mechanism of a standard for operating in a simple way on the territories where urban growth occurs today and which often lack, not only the right tools for the setting of standards, but also adequate urban planning legislation.

As a contribution to this debate, this work raises the problem of urban development not only as a matter of magnitude but also of the adopted urban model, where design, that is, morphology and urban standards play a fundamental role in the achievement of objectives of sustainability.

For the setting of this standard, not only determines the territory of the binomial free/occupied area, derived from the implantation on the territory of the building mass and which has been referred in some forums, but is taken as a basis and reference all the principles that determine sustainability in urban growth.

With this aim, a selection of key indicators such as density, diversity of uses, open public space and the treatment of free building surfaces are analyzed, allowing *the urban model to be optimized with sustainability criteria,* being its application very simple and replicable in any territory.

4.1 Density

The morphological character of each space unit in the city is determined by defining characteristics of how urban extension has occurred, one of the most relevant being density or compactness. This characteristic is indicative of the relationship between built space and free space. Both spaces are closely related since it is the built space that demands free space to achieve the proper development of all the activities that urban life implies (Moliní et al. 2012).

The density in the cities is expressed by an index that expresses the relation between the number of inhabitants, dwellings or built surface and the surface of soil that occupies. This index therefore relates the morphology of the building mass and the way in which it occupies the territory and influences very significantly in the conformation of the urban structure of the city and in the operation of the same, giving rise to the compact city and to the scattered city or urban sprawl.

Adequate density is of paramount importance for the optimum functioning of the city in terms of sustainability in order not to incur disproportionate cost levels, and an essential condition for achieving diversity, urban complexity and social interaction, especially when accompanied by a variety of uses and building typologies.

From the point of view of sustainability, the high density of the compact city is more advisable since they have a smaller impact on the territory and the environment, due to their less occupation of land and the concentration of activities avoiding, to a certain extent, displacements and therefore consumption of energy, as shown Fig. 4 in gasoline use per capita.

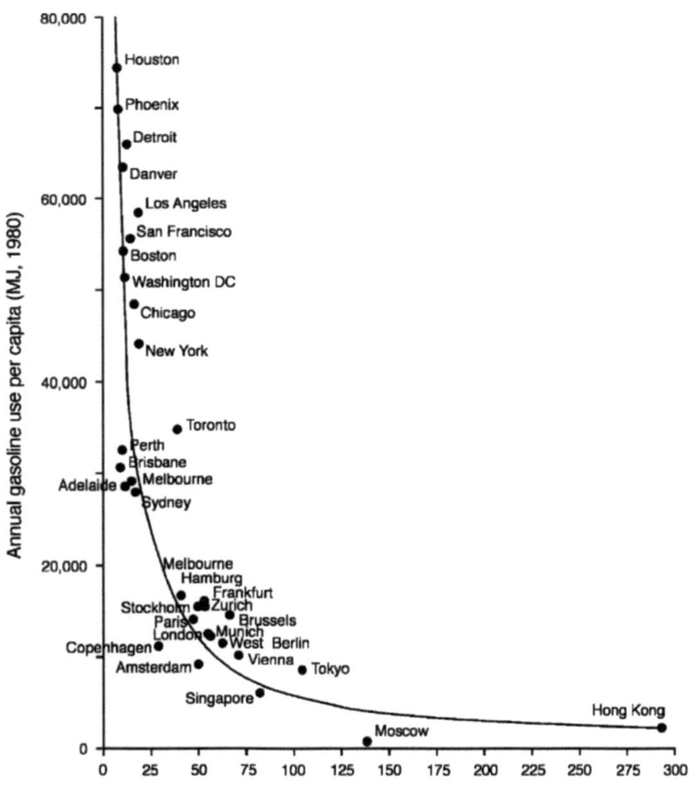

Fig. 4 Gasoline use per capita versus urban density 1980 *Source* Mindali et al. (2004), from Newman and Kenworthy 1989

In Figs. 5 and 6, it is shown Houston, the city with the lowest density of Fig. 4, with less than 15 per/ha. As seen in ortophotography and streetview, it is an urban fabric of single family housings of one floor, with a lot of green areas and roads network.

Figures 7 and 8 are samples of Vienna fabric, a compact city in the middle of the graphic of Fig. 4. This city has got a density about 75 per/ha, with a low consumption of gasoline per capita. The urban fabric is composed of multifamiliar housing among two and five floors, with a road network contained.

Finally in the selection of urban fabrics, it is shown in Figs. 9 and 10 Hong Kong, the city with the highest density of Fig. 4, with 300 per/ha, and one of the lowest consumption of gasoline per capita. The skyline is totally different to other examples, and it is composed of skycrapers.

Compact cities are also more economically sustainable due to their lower implementation and maintenance costs. However, to the compactness of cities, it is necessary to set limits so as not to incur the collapse of infrastructure and urban services.

Density in cities implies proximity to urban uses and functions, and facilitates contact, exchange and communication among citizens, issues that constitute the essence of the city. This proximity and use of the urban elements define a model of occupation of the territory and a degree of urban sustainability.

The density of cities can be defined, first and taking into account that the inhabitant is the protagonist of the same, as the relationship between the number of inhabitants and the area they occupy. This density is very used in geography, it is not the most determinant for the conformation of the urban form since the number of inhabitants in the city is fluctuating and varies continuously.

Only the permanent inhabitants of the cities develop their social life in the spaces they occupy, for it will be determining the occupation of a dwelling and this is how another form of defining the density in the cities is generated, the result of dividing the number of housing and the area they occupy. However, this relationship only characterizes the parts of the city with a predominantly residential use and, in any

Fig. 5 Houston, United States: Ortophotography and Streetview *Source* Google Earth, https://www.google.es/intl/es/earth/

Fig. 6 Houston, United States: Ortophotography and Streetview *Source* Google Earth, https://www.google.es/intl/es/earth/

Fig. 7 Vienna, Austria: Ortophotography and Streetview *Source* Google Earth, https://www.google.es/intl/es/earth/

case, the result obtained is very dependent on the size of the dwellings and the ratio of inhabitants per dwelling.

The third way of defining the density in the cities, and is the one that will interest to characterize the compactness of the building mass and its morphology in the space occupied by it, regardless of the uses and the housing model of the inhabitants in the dwellings, Is the one that relates the constructed square meters and the surface in which they are developed. In gross or absolute terms, the surface on which the built square meters are developed would result from adding the surface area occupied by the building mass and the free surface that remains between it and that is necessary for its optimal functioning. In this way of measuring the density is where it is clearly seen the close relationship that exists between the buildability, the occupied floor surface and the free floor surface. In Fig. 11, it is shown these parameters.

Fig. 8 Vienna, Austria: Ortophotography and Streetview *Source* Google Earth, https://www.google.es/intl/es/earth/

Fig. 9 Hong Kong: Ortophotography and Streetview *Source* Google Earth, https://www.google.es/intl/es/earth/

If we start from a representative piece of a city, for example of one hectare, in dimensional terms of a square of 100 × 100 m, applying the parameters 50–50, would result in an occupied surface of 5000 m^2 and another one of free surface. Taking as a reference a building typology of two plants, not to increase construction costs and not to collapse the city with building masses in height, would result in 10,000 m^2 of buildability (5000 m^2 of occupied surface × two floors). Therefore, the resulting density, starting from the initial premise of binomial 50–50 and taking as reference an optimum building of two plants, is defined by the ratio of total buildability to total area, both of which occupies and of the Which generates as free, that is to say 10,000 m^2 buildable between 10,000 m^2 of surface, 1 m^2e/m^2s, gross density sanctioned as maximum by many urban regulations like in Spain case.

This density, in a residential neighborhood, and translated to dwellings and inhabitants, would give rise to densities of 100 dwellings and 300 inhabitants per

Fig. 10 Hong Kong: Ortophotography and Streetview *Source* Google Earth, https://www.google.es/intl/es/earth/

Fig. 11 Density concept: scheme *Note* In this scheme the density is 1, because there is the same built area that of soil area

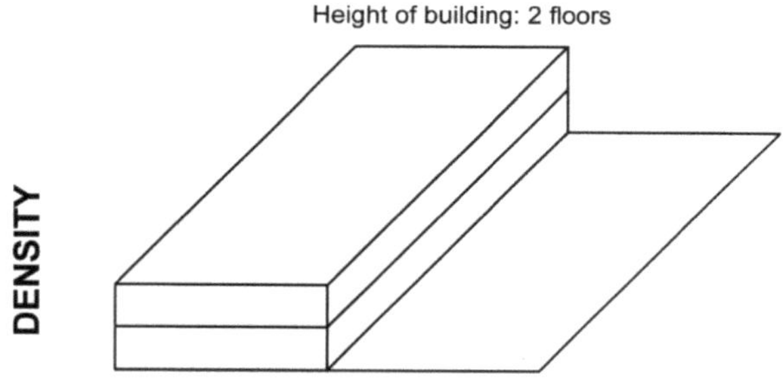

Table 3 Existing relationship between densities and number of plants taking as reference parameter 50–50 (occupied-free) and a residential neighborhood with dwellings of 100 m^2 with a ratio of 3 inhabitants per dwelling

	Density parameters		
Number of floors	Gross buildable area	Dwellings per hectare	Inhabitants per hectare
1	0.5	50	150
2	1	100	300
3	1.5	150	450
4	2	200	600
5	2.5	250	750

hectare without taking into account the built surface destined to endowments and taking as reference houses of 100 m^2 of built surface and 3 inhabitants per housing.

It is thus evidenced in this form of operation that the density, once taken the premise 50–50, is related to the height of the buildings. If we perform the same

Table 4 Densities of population in some cities of the world according to the report of the center of urban studies American

Rank	Geography	Urban area	Population estimate	Land area (per km^2)	Population (per km^2)	Density (Per ha)
1	Bangladesh	Dhaka	16,235,000	368	44,117	441
6	China	Hong Kong	7,280,000	285	25,544	255
58	Colombia	Bogota	9,520,000	562	16,940	169
111	Afganistan	Kabul	3,650,000	259	14,093	141
172	India	Delhi	3,240,000	272	11,912	119
246	Turkey	Istambul	13,520,000	1360	9941	99
565	China	Beijing	20,390,000	3937	5179	52
757	France	Paris	10,870,000	2845	3821	38
945	United States	Chicago	9,185,000	6856	1340	13

Source Demographia 2016

operation for cities of more than two plants we would obtain the following results of Table 3.

That the city with In Table 4, data show that the 50–50 ratio leads to cities that are too dense, taking into account the highest population density in the world is Dhaka (Bangladesh) with 441 inhab/ha, according to the report of the US urban studies center Demographia for the year 2016.

Any increase in density, on the basis of increasing the number of plants, should be directed, therefore, towards a greater release of the occupied soil. And any decrease in density should be linked to the incorporation of other uses in the building mass as the facilities, as it will be exposed in the following section regarding the diversity of uses.

Therefore, to any binomial in which the relationship between the occupied surface and the free surface is established, the density parameter must be introduced in square meters of building land per square meter of soil for definition as standard.

For example, in Table 5 it is shown how initial standards translate into derived standards. As a kick-off point, it is chosen 10,000 m^2 of land (1 ha) and an initial standard of occupied surface of 50% of the land, with a residential building of 2 floors.

If in a residential neighborhood with a maximum density or gross build ability of 100 m^2 of building per square meter of floor, for housing of 100 m^2 built and an average of 3 inhabitants per dwelling, we would obtain standards derived from a 50% free surface, 100 dwellings and 300 inhabitants, an operation that could be put in reverse.

Table 5 Example of initial standards and derived ones related to density in cities

Initial surface	10,000	m^2			
Initial standards			Derived standards		
Density					
Occupied floor area	50	%	Surface free of building	50	%
Main use		Residential			
Density (gross buildable area)	1	m^2/m^2s			
Average dwelling area	100	m^2	Total dwellings per hectare	100	dwellings
Inhabitants per dwelling	3	inhabitants	Total inhabitants per hectare	300	inhab.

4.2 *Diversity of Uses*

The cities are based on a specialization by uses of the areas in which they are developed (residential, tertiary, commercial, industrial, facilities, to mention the most important). These uses are those that characterize each of the parts of the city, and the relations that occur between these parts and the uses that are developed in them are determinants for the urban sustainability.

In each of the parts in which the city can be divided, in detail, another variety of uses occurs on a scale more decisive for sustainability, the local or neighborhood scale. In residential areas, reserving nearby spaces for commercial premises, offices or other compatible uses is essential in order to generate a certain density of activity and therefore a reduction in travel and an increase in the probability of exchange and contact between people. The practice of zoning, that is to say, the mono-functional sectors, as much residential as of activity, like the great commercial surfaces, generate a high number of displacements in motor vehicle and they dis-articulate the social interaction.

The mixture of functions and compatible urban uses in the same urban space, for the satisfaction of daily needs by the population, contributes to the functional diversity and the self-contained mobility, especially motorized by its influence on climate change, And therefore to sustainability.

The regulation to build, in those municipalities that have regulations, will define/allow/enforce the uses that can be implemented in the plots and for this they will define a main use as well as compatible and prohibited uses. When this type of regulation exists, the compatible uses are those uses subordinated to the principal that, in some cases, contribute to its correct functioning and, in others, complement it.

As is sometimes the case with regard to the compatibility of uses in town planning, the power to decide the uses that are implemented in each of the urban areas, except in exceptional cases, correspond exclusively to the owner-developer

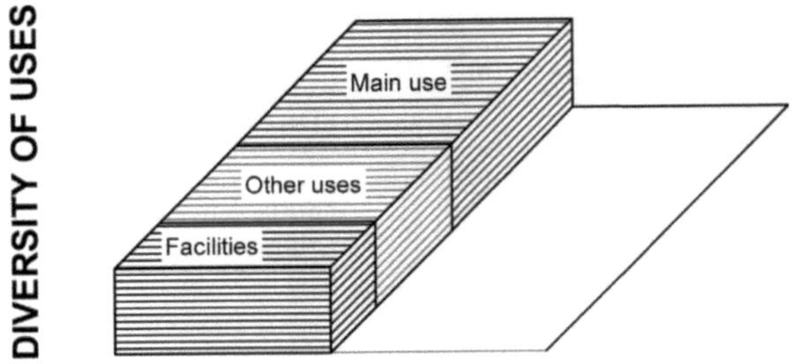

Fig. 12 Diversity of uses concept: scheme *Note* Mixed functions and compatible urban uses in the same urban space, for the satisfaction of daily needs of the population. It contributes to functional diversity and self-containment in mobility, especially motorized

Table 6 Range of existence of facilities of the different types of city and with sustainable population according to the density of dwellings

Density/buildability	Residential		Medium		Central		Pedestrian	
	Bg	Bn	Bg	Bn	Bg	Bn	Bg	Bn
25 dwe/ha	0.29	0.38	0.34	0.41	0.38	0.44	0.38	0.41
50 dwe/ha	0.59	1.12	0.67	1.05	0.76	1.04	0.76	0.89
75 dwe/ha	0.88	3.68	1.00	2.21	1.13	1.91	1.13	1.47
100 dwe/ha	1.17	26.00	1.34	4.90	1.51	3.30	1.51	2.18
125 dwe/ha			1.68	18.11	1.89	5.88	1.89	3.06
150 dwe/ha					2.27	12.29	2.27	4.20
175 dwe/ha					2.64	53.12	2.64	5.71

Source Hernández Aja 1997. *Note* Bg: Index of gross buildability on foreclosable parcels in m^2/m^{2}s Bn: Index of net buildability on foreclosable parcels in m^2/m^{2}s

and therefore Both will be subject to the economic profitability of the different uses, thus undermining the diversity of uses in terms of environmental sustainability.

The first use to be made in any urban design process and at a stage where there is an estimate of the population density is the one that covers the basic necessities derived from the life in the cities, that is to say the facilities, whether educational, health, recreational, cultural or other, whether public or private, as shown in Fig. 12. The enumerated uses consume building mass and must be understood to be included, therefore in the space occupied by the building, thus lowering the resulting density of urban areas.

The optimal urban standard reserved for endowments should be established on the basis of comprehensive and complex studies on the population. Perhaps this is the standard that must be more open since cultural facts and geographical features are involved in the needs of the population.

The optimal urban standard reserved for endowments should be established on the basis of comprehensive and complex studies on the population. Perhaps this is the standard that must be more open since cultural facts and geographical features are involved in the needs of the population.

Table 7 Ratios of suitability for building destined to the commercial, tertiary and industrial uses per worker

Use		Ratio	Population	Active population		Suitability for building	Ratio
	%	m^2/worker	inhab.	workers		m^2 built area for other uses	% of suitability for building
Commercial	30	25	300	150	45	1.125	11.25
Tertiary	50	25			75	1.875	18.75
Industrial	20	40			30	1.200	12.00
	100						

Table 8 Initial standards and derived ones related to the diversity of uses in cities

Initial standards			Derived standards		
Diversity of uses					
Active population	50	%	Number of workers	150	Workers
Commercial use	30	%			
	25	m^2/worker	Commercial buildability	1125	m^2
Tertiary use	50	%	Tertiary buildability		
	25	m^2/worker		1875	m^2
Industrial use	20	%			
	40	m^2/worker	Industrial buildability	1200	m^2

Fig. 13 Levittown, Pennsylvania (United States): Ortophotography and Streetview *Source* Google Earth, https://www.google.es/intl/es/earth/

A selection of range of desirable existence of facilities of the different types of city and with sustainable population according to the density of dwellings, is shown in Table 6.

Once part of the building is dedicated to meeting the needs of urban life, the following uses that must be included in any process of urban design in the building mass are those that are linked to work and commerce, are all lucrative uses: Industrial, tertiary and commercial. The spatial implementation of these uses, especially those linked to work (industrial and tertiary offices) is closely linked to the economic base of the city, but in any case must be included in urban design in a different way, mixing them together Unless there is incompatibility between them, as can happen with the industrial compared to the residential.

As a reference value we have considered the space needed to cover the demands of employment in an area, using the parameters corresponding to the Comunidad de Madrid (Madrid Region), with 3 inhabitants per dwelling and an active population

Fig. 14 Levittown, Pennsylvania (United States): Ortophotography and Streetview *Source* Google Earth, https://www.google.es/intl/es/earth/

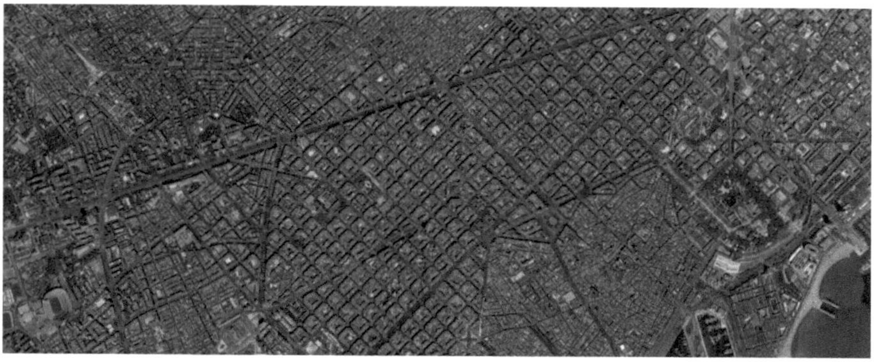

Fig. 15 Barcelona, Spain: Ortophotography and Streetview *Source* Google Earth, https://www.google.es/intl/es/earth/

rate of 50%. The distribution of employment by sector and area per job would be: for the commercial 30% with 25 m^2 built by employment; For the tertiary 50% with 25 m^2 built by employment; and for industrial use 20% with 40 m^2 buildable per job (Hernández Aja 2011), as shown in Table 7.

Its application to an area in which we know the number of inhabitants that we live in would give us the demand of space for lucrative uses destined to the non-residential area (tertiary, commercial, industrial, etc.).

For example, for the binomial 50–50 and a maximum density resulting from 100 m^2 of building per square meter of soil, 100 houses per hectare and 300 inhabitants per hectare, taking as reference a residential neighborhood with dwellings of 100 m^2 of surface Built and 3 inhabitants per dwelling, there are standards derived from 1125 m^2 networks and reserve of open publicbuilt for commercial use and 1875 m^2 built for tertiary use, i.e. a reference value of

Fig. 16 Barcelona, Spain: Ortophotography and Streetview *Source* Google Earth, https://www.google.es/intl/es/earth/

approximately 30% of the total building area (10,000 m^2e) for others uses to generate the necessary diversity for sustainability, taking as reference the number of inhabitants and the active population that it represents, as shown in Table 8.

For showing the importance of diversity of uses, two opposite examples have been shown: Levittown, Pennsilvania, in Figs. 13 and 14, and Barcelona, Spain, in Figs. 15 and 16. Although both urban fabrics area planned (because if not, the comparison were more difficult) the result it cannot be more different.

In one hand, Levittown represents a derived version of garden city, but without the important issue of garden and green areas, and with a monofunctional urban fabric of single familiar houses of one floor. The other uses as commercial ones are located apart from the fabric creating a shopping mall with a parking lot. Private car is compulsory for the mobility of this kind of city.

In the other hand, Barcelona represents one of the best responses to the extension of the cities, with a mixed-uses city, housing or residential, facilities,tertiary and some kinds of industrial uses are working together to improve the quality of habitat.

4.3 *Open Public Space*

The free floor not occupied by the building forms a continuum that must be structured, qualified and at least partially reserved to the public domain. It is the local or supralocal governments that must have the disposition and command of most of the land left free without building to be able to implement the ordination and organization necessary and sufficient for the proper functioning of the city in global and structuring terms, for purposes of its sustainability.

To achieve this task, it is necessary to qualify this free and public soil in two large groups, such as road, street or plaza, i.e. within the system or the communications network, and as urban park or garden, i.e. within the system or network of

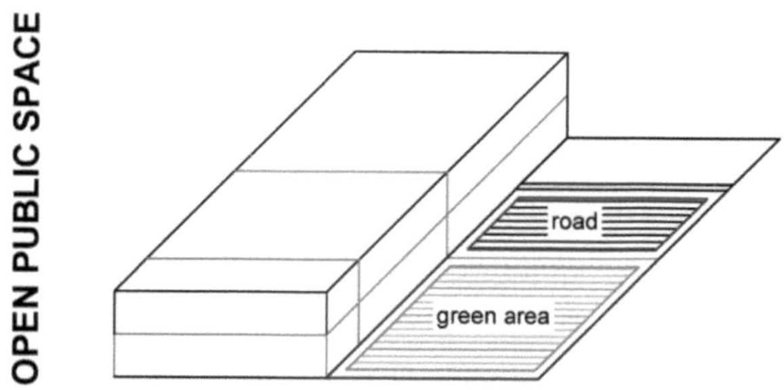

Fig. 17 Open public space concept: *green* areas, road networks and reserve of open public space: scheme *Note* Soil in green: green areas; Soil in *grey*: roads, platforms and squares; Soil in *white*: Reservation of free private or public land for ensuring the ventilation of the building mass. It is obtained by applying the position conditions of the building on the plot (setbacks, maximum occupation, etc.). Reserve of open public space to guarantee the functioning of the city (Color figure online)

Fig. 18 Cartagena de Indias, Colombia: Ortophotography and Streetview *Source* Google Earth, https://www.google.es/intl/es/earth/

green areas, where the treatment of the surface of the ground is going to be determinant and in which the different facilities uses that do not require buildability, like a sports tracks in the open air can be developed.

Regarding the standard for the road, the minimum essential for the operation of the city and for the shaping of the urban fabric must be considered based on dividing the building mass into lots.

Regarding the defining standard of the quantification of green areas, the World Health Organization proposes a fairly moderate standard of 9 m^2 of green areas per inhabitant (World Health Organization 2010). This standard does not refer to the free ground surface of the building but to that which has a green treatment. This green layer in cities is the most natural method of reducing CO_2 emissions, the most present greenhouse gas in the atmosphere (Walker et al. 2015).

If we carry these considerations to our example we would have that for binomial 50–50 (built/free land) and a maximum density resulting from 100 m^2 of building

Fig. 19 Cartagena de Indias, Colombia: Ortophotography and Streetview *Source* Google Earth, https://www.google.es/intl/es/earth/

Fig. 20 Nairobi, Kenia: Ortophotography and Streetview *Source* Google Earth

per square meter of soil, 100 houses per hectare and 300 inhabitants per hectare, would result in a minimum road of 25 m^2 per dwelling (Hernández Aja 2011), that is to say 2500 m^2 of land destined to road, leaving the 2500 m^2 of remaining ground destined to green zones, an operation that yields a standard of 8.3 m^2 of green areas per inhabitant, very close to that recommended by WHO (Fig. 17).

In Fig. 18 and 19, it is shown informal settlements in Cartagena de Indias, Colombia, where open public space is undersized, with poor quality, overall, and without a structure that allows a lecture of the neighbourhood and city.

In Figs. 20 and 21, it is shown Nairobi, with a very important urban growth in recent years, where live together formal and informal settlements, and where open public space has got a better size than the previous example, but without a structure that answers to the full city, really metropolitan area or conurbation.

Fig. 21 Nairobi, Kenia: Ortophotography and Streetview *Source* https://www.google.es/intl/es/earth/ and http://4.bp.blogspot.com/_C6bSE7QLZEc/TSIlB6vkYxI/AAAAAAAAAt8/oPeWVMG-pKs/s1600/Nairobi%2Bphoto.jpg

4.4 Treatment of Surface Free of Building

Each urban fabric can be analyzed from the point of view of density, diversity of uses and reserve of open public space, but not only these indicators are those that define in terms of sustainability the adopted urban model.

The construction of the city implies a transformation of the important natural environment, the free space in a process of urbanization represents that part of the city that will not be occupied by the building mass and that will allow the city to function when being configured as space of circulation and of stay. This free space, although it is not built, can be urbanized due to the alteration of the surface layer becoming an inert soil (without activity) from the environmental point of view.

The surface free of building, that is to say not occupied by the buildings, can be treated in the city of very different forms. The treatment of the surface of this soil will imply environmental consequences related to permeability, recharge of aquifers, environmental humidity and therefore to urban sustainability.

The treatment of the surface free of building will depend fundamentally on its function. There will be a public space, free of building, made up of the roads where the treatment will turn the soil into opaque or occupied, and other spaces made up of free spaces and green areas that may have more permeable treatments. There is also a public space destined to the facilities where the building mass will occupy part of the ground being the other part subject to diverse treatments according to the function that derives from these facilities (Fig. 22).

Fig. 22 Treatment of free of building surface: scheme *Note* Environmental consequences related to permeability, recharge of aquifers, etc

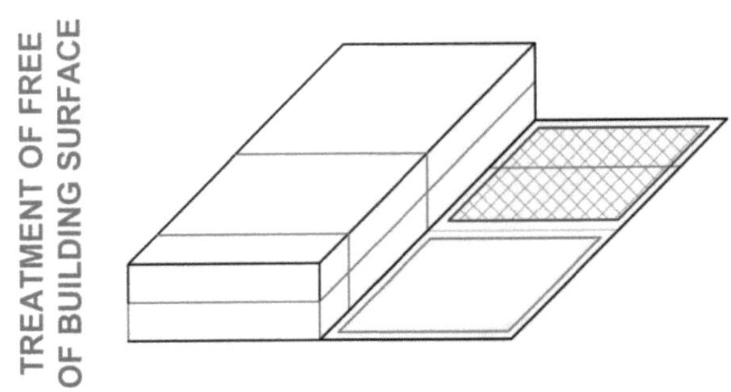

Fig. 23 Energy exchange of materials in soil: radiation, evapotranspiration, convection, irradiance and conduction *Source* Hernández Aja 2013

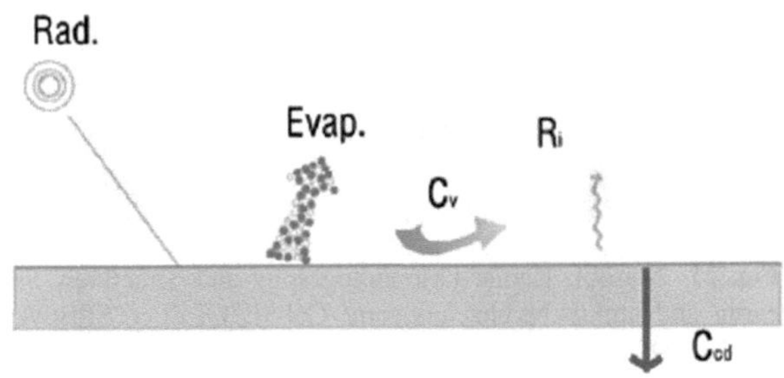

The private space will be characterized by space occupied in plant by the building and by the free space. The private space not occupied by the building mass or free space, from the environmental point of view, will contribute as well as the public space not built to the urban sustainability although linked to a private property regime.

The provision on the treatment of private free spaces has been regulated in urban regulations in order to make a greater contribution to urban sustainability by private property. As an example, in the case of the city of Madrid, Spain, there is a regulation called surface conditioning of the private open spaces of Masterplan of Madrid (Article 6.10.20) that establishes that 50% of the area not occupied by the building must be landscaped with vegetation. It allows, however, that this space be occupied by the building in a low level situation, prescribing for this possibility the obligation of introducing a layer of minimum vegetal soil of 60 cm.

This type of measures is desirable in current context of urban microclimate. For example, northern hemisphere urban areas annually have an average of 12% less solar radiation, 8% more clouds, 14% more rainfall, 10% more snowfall, and 15% more thunderstorms than their rural counterparts. Urban pollutant concentrations can be 10 times higher than those of the 'clean' atmosphere and air temperatures can be on the average 2 °C higher, effect that is called urban areas heat island (Taha 1997).

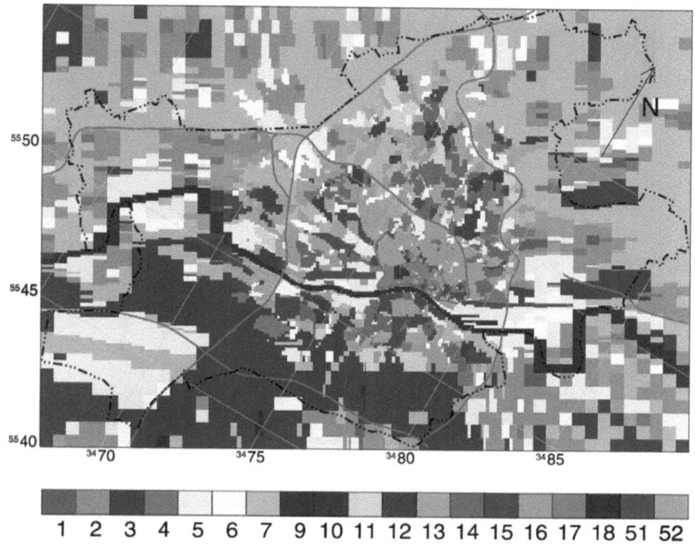

Fig. 24 Map of the land use: 1—settlement (dense), 2—settlement (sparse), 3—forest, 4—park, 5—commercial and industrial zone (dense), 6—commercial and industrial zone (sparse), 7—open space, 9—water, 10—financial district, 11—single- and multifamily residential (sparse), 12—village and single-family residential (dense), 13—terraced housing estate, 14—terrace houses (middle), 15—terrace houses (dense), 16—tenement block residential, 17—city, 18—medieval city, 51—tracks, 52—allotment *Source* Früh et al. 2011, 175

In this context, the type of energy exchange of materials in soils, as radiation, evapotranspiration, convection, irradiance and conduction, depend on the treatment of surface free of building, that is, the selection of materials. And this accumulated process of selection of materials, furthermore of building mass, is one of the causes of the processes described above (Fig. 23).

Especially important are albedo, which is the effect by which materials reflect sunlight and helps prevent temperatures from rising, evapotranspiration, and of course anthropogenic heat (Taha 1997).

Results from meteorological simulations suggest that cities can feasibly reverse heat islands and offset their impacts on energy use simply by increasing the albedo of roofing and paving materials and reforesting urban areas (Taha 1997).

Urban areas heat island, effect by which urban areas have higher temperature than counterparts or hinterland, is a phenomenon that is going to increase with climate change. For all of this, the inclusion of treatment as a standard of sustainability is key at present and its importance will increase in the coming years.

In Fig. 24, map of the land use is shown for the case of Frankfurt. The higher fraction of buildings with a higher degree of inhomogeneity is situated in the city center. The two longish commercial and industrial zones (SW of the domain) show the Rhein-Main airport with the runways in between the two building structures (Früh et al. 2011).

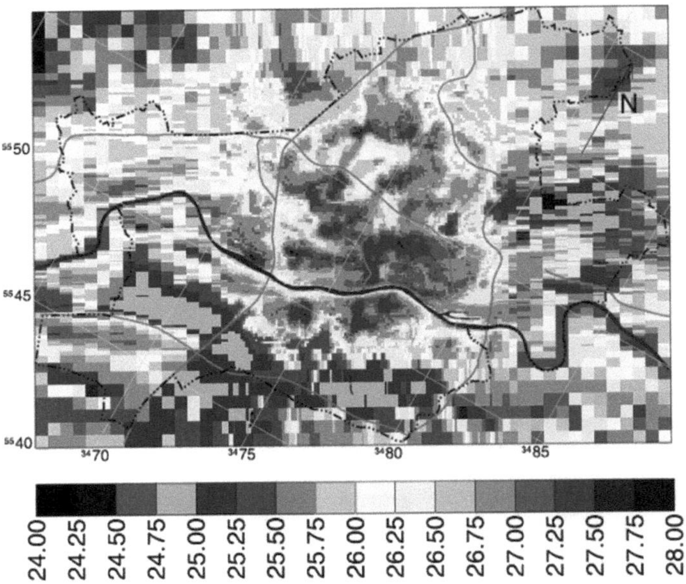

Fig. 25 Map of T_{max} (°C) from CUB_{int} at 5-m *Source* Früh et al. 2011, 178

Figure 25 shows the map of the daily maximum temperature. It can be seen the consistency of both data.

In addition, authors of these maps research estimated the impact of climate change on the climate indices. According to the analysis of the regional climate projections from four ensemble members, they found that the annual mean number of summer days will increase by 5–32 days at the 90% significance level in Frankfurt for 2021–50 relative to 1971–2000 (Früh et al. 2011).

Another important point to follow is that the outcome of climate research has to be linked more explicitly to the objectives of sustainable settlement, since settlement planning is a key aspect of sustainability (Mills 2006).

5 Limits and Origin of Urban Standards: The Citizen Needs as a Measure Unit

Of the four standards analyzed, determinants for sustainability and that are reviewed in this work, the first two density and diversity of uses, would form part of the occupied land and try to characterize and design the urban form of the built volume, what is its volume and how many different uses are included in that volume.

The other two issues, the open public space (including green areas, road networks and reserve of soil for facilities) and the treatment of surface free of building, would form part of the soil not occupied by the building and try to guarantee the operation of the city and its sustainability from an environmental point of view.

Urban density and diversity of uses, therefore, seek the most efficient relation between the resources used, basically the natural resources—especially the soil—and the objectives achieved, to provide a functional and healthy habitat to the inhabitants in which they can develop the largest number of activities linked to urban life with the least possible consumption of resources.

The characterization of open public space seeks the most efficient operation of the city in terms of displacement and stay without damaging to excess the environment in its conversion to urban environment.

The establishment of optimal standards is intimately related to the concept of urban efficiency and therefore sustainability, this efficiency involves environmental, economic and social aspects.

From the environmental point of view, the least possible impact on the environment must be achieved, consuming the least amount of resources and energy, generating the least possible amount of waste and minimizing pollution. Taking the concept of density, this environmental efficiency is closely related to the occupation of the land by urbanization, not only the occupation produced by the buildings but also that produced by non-permeable paving materials such as usually used in roadways.

From the economic point of view, cities must be economically viable in their operation so as not to compromise the Public Treasuries affected by the implementation and maintenance thereof. The cost of urbanization is linked to the density, open public space and the treatment of the surface free of building. In this last standard, it is not the same cost asphalting than gardening, but in a more accentuated way the costs will be linked to the maintenance of these free floor surfaces. It is here that viability must be seen from the point of view of sustainability.

And from the social side, cities will be required to respond to social demands by improving the quality of life of the population by fostering social interaction.

Some decades ago, Le Corbusier introduced Modulor (Le Corbusier 1961, 2004) as a system of measures in the search for a mathematical relationship between the measures of man and nature.

The way of transforming these indicators into urban standards, that are also related to the idiosyncrasy of the society is to take the needs of the citizen, not only current but future, as a unit of measure. *The citizen is at the center of the discourse of urban sustainability* in order to achieve greater and better development and wellbeing. Urban design must do a lecture of citizen needs for achieving a proportion between land uses and human activities, including not only nowadays situation but also prospects for improvement, for example, land for urban facilities although financing for construction is not yet available, or soil protection previewing different climate change scenarios.

6 Conclusions

The soil is a basic natural resource, not renewable, and this implies it must be assessed in all urban extension actions, the occupation of the land by urbanization, its necessity, suitability and opportunity, because of principles of irreversibility and uncertainty.

It is shown in this work the need to include a selection of determinants for sustainability, among which are found as fundamental those indicated in the text: density, diversity of uses, open public space and treatment of free of building surface. The debate should result in the agreement of internationally accepted basic standards as has happened with the standard of green zones set by the WHO. At the same time, flexible standards through the fixing of forks or minimum/maximum, so that they are applicable to the social, geographical and economic characteristics of the different territories.

In addition, it is observed that some of the indicators used are more decisive than others, as is treated in the case of density, and for that reason a weight should be considered in an aggregated index of sustainability.

References

Brenner, N., Marcuse, P., & Mayer, M. (Eds.). (2012). *Cities for people, not for profit: Critical urban theory and the right to the city*. Routledge.

Carrión, F., Cohen, M., Borja, J., Corti, M., Gutman, M., Pirez, P. et al. (2016). Manifiesto de Quito: Hacia un Habitat 3 Alternativo. Aprobado el 20 de octubre de 2016. Seminario Habitat 3 Alternativo. https://www.flacso.edu.ec/flax15/_upload/habitatIII/Manifiesto_Quito.pdf. Accessed 31 December 2016).

Chen, M., Zhang, H., Liu, W. & Zhang W (2014). The global pattern of urbanization and economic growth: evidence from the last three decades. *PLoS ONE, 9*(8), e103799. doi:10.1371/journal.pone.0103799. Accessed 2 September 2016).

Corbusier, L. (1961). *El Modulor: ensayo sobre una medida armónica a la escala humana, aplicable universalmente a la arquitectura ya la mecánica*. Poseidón.

Corbusier, L. (2004). *The Modulor and Modulor 2* (Vol. 1). Springer Science & Business Media.

Demographia. (2016). Demographia world urban areas 12th, annual edition: 2016:04, www.demographia.com/db-worldua.pdf. Accessed 2 September 2016.

Delgado-Jiménez, A. (2015). Del derecho a la ciudad al reto de la ciudad sostenible e inteligente: la ciudad sostenible al servicio del derecho a la ciudad y la ciudad inteligente al servicio de la ciudad sostenible. *Revista de Derecho Urbanístico,* ISSN 1139–4978, núm. 300, Madrid, septiembre-octubre, 107–134.

Delgado-Jiménez, A. (2016a). La COP21 ante la escala local: ciudades, comunidades locales y pueblos indígenas en transformación. *Revista Ciudad Sostenible, 2016*(26), 6–8.

Delgado-Jiménez, A. (2016b). Año clave para la vivienda y desarrollo urbano sostenible con Hábitat III, Nebrija Actualidad, http://www.nebrija.com/medios/actualidadnebrija/2016/07/07/2016-ano-clave-la-vivienda-desarrollo-urbano-sostenible-habitat-iii/. Accessed on 2/09/2016.

Früh, B., Becker, P., Deutschländer, T., Hesserl, J-D., Kossmann, M., Mieskes, I. et al. (2011). Estimation of climate-change impacts on the urban heat load using an urban climate model and regional climate projections. *Journal of Applied Meteorology and Climatology, American Meteorological Society*, *50*, 167–184, doi:10.1175/2010JAMC2377.1.

Gollin, D. (University of Oxford), Jedwab, R. (GWU), & Vollrath, D. (Houston). (2015). Urbanization with and without Industrialization, WB-GWU Conference on Urbanization and Poverty Reduction.
Gollin, D. (University of Oxford), Jedwab, R. (GWU), & Vollrath, D. (Houston). (2016). Urbanization with and without Industrialization, *Journal of Economic Growth, 21*(1), 35–70. First online 13 November 2015.
Harvey, D. (2003). The right to the city. *International Journal of Urban and Regional Research, 27*(4), 939–941.
Harvey, D. (2008). The right to the city. *The City Reader, 6,* 23–40.
Hernández Aja, A. (1997). *La ciudad de los ciudadanos*. Madrid: Ministerio de Fomento.
Hernández Aja, A. (2011). Análisis de los estándares de calidad urbana en el planeamiento de las ciudades españolas. Cuadernos de investigación urbanística, (11).
Hernández Aja, A. (coord.). (2013). Manual de diseño bioclimático urbano. Recomendaciones para la elaboración de normativas urbanísticas. Redacción: José Fariña, Victoria Fernández, Miguel Ángel Gálvez, Agustín Hernández y Nagore Urrutia. Colaboradoras: Carolina Astorga e Itxaso Ceberio. Coordinación editorial y traducción al portugués: Artur Gonçalves, Antonio Castro y Manuel Feliciano. Bragança [Portugal]: Instituto Politécnico de Bragança, 2013. ISBN: 978-972-745-157-9.
Investopedia. (2016). How does industrialization lead to urbanization? http://www.investopedia.com/ask/answers/041515/how-does-industrialization-leadurbanization.asp#ixzz4Ekraa4mk. Accessed 2 September 2016.
Lefebvre, H. (1972). *Le droit a la ville suivi de l'Espace et politique*, 1972. Paris: Ed. Anthropos [Edición original (1968) *Le Droit à la ville*, Paris: Anthropos].
Marcuse, P. (2009). From critical urban theory to the right to the city. *City, 13*(2–3), 185–197.
Matthews, H. D., Graham, T. L., Keverian, S., Lamontagne, C., Seto, D., & Smith, T. J. (2014). National contributions to observed global warming. *Environmental Research Letters, 9*(1), 014010.
Mattioli, D., & Elorza, A. L. (2016). Críticas al habitat III. Perspectivas y reflexiones en torno a las ciudades. *Vivienda y Ciudad, 3,* 93–96.
McCauley, S., Murray, K., & Felt, J. (2016). How did the industrial revolution affect the modern world? The effects of Industrialization, https://sites.google.com/site/5effectsofindustrialization/effects-in-the-1800s-1900s. Accessed 2 September 2016.
Mills, G. (2006). Progress toward sustainable settlements: A role for urban climatology. *Theoretical and Applied Climatology, 84*(1–3), 69–76.
Mindali, O., Raveh, A., & Salomon, I. (2004). Urban density and energy consumption: A new look at old statistics. *Transportation Research Part A: Policy and Practice, 38*(2), 143–162.
Moliní, F., & Salgado, M. (2012). Los impactos ambientales de la ciudad de baja densidad en relación con los de la ciudad compacta. Biblio 3 W. Revista Bibliográfica de Geografía y Ciencias Sociales, (958).
Pahl, R. E. (1970). *Whose city? and other essays on sociology and planning*. Longmans.
Population Reference Bureau. (2016). 2016 world population data sheet. Population Reference Bureau. Retrieved September 15, 2016. http://www.prb.org/Publications/Datasheets/2016/2016-world-population-data-sheet.aspx. Accessed 2 October 2016.
Purcell, M. (2002). Excavating Lefebvre: The right to the city and its urban politics of the inhabitant. *GeoJournal, 58*(2–3), 99–108.
Sassen, S. (1996). Whose city is it? *Traditional Dwellings and Settlements Review, 8*(1), 11–11.
Sassen, S. (1999). Whose city is it? Globalization and the formation of new. *Sustainable cities in the 21st century, 145.*
Taha, H. (1997). Urban climates and heat islands: albedo, evapotranspiration, and anthropogenic heat. *Energy and Buildings, 25*(2), 99–103.
United Nations. (2015). Department of economic and social affairs, population division. World Population Prospects: The 2015 Revision, Key Findings and Advance Tables. 2015. Working Paper No. ESA/P/WP.241. https://esa.un.org/unpd/wpp/Publications/Files/Key_Findings_WPP_2015.pdf. Accessed 2 September 2016.

United Nations. (2016). The new urban agenda. Draft outcome document of the United Nations Conference on Housing and Sustainable Urban Development (Habitat III). Conference on Housing and Sustainable Urban Development (Habitat III), Quito, 17–20 October 2016. Adoption of the final outcome of the Conference, http://es.unhabitat.org/. Accessed 31 December 2016.

UN-HABITAT (2016). Climate change. http://unhabitat.org/urban-themes/climate-change/?noredirect=en_US. Accessed 2 September 2016.

Uhel, R. (2006). Urban sprawl in Europe: the ignored challenge. *EAA Report, 10*(2), 56.

Walker, F. B., Fernández, P. W., & Freitas, J. M. (2015). Modelo de cálculo de áreas verdes en planificación urbana desde la densidad habitacional. *Urbano, 10*(15), 97–101.

World Health Organization. (2010). Urban planning, environment and health: From evidence to policy action, http://www.euro.who.int/__data/assets/pdf_file/0004/114448/E93987.pdf?ua=1. Accessed 22 December 2016.

MIX
Papier aus verantwortungsvollen Quellen
Paper from responsible sources
FSC® C105338

If you have any concerns about our products,
you can contact us on
ProductSafety@springernature.com

In case Publisher is established outside the EU,
the EU authorized representative is:
Springer Nature Customer Service Center GmbH
Europaplatz 3, 69115 Heidelberg, Germany

Printed by Libri Plureos GmbH
in Hamburg, Germany